Oxidative Stress in Applied Basic Research and Clinical Practice

Editor-in-Chief
Donald Armstrong

For further volumes:
http://www.springer.com/series/8145

Note from the Editor-in-Chief

All books in this series illustrate point-of-care testing and critically evaluate the potential of antioxidant supplementation in various medical disorders associated with oxidative stress. Future volumes will be updated as warranted by emerging new technology, or from studies reporting clinical trials.

Donald Armstrong
Editor-in-Chief

Hirokazu Tsukahara • Kazunari Kaneko
Editors

Studies on Pediatric Disorders

Humana Press

Editors
Hirokazu Tsukahara
Department of Pediatrics
Okayama University Graduate School of Medicine, Dentistry and Pharmaceutical Sciences
Okayama, Japan

Kazunari Kaneko
Department of Pediatrics
Kansai Medical University
Hirakata-shi
Osaka, Japan

ISSN 2197-7224 ISSN 2197-7232 (electronic)
ISBN 978-1-4939-0678-9 ISBN 978-1-4939-0679-6 (eBook)
DOI 10.1007/978-1-4939-0679-6
Springer New York Heidelberg Dordrecht London

Library of Congress Control Number: 2014937525

Printed on acid-free paper

Humana Press is a brand of Springer
Springer is part of Springer Science+Business Media (www.springer.com)

Contents

Contributors

Koji Abe, M.D., Ph.D. Department of Neurology, Okayama University Graduate School of Medicine, Dentistry and Pharmaceutical Sciences, Okayama, Japan

Saurabh Aggarwal The Pulmonary Disease Program, Vascular Biology Center, Georgia Regents University, Augusta, GA, USA

Stephen M. Black, Ph.D. The Pulmonary Disease Program, Vascular Biology Center, Georgia Regents University, Augusta, GA, USA

Gediminas Cepinskas, D.V.M., Ph.D. Centre for Critical Illness Research, London, ON, Canada

Francesco Chiarelli Department of Pediatrics, University of Chieti, Chieti, Italy

Valentina Chiavaroli Department of Pediatrics, University of Chieti, Chieti, Italy

Michael F. Cohen, Ph.D. Department of Biology, Sonoma State University, Rohnert Park, CA, USA

Allan Doctor, M.D. Departments of Pediatrics and Biochemistry, Washington University in Saint Louis, School of Medicine, St. Louis, MO, USA

Douglas D. Fraser, M.D., Ph.D. Pediatric Critical Care Medicine, Translational Research Centre, Children's Health Research Institute and Centre for Critical Illness Research, Western University, London, Ontario, Canada

Tohru Fukai, M.D., Ph.D. Departments of Medicine (Section of Cardiology) and Pharmacology, Center for Cardiovascular Research, University of Illinois at Chicago, Chicago, IL, USA

Jon Fukuto, Ph.D. Department of Chemistry, Sonoma State University, Rohnert Park, CA, USA

Christine M. Gross The Pulmonary Disease Program, Vascular Biology Center, Georgia Regents University, Augusta, GA, USA

Aurelia Guarini Department of Pediatrics, University of Chieti, Chieti, Italy

Kenji Hamaoka, M.D., Ph.D. Department of Pediatric Cardiology and Nephrology, Graduate School of Medical Science, Kyoto Prefectural University of Medicine, Kyoto, Japan

Masaharu Hayashi, M.D. Department of Brain Development and Neural Regeneration, Tokyo Metropolitan Institute of Medical Science, Tokyo, Japan

Li-Tung Huang Department of Pediatrics, Kaohsiung Chang Gung Memorial Hospital and Chang Gung University College of Medicine, Kaohsiung, Taiwan

Seiji Ito Department of Medical Chemistry, Kansai Medical University, Hirakata, Japan

Kazunari Kaneko, M.D., Ph.D. Department of Pediatrics, Kansai Medical University, Hirakata-shi, Osaka, Japan

Tayo Katano Department of Medical Chemistry, Kansai Medical University, Hirakata, Japan

Masayuki Kubo, Ph.D. Department of Public Health, Okayama University Graduate School of Medicine, Dentistry and Pharmaceutical Sciences, Okayama, Japan

Takashi Miida, M.D., Ph.D. Department of Clinical Laboratory Medicine, Juntendo University School of Medicine, Tokyo, Japan

Kousuke Minami, Ph.D. The International Center for Materials Nanoarchitectonics (MANA), National Institute for Materials Science (NIMS), Ibaraki, Japan

Rie Miyata, M.D. Department of Brain Development and Neural Regeneration, Tokyo Metropolitan Institute of Medical Science, Tokyo, Japan

Angelika Mohn Department of Pediatrics, University of Chieti, Chieti, Italy

Tsuneo Morishima, M.D., Ph.D. Department of Pediatrics, Okayama University Graduate School of Medicine, Dentistry and Pharmaceutical Sciences, Okayama, Japan

Yoshiharu Nagaoka, M.D. Department of Pediatrics, Okayama University Graduate School of Medicine, Dentistry and Pharmaceutical Sciences, Okayama, Japan

Hironori Nagasaka, M.D. Department of Pediatrics, Takarazuka City Hospital, Takarazuka, Japan

Yuji Naito, M.D., Ph.D. Department of Molecular Gastroenterology and Hepatology, Graduate School of Medical Science, Kyoto Prefectural University of Medicine, Kyoto, Japan

Eisei Noiri, M.D., Ph.D. Nephrology and Endocrinology, Hemodialysis and Apheresis, The University of Tokyo Hospital, Tokyo, Japan

Keiki Ogino, M.D., Ph.D. Department of Public Health, Okayama University Graduate School of Medicine, Dentistry and Pharmaceutical Sciences, Okayama, Japan

Tsuyoshi Oikawa, Ph.D. CMIC Holdings Co. Ltd., Tokyo, Japan

Eric Patterson, Ph.D. Centre for Critical Illness Research, London, ON, Canada

Ruslan Rafikov The Pulmonary Disease Program, Vascular Biology Center, Georgia Regents University, Augusta, GA, USA

Stephen Rogers, Ph.D. Departments of Pediatrics and Biochemistry, Washington University in Saint Louis, School of Medicine, St. Louis, MO, USA

Pavel Rossner Jr. The Institute of Experimental Medicine ASCR, Academy of Sciences of the Czech Republic, Prague, Czech Republic

Takeyori Saheki Institute of Resource Development and Analysis, Kumamoto University, Kumamoto, Japan

Ola Didrik Saugstad, M.D. Department of Pediatric Research, Oslo University Hospital, Oslo, Norway

Fukashi Serizawa, M.D., Ph.D. Centre for Critical Illness Research, London, ON, Canada

Manuel Silva, M.D. Departments of Pediatrics and Biochemistry, Washington University in Saint Louis, School of Medicine, St. Louis, MO, USA

Radim J. Sram The Institute of Experimental Medicine ASCR, Academy of Sciences of the Czech Republic, Prague, Czech Republic

Varadarajan Sudhahar Departments of Medicine (Section of Cardiology) and Pharmacology, Center for Cardiovascular Research, University of Illinois at Chicago, Chicago, IL, USA

Jesse Brown Veterans Affairs Medical Center, Chicago, IL, USA

Takeshi Sugaya, Ph.D. CMIC Holdings Co. Ltd., Tokyo, Japan

Department of Nephrology and Hypertension, Internal Medicine, St. Marianna University School of Medicine, Kanagawa, Japan

You-Lin Tain Department of Pediatrics, Kaohsiung Chang Gung Memorial Hospital and Chang Gung University College of Medicine, Kaohsiung, Taiwan

Tomohisa Takagi, M.D., Ph.D. Department of Molecular Gastroenterology and Hepatology, Graduate School of Medical Science, Kyoto Prefectural University of Medicine, Kyoto, Japan

Naoyuki Tanuma, M.D. Department of Brain Development and Neural Regeneration, Tokyo Metropolitan Institute of Medical Science, Tokyo, Japan

Shoji Tsuji, M.D., Ph.D. Department of Pediatrics, Kansai Medical University, Hirakata-shi, Osaka, Japan

Hirokazu Tsukahara, M.D., Ph.D. Department of Pediatrics, Okayama University Graduate School of Medicine, Dentistry and Pharmaceutical Sciences, Okayama, Japan

Kazuhiko Uchiyama, M.D., Ph.D. Department of Molecular Gastroenterology and Hepatology, Graduate School of Medical Science, Kyoto Prefectural University of Medicine, Kyoto, Japan

Naoko S. Watanabe, M.Sc. Faculty of Science, University of the Ryukyus, Nakagami-gun, Okinawa, Japan

Michael S. Wolin Department of Physiology, New York Medical College, Valhalla, NY, USA

Akihiro Yachie, M.D., Ph.D. Department of Pediatrics, School of Medicine, Graduate School of Medical Sciences, Kanazawa University, Kanazawa, Japan

Tomoyo Yahata Department of Pediatric Cardiology and Nephrology, Graduate School of Medical Science, Kyoto Prefectural University of Medicine, Kyoto, Japan

Hideo Yamasaki, Ph.D. Faculty of Science, University of the Ryukyus, Nakagami-gun, Okinawa, Japan

Toru Yamashita Department of Neurology, Okayama University Graduate School of Medicine, Dentistry and Pharmaceutical Sciences, Okayama, Japan

Masato Yashiro, M.D., Ph.D. Department of Pediatrics, Okayama University Graduate School of Medicine, Dentistry and Pharmaceutical Sciences, Okayama, Japan

Part I
General Topics

Chapter 1
Oxygen and Oxidative Stress in the Newborn

Ola Didrik Saugstad

Abbreviations

ARE	Antioxidant response element
BPD	Bronchopulmonary dysplasia
CEBP	CCAAT/enhancer-binding protein
FiO_2	Fraction of oxygen
GSH	Glutathione
GSSG	Oxidized glutathione
HIE	Hypoxic ischemic encephalopathy
HIF	Hypoxia inducible factor
IGF-1	Insulin-like growth factor-1
NO	Nitric oxide
Nrf2	Nuclear factor (erythroid-derived 2)-like 2
OS	Oxidative stress
PG	Prostaglandin
ROP	Retinopathy of prematurity
ROS	Reactive oxygen species
SaO_2	Arterial oxygen saturation
S_pO_2	Oxygen saturation measured by pulse oximetry
SOD	Superoxide dismutase
VEGF	Vascular endothelial growth factor

O.D. Saugstad, M.D. (✉)
Department of Pediatric Research, Oslo University Hospital, Rikshospitalet,
PB 4950 Nydalen, Oslo 0424, Norway
e-mail: odsaugstad@rr-research.no

H. Tsukahara and K. Kaneko (eds.), *Studies on Pediatric Disorders*, Oxidative Stress in Applied Basic Research and Clinical Practice, DOI 10.1007/978-1-4939-0679-6_1,

1.1 Introduction

Hyperoxia induces toxic effects in several organs. The German obstetrician Jörg already in 1835 reported an inflammatory response of the small airways to oxygen administration [1]. The association between retinopathy of prematurity (ROP) and oxygen has been known since the publication of Campbell in 1951 [2]. Immediately after the first description of bronchopulmonary dysplasia (BPD) by Northway et al. in 1967 [3], these authors linked the condition to oxygen toxicity. A decade earlier Gerschman and her colleagues had introduced the free radical theory explaining oxygen toxicity [4]. Already in early 1950s oxidative stress (OS) per se was linked to neonatal disease, primarily hemolysis when the newborn erythrocyte membrane was shown to be less resistant to H_2O_2 than the adult one [5]. Vitamin E was tested against a number of neonatal conditions, however, without good results. It has been known since 1973 that oxygen radicals are important in antibacterial defense through oxidative burst of leukocytes [6]. Twenty-five years ago it was also understood that oxygen radicals contribute to vasoregulation of several organs, therefore exerting physiological functions [7].

A leap forward in the understanding of the pathogenesis of these conditions occurred in the 1980s when the concept of OS in preterm infants was focused. A major breakthrough was in fact achieved when it was understood that OS is not related to oxygen treatment only. For instance, the understanding that inflammation and OS are two sides of the same coin was extremely useful for understanding OS-related conditions, especially BPD. Free radical production theoretically could explain the pathogenesis of a wide variety of conditions in medicine. In 1988 we coined the term “oxygen radical disease of the newborn” which implicated that ROP, BPD, and some other conditions as necrotizing enterocolitis, and intraventricular hemorrhage, may have a common pathogenesis via free radicals but different manifestations due to which organ is mostly affected [8]. Since then a large body of studies and literature have accumulated and it has been shown that OS is associated with these conditions as well as others such as periventricular leucomalacia and patent ductus arteriosus. Recently genes related to BPD and ROP have been studied and more such data will come [9, 10].

1.2 Effects of Hyperoxia

Through the evolution the body tightly regulated its defense against low oxygen states. A master molecule, the hypoxia inducible factor (HIF)-1α is activated during hypoxia and transcribes a large number of genes which defend the organism against hypoxia. Among these are genes related to angiogenesis, erythropoiesis, increased breathing and glucose uptake resulting in reduced oxygen consumption and increased oxygen delivery. In normoxia and hyperoxia HIF-1α is turned off and degraded. For review, see [11].

Oxidative defense mechanisms have also been developed through evolution, as antioxyenzymes and mitochondria which convert oxygen to water in the respiratory chain. Recently it was shown that *Drosophila melanogaster* breeded in, for instance, 90 % oxygen for 13 generations could live, develop, and reproduce. The body weight increased, also with increased wing area, reaching a maximal weight in 70 % oxygen [12].

1.3 Oxygen-Free Radicals

Oxygen can only receive single electrons with antiparallel spin to complete electron pairings. By feeding oxygen with one electron at a time, for instance, from iron the oxygen molecule is stabilized, this phenomenon explains the high affinity of iron to oxygen and the production of rust. During oxidative phosporylation in the mitochondria, single electrons escape and join with 1–2 % of the total oxygen consumed by the cells to form superoxide radicals; however, during physical exercise this may increase to 10 %. By adding 2, 3, and subsequently 4 electrons hydrogen peroxide, the hydroxyl radical and finally water are formed, respectively.

Oxygen radicals or reactive oxygen species (ROS) have a number of actions and oxidize free fatty acids, proteins, and DNA. They have important physiologic properties as in the defense against microbes. They are also signaling substances, and redox processes probably are important for controlling growth and development [13].

1.4 Significance of Glutathione in Fetus and Preterm Infants

Glutathione (GSH) is perhaps the most abundant antioxidant, and it can also regenerate other antioxidants. It has an antiperoxide activity, can be a radical scavenger, and is present in high concentrations (molars) in the cytosol of mammalian cells but is constantly degraded and resynthesized. Its availability is limited in the newborn, especially the preterm infant, and this is related to cell death. GSH is a tripeptide (γ-glutamylcysteinyl-glycine) containing a sulfhydrol also called a thiol. GSH can react with superoxide and hydroxyl radicals and in the presence of GSH peroxidase donate an electron to H_2O_2 and organic peroxides and itself becomes oxidized (GSSG). At high pH, GSH is spontaneously converted to oxidized GSH (GSSG).

GSH is not only an antioxidant but also an important detoxifying agent. It is therefore of importance to keep GSH high and GSSG low. Preterm infants with a low GSH therefore not only have a lowered antioxidant defense but also a lowered capacity for detoxification. For review, see [14].

1.5 Oxidative Stress and Transcription Factors

The last years it has been better understood that OS may activate transcription factors which may transcribe genes that are related to inflammation, apoptosis, as well as OS. Today we recognize a list of transcription factors, reviewed by Wright and Dennery [14], in addition to HIF-1α that play a role in hyperoxia [15]. Nuclear factor (erythroid-derived 2)-like 2 (Nrf2) is activated by hyperoxia and activates antioxidant response element (ARE). This regulates detoxifying and antioxidant enzymes and increases expression of antioxidant enzymes. It is cytoprotective in type II cells of the lung and ameliorates O_2-induced lung injury in mice. AP-1 controls genes regulating apoptosis, inflammation, and OS. NF-κB activates genes regulating apoptosis, inflammation, and OS. It is activated by endotoxins and OS via toll-like receptors in the cell membrane. P53 regulates expression of target genes related to cell cycle arrest, cell death, and DNA repair. CCAAT/enhancer-binding protein (CEBP) regulates cell proliferation and tissue development and is increased in the lung of rats exposed to hyperoxia. STATs are polypeptides participating in signaling pathways and may be protective to hyperoxia by induction of heme-oxygenase, which is a highly inducible cytoprotective enzyme following exposure to hyperoxia.

1.6 Oxidative Stress as Physiologic Regulators

About 25 years ago, it was understood that free radicals are not only toxic but may play important roles in regulating normal physiologic processes. Thus it became clear that OS is related to vasoregulatory control of several vascular beds, as in the lung and brain. It was also shown that oxygen radicals dilate the lamb ductus arteriosus probably through stimulation of PGE_2 [16]. This indicates that OS may be involved in regulation of the perinatal circulation.

ROS are now identified as signaling molecules. They affect signal transduction by stimulating, for instance, Ca^{++} signaling and protein phosphorylation. This induces biological processes such as gene expression. Antioxidants block receptor-mediated signal transduction [17].

1.7 Antioxidant Defense in Fetal Life. Why the Premature Infant is Exposed to Oxidative Stress

The embryonic and fetal development occurs in a hypoxemic environment. This is a highly interesting observation and indicates that redox processes are of importance in regulating embryogenesis. In the fetus the oxygen saturation of blood is around 50–60 %. An important question therefore is what the developmental consequences are when the redox status is changed in immature infants treated with oxygen to achieve a higher oxygen tension than in fetal life. This also shuts down HIF-1α in these babies. To be born means that the oxidative load is sharply increased. At the same time, the oxygen demands increase abruptly. The term baby in most cases easily adapts to this transition but for the preterm infant the intra- to extra-uterine transition is not without risks.

Several OS pathways have been defined which include antioxidant enzymes, antioxidant reactions at the mitochondrial level. These include thioredoxins and ubiquinones. They are important antioxidants that are located in the inner mitochondrial membrane scavenging ROS and other free radicals. Thioredoxins as peroxiredoxins contribute to remove hydrogen peroxide from the cells and this antioxidant system which is present in the lung of baboons in the last third part of gestation does not seem to be increased towards term [18]. In humans the thioredoxin system was not detected in week 17th; however, in the neonatal lung levels were similar as in adult lungs. This indicates that the premature infant in spite of a lowered antioxidant enzymatic defense has some well-developed antioxidant defense systems.

In spite of that the intracellular defense against OS is poorly developed in fetal life and the premature infant is often for various reasons exposed to high concentration of supplemental oxygen. A new momentum in the understanding of OS in the perinatal period was added when Frank and his coworkers in 1984 showed that antioxidant enzyme activity in the rat lungs are low in fetal life indicating a lowered antioxidant defense in premature infants [19]. The final days in utero the increase was between 100 and 200 %. This maturation of antioxyenzymes paralleled the maturation of pulmonary surfactant. In a subsequent article, Frank and Sosenko showed that term newborn rabbits increase their antioxidant enzyme activities as response to a hyperoxic challenge. By contrast, premature rabbits did not have such a response [20]. This means that premature rabbits are vulnerable to hyperoxia exposure for at least two reasons, (1) they have a lowered defense and (2) they are not able to mount a defense during a hyperoxic challenge.

In the human brain parietal white matter, Folkerth et al. have shown that the expression of CuZn superoxide dismutase (SOD), MnSOD, and catalase increase during the last half part of pregnancy [21]. In the human developing lung varying results have been found. However, Kaartenaho-Wiik et al. found both CuZn and MnSOD as well as rate limiting enzymes for GSH synthesis are present in bronchial epithelium already from gestational week 17th, that is, in the canalicular stage. Catalase was negative or only weakly positive until gestational week 31 [22]. It therefore seems clarified that the human fetus has SOD and GSH present in the lung at an early stage of development. Whether SOD increases towards term in the human as in animal models is possible but not entirely clear. However, it seems that catalase is low until the last 10 weeks of gestation and increases towards term. Both steroids and endotoxins increase fetal levels of antioxidants.

The fetus and premature infant is also susceptible to inflammations and infections that lead to an increased OS. The premature, especially the sick premature infant, often exhibits free iron in tissues which may enhance the production of hydroxyl radicals through Fenton chemistry [23].

It therefore became clear that premature infants are more exposed to high OS not only because they are (1) exposed to oxygen therapy but also because they (2) are exposed to inflammation (3) have a poorly developed defense against OS, and (4) often have free iron in tissues which triggers production of hydroxyl radicals [24]. This knowledge contributed to a new attention and approach to prevent oxygen radical-related conditions in the newborn or the so-called oxygen radical disease of the newborn. However, since therapies with different antioxidants so far have not been very useful, the most rational approach to reduce OS is to control oxygen supplementation and prevent inflammation. Other potential OS promoting factors as blood transfusions and infusion of parenteral nutrition should also be taken into consideration.

1.8 Effects of Oxidative Stress on the Newborn Brain, Lungs, and Eyes

1.8.1 The Brain

The neonatal brain is susceptible to OS because of its high content of polyunsaturated free fatty acids, its low antioxidative defense, the presence of free iron, exposure to oxygen therapy, and inflammations. Both neuronal and inducible nitric oxide (NO) synthases are high in the developing brain. Hypoxia activates NMDA receptors which lead to calcium influx and subsequent calmodulin activation of neuronal NO synthase. The formation of NO leads to formation of peroxynitrite. This may initiate lipid peroxidation, but also exerts neuroprotection by inducing vasodilatation, angiogenesis, and inhibition of platelet aggregation. Through its inhibition of cytochrome C release from mitochondria peroxyntrite also has antiapoptotic actions. Activated microglia release both reactive oxygen and nitrogen species. For review, see [25].

The immature and pre-oligodendrocytes are more vulnerable to OS than the mature oligodendrocytes. In rodents it seems to be a sensitive window the first week of life when the brain is more easily injured by hyperoxia. For instance, in newborn rats at day 7 exposed to hyperoxia for 24 h, there is induction of neuronal degeneration and apoptosis. A few days later the brain is not so vulnerable to such exposure [26].

1.8.2 The Lungs

The lungs are directly affected by hyperoxia. A number of studies have shown that babies who develop BPD also have elevated markers of OS at an early stage before the diagnosis of BPD is formally set. These markers reflect both protein and lipid peroxidation. Recently we described that genes in oxidative phosphorylation are downregulated in the newborn mice lung after exposure to hyperoxia following a period of hypoxia. Hyperoxic reoxygenation affects pathways regulating cell growth and survival. DNA-damage-responsive genes are restricted to reoxygenation with 100 % oxygen. These findings indicate that cell cycling is reduced and inflammation is enhanced by hyperoxia [27].

A close link between inflammation and OS of the lung has been established [28, 29]. The mechanisms triggering hyperoxia-induced lung injury was recently summarized by Bhandari [30]. It is characterized by an influx of inflammatory cells, increased pulmonary permeability, apoptosis, or cell necrosis leading to endothelial and epithelial cell death. Inflammatory cells produce cytokines and chemoattractants. Vascular endothelial growth factor (VEGF) is initially increased and then depressed. In animal models hyperoxia induces disruption of the alveolar capillary unit and increases vascular permeability. The lung pathological characteristics found in several animal models are similar to those found in BPD in preterm infants.

We recently performed a whole genome study in preterm babies <32 weeks developing OS-related conditions as BPD [9] and ROP [10]. For the BPD study, 111 newborns were included. The mean birth weight was 1,029 g (SD: 290), and the mean gestational age was 27.8 weeks (SD: 2.5). Blood samples were drawn from the study participants on the 5th, 14th, and 28th day of life. The infants were divided into two groups: BPD ($n=68$) and control ($n=43$). Overall 2,086 genes were differentially expressed on the day 5, only 324 on the day 14, and 3,498 on the day 28. Based on pathway enrichment analysis, we found that the cell cycle pathway was up-regulated in the BPD group. The activation of this pathway does not seem to be related with the maturity of the infant. Four pathways related to inflammatory response were continuously on the 5(th), 14(th), and 28(th) day of life downregulated in the BPD group. However, the expression of these genes depended on both immaturity and disease severity. The most significantly downregulated pathway was the T cell receptor signaling pathway. The results of this whole genome expression study revealed alteration of the expression of nearly 10 % of the genome in BPD patients.

1.9 Retinopathy of Prematurity

Today we know that ROP is more frequent in preterm infants if arterial oxygen saturation (SaO_2) is kept higher than 91–95 % compared to 85–89 %. Fluctuations in SaO_2 especially in the high SaO_2 ranges should be avoided. The fundamental studies of Smith and her colleagues showed that VEGF plays an important role in the development of ROP. Hyperoxia, by inhibiting VEGF, leads to a halt in vessel growth and subsequently a relative hypoxia in the retina. Hypoxia then triggers an increase of VEGF and uncontrolled neovascularization occurs. It has been shown that IGF-1 receptor increases VEGF's activation of the mitogen-activated protein kinase system leading to angiogenesis. Low IGF-1 therefore seems to be a prerequisite for the halt in vascularization in the first phase of ROP; this growth factor rises quickly in the third trimester and falls precipitously after premature birth and a low level has been related to severe ROP. Thus a close, however complex relationship between hyperoxia and triggering of ROP has been established [31, 32].

We performed a whole genome study in preterm infants <32 weeks gestational age who developed ROP [10]. Overall, 794 genes were differentially expressed on the 5th day of life, 1,077 on the 14th day of life, and 3,223 on the 28th day of life. In each of the three time points during the first month of life, more genes were under-expressed than overexpressed in the ROP group. Fold change, which was used in the analysis of gene expression data, ranged between 1.0 and 1.5 in the majority of genes differentially expressed. Pathway enrichment analysis revealed that genes in four pathways related to inflammatory response were consistently downregulated.

1.10 Oxygen Therapy of Newborn Infants

Hyperoxygenation may occur both during resuscitation and in the first weeks of life. During procedures such as suctioning of the endotracheal tube, there has been a tradition to elevate FiO_2 both before and after. Following apneas and desaturation episodes FiO_2 has often been increased. After tracheal instillation of natural surfactant, SaO_2 and PaO_2 rise quickly if FiO_2 is not turned down appropriately. It has been shown that a too high PaO_2 after surfactant instillation may be detrimental. In addition, premature infants treated with NO and oxygen may generate the toxic peroxynitrite radical.

Term infants resuscitated with 100 % oxygen in the delivery room when compared to those given air had elevation of OS markers several weeks after birth [33]. Term babies in need of resuscitation who had a high oxygen tension at entrance into the neonatal intensive care unit had developed more HIE and had a higher need for hypothermia treatment when compared to those who had a normoxic condition at transferal [34]. Preterm infants <29 weeks of gestation resuscitated in the delivery room with 90 % oxygen had increased OS for days and weeks when compared to those resuscitated with 30 % oxygen [35]. We presently recommend that preterm infants <33 weeks gestation could be resuscitated initially with 21–30 % oxygen.

In preterm infants <28 weeks of gestation who were aimed to reach an SpO_2 target of 91–95 % when compared to 85–89 % had lower mortality and necrotizing enterocolitis but higher survival [36].

1.11 Antioxidant Therapy

Until now antioxidant therapy has not been successful in treating or preventing BPD or ROP. There have been two studies with SOD. The first one with bovine SOD and the second with human recombinant SOD. This last study was a randomized North

American study including 302 premature infants (600–1,200 g birth weight) treated with exogenous surfactant at birth for respiratory distress syndrome. They were randomized to receive either intratracheal recombinant human SOD (CuZnSOD, 5 mg/kg in 2 mL/kg saline) or placebo every 48 h (as long as intubation was required) for up to 1 month of age. There was no difference in the development of BPD at 28 days of life or 36 weeks' postmenstrual age. At 1 year corrected age, SOD-treated infants had significantly less episodes of wheezing or other respiratory illnesses requiring treatment with asthma medication as well as decrease in emergency department visits and hospitalizations [37]. However, when the authors analyzed the data regarding SOD, they found a significant reduction in severe ROP in babies with gestational age <26 weeks who had been treated with SOD [38].

1.12 Conclusion

ROP and BPD were already in the 1950s and 1960s associated with oxygen toxicity. In the recent decades, it has been understood more clearly how OS is involved in the development of BPD, ROP, and other OS-related newborn conditions. The close linkage found between inflammation and BPD and how this affects OS has further increased our insight in the pathogenesis of this condition. The emphasis on OS and a possible relation to BPD was one rationale for testing out how different SaO_2 effects the development of BPD. Today it seems established that SaO_2 should not exceed certain limits for instance 90–95 %, the optimal range of SaO_2 in these babies is, however not yet established. Antioxidant therapies have so far not clearly demonstrated a benefit. Hyperoxic resuscitation triggers long-term augmentation of OS both in term and preterm infants, and therefore should be avoided, if possible. Whole genome studies indicate that hyperoxia induces inflammation and reduces cell cycling in the newborn.

Term newborn infants in need of resuscitation at birth should be started with air. Preterm infants <33 weeks should be started with 21–30 % oxygen. Target of SpO_2 in immature infants <28 weeks should be between 90 and 95 %.

References

1. Obladen M (2009) History of neonatal resuscitation. Part 2: oxygen and other drugs. Neonatology 95:91–96
2. Campbell K (1951) Intensive oxygen therapy as a possible cause of retrolental fibroplasia: a clinical approach. Med J Aust 2:48–50
3. Northway WH Jr, Rosan RC, Porter DY (1967) Pulmonary disease following respirator therapy of hyaline-membrane disease. Bronchopulmonary dysplasia. N Engl J Med 276:357–368
4. Gerschman R, Gilbert D, Nye SW, Dwyer P, Fenn WO (2001) Oxygen poisoning and X-irradiation: a mechanism in common. 1954. Nutrition 17:162
5. Bracci R, Benedetti PA, Ciambellotti V (1970) Hydrogen peroxide generation in the erythrocytes of newborn infants. Biol Neonate 15:135–141

6. Babior BM, Kipnes RS, Curnutte JT (1973) Biological defense mechanisms. The production by leukocytes of superoxide, a potential bactericidal agent. J Clin Invest 52:741–744
7. Wei EP, Christman CW, Kontos HA, Povlishock JT (1985) Effects of oxygen radicals on cerebral arterioles. Am J Physiol 248:H157–H162
8. Saugstad OD (1988) Hypoxanthine as an indicator of hypoxia: its role in health and disease through free radical production. Pediatr Res 23:143–150
9. Pietrzyk JJ, Kwinta P, Wollen EJ, Bik-Multanowski M, Madetko-Talowska A, Günther CC, Jagła M, Tomasik T, Saugstad OD (2013) Gene expression profiling in preterm infants: new aspects of bronchopulmonary dysplasia development. PLoS One 8:e78585
10. Pietrzyk JJ, Kwinta P, Bik-Multanowski M, Madetko-Talowska A, Jagła M, Tomasik T, Mitkowska Z, Wollen EJ, Nygård S, Saugstad OD (2013) New insight into the pathogenesis of retinopathy of prematurity: assessment of whole-genome expression. Pediatr Res 73(4 Pt 1): 476–483
11. Maltepe E, Saugstad OD (2009) Oxygen in health and disease: regulation of oxygen homeostasis—clinical implications. Pediatr Res 65:261–268
12. Zhao HW, Zhou D, Nizet V, Haddad GG (2010) Experimental selection for Drosophila survival in extremely high O2 environments. PLoS One 5:e11701
13. Hensley K, Robinson KA, Gabbita SP, Salsman S, Floyd RA (2000) Reactive oxygen species, cell signaling, and cell injury. Free Radic Biol Med 28:1456–1462
14. Wright CJ, Dennery PA (2009) Manipulation of gene expression by oxygen: a primer from bedside to bench. Pediatr Res 66:3–10
15. Saugstad OD (2005) Oxidative stress in the newborn—a 30-year perspective. Biol Neonate 88:228–236
16. Clyman RI, Saugstad OD, Mauray F (1989) Reactive oxygen metabolites relax the lamb ductus arteriosus by stimulating prostaglandin production. Circ Res 64:1–8
17. Coalson JJ (2003) Pathology of new bronchopulmonary dysplasia. Semin Neonatol 8:73–81
18. Das KC (2005) Thioredoxin and its role in premature newborn biology. Antioxid Redox Signal 7:1740–1743
19. Frank L, Groseclose EE (1984) Preparation for birth into an O2-rich environment: the antioxidant enzymes in the developing rabbit lung. Pediatr Res 18:240–244
20. Frank L, Sosenko IR (1991) Failure of premature rabbits to increase antioxidant enzymes during hyperoxic exposure: increased susceptibility to pulmonary oxygen toxicity compared with term rabbits. Pediatr Res 29:292–296
21. Folkerth RD, Haynes RL, Borenstein NS, Belliveau RA, Trachtenberg F, Rosenberg PA, Volpe JJ, Kinney HC (2004) Developmental lag in superoxide dismutases relative to other antioxidant enzymes in premyelinated human telencephalic white matter. J Neuropathol Exp Neurol 63:990–999
22. Kaarteenaho-Wiik R, Kinnula VL (2004) Distribution of antioxidant enzymes in developing human lung, respiratory distress syndrome, and bronchopulmonary dysplasia. J Histochem Cytochem 52:1231–1240
23. Buonocore G, Zani S, Sargentini I, Gioia D, Signorini C, Bracci R (1998) Hypoxia-induced free iron release in the red cells of newborn infants. Acta Paediatr 87:77–81
24. Saugstad OD (2001) Is oxygen more toxic than currently believed? Pediatrics 108:1203–1205
25. Haynes RL, Baud O, Li J, Kinney HC, Volpe JJ, Folkerth DR (2005) Oxidative and nitrative injury in periventricular leukomalacia: a review. Brain Pathol 15:225–233
26. Gerstner B, DeSilva TM, Genz K, Armstrong A, Brehmer F, Neve RL, Felderhoff-Mueser U, Volpe JJ, Rosenberg PA (2008) Hyperoxia causes maturation-dependent cell death in the developing white matter. J Neurosci 28:1236–1245
27. Wollen EJ, Sejersted Y, Wright MS, Bik-Multanowski M, Madetko-Talowska A, Günther CC, Nygård S, Kwinta P, Pietrzyk JJ, Saugstad OD (2013) Transcriptome profiling of the newborn mouse lung after hypoxia and reoxygenation: hyperoxic reoxygenation affects mTOR signaling pathway, DNA repair, and JNK-pathway regulation. Pediatr Res 74:536–544
28. Merritt TA, Cochrane CG, Holcomb K, Bohl B, Hallman M, Strayer D, Edwards DK 3rd, Gluck L (1983) Elastase and alpha 1-proteinase inhibitor activity in tracheal aspirates during

respiratory distress syndrome. Role of inflammation in the pathogenesis of bronchopulmonary dysplasia. J Clin Invest 72:656–666
29. Thomas W, Speer CP (2014) Chorioamnionitis is essential in the evolution of bronchopulmonary dysplasia—the case in favour. Paediatr Respir Rev 15(1):49–52
30. Bhandari V (2010) Hyperoxia-derived lung damage in preterm infants. Semin Fetal Neonatal Med 15:223–229
31. Hellström A, Smith LE, Dammann O (2013) Retinopathy of prematurity. Lancet 382: 1445–1457
32. Hård AL, Smith LE, Hellström A (2013) Nutrition, insulin-like growth factor-1 and retinopathy of prematurity. Semin Fetal Neonatal Med. pii: S1744-165X(13)00007-3. doi: 10.1016/j.siny.2013.01.006
33. Vento M, Asensi M, Sastre J, Lloret A, García-Sala F, Viña J (2003) Oxidative stress in asphyxiated term infants resuscitated with 100% oxygen. J Pediatr 142:240–246, Erratum in: J Pediatr. 2003;142:616
34. Kapadia VS, Chalak LF, DuPont TL, Rollins NK, Brion LP, Wyckoff MH (2013) Perinatal asphyxia with hyperoxemia within the first hour of life is associated with moderate to severe hypoxic-ischemic encephalopathy. J Pediatr 163:949–954
35. Vento M, Moro M, Escrig R, Arruza L, Villar G, Izquierdo I, Roberts LJ II, Arduini A, Escobar JJ, Sastre J, Asensi MA (2009) Preterm resuscitation with low oxygen causes less oxidative stress, inflammation, and chronic lung disease. Pediatrics 124:e439–e449
36. Saugstad OD, Aune D (2013) Optimal oxygenation of extremely low birth weight infants: a meta-analysis and systematic review of the oxygen saturation target studies. Neonatology 105:55–63
37. Davis JM, Parad RB, Michele T, Allred E, Price A, Rosenfeld W, North American Recombinant Human CuZnSOD Study Group (2003) Pulmonary outcome at 1 year corrected age in premature infants treated at birth with recombinant human CuZn superoxide dismutase. Pediatrics 11:469–476
38. Parad RB, Allred EN, Rosenfeld WN, Davis JM (2012) Reduction of retinopathy of prematurity in extremely low gestational age newborns treated with recombinant human Cu/Zn superoxide dismutase. Neonatology 102:139–144

Chapter 2
Reactive Oxygen Species and Nitric Oxide in Vascular Function

Michael S. Wolin

Abbreviations

ecSOD	Extracellular superoxide dismutase (SOD3)
eNOS	Endothelial nitric oxide synthase (NOS3)
ETC	Electron transport chain
H_2O_2	Hydrogen peroxide
HPV	Hypoxic pulmonary vasoconstriction
NADH	Nicotinamide adenine dinucleotide (reduced)
NADPH	Nicotinamide adenine dinucleotide phosphate (reduced)
NAD(P)H	NADH and NADPH
NO	Nitric oxide
iNOS	Inducible or inflammatory nitric oxide synthase (NOS2)
nNOS	Neuronal nitric oxide synthase (NOS1)
NOS	Nitric oxide synthase
Nox	NADPH oxidase
Nrf2	Nuclear factor (erythroid-derived 2)-like 2
O_2^-	Superoxide anion
PG	Prostaglandin
PKG	Protein kinase G
RNS	Reactive nitric oxide-derived species
ROS	Reactive oxygen species
SOD	Superoxide dismutase

M.S. Wolin (✉)
Department of Physiology, New York Medical College,
Valhalla, NY 10595, USA
e-mail: mike_wolin@nymc.edu

H. Tsukahara and K. Kaneko (eds.), *Studies on Pediatric Disorders*, Oxidative Stress in Applied Basic Research and Clinical Practice, DOI 10.1007/978-1-4939-0679-6_2,

2.1 Roles for Reactive Oxygen Species and Nitric Oxide in Vascular Function

The vasculature can be exposed to sources of reactive oxygen species (ROS) originating from several different oxidases located in key cell types such as endothelium, smooth muscle, fibroblasts, and inflammatory cells that are present in the vessel wall. While the endothelium is a prominent source of nitric oxide (NO) derived from endothelial NO synthase (eNOS) associated with the physiological regulation of vascular function, other cell types (e.g., nerves) and forms of nitric oxide synthase (NOS) (e.g., neuronal and inflammation-induced) can be sources NO influencing vascular function. Cells that surround the microcirculation of tissues and components of blood can also be sources of ROS and NO that control vascular function. Early research on vascular regulation by ROS evolved from several different directions. Studies on the pathophysiology of conditions such as ischemia/reperfusion, hypertension, diabetes, complications of pregnancy, atherosclerosis, inflammation, adult respiratory distress syndrome identified conditions where ROS appeared to be influencing vascular function. In addition, the investigation of processes involved in the regulation of vascular smooth muscle by the endothelium and hypoxia had major roles in beginning to define the signaling aspects of these species in the regulation of vascular tone. The reaction of the superoxide anion with NO helped with the initial identification of the biological importance of NO in the context of its role as a key endothelium-derived relaxing factor [11, 20]. In addition, this reaction is also a major contributing factor in the progression of most vascular diseases. This chapter will consider sources and processes through which NO and ROS influence vascular function. Figure 2.1 shows how enzymes producing and metabolizing ROS are involved in initiating signaling processes.

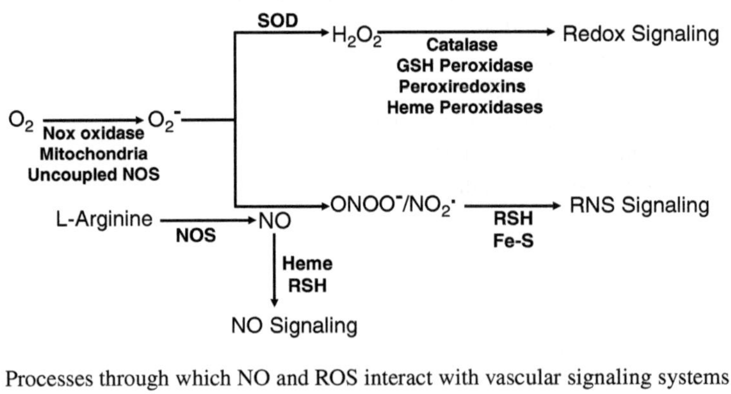

Fig. 2.1 Processes through which NO and ROS interact with vascular signaling systems

2.2 Origins of Reactive Oxygen and Nitric Oxide-Derived Species in the Vasculature

The enzymatic reactions producing ROS often begin with reactions that transfer one or two electrons from sources such as nicotinamide adenine dinucleotide phosphate (NADPH) or nicotinamide adenine dinucleotide (NADH) to molecular oxygen (O_2)-generating superoxide anion (O_2^-) and hydrogen peroxide (H_2O_2), respectively. NOS enzymes produce NO from L-arginine. These enzymes have distinctive cellular and subcellular localizations, and many very specific mechanisms regulating their expression, activity, and subcellular localization that appear to be important components of processes they regulate. As will be discussed in other sections, the conversion of superoxide anion to H_2O_2 by the various forms of superoxide dismutase (SOD) is also a major factor contributing to the origins of peroxide (and in preserving NO). Table 2.1 lists some of the properties of systems generating NO and ROS that are associated with the regulation of vascular function.

2.2.1 *NOS as a Generator of NO and Superoxide*

There are three major forms of NOS enzymes which produce NO from L-arginine, NADPH, and oxygen [17]. NOS1 or nNOS was first described as a neuronal NOS regulated by intracellular calcium. NOS2 or iNOS was identified as an inducible enzyme associated with inflammatory cells and responses. NOS3 or eNOS was discovered as receptor regulated, calcium-activated endothelial source of NO involved in endothelium-dependent relaxation. The calcium regulated forms of NOS seem to become uncoupled from NO generation associated with the production superoxide in place of NO. A depletion or oxidation of the tetrahydrobiopterin cofactor of NOS and/or the lack of availability of L-arginine seem to be some of the factors that promote uncoupling [16]. This will create situations where cells such as endothelium are exposed to increased superoxide, which will either react with any NO that is available or be converted to H_2O_2 by the SOD enzymes. Thus, the availability of NO will decrease in the presence of superoxide associated with the formation and biological effects of increased reactive NO-derived species and/or H_2O_2

2.2.2 *Nox Oxidases as Sources of Superoxide and Peroxide*

Nox oxidases are important sources of ROS that influence the regulation of vascular function [13]. The phagocytic cell gp91phox or Nox2 was the first well defined oxidase in this family of NAD(P)H oxidases. This oxidase contains membrane-bound gp91phox (Nox2) and p22phox subunits which are activated to produce superoxide by signaling mechanisms controlling the binding of cytosolic subunits.

Table 2.1 Regulation of enzymatic sources of NO and ROS in the vasculature

Enzyme	Properties	Regulation
NOS	Membrane-bound endothelial cell eNOS, neuronal nNOS, and increasing the expression of inducible iNOS in vascular smooth muscle can be sources of vasodilator levels of NO	The ability of the endothelium to generate NO (or ROS) is stimulated by receptors (e.g., acetylcholine, adenosine) and the shear force of flow through increased calcium and multiple additional signaling mechanisms. Nerve stimulation of nNOS by mechanisms similar to eNOS and inflammatory signaling increasing the expression of iNOS also increase vascular NO
	Uncoupling of eNOS allows it to generate superoxide in place of NO, which can be associated with vasodilator levels of peroxide	Oxidation or loss of tetrahydrobiopterin and a deficiency in L-arginine promote uncoupling of eNOS (and nNOS)
NAD(P)H oxidases (Nox)	Nox1 is a membrane-bound oxidase which also contains a p22phox subunit. Its expression is often associated with vascular smooth muscle proliferation	Nox1 expression is enhanced by growth factors, and its activity is stimulated by phosphorylation of p47phox or increased expression of NoxO1 subunits, and by oxidant signaling activation of rac1
	Nox2 has a p22phox subunit and it appears to be an important oxidase in vascular smooth muscle that may participate in the secretion of superoxide	Nox 2 is stimulated by p47phox or increased and by oxidant signaling activation of rac1
	Nox4 is a constitutively active oxidase in multiple vascular cell types containing a p22phox subunit. This oxidase seems to be most closely associated with hydrogen peroxide generation. It appears to have a nuclear and microsomal localization in vascular smooth muscle cells	Nox4 expression is stimulated by agents such as TGFbeta. The level of expression of the Nox4 subunit appears to be key regulator of the activity of this oxidase
	Nox5 is a human vascular oxidase	The activity of Nox5 appears to be regulated by intracellular calcium levels

Mitochondria	Mitochondria appear to generate superoxide primarily from regions in Complexes I and III of the electron transport chain (ETC). Peroxide may be the dominant ROS released from mitochondria	Increased mitochondrial NADH and electron density appear to promote superoxide generation by Complex I and conditions increasing the semiquinone form of Coenzyme Q may control superoxide generation from Complex III. Changes in the redox of sites such as thiols and expression of ETC protein subunits can modulate superoxide generation
Xanthine oxidase	This enzyme is enriched in microvascular endothelium, and it normally uses hypoxanthine and xanthine to function as a dehydrogenase that reduces NAD to NADH. When converted to an oxidase, it generates both superoxide and hydrogen peroxide	Hypoxia, activation of P38-MAP kinase and conditions promoting thiol oxidation and proteolysis appear to stimulate xanthine oxidase activity. Limitations in tissue energy metabolism increase oxidase activity by promoting the availability of hypoxanthine. NADH availability may also drive NADH oxidase activity
Cyclooxygenase	These enzymes have a peroxidase activity which appears to generate radicals from NADH and NADPH, which react with oxygen to generate superoxide. It also generates and consumes the peroxide PGG_2 in the process of generating PGH_2	Peroxide stimulates the release of the arachidonic acid substrate and it stimulates and maintains the prostaglandin (PG) and superoxide generating activity of cyclooxygenase enzymes
Cytochrome P450	The endothelial cell form of this enzyme in some vascular segments can potentially generate vasodilator levels of peroxide	Some endothelial cell receptors may activate the production of ROS by cytochrome P450

Some of the key properties of Nox oxidases in vascular tissue are summarized in Table 2.1. While the phagocytic oxidase appears to use only cytosolic NADPH for generating superoxide, the availability of NADPH (and perhaps NADH) may be factors in modulating vascular Nox oxidase activity [13, 28]. Since lactate appears to increase superoxide and it can elicit vasodilation through peroxide, increased cytosolic NADH as a substrate for oxidases may be a factor in promoting ROS-mediated responses. Vascular smooth muscle cells may also have the ability to secrete superoxide from oxidases such as Nox1 or Nox2 when these oxidases are localized in the plasma membrane region [29]. The diversity of mechanisms controlling the subunit expression, subcellular localization, and activation of Nox oxidases in vascular smooth muscle and other cells present in blood vessels enables these oxidases to have numerous roles in the regulation of many aspects of vascular physiology and pathophysiology.

2.2.3 Mitochondria as Sources of ROS

Early observations led to the recognition that the redox status of components of the mitochondrial electron transport chain (ETC) were key factors in controlling the generation of superoxide and/or H_2O_2, and that vascular mitochondria appear to secrete peroxide [2, 25, 28]. However, the actual mechanisms controlling ROS generation by mitochondria appear to be influenced by many poorly defined processes summarized in Table 2.1. Hypoxia may influence these factors in ways that promote increased ROS generation as a dominating factor over limitations in the availability of oxygen for ROS production [25]. Studies in vascular disease models suggest that alterations in the expression and function of ETC components and ROS metabolizing enzymes, and mitochondrial membrane potential also appear to be major factors in controlling regulation by mitochondria-derived ROS [5].

2.2.4 Xanthine Oxidase, Cyclooxygenase, and Other Sources of ROS Regulating Vascular Function

A variety of additional cellular enzymes appear to be sources of ROS, however, only a few of these sources have been documented to be factors in vascular regulation. Some of the properties of ROS generation by these oxidases included in this section are described in Table 2.1. It appears that xanthine oxidase activity can be an important source of ROS generation by microvascular endothelium under some pathophysiological conditions, such as ischemia or hypoxia [4]. The initial enzyme generating prostaglandins, cyclooxygenase also appears to be a source of ROS influencing vascular function [12]. There are other sources of ROS generation by the vasculature such as cytochrome P450 and monoamine oxidase. The specific roles of the sources of ROS included in this section in vascular regulation need to be better defined.

2.3 Chemical and Metabolic Properties of Reactive Oxygen and Nitric Oxide-Derived Species

The properties of chemical reactions associated with individual ROS and NO are major factors in the removal of these species and their conversion to other species or redox events that participate in the control of signaling mechanisms associated physiological and pathophysiological aspects of vascular regulation. Many of these properties are included in Table 2.2.

Table 2.2 Metabolic properties of NO and ROS that potentially participate in vascular regulation

Species	Properties	Origins of regulatory actions
Superoxide	Superoxide anion is a negatively charged free radical formed by a one-electron reduction of oxygen	• SOD enzymes convert superoxide to peroxide • Superoxide reacts with NO to form $ONOO^-$ • Superoxide disrupts iron–sulfur (Fe–S) centers • Superoxide reacts efficiently with catecholamines (e.g., epinephrine and norepinephrine)
Peroxide	Oxidases generate peroxide either by a two-electron reduction of oxygen or from superoxide via the activity of SOD enzymes	• Peroxide metabolism by catalase, glutathione peroxidase, peroxiredoxins, or heme peroxidases can initiate interactions with regulatory systems by producing oxidized glutathione and thioredoxin, and reactive oxidized forms iron
NO	NO is a dissolved membrane permeable free radical gas	• NO binds ferrous hemoproteins (e.g., guanylate cyclase) • NO reacts with superoxide to form peroxynitrite • NO reacts with oxyhemoglobin to form nitrate and methemoglobin • NO reacts with oxygen to form nitrogen dioxide, and subsequently nitrite and nitrate
Peroxynitrite ($ONOO^-$)	$ONOO^-$ is an unstable anion which generates nitrogen dioxide and reactive radicals	• $ONOO^-$ oxidizes or nitrates tetrahydrobiopterin, thiols, unsaturated fatty acids, tyrosine sites, etc. • Nitrogen dioxide also oxidizes and nitrates thiols and unsaturated fatty acids, to generate molecules which often slowly release NO
Reactive species	Peroxide reactions with iron and peroxynitrite generate reactive radicals with hydroxyl radical-like reactivities	• These reactions may elicit biological effects through promoting multiple oxidation and lipid peroxidation reactions which may have regulatory effects through carbonyl formation

2.3.1 Properties of NO and Its Oxidized Species

Nitric oxide is a dissolved gas with hydrophobic properties which allow it to diffuse efficiently between cells in the vascular region until it is removed by interactions with biological molecules [17, 22]. The unpaired electron or free radical aspect of NO allows it to react extremely efficiently with other molecules with unpaired electrons such as superoxide anion. The reaction with superoxide is an extremely efficient mechanism for both the removal of NO and formation of peroxynitrite anion ($ONOO^-$). NO also either binds or reacts with some transition metal centers, and its interactions with iron containing molecules seems to be involved in some of the better understood roles it has in signaling and in its removal. For example, some of the most sensitive sites for biological regulation by NO include the hemoproteins soluble guanylate cyclase and cytochrome oxidase in the mitochondrial ETC. The binding of NO to hemoglobin in erythrocytes generates methemoglobin and nitrate (NO_3^-), and this may be a major mechanism for the removal of NO as it enters the circulation. The spontaneous aerobic decomposition of NO results in nitrogen dioxide ($NO_2^{\bullet}$) formation, and this also generates nitrite (NO_2^-) and nitrate. Species such as peroxynitrite and nitrogen dioxide are viewed as reactive nitric oxide-derived species (RNS).

Peroxynitrite is an unstable molecule which appears to participate in both signaling and pathophysiological processes [21]. Peroxynitrite readily generates highly reactive oxidants such as nitrogen dioxide and carbon dioxide radical. The species formed from peroxynitrite are associated with reactions such as oxidation and/or nitration of thiols, unsaturated fatty acids, tyrosine, and urate. At low levels of RNS, these reactions may function in vascular signaling-related processes through mechanisms such as modifying protein thiols, generating nitrated lipids and species which slowly regenerate NO. Whereas, as the levels of RNS increase, they are likely to drive pathophysiological processes through actions such as inactivating enzymes by nitrating protein tyrosine sites, oxidizing thiols, and depleting antioxidant systems such as mitochondrial or Mn-SOD.

2.3.2 Properties of Superoxide and Its Metabolism by Various Forms of SOD

The free radical side of superoxide results in it reacting very rapidly with NO, and for it to react with itself generating both oxygen and H_2O_2. This reaction is accelerated by SOD enzymes present in the intracellular (Cu, Zn-SOD), mitochondrial matrix (Mn-SOD), and extracellular (ecSOD) regions. While only a limited number of cell types appear to secrete ecSOD, vascular smooth muscle is an important source of ecSOD, and the levels of this enzyme appear to be very influential in controlling regulation by NO and the balance between some physiological and pathophysiological aspects of vascular regulation through interactions listed in Table 2.2.

2.3.3 *Properties of Hydrogen Peroxide and Various Aspects of Its Metabolism*

H_2O_2 is a rather stable neutral molecule which appears to be transported in a manner similar to water. While some oxidases directly generate H_2O_2, the metabolism of superoxide by SOD enzymes is a major source of its generation. There are several enzymes which metabolize peroxide (See Table 2.2), and its metabolism by these enzymes located in various cellular regions and organelles may be some of the most sensitive processes used in selectively influencing specific signaling mechanisms controlled by peroxide [28]. Oxidized glutathione (GSH) and thioredoxin (Trx), generated by peroxide metabolizing enzymes such as GSH peroxidases and peroxiredoxins, have important roles in controlling the redox status of thiols on proteins in signaling-related processes. There are other heme peroxidases, such as cyclooxygenase myeloperoxidase, which promote regulation as a result of consuming peroxide. There is also a circulating form of glutathione peroxidase which can potentially influence vascular function by removing extracellular peroxide. While this extracellular enzyme can utilize various thiols such as cysteine to metabolize peroxide, the low levels of thiols in the extracellular environment might be a factor influencing the function of this system. These various peroxide metabolizing enzymes function to both maintain low levels of peroxide and selectively regulate signaling systems controlling vascular function through the oxidized factors generated during the consumption of peroxide.

Peroxide reactions with iron and mediators such as peroxynitrite and nitrogen dioxide are sources of highly reactive chemical species that promote processes such as lipid modifications associated with oxidation, nitration and autooxidation, and protein modifications such as tyrosine nitration, carbonyl formation, and irreversible oxidations of cysteine thiols. Many of the antioxidants found in nature scavenge some of the reactive species discussed in this section, without markedly influencing the signaling actions of NO, superoxide, and peroxide.

2.4 Origins of Interactions of Reactive Oxygen and Nitric Oxide Species with Vascular Signaling Systems

The local concentrations and chemical interactions of each individual reactive oxygen and NO species greatly determines their influence on signaling systems controlling vascular function through processes such as the chemical or redox interactions described in the previous section. This section considers interactions summarized in Table 2.3 with some of the better understood systems that appear to be directly regulated by individual reactive oxygen or NO species.

Table 2.3 Key targets of direct interactions of NO and ROS with vascular regulatory systems

Species	Regulatory interaction
NO	• NO readily binds the Fe^{2+} heme of sGC and stimulates cGMP generation and PKG signaling • NO readily reversibly binds a heme site in cytochrome oxidase of mitochondria in competition with oxygen associated with a reversible inhibition of respiration and perhaps oxygen sensing • NO regulates signaling by nitrosation of thiols (RSNO) on proteins
Superoxide	• The reaction of superoxide with NO attenuates its regulatory effects and shifts regulation to the actions of $ONOO^-$ • Superoxide disrupts Fe–S centers associated with inhibiting mitochondrial respiration and perhaps altered aspects of Fe metabolism
Peroxide	• The heme peroxidase-like metabolism of peroxide by cyclooxygenase promotes the generation vasodilator and/or vasoconstrictor prostaglandins • The metabolism of peroxide by catalase can stimulate sGC, promoting cGMP-stimulated PKG signaling • Peroxide metabolism through peroxidases such as myeloperoxidase can convert nitrite to reactive nitrogen dioxide-like species which promote nitration of protein tyrosines • The metabolism of peroxide by peroxiredoxins or glutathione peroxidases may be involved in the activation of multiple vascular regulatory mechanisms such as vasodilation by dimerization–activation of PKG, stimulation of vasoconstriction through systems including PKC, rho kinase, ERK and P38 MAP kinases, and multiple aspects of systems regulating of intracellular calcium. Peroxide metabolism by these pathways may also regulate multiple processes through inhibiting tyrosine phosphatases
$ONOO^-$ (and nitrogen dioxide)	• $ONOO^-$ disrupts Fe–S centers associated with irreversible inhibition of mitochondrial respiration, and perhaps altered aspects of Fe metabolism • $ONOO^-$ inhibits tyrosine phosphatases • Prostaglandin I_2 synthase is inactivated by $ONOO^-$ via nitration of an active site tyrosine • Oxidation and nitration of unsaturated fatty acids generates vasoactive lipid mediators
NADP/NADPH redox	• Oxidation of cytosolic NADPH by decreased biosynthesis or increased ROS-associated consumption can coordinate multiple mechanisms promoting vasodilation by increasing cytosolic NADP/NADPH ratios. Vasoconstriction can result from decreasing NADP/NADPH ratios

2.4.1 Roles for Prostaglandins in the Vasoactive Effects of ROS

One of the earliest observed actions of peroxide on vascular and microvascular preparations was that stimulation of the generation of prostaglandins was often a prominent factor in the vasodilator or vasoconstrictor responses that were observed.

It appears that both the release of arachidonic acid and the heme peroxidase activity of cyclooxygenase used to generate prostaglandin H_2 are very sensitive to stimulation by peroxide [27]. The vasodilator prostaglandin I_2 is often the dominant endothelium-derived prostaglandin generated under physiological conditions from prostaglandin H_2. However, the enzyme generating this prostaglandin appears to be very susceptible to inactivation by oxidant stress conditions associated with pathophysiology. This can change the profile of prostaglandins generated by the endothelium towards vasoconstrictors such as prostaglandin H_2. The prostaglandin I_2 synthase enzyme has a tyrosine group which is one of the most sensitive sites for nitration by peroxynitrite, and tyrosine nitration results in the inactivation of this enzyme. Peroxide stimulation of prostaglandins from sources such as the activation of platelets generating the vasoconstrictor thromboxane A_2 and prostaglandin generation by vascular smooth muscle cells can also be important factors in vascular regulation by ROS. Thus, prostaglandins can be key factors in the vasoactive actions of reactive oxygen and NO-derived species.

2.4.2 *Vascular Smooth Muscle Regulation Associated with cGMP, a Potential Coordinator of Multiple Processes Contributing to Vascular Relaxation*

It is well established that the stimulation of protein kinase G (PKG) by cGMP regulates vascular smooth muscle relaxation though coordinating the activation of multiple mechanisms that either lower intracellular calcium and/or inhibit the actions of calcium [15]. There are multiple redox-associated processes summarized in Table 2.3 which have major effects on this regulatory system [19]. Stimulation of the generation of cGMP by sGC seems to be the most sensitive target for the actions of NO as a result of the affinity of the Fe^{2+} or ferrous heme of sGC for NO being in the low nanomolar concentration range. Any significant source of superoxide attenuates sGC stimulation through scavenging NO. The heme of sGC also appears to readily undergo oxidation to the ferric of Fe^{3+} form of heme, which does not support sGC stimulation by NO. Oxidation of the heme of sGC also appears to result in a protesomal degradation that can deplete sGC [23].

A peroxide has several regulatory interactions with the sGC/PKG systems which are listed in Table 2.3 [19]. For example, a peroxide-elicited mechanism of activating PKG-1alpha in a cGMP-independent manner by thiol oxidation causing dimerization of its two subunits has recently been identified and found to function as an important vasodilator mechanism in vascular smooth muscle. Thus, the activity of PKG is extremely sensitive to the influence of NO, superoxide, peroxide, and processes controlling protein thiol redox. Redox-related alterations in cGMP-associated signaling are likely to be major factors in vascular regulation due to the many coordinating roles PKG has in controlling vascular function.

2.4.3 The Subcellular Organization of Thiol and NAD(P)H Redox Systems Enables Them to Participate in Controlling Multiple Aspects of Vascular Function

The ability of ROS and RNS to regulate multiple vasodilator, vasoconstrictor, and other pathophysiological processes originates from the large number of cellular regulatory systems that have one or more key protein components which are regulated by changes in thiol redox centers. Changes in thiol or sulfur redox can be initiated by either oxidant species or as a result of decreased activity of systems maintaining their reduced status. Some of the most common enzymes directly controlling protein thiol redox are thioredoxin and glutathione redox regulation of glutaredoxin [8]. Control of the redox status of these systems through oxidation by peroxide metabolizing enzymes or reduction by NADPH-dependent thioredoxin and glutathione reductases in localized subcellular regions is likely to have a major influence on individual-specific cellular regulatory systems. Reactive species such as peroxynitrite and nitrogen dioxide can also directly oxidize protein thiols. The control of protein phosphorylation signaling by systems such as tyrosine kinases are often enhanced by oxidation of thiols present in the active sites of tyrosine phosphatases. Most potassium channels have thiols that are thought to regulate channel opening in a redox-dependent manner. Calcium regulation by the sacroplasmic reticulum can be controlled by thiol redox changes in systems such as the SERCA pump. Thus, these and many additional thiol redox-regulated cellular systems (See Table 2.3) are controlled by the influence of reactive oxygen and NO-derived species. It is also important to note that control of the redox status of NADPH in subcellular regions is also potentially a major factor in the signaling that is observed because NADPH maintains the reduced forms of Trx and glutathione. It seems multiple biological regulatory processes can be selectively controlled by differences in subcellular interactions related to ROS, and NAD(P)H and thiol redox [18, 25, 28]. These poorly understood interactions potentially participate in aspects of vascular regulation ranging from the control of contractile function to processes associated with remodeling.

2.5 Physiological Mechanisms of Vascular Regulation That Appear to Involve Reactive Oxygen and Nitric Oxide-Derived Species

This section considers how changes in multiple reactive oxygen and NO species can participate in some of the better understood physiological vascular regulatory systems. Many aspects of the physiological status of the cells in the vessel wall, such as exposure to vasoactive agents, pressure, flow, hypoxia, mediators of inflammation and adaptive remodeling, may also help determine the properties of signaling mechanisms which can be regulated.

2.5.1 Roles for Reactive Oxygen and NO Species in Endothelial Regulation of Vascular Function by Vasoactive Agents, Flow and Pressure

The actions of vasoactive substances, shear force of flow and impact of increased pressure often have effects on endothelium associated with regulation by NO and ROS [14]. Receptors on endothelium which promote activation of signaling mechanisms such as increased intracellular calcium and/or oxidase activation promote the release of mediators such as NO, H_2O_2, prostaglandins, and cytochrome P450-derived eicosanoids. Increases in the shear force of flow and elevated blood pressure often activate endothelium in a manner that also releases some of these mediators, associated with flow-induced vasodilation. At high levels of flow-induced shear forces on endothelium and/or in the presence of elevated pressure, increased generation of reactive oxygen and NO-derived reactive species begins to change the profile of mediators released from endothelium. For example, these agents can uncouple eNOS and shift the prostaglandins generated to vasoconstrictors.

2.5.2 Roles for Redox Processes in Vascular Smooth Muscle Vasodilator and Constrictor Mechanisms Activated by Vasoactive Agents and Increased Pressure

The stimulation of receptors promoting vasoconstriction and exposure of vascular tissue to increased stretch or pressure has been shown to often be associated with oxidase activation and the stimulation of force enhancing mechanisms such as protein kinase C, rho kinase, and ERK and/or P38 MAP kinases [9, 24]. As discussed in Sect. 2.4, there are also multiple redox controlled signaling mechanisms promoting vasodilation. However, very limited information is available on vascular smooth muscle receptor-mediated vasodilator mechanisms mediated by stimulation of increased ROS generation. Vasoconstrictors have also been shown to stimulate glucose-6-phosphate dehydrogenase activity, and this should be associated with decreased vasodilation as a result of increased cytosolic NADPH/NADP ratios [3]. Increased stretch or pressure has been shown to activate oxidases in multiple cell types in the vessel wall, including vascular smooth muscle. This action of stretch or pressure is also associated with stimulation of force enhancing signaling mechanisms discussed above in this section.

2.5.3 *Roles for Redox Processes in Vascular Mechanisms of Sensing Hypoxia*

The pulmonary circulation responds to decreased availability of oxygen through a hypoxic pulmonary vasoconstriction (HPV) response which is designed to match ventilation to perfusion. There is much evidence that the oxygen sensor in pulmonary arteries producing the HPV response appears to originate primarily from redox changes in pulmonary arterial smooth muscle. However, there is much controversy in the direction of changes in ROS caused by hypoxia and in the signaling mechanisms by which hypoxia controls contractile function. Some of the mechanisms proposed for the pulmonary artery HPV response include hypoxia decreasing voltage-regulated potassium channel opening by mitochondrial-derived ROS [18], or decreased Nox4-derived peroxide stimulation of both sGC/cGMP and thiol oxidation-elicited dimerization of PKG mechanisms [19]. In contrast, Schumaker's studies suggest hypoxia increases mitochondrial ROS promoting pulmonary artery contraction via cytosolic oxidation processes increasing intracellular calcium [25]. In contrast, in various segments of the systemic circulation, hypoxia usually increases blood flow to help match oxygen delivery to the metabolic needs of the tissue being perfused. While metabolic mediators released from the tissues and the endothelium may be factors in this response [7], hypoxic sensing by ROS and/or redox-associated mechanisms in systemic arterial smooth muscle may be a factor in regulating the increase in blood flow that is observed. While hypoxia also appears to modulate ROS in the systemic arteries, often in a manner similar to pulmonary arteries [26], our studies in bovine coronary arterial smooth muscle suggest that the relaxation to hypoxia seems to originate from an ROS-independent oxidation of cytosolic NADPH, and dimerization of PKG may be a factor in promoting relaxation to hypoxia under these conditions [19]. In contrast, hypoxia appears to promote increased NADPH in bovine pulmonary arterial smooth muscle associated with decreased PKG dimerization and activation [10]. Further studies are needed to define the dominant mechanisms involved in vascular regulation by hypoxia.

2.6 Pathophysiological Changes in Vascular Regulation Associated with Alterations in Regulation by Reactive Oxygen and Nitric Oxide-Derived Species

This section considers aspects of how changes in oxidant signaling and stress participate in the progression and/or attenuation of vascular disease processes. Many pediatric and adult disease processes are associated with prominent roles for increased generation of ROS and their interactions with multiple vascular regulatory systems shown in Fig. 2.2 that attenuate regulation by NO and increase vascular contraction and reactivity.

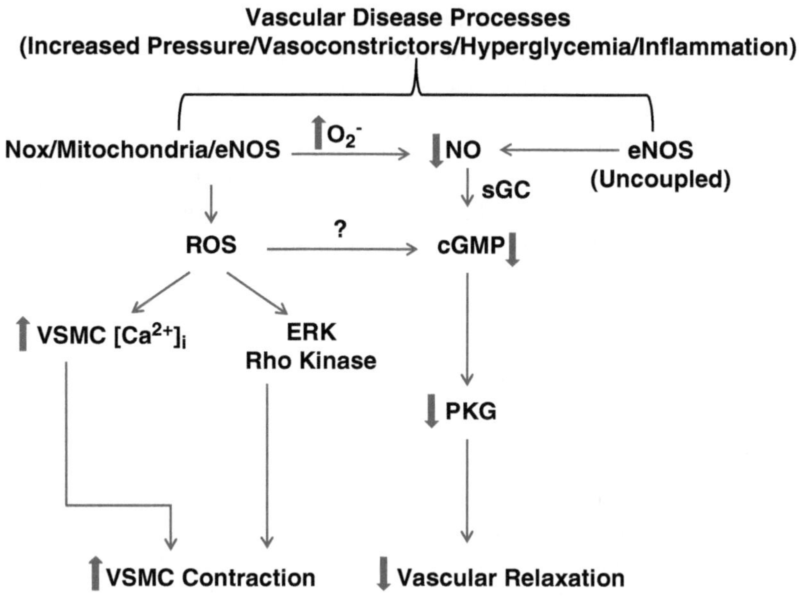

Fig. 2.2 Vascular disease processes activate various oxidases to promote signaling mechanisms that increase vascular contraction and reactivity

2.6.1 Processes Associated with Disease Progression

One of the most well-documented processes occurring in many pathophysiological situations is superoxide impairing the actions of NO, and promoting the consequences of increased reactive peroxynitrite-derived species [21]. As previously discussed, some of the major consequences of increased peroxynitrite formation in endothelium include a loss of tetrahydrobiopterin promoting eNOS uncoupling and a nitration-induced inhibition of PGI_2 synthase resulting in a shifting prostaglandins generated towards agents such as PGH_2 and thromboxane A_2 promoting vasoconstriction and thrombosis. Since superoxide also readily reacts with and inactivates catecholamines such as norepinephrine and epinephrine, this process can influence vascular function in many pathophysiological situations [27]. Both superoxide and peroxynitrite disrupt iron–sulfur centers, and this could alter key aspects of mitochondrial function such as electron transport and heme biosynthesis in ways that either regulate vascular function or the release of hypoxia-associated metabolic regulators of blood flow. Many of the mediators involved in disease processes activate vascular oxidases in the endothelium and/or vascular smooth muscle. Some examples of the most extensively studied agents or conditions associated with oxidase activation include angiotensin II, endothelin, elevated blood pressure, increased shear stress, ischemia/reperfusion, and inflammation. Metabolic aspects of disease processes such as elevated glucose, low-density lipoprotein, and homocysteine, and chronic hypoxia may also have marked effects on vascular oxidase activities. As discussed in Sect. 2.2, ROS generation by oxidases can be increased by processes including cell signaling-associated mechanisms, increased expression of oxidase subunits, and ROS promoting the activation of additional oxidase mechanisms. There are many studies describing specific signaling mechanisms activating individual sources of ROS which are linked to adaptive physiological or pathophysiological vascular regulatory processes influencing vascular function and remodeling. Initial studies by Harrison related to Nox oxidases causing eNOS uncoupling have helped highlight the importance and complexity of the roles for ROS-elicited oxidant-induced activation of additional sources of ROS in disease mechanisms [6]. When eNOS is uncoupled, it can become an endothelial source of vasodilator levels of H_2O_2. However, peroxide is likely to differ from NO in many of the additional vascular remodeling-related processes it regulates. Thus, conditions associated with the progression of disease processes appear to increase the generation of ROS by various vascular oxidases in a manner which is often associated with a loss of NO and/or increased generation of reactive NO-derived species. Under these conditions, many processes become activated in addition to those controlling vascular contractile function, such as inflammation, remodeling, and thrombosis.

2.6.2 Protective Mechanisms Activated by Oxidant Signaling

Oxidant-associated vascular disease processes may also markedly alter the expression of components of antioxidant and signaling systems which can either be protective or part of the sequence of pathophysiological processes. For example, oxidant conditions often activate the expression of heme oxygenase-1 (HO-1), a system which has protective actions associated with removal of heme and the generation of carbon monoxide and biliverdin (and its conversion to bilirubin) [1]. The availability of heme enhances the pathophysiology of ROS through generating highly reactive hydroxyl radical-like species. Carbon monoxide has multiple actions on systems such as ROS and NO generation and potassium channel regulation which can be associated with both vasodilation and anti-inflammatory protective effects. Bilirubin is both a potent antioxidant and an inhibitor of Nox oxidase activation. Electrophilic molecules derived from oxidant reactions can promote the expression of multiple antioxidant enzymes through regulators of gene expression such as Nrf-2. Some signaling systems such as NO have effects on the enhancing the expression of other systems such as ecSOD which function to protect NO signaling. Thus, feedback activation of protective systems is often part of the progression of vascular disease processes.

2.6.3 *Antioxidant-Associated Therapeutic Approaches*

Certain aspects of oxidant regulation in vascular disease processes can have beneficial therapeutic effects. The delivery of SOD or agents with SOD activity can have marked beneficial effects through lowering the levels of superoxide and preventing its ability to scavenge NO and generate peroxynitrite, liberate iron from iron–sulfur centers, or attenuate NO biosynthesis through depletion of tertahydrobiopterin. Agents targeting the biosynthesis or actions on specific receptors promoting oxidase activation such as systems associated with angiotensin II and endothelin-1 can have major beneficial effects in treating vascular disease processes such as various forms of systemic and pulmonary hypertension. A consequence of statin drugs inhibiting the biosynthesis of cholesterol can include preventing the isoprenylation of rac-1 a key component promoting Nox-2 oxidase activation. While most dietary antioxidants have rather weak superoxide and peroxide scavenging activity, they may possess scavenging activities for reactive species generated from NO and peroxynitrite. This profile of actions would allow important basal regulatory actions of NO and ROS that participate in normal physiological processes such as endothelial regulation of vascular tone and oxygen sensing. However, some of the pathophysiological regulatory actions of reactive oxygen and NO-derived species associated with their chemical reactivity could be attenuated by dietary antioxidants.

2.7 Concluding Remarks

There are many similarities in how ROS are involved in altering vascular function across diseases and vascular segments. Increased pressure, excessive hypoxia, hyperglycemia, and disease-associated mediators often activate oxidases in ways that promote common effects. Endothelial dysfunction, impairment of vasodilator mechanisms, such as a loss of NO and its ability to stimulate cGMP, enhanced activation of vasoconstriction, loss of vascular regulation by hypoxia, enhanced remodeling, etc. can all be caused by increased ROS. However, due to the multifaceted roles ROS have in vascular regulation, the identification and therapeutic targeting of the specific processes stimulating ROS generation seems to often be more beneficial than antioxidant approaches for scavenging specific ROS or RNS.

Acknowledgments Recent studies from the authors' laboratory have been funded by USPHS grants HL031069, HL043023, HL066331, and HL115124.

References

1. Abraham NG, Kappas A (2008) Pharmacological and clinical aspects of heme oxygenase. Pharmacol Rev 60:79–127
2. Addabbo F, Montagnani M, Goligorsky MS (2009) Mitochondria and reactive oxygen species. Hypertension 53:885–892

3. Ata H, Rawat DK, Lincoln T, Gupte SA (2011) Mechanism of glucose-6-phosphate dehydrogenase-mediated regulation of coronary artery contractility. Am J Physiol Heart Circ Physiol 300:H2054–H2063
4. Boueiz A, Damarla M, Hassoun PM (2008) Xanthine oxidoreductase in respiratory and cardiovascular disorders. Am J Physiol Lung Cell Mol Physiol 294:L830–L840
5. Dai D-F, Rabinovitch PS, Ungvari Z (2012) Mitochondria and cardiovascular aging. Circ Res 110:1109–1124
6. Dikalova AE, Góngora MC, Harrison DG, Lambeth JD, Dikalov S, Griendling KK (2010) Upregulation of Nox1 in vascular smooth muscle leads to impaired endothelium-dependent relaxation via eNOS uncoupling. Am J Physiol Heart Circ Physiol 299:H673–H679
7. Duncker DJ, Bache RJ (2008) Regulation of coronary blood flow during exercise. Physiol Rev 88:1009–1086
8. Forman HJ, Fukuto JM, Torres M (2004) Redox signaling: thiol chemistry defines which reactive oxygen and nitrogen species can act as second messengers. Am J Physiol Cell Physiol 287:246–256
9. Gupte SA, Kaminski PM, George S, Kouznestova L, Olson SC, Matthew R, Hintze TH, Wolin MS (2009) Peroxide generation by $p47^{phox}$-Src activation of Nox2 has a key role in protein kinase C-induced arterial smooth muscle contraction. Am J Physiol Heart Circ Physiol 296:H1048–H1057
10. Gupte RS, Rawat DK, Chettimada S, Cioffi DL, Wolin MS, Gerthoffer WT, McMurtry IF, Gupte SA (2010) Activation of glucose-6-phosphate dehydrogenase promotes acute hypoxic pulmonary artery contraction. J Biol Chem 285:19561–19571
11. Ignarro LJ, Byrns RE, Buga GM, Wood KS, Chaudhuri G (1988) Pharmacological evidence that endothelium-derived relaxing factor is nitric oxide: use of pyrogallol and superoxide dismutase to study endothelium-dependent and nitric oxide-elicited vascular smooth muscle relaxation. J Pharmacol Exp Ther 244:181–189
12. Kukreja RC, Kontos HA, Hess ML, Ellis EF (1986) PGH synthase and lipoxygenase generate superoxide in the presence of NADH or NADPH. Circ Res 59:612–619
13. Lassègue B, San Martín A, Griendling KK (2012) Biochemistry, physiology, and pathophysiology of NADPH oxidases in the cardiovascular system. Circ Res 110:1364–1390
14. Li A-M, Shah AM (2004) Endothelial cell superoxide generation: regulation and relevance for cardiovascular pathophysiology. Am J Physiol Regul Integr Comp Physiol 287:R1014–R1030
15. Lincoln TM, Dey N, Sellak H (2001) Invited review: cGMP-dependent protein kinase signaling mechanisms in smooth muscle: from the regulation of tone to gene expression. J Appl Physiol 91:1421–1430
16. Luiking YC, Ten Have GAM, Wolfe RR, Deutz NEP (2012) Arginine de novo and nitric oxide production in disease states. Am J Physiol Endocrinol Metab 303:E1177–E1189
17. Moncada S, Palmer RM, Higgs EA (1991) Nitric oxide: physiology, pathophysiology, and pharmacology. Pharmacol Rev 43:109–142
18. Moudgil R, Michelakis ED, Archer SL (2005) Hypoxic pulmonary vasoconstriction. J Appl Physiol 98:390–403
19. Neo BH, Kandhi S, Ahmad M, Wolin MS (2010) Redox regulation of guanylate cyclase and protein kinase G in vascular responses to hypoxia. Respir Physiol Neurobiol 174:259–264
20. Palmer RM, Ferrige AG, Moncada S (1987) Nitric oxide release accounts for the biological activity of endothelium-derived relaxing factor. Nature 327(6122):524–526
21. Pacher P, Beckman JS, Liaudet L (2007) Nitric oxide and peroxynitrite in health and disease. Physiol Rev 87:315–424
22. Pryor WA, Houk KN, Foote CS, Fukuto JM, Ignarro LJ, Squadrito GL, Davies KJA (2006) Free radical biology and medicine: it's a gas, man! Am J Physiol Regul Integr Comp Physiol 291:R491–R511
23. Stasch J-P, Schmidt PM, Nedvetsky PI, Nedvetskaya TY, Arun Kumar HSA, Meurer S, Deile M, Taye A, Knorr A, Lapp H, Müller H, Turgay Y, Rothkegel C, Tersteegen A, Kemp-Harper B, Müller-Esterl W, Schmidt HHHW (2006) Targeting the heme-oxidized nitric oxide receptor for selective vasodilatation of diseased blood vessels. J Clin Invest 116:2552–2561

24. Ungvari Z, Wolin MS, Csiszar A (2006) Mechanosensitive production of reactive oxygen species in endothelial and smooth muscle cells: role in microvascular remodeling? Antioxid Redox Signal 8:1121–1129
25. Waypa GB, Schumacker PT (2005) Hypoxic pulmonary vasoconstriction: redox events in oxygen sensing. J Appl Physiol 98:404–414
26. Waypa GB, Marks JD, Guzy R, Mungai PT, Schriewer J, Dokic D, Schumacker PT (2010) Hypoxia triggers subcellular compartmental redox signaling in vascular smooth muscle cells. Circ Res 106:526–535
27. Wolin MS (2000) Interactions of oxidants with vascular signaling systems. Arterioscler Thromb Vasc Biol 20:1430–1442
28. Wolin MS, Ahmad M, Gupte SA (2005) Oxidant and redox signaling in vascular oxygen sensing mechanisms: basic concepts, current controversies, and potential importance of cytosolic NADPH. Am J Physiol Lung Cell Mol Physiol 289:L159–L173
29. Zhang G, Zhang F, Muh R, Yi F, Chalupsky K, Cai H, Li PL (2007) Autocrine/paracrine pattern of superoxide production through NAD(P)H oxidase in coronary arterial myocytes. Am J Physiol Heart Circ Physiol 292:H483–H495

Chapter 3
Nitrite-Dependent Nitric Oxide Production Pathway: Diversity of NO Production Systems

Hideo Yamasaki, Naoko S. Watanabe, Jon Fukuto, and Michael F. Cohen

Abbreviations

AsA	Ascorbate
DHA	Dehydroascorbate
dNiR	Dissimilatory nitrite reductase
eNOS	Endothelial nitric oxide synthase
HO1	Heme oxygenase-1
iNOS	Inducible nitric oxide synthase
MDA	Monodehydroascorbate
NADPH	Nicotinamide adenine dinucleotide phosphate
NiR	Nitrite reductase
NO	Nitric oxide
NOS	Nitric oxide synthase
NR	Nitrate reductase
RNS	Reactive nitrogen species
ROS	Reactive oxygen species
US EPA	United States Environmental Protection Agency
XOR	Xanthine oxidoreductase

H. Yamasaki, Ph.D. (✉) • N.S. Watanabe, M.Sc.
Faculty of Science, University of the Ryukyus, Nishihara, Nakagami-gun,
Okinawa 903-0213, Japan
e-mail: yamasaki@sci.u-ryukyu.ac.jp

J. Fukuto, Ph.D.
Department of Chemistry, Sonoma State University, Rohnert Park, CA 94928, USA

M.F. Cohen, Ph.D.
Department of Biology, Sonoma State University, Rohnert Park, CA 94928, USA

H. Tsukahara and K. Kaneko (eds.), *Studies on Pediatric Disorders*, Oxidative Stress in Applied Basic Research and Clinical Practice, DOI 10.1007/978-1-4939-0679-6_3,

3.1 Introduction

Nitrite has long been recognized for its toxicity as the causative agent of methemoglobinemia, also called cyanosis or "blue baby syndrome." The condition was first reported by pediatric resident Hunter Comly [1] in rural areas having nitrate-contaminated groundwater. The ingested nitrate is reduced by bacteria of the gastrointestinal tract to nitrite. Once in the bloodstream nitrite oxidizes hemoglobin (Fe^{2+}) to methemoglobin (Fe^{3+}), giving rise to the characteristic blue-hued skin. Infants under 6 months of age are particularly prone to methemoglobinemia since they have a low activity of circulating methemoglobin reductase [2]. Later, the use of nitrite in curing meat drew concern due to the formation of mutagenic nitrosamines during cooking [3] although extensive subsequent animal and epidemiological studies have not indicated that nitrite in meat leads to carcinogenesis [4].

More recently we have come to appreciate other, often beneficial, aspects of nitrite as a nitric oxide (NO) precursor in a range of physiological functions, which have recently been reviewed by Lundberg et al. [4] and are covered in Chap. 2 of this volume. At room temperature and pressure, pure NO is a gaseous free radical. However, as a biological mediator, it is an aqueous solute that exhibits versatile functions not only in humans but probably in all organisms [5–7].

Figure 3.1 illustrates major sources of nitrite for NO synthesis in humans. Unlike L-arginine-dependent NO production, which is a substrate-specific reaction in a local microenvironment, NO production from nitrite involves multiple routes

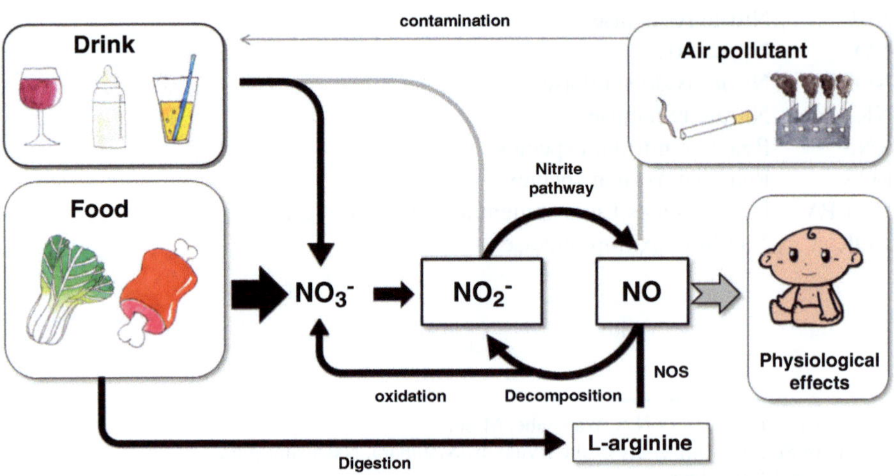

Fig. 3.1 Exogenous and endogenous sources of nitrite. Physiologically available nitrite is supplied mostly as nitrate contained in dietary foods and drinks. Leafy vegetables and seaweeds are major sources of nitrate. Nitrite is also obtained from processed food such as meat products and, in some cases, contaminated groundwater. In addition to these exogenous sources, nitrite can be endogenously synthesized as an oxidation product of NO produced by NO synthase (NOS). The relationship between nitrite and NO shows circularity

(exogenous and endogenous) and mechanisms (chemical and enzymatic). Nitrite and nitrate are oxidation products of NO synthesized by nitric oxide synthase (NOS). The relationship between nitrite and NO shows a circularity in terms of NO production, giving rising to the "chicken and egg" issue [8]. The multiplicity and circularity characteristic of nitrite may lead current researchers to a state of confusion. To help integrate new findings into our body of knowledge, in this chapter we summarize our current understanding of this alternative route for NO production from a biological perspective.

3.2 Dietary Sources of Nitrate and Nitrite

Nitrite is an inorganic nitrogen compound that exists as ionic form in a solution (NO_2^-; the p*K*a of nitrous acid, HNO_2, is approximately 3.2). Although nitrite is an intermediate metabolite in the plant nitrate assimilation pathway that synthesizes amino acids, nitrite occurs only at very low to undetectable levels in plants [9, 10]. Thus, natural foods such as leafy vegetables and seaweeds provide little nitrite but high levels of nitrate [11], a possible health benefit of Japanese and Mediterranean diets [12–14]. Bioavailable nitrite is mostly supplied exogenously via nitrate contained in foods but nitrite itself may also be obtained from processed foods or cured meats.

3.2.1 Vegetables

Vegetables are major dietary sources of nitrate. More than 80 % of nitrate ingested can be attributed to vegetables [11]. Absorbed by roots and translocated through the plant vascular system, nitrate can become highly concentrated in plant tissues, even to the point of crystallization [15]. Nitrate concentrations vary widely between plant species [9, 10] and even within the same tissue types of the same species, most likely due to differences in nitrate fertilization during cultivation [16]. A recent survey found a hundredfold variation in average nitrate levels in cooked greens, ranging from 4,850 mg kg^{-1} in English spinach to 48 mg kg^{-1} in iceberg lettuce [9].

3.2.2 Meat Curing Ingredients

Nitrate and nitrite are important meat preservatives, having a long history going back as far as 5,000 years throughout civilizations worldwide [17]. The original "curing" agent saltpeter (potassium nitrate) relies on bacterial conversion of nitrate to nitrite within the meats [18]. Since the late nineteenth century sodium nitrite instead of nitrate has been directly used to certain meat products, such as ham, bacon, and sausage [19]. The U.S. Federal Code of Regulations (21CFR

172.170 and 172.175) requires levels of nitrate and nitrite not to exceed 365 and 146 mg kg^{-1} in the finished product, respectively, similar to regulations prescribed by other countries.

3.2.3 Drinking Water

Water supplies, especially near agricultural areas, are often contaminated with nitrate derived from crop fertilizer. In many countries the content of nitrate in drinking water is regulated. A limit of 44 mg nitrate L^{-1} has been set by the United States Environmental Protection Agency (US EPA) for drinking water, for example [20]. In addition to direct contamination, nitric oxides (NOx) in industrial air pollution indirectly increase nitrate and nitrite content of stock water (Fig. 3.1).

3.2.4 Human Breast Milk

In the context of pediatrics, milk should be discussed separately from drinking water. Milk is a highly evolved product designed by millions of years of natural selection to maximize survival of newborns. Mammals secrete nitrate into milk at levels ranging from 1.4 to 11.2 mg nitrate L^{-1} [11, 21, 22]. The highest levels of nitrate are released 2–5 days postpartum. Nitrite is typically only found in milk at barely or non-detectible levels except early after birth, with reports of 2.1 mg L^{-1} 2–5 days postpartum and 0.8 mg L^{-1} 1–3 days postpartum [23]. The provisioning of nitrate and nitrite in milk at levels substantially exceeding that found in plasma [11] implies a physiological benefit of these components. The inability to produce nitrite due to the low level of nitrate reductase (NR) activity in the mouths of infants [24] explains the inclusion of nitrite in early breast milk. Mammals also secrete into milk substantial nanomolar levels of phytochemicals, including flavonoids, which are reported to enhance bioconversion of nitrite to NO [25].

3.3 Nitrite as a Degradation Product of NO, an Endogenous Source of Nitrite

The presence of endogenous nitrite formation in human body was first noted by Mitchell and coworkers a century ago [26] but the endogenous source of this nitrite remained unknown until the 1980s [27]. Up until then it had been known that exogenously supplied nitrate can be converted to nitrite by the nitrate reductase activity of oral bacteria (3.1).

$$NO_3^- + 2e^- + 2H^+ \rightarrow NO_2^- + H_2O \tag{3.1}$$

In addition to exogenous dietary sources of nitrate intake, recent studies have confirmed that nitrite can be generated through the oxidation of NO produced by the L-arginine-dependent NOS pathway (see Chap. 2), thereby explaining the mechanism for endogenous nitrite generation that had been unknown for over 60 years.

In the presence of O_2, NO can be oxidized (3.2).

$$4NO + O_2 + 2H_2O \rightarrow 4NO_2^- + 4H^+ \tag{3.2}$$

Some proteins such as ceruloplasmin are reported to facilitate oxidation of NO to nitrite [28]. Nitrite is formed stoichiometrically by autoxidation of NO, thereby allowing the widely available Griess assay to be applied for measurement of NOS activity [29].

3.4 Nitrite Pathway: An NOS-Independent NO Production System

Recent studies have provided strong evidence for the presence of L-arginine-independent NO-generating systems that are as important as the NOS system, especially under hypoxic conditions. Contrary to the view of nitrite simply as a "waste" product of the NOS reaction, it is now understood that nitrite is an alternative source of NO in our body. For simplification we will refer to this newly recognized NOS-independent NO-generating pathway hereafter as "the nitrite pathway" [8].

3.5 Chemical Mechanisms for NO Production from Nitrite

In contrast to enzymatic NO synthesis with L-arginine, the basic chemistry of NO production from nitrite is rather simple: one-electron reduction of nitrite produces NO. Because of this simplicity, multiple routes and mechanisms, including enzymatic and nonenzymatic (or chemical) reactions, are possible.

3.5.1 Acid (Proton)

Nitrite can be converted to NO without the aid of enzymes, an important feature not observed of L-arginine-dependent NO production. Nitrite ion (NO_2^-) exists in equilibrium with its conjugate acid nitrous acid (HNO_2) (3.3). Formation of the HNO_2 anhydride gives dinitrogen trioxide (N_2O_3) (3.4) and NO is then generated through decomposition of N_2O_3 (3.5).

$$NO_2^- + H^+ \leftrightarrow HNO_2 \tag{3.3}$$

$$2HNO_2 \leftrightarrow N_2O_3 + H_2O \tag{3.4}$$

$$N_2O_3 \leftrightarrow NO + NO_2 \tag{3.5}$$

Spontaneous NO production from nitrite is prominent under acidic conditions due to the HNO_2 p*K*a of 3.1–3.5. In mammalian systems, gastric fluid fits this requirement. Within hours after the clearance of slightly alkaline ammonic fluid from the stomach its contents become acidified to pH 1–3, sufficiently low to drive nonenzymatic production of NO from nitrite [11]. It needs to be stressed, however, that the formation of N_2O_3 from HNO_2 is second order and will only be relevant under conditions of high NO_2^- and H^+ concentrations. Since spontaneous NO production from nitrite under acidic conditions is relatively slow [30], it may have minimal physiological relevance.

3.5.2 *Reductants (Antioxidants)*

A large amount of NO can be generated by the reduction of acidified nitrite solution in the presence of reductants. It has long been known that NO can be generated by the reduction of nitrite in the presence of iodide or ferrocyanide in acidified solution and historically this reaction was used to quantify nitrate in urine [26]. This chemical production of NO is almost stoichiometric, thereby being widely used as a convenient method to calibrate NO concentrations in a solution even today [31]. A more biologically relevant reductant, ascorbate (AsA), an essential antioxidant in vegetables [32], has also been reported to induce chemical NO production from nitrite [33, 34]. One-electron reduction of nitrite produces NO and dehydroascorbate (DHA) through the formation of monodehydroascorbate (MDA) radical:

$$2HNO_2 + AsA \rightarrow 2NO + DHA + 2H_2O \tag{3.6}$$

Figure 3.2a shows pH dependence of chemical NO production from nitrite with potassium iodide (KI) and AsA. Both dependences clearly follow the Henderson–Hasselbalch equation when the p*K*a of HNO_2 is assumed to be 3.2 [6], indicating the requirement for acidic conditions (or HNO_2). In daily diets, phytochemicals such as polyphenols of vegetables are also strong antioxidants [35, 36] and can act, in addition to AsA, to reduce nitrite to NO [37].

3.6 Enzymatic Mechanisms for NO Production from Nitrite

Under hypoxic conditions, with lower competition for binding sites by O_2, enzymatic reduction of nitrite to NO takes place more readily thereby allowing for production of NO without the need for the NOS activity, which requires O_2. Accordingly, organisms commonly recognize NO as a signal of hypoxic conditions [38, 39]. Previously, enzymatic NO production from nitrite was only appreciated in denitrifying and nitrifying bacteria (Fig. 3.3) with little seeming relevance to human physiology. Recent investigations have revealed that multiple enzymes or proteins in mammalian systems are also capable of NO production from nitrite. Interestingly, many of these are well-known classical enzymes and proteins.

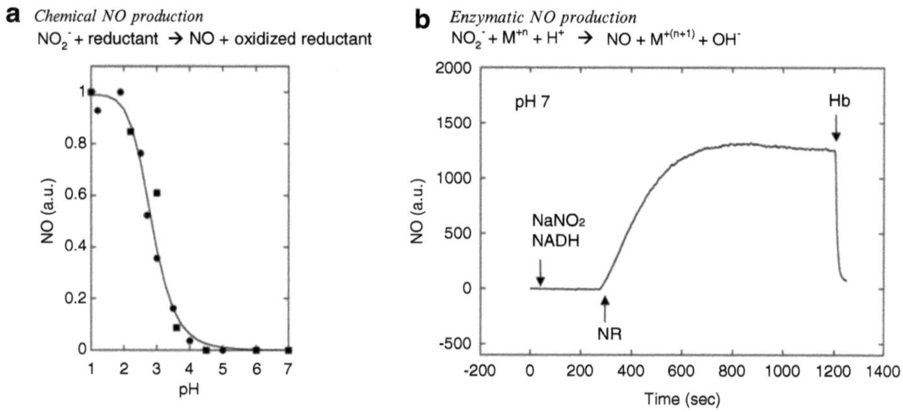

Fig. 3.2 NO production from nitrite. Chemical NO production with nitrite strongly depends on pH (**a**). Since HNO_2 (nitrous acid) is required to produce NO, the pH dependence follows the Henderson–Hasselbalch equation. In the presence of a reductant such as ascorbate, the rate of NO production becomes much faster. Whatever the reductants (*filled circle*, ascorbate; *filled square*, KI), the pH dependence shows identical *curves*. At a neutral pH, chemical NO production is not significant even in the presence of the reductant NADH (**b**). By adding enzyme, e.g., nitrate reductase (**b**), nitrite is rapidly converted to NO in a solution. NO was electrochemically detected with an NO-specific electrode. *M* metal-containing enzyme, *NR* nitrate reductase, *Hb* bovine hemoglobin. The graphs are redrawn after Yamasaki (2000) [6] and Yamasaki and Sakihama (2000) [101] for (**a**) and (**b**), respectively

3.6.1 Hemoglobin and Myoglobin

During hypoxia, deoxyhemoglobin in erythrocytes [40] and deoxymyoglobin in myocardial cells [41] can reduce nitrite to NO to form methemoglobin and metmyoglobin, respectively. In isolated erythrocytes the intracellular reduction of nitrite has been shown not to be limited by transport of nitrite across the membrane, which occurs through a combination of simple diffusion of HNO_2 and facilitated diffusion [42]. Recently, similar NO-producing nitrite reductase (NiR)-like activities have been found in neuroglobin [43], cytoglobin [44], and plant hemoglobins [45]. It appears that such NiR-like activity is conserved throughout these hemoglobin families [46] but it remains to be convincingly demonstrated whether the kinetic constraints on the reaction observed in purified systems are overcome under physiological conditions.

3.6.2 Xanthine Oxidoreductase

Xanthine oxidoreductase (XOR) functions in the catabolism of purines in organisms across the phylogenic spectrum. Under hypoxic conditions XOR can act as an NiR [47, 48], utilizing xanthine as an electron donor and at nitrite concentrations

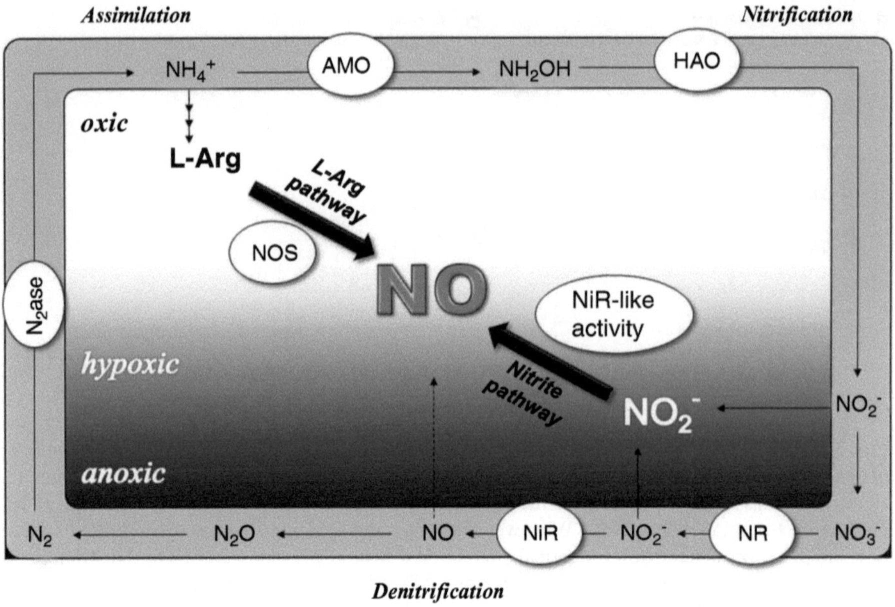

Fig. 3.3 Two pathways for enzymatic NO production in the bacterial nitrogen cycle. Various bacterial metabolisms can perform all the steps of the biological nitrogen cycle, processing nitrogen through oxidation states ranging from 3– (e.g., ammonia) to 5+ (e.g., nitrate). Nitrate is formed in nature primarily by nitrification, the sequential oxidation of ammonia by ammonia- and nitrite-oxidizing bacteria as a means to obtain energy. Under anaerobic conditions some bacteria are capable of inducing protein complexes that act in series, in a process termed denitrification, to reduce nitrate to N_2 gas. A portion of the nitrogen flowing through both nitrification and denitrification is released as NO. In animals, some nitrate ingested from foods is converted only to nitrite by bacteria in the mouth and gut. Nitrite produced from nitrate can be used to form NO either by nonenzymatic (chemical) or by enzymatic mechanisms using reductants. This nitrate–nitrite–NO pathway (nitrite pathway) complements the L-arginine-dependent NO production pathway driven by the NOS enzymes. An important difference between the two NO-producing systems is the oxygen requirement. Selected enzymes shown (in *ovals*) are: *N_2ase* nitrogenase, *AMO* ammonia monooxygenase, *HAO* hydroxylamine oxidoreductase, *NiR* nitrite reductase, *NR* nitrate reductase, *NOS* nitric oxide synthase

(5–40 μM) that are typically found in myocardial tissue [49]. Together, both XOR and deoxymyoglobin appear to be responsible for the majority of nitrite-responsive vasodilation activity although their apparent relative contributions vary among studies [41, 50, 51].

3.6.3 *Mitochondrial Electron Transport*

Under hypoxic conditions the electron transport systems of mitochondria throughout the Eukarya domain show evidence of their denitrifying heritage by reducing nitrite to NO at complex IV [52, 53], as well as at complex III and cytochrome *c*

[54]. Exposure to nitrite also dampens mitochondrial respiratory activity during anoxia, presumably by *S*-nitrosation of complex I, and thereby lowers production of damaging reactive oxygen species (ROS) by mitochondria during reperfusion [13].

3.6.4 Dissimilatory Nitrite Reductase

NO is a product of dissimilatory nitrite reductase (dNiR) in both denitrifying bacteria and ammonia-oxidizing bacteria [55]. NO released by these bacterial pathways has local effects on neighboring organisms (see below) and constitutes a major biogeochemical source of nitrogen oxides that return to terrestrial and marine environments [56]. Only a fraction of the NO formed in denitrifying bacteria diffuses from cells. The majority is a substrate for NO reductase, which is the evolutionary antecedent of the terminal oxygen-reducing cytochrome *c* oxidase in mitochondria and its bacterial relatives [52]. In the ancient atmosphere, before cyanobacterial photosynthesis generated the oxic atmosphere, NO was probably the predominant terminal electron acceptor for respiratory bacteria [57]. Thus, modern metabolic strategies for handling O_2, including cytochrome oxidases and globin proteins, are adaptions built on an NO-metabolizing framework [58].

3.6.5 Nitrate Reductase

Nitrate reductase (NR) can reduce nitrite to NO in plants [59, 60], fungi [6], and bacteria [61, 62] under high nitrite concentrations. Analogous to leukocyte inducible NOS (iNOS) activity in the innate immunity of animals [63], production of NO by nitrate reductase in plants stimulates expression of pathogen defense genes [64–67]. Using nicotinamide adenine dinucleotide phosphate (NADPH) as an electron donor, molybdenum (Mo^{2+}) at the reaction center appears to reduce nitrite, similar to the reaction catalyzed by XOR [68].

3.7 Other Key Reactions of Nitrite

It is evident that nitrite serves as an alternative NO source. Because of the unique chemistry of nitrite, some reactions should be considered separately from those of NO.

3.7.1 HNO Production from Nitrite

HNO, or nitroxyl, the one-electron reduction product of NO, reacts with heme proteins to exert important physiological effects that are unique from those of NO [69–71]. HNO is produced naturally within organisms but the biological mechanism of its formation remains unknown largely due to its short half-life. Based on the results of in vitro experiments it is conceivable that endogenous production of HNO could proceed by the reaction of nitrite with AsA [72] and subsequent decay of *O*-nitroascorbate to DHA and HNO [73]. Thus, nitrite may also act as a precursor to HNO in addition to NO.

3.7.2 Nitration and Nitrosation

The reaction between NO and the ROS superoxide (O_2^-) produces peroxynitrite ($ONOO^-$), a potentially toxic reactive nitrogen species (RNS). In the presence of catalytic CO_2, $ONOO^-$ nitrates the amino acid tyrosine to form nitro-tyrosine [74]. Nitration ($-NO_2$) of biomolecules can also involve nitrite itself through distinctive routes and mechanisms [75]. In macrophages, myeloperoxidase catalyzes the formation of nitro-tyrosine with nitrite and hydrogen peroxide (H_2O_2) [76].

In the oxidation of NO, N_2O_3 is generated. Reactions between N_2O_3 and biomolecules including amino acids result in nitrosation (–NO) of the residues at nucleophilic sites. The reaction is catalyzed by thiocyanate contained in the cabbage family and is inhibited by antioxidants including AsA [77]. Since the 1850s, the nitrosation reaction that produces nitrosoamines has attracted much attention as a potential carcinogenic effect of nitrite [11]. Although there has been long debate on the potential toxicity of nitrite in carcinogenesis, extensive animal as well as epidemiological studies have suggested that nitrite in diets does not in fact lead to carcinogenesis in normal conditions [4].

3.8 Bacterial Associations with the Nitrite Pathway

To fully grasp the function of nitrite pathway in vivo, knowledge of human physiology must be complemented with an understanding of nitrite processing by the human microbiome.

3.8.1 Nitrate Reduction by Gastrointestinal Bacteria

Approximately, one-third to one-half of nitrate ingested by mammals exits the body through bacterial denitrification to gas as well as via assimilation into bacterial biomass that is ultimately eliminated in feces [11]. Some bacteria that colonize the lumen of the gastrointestinal tract are capable of switching from aerobic to nitrate respiration, reducing nitrate to N_2 but often releasing a substantial portion of N from the pathway as NO [78]. Certain others, while in the nitrate respiratory mode, produce only nitrate reductase, thereby releasing nitrite for intestinal absorption or acid-mediated conversion to NO [4, 79].

NO production via the nitrite pathway is a common trait of the commensal bacteria of the intestine, especially among the *Lactobacillus* and *Bifidobacterium* genera [80]. Though some commensal bacteria consume NO, overall there is net production of NO by bacteria in the gastrointestinal tract [81, 82], which helps to explain the significant loss of intestinal NO emission in antibiotic-treated neonates compared to their untreated counterparts [81].

Of the nitrate absorbed from the intestine approximately one-quarter is returned to the upper gastrointestinal tract via saliva, presumably to permit reduction of nitrate to nitrite by mouth flora. Some of this nitrite will undergo acid-mediated reduction to NO in the stomach while the rest is absorbed into the bloodstream. Interestingly this nitrite-production activity in the mouth appears to exert a significant blood pressure-lowering effect on the host. Subjects that lowered their oral nitrite-producing activity by 90 % through use of an antibacterial mouthwash showed 2–3.5 mg Hg increases in systolic and diastolic blood pressure that correlated with a 25 % reduction in plasma nitrite levels [83], consistent with previous studies showing that dietary supplementation with nitrate lowers blood pressure [84].

These findings linking nitrite levels with blood pressure may imply a mechanism for the association between breastfeeding and lower systolic and diastolic pressure sustained into adulthood identified in a meta-analysis study [85] and thereby explain the observation of particularly high levels of nitrate in early postpartum milk. Nitrate in milk could conceivably favor the colonization of the mouth by nitrate reducing bacteria [24]; it would be interesting to learn whether the low levels of nitrate in certain baby formulas alter the ultimate composition of the microbial mouth flora.

3.8.2 Nitrite-Derived NO in Biofilm Dispersal: Ecological Aspects

Biofilms are multispecies assemblages of microbes that display coordinated metabolism and development and resilience in the face of physiochemical stresses and antibiotic treatment [86]. They are associated with increased virulence of a variety of infections that are particularly baneful to pediatric medicine, including dental carries and upper respiratory tract and medical device-associated infections. The denitrifying bacterium *Pseudomonas aeruginosa* is a common member of pathogenic biofilm communities [87–89] that releases NO under denitrifying conditions. NO has properties ideally suited for signaling within biofilms: it is membrane diffusible and can interact with several cellular targets including heme groups, iron sulfur clusters, and thiols [90], allowing it to exert a variety of post-transcriptional influences.

3.9 Behind the Multiplicity in a Simple System: Evolutional Aspects

Looking back on a long research history, one can notice that our society's recognition of the simple molecule nitrite (or nitrate) has changed over time. There have been contradictory findings and interpretations regarding this ubiquitous molecule. Two opposed effects of nitrite (carcinogenic or therapeutic) may lead nonspecialists to a state of confusion: is it good or bad? Moreover, the circularity of the relationship between nitrite and NO gives rise to the chicken and egg issue: substrate or product? For understanding the nature of ubiquitous and essential biomolecules, including O_2, NO, and nitrite, knowing their fine "balance" is more valuable than strict categorization [4, 8, 91]. As Fritjof Capra introduced the parallelism between modern physics and eastern philosophies [92], the application of "Yin–Yang" (shadow and light in a unity) philosophy may be of help in illustrating the opposing behaviors of the key players in oxidative stress [35] (Fig. 3.4).

The diversity in the effects of nitrite and NO can also be explained from an evolutionary perspective. High concentrations of NO arising from lightening the ancient atmosphere [93] and the consequently formed NO_2^- (reaction 3.2) would have bound iron, inhibiting electron transport processes, while the species N_2O_3 and nitrogen

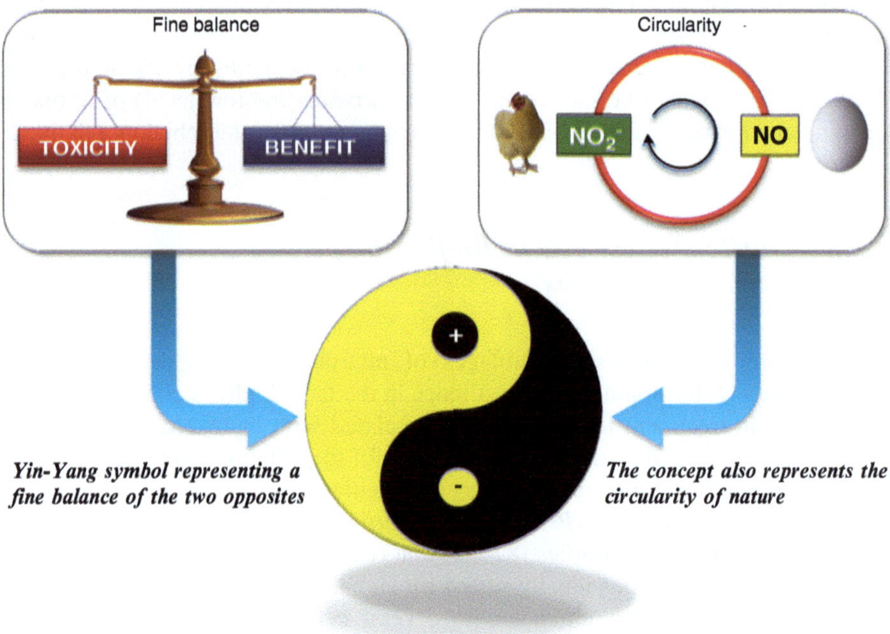

Fig. 3.4 Philosophical aspect of nitrite's dualism. Yin–Yang philosophy is one of the oldest concepts in Chinese medicine and Chinese food therapy. The philosophy is often helpful to understand contradictory features (good and bad) of ubiquitous molecules including nitrite

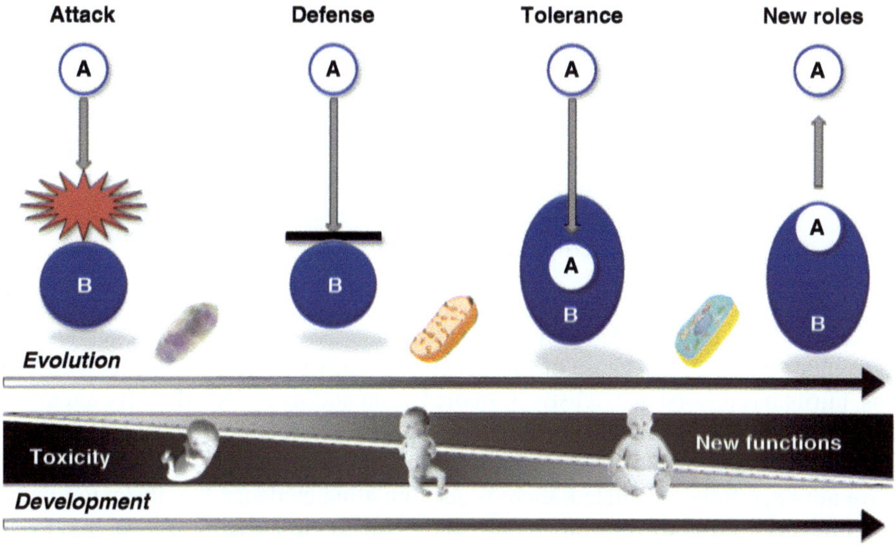

Fig. 3.5 Evolutional aspect of nitrite's dualism. This illustration shows conceptual steps representing how living organisms have evolved to cope with unfavorable exogenous chemicals. Suppose the element A is harmful to the element B. At the beginning, B suffers due to the attack by A (attack stage). Then, B develops protection mechanisms (defense stage) and further evolves tolerance mechanisms (tolerance stage). Finally, A plays new roles for B (new roles stage). Eventually, B is able to actively produce A to utilize it. This concept can account for the dualism of ROS, RNS, H_2S, and even for biological interactions. The figure is redrawn from Yamasaki (2005) [8]

dioxide (NO_2), formed in reactions (3.4) and (3.5), respectively, would have nitrosated proteins, causing general toxicity. Tolerance would have been gained through evolution of means to reduce NO_2^- and NO to N_2O or N_2 gas, in conjunction with deployment of cytosolic thiols, as in glutathione. It is conceivable that the need for tolerance to nitrogen oxides drove the evolution of their corresponding reductases. Subsequent development of protein complexes capable of linking proton pumping with the reduction of NO_2^- and NO would have allowed cells to take advantage of the positive redox potential of these compounds and thereby extract more cellular energy from their electron donors [57]. Free atmospheric NO would have plummeted with the rise in the atmosphere of O_2 resulting from oxygenic photosynthesis; NO could then serve as a signal of hypoxic conditions. Cells already well adapted to NO would find the need to synthesize it for certain physiological purposes, leading to the evolution of NOSs.

Figure 3.5 is generalized model for how biological relationships of two elements change in the process of evolution and development [8]. Suppose that in an early stage hostile elements (invaders) solely cause harm to living organisms (stage 1; attack). Later, living organisms evolve protective survival mechanisms (stage 2; defense). After the acquisition of the protecting mechanism, invaders may exist inside and may even function for mutual benefit (stage 3; tolerance). At the final stage, living organisms actively reproduce the past-invader to provide new functions (stage 4; new roles). This sequential change in relationship can be found in many biological interactions, e.g., pathogenesis, parasitism, and symbiosis. Human beings are snapshots of life's evolution, encompassing both "primitive" and "advanced" relationships and mechanisms that explain the dualism of simple molecules such as O_2, H_2S, and NO [8].

3.10 Implications for Therapeutic Application of Nitrite

It is evident that nitrite can supply NO to blood and tissues through endogenous and exogenous pathways. Because of the difficulty in directly treating patients with NO gas, nitrite has now been extensively studied as an alternative NO delivery agent for stroke, myocardial ischemia, hypertension, and transplantation therapy and for promoting angiogenesis [94]. Although the effects of nitrite therapy on those disorders have not been fully elucidated, there is accumulating evidence that the cytoprotective effects of nitrite therapy are mostly ascribable to the action of NO derived from nitrite and are independent of endothelial NOS (eNOS) and heme oxygenase-1 (HO1) activities [94].

It should be noted again that nitrite and nitrate are found at a high concentration (submillimolar) in postpartum breast milk [95]. In particular, colostrum contains the highest amount of nitrite [23]. Accordingly, a concentration of NO in stomach gas is higher than 7 ppm between postpartum days 2 and 5, which is not observed in neonates fed on low-nitrite formulas [96]. The gastrointestinal tract of the neonates must be sterile until the successful colonization of commensal bacteria that originate from the mother and environments [23]. It is most likely that a high concentration of nitrite in the early postpartum period is needed until the gut microbial flora establishes to metabolize nitrate as the substrate [23].

It appears that an early exposure to nitrite influences subsequent disease risk of children. Breastfeeding has been shown to reduce the risks of the infant developing asthma and allergies [97] as well as childhood acute leukemia [98]. Cardiovascular disease risk is also reduced through the reduction of obesity, blood pressure, and cholesterol. Interestingly, breastfeeding during infancy is associated with a reduction in risk of ischemic cardiovascular disease later in life [99]. It is well appreciated that immunoglobulins transferred from mother to neonate are one of the strong benefits of colostrum. Likewise it appears that human breast milk may supplement insufficient NO production capacity in neonates by providing nitrite orally, a new beneficial role of human breast milk.

3.11 Future Prospects

Nitrate–nitrite–NO metabolism in mammals has been scientifically investigated for more than 100 years. The long-standing question—what is the endogenous source of nitrate and nitrite?—can be now answered: it is the NO produced by host NOS systems and commensal bacterial activities. Autoxidation of NO to nitrite and nitrate can thus be understood to constitute a simple mechanism for recycling NO-generating capacity for the nitrite pathway.

As foods and drinking water change bacterial flora in gut [100], cytoprotective effects of nitrite may be altered among individuals depending on health conditions, age, daily diet, and local food customs. To determine an appropriate dose of nitrite for an individual patient, finding the balance point for attaining the maximal beneficial effect will be necessary. Although development of such personalized medicine must wait for future research, integration of our diverse knowledge of nitrite will be necessary to find the "Yin–Yang" balance of this essential inorganic molecule.

Acknowledgements Work in the authors' laboratory (H.Y.) was supported by a Grant-in-Aid for Scientific Research (B) from Ministry of Education, Culture, Sports, Science and Technology (MEXT), Japan.

References

1. Comly HH (1945) Cyanosis in infants caused by nitrates in well water. JAMA 129(2): 112–116. doi:10.1001/jama.1945.02860360014004
2. Knobeloch L, Salna B, Hogan A, Postle J, Anderson H (2000) Blue babies and nitrate-contaminated well water. Environ Health Perspect 108:675–678
3. Wolff IA, Wasserman AE (1972) Nitrates, nitrates & nitrosamines. Science 177:15–19
4. Lundberg JO, Weitzberg E, Gladwin MT (2008) The nitrate-nitrite-nitric oxide pathway in physiology and therapeutics. Nat Rev Drug Discov 7(2):156–167
5. Cohen MF, Yamasaki H (2003) Involvement of nitric oxide synthase in sucrose-enhanced hydrogen peroxide tolerance of *Rhodococcus* sp. strain APG1, a plant-colonizing bacterium. Nitric Oxide 9:1–9
6. Yamasaki H (2000) Nitrite-dependent nitric oxide production pathway: implications for involvement of active nitrogen species in photoinhibition in vivo. Philos Trans R Soc Lond B Biol Sci 355:1477–1488
7. Bouchard JN, Yamasaki H (2008) Heat stress stimulates nitric oxide production in *Symbiodinium microadriaticum*: a possible linkage between nitric oxide and the coral bleaching phenomenon. Plant Cell Physiol 49(4):641–652
8. Yamasaki H (2005) The NO world for plants: achieving balance in an open system. Plant Cell Environ 28:78–84
9. Hsu J, Arcot J, Alice Lee N (2009) Nitrate and nitrite quantification from cured meat and vegetables and their estimated dietary intake in Australians. Food Chem 115(1):334–339. doi:10.1016/j.foodchem.2008.11.081
10. Hord NG, Tang Y, Bryan NS (2009) Food sources of nitrates and nitrites: the physiologic context for potential health benefits. Am J Clin Nutr 90(1):1–10. doi:10.3945/ajcn.2008.27131
11. L'hirondel J, L'hirondel JL (2002) Nitrate and man: toxic, harmless or beneficial? CABI Publishing, Oxon
12. Lundberg JO, Feelisch M, Björne H, Jansson E, Weitzberg E (2006) Cardioprotective effects of vegetables: Is nitrate the answer? Nitric Oxide 15(4):359–362. doi:10.1016/j.niox.2006.01.013
13. Shiva S, Sack MN, Greer JJ, Duranski M, Ringwood LA, Burwell L, Wang X, MacArthur PH, Shoja A, Raghavachari N, Calvert JW, Brookes PS, Lefer DJ, Gladwin MT (2007) Nitrite augments tolerance to ischemia/reperfusion injury via the modulation of mitochondrial electron transfer. J Exp Med 204(9):2089–2102. doi:10.1084/jem.20070198

14. Sobko T, Marcus C, Govoni M, Kamiya S (2010) Dietary nitrate in Japanese traditional foods lowers diastolic blood pressure in healthy volunteers. Nitric Oxide 22(2):136–140
15. Wright MJ, Davison KL (1964) Nitrate accumulation in crops and nitrate poisoning in animals. Adv Agron 16:197–247
16. Zhen R, Leigh R (1990) Nitrate accumulation by wheat (*Triticum aestivum*) in relation to growth and tissue N concentrations. Plant and Soil 124(2):157–160. doi:10.1007/bf00009253
17. Gladwin MT, Schechter AN, Kim-Shapiro DB, Patel RP, Hogg N, Shiva S, Cannon RO, Kelm M, Wink DA, Espey MG (2005) The emerging biology of the nitrite anion. Nat Chem Biol 1(6):308–314
18. Keeton JT (2011) History of nitrite and nitrate in food. In: Bryan NS, Loscalzo J (eds) Nitrite and nitrate in human health and disease. Springer, New York, pp 69–84
19. Honikel K-O (2008) The use and control of nitrate and nitrite for the processing of meat products. Meat Sci 78(1–2):68–76. doi:10.1016/j.meatsci.2007.05.030
20. US EPA (2009) National primary drinking water regulations. Document no. EPA 816-F-09-004
21. Manning PB, Coulter ST, Jenness R (1968) Determination of nitrate and nitrite in milk and dry milk products. J Dairy Sci 51(11):1725–1730
22. Somogyi A, Beck H (1993) Nurturing and breast-feeding: exposure to chemicals in breast milk. Environ Health Perspect 101(suppl 2):45–52
23. Hord NG, Ghannam JS, Garg HK, Berens PD, Bryan NS (2011) Nitrate and nitrite content of human, formula, bovine, and soy milks: implications for dietary nitrite and nitrate recommendations. Breastfeed Med 6(6):393–399
24. Kanady JA, Aruni AW, Ninnis JR, Hopper AO, Blood JD, Byrd BL, Holley LR, Staker MR, Hutson S, Fletcher HM, Power GG, Blood AB (2012) Nitrate reductase activity of bacteria in saliva of term and preterm infants. Nitric Oxide 27(4):193–200, doi: http://dx.doi.org/10.1016/j.niox.2012.07.004
25. Song BJ, Jouni ZE, Ferruzzi MG (2013) Assessment of phytochemical content in human milk during different stages of lactation. Nutrition 29(1):195–202, doi: http://dx.doi.org/10.1016/j.nut.2012.07.015
26. Mitchell H, Shonle H, Grindley H (1916) The origin of the nitrates in the urine. J Biol Chem 24(4):461–490
27. Hibbs J Jr, Taintor RR, Vavrin Z (1987) Macrophage cytotoxicity: role for L-arginine deiminase and imino nitrogen oxidation to nitrite. Science 235(4787):473–476
28. Shiva S, Wang X, Ringwood LA, Xu X, Yuditskaya S, Annavajjhala V, Miyajima H, Hogg N, Harris ZL, Gladwin MT (2006) Ceruloplasmin is a NO oxidase and nitrite synthase that determines endocrine NO homeostasis. Nat Chem Biol 2(9):486–493
29. Arita NO, Cohen MF, Tokuda G, Yamasaki H (2006) Fluorometric detection of nitric oxide with diaminofluoresceins (DAFs): applications and limitations for plant NO research. In: Lamattina L, Polacco JC (eds) Nitric oxide in plant growth, development and stress physiology, Springer book series: plant cell monographs. Springer, Heidelberg, pp 269–280
30. Miles AM, Wink DA, Cook JC, Grisham MB (1996) Determination of nitric oxide using fluorescence spectroscopy. Methods Enzymol 268:105–120
31. Schmidt K, Mayer B (1998) Determination of NO with a Clark-type electrode. In: Titheradge MA (ed) Nitric oxide protocols. Humana Press, Totowa, pp 101–109
32. Asada K (2000) The water-water cycle as alternative photon and electron sinks. Philos Trans R Soc Lond B Biol Sci 355(1402):1419
33. Weitzberg E, Lundberg J (1998) Nonenzymatic nitric oxide production in humans. Nitric Oxide 2(1):1–7
34. Evans HJ, McAuliffe C (1956) Identification of NO, N_2O, and N_2 as products of the nonenzymatic reduction of nitrite by ascorbate or reduced diphosphopyridine nucleotide. In: McElroy WD, Glass B (eds) Inorganic nitrogen metabolism. Johns Hopkins Press, Baltimore, pp 189–197
35. Sakihama Y, Cohen MF, Grace SC, Yamasaki H (2002) Plant phenolic antioxidant and prooxidant activities: phenolics-induced oxidative damage mediated by metals in plants. Toxicology 177(1):67–80

36. Yamasaki H, Uefuji H, Sakihama Y (1996) Bleaching of the red anthocyanin induced by superoxide radical. Arch Biochem Biophys 332(1):183–186
37. Peri L, Pietraforte D, Scorza G, Napolitano A, Fogliano V, Minetti M (2005) Apples increase nitric oxide production by human saliva at the acidic pH of the stomach: a new biological function for polyphenols with a catechol group? Free Radic Biol Med 39(5):668–681
38. Dijkers PF, O'Farrell PH (2009) Dissection of a hypoxia-induced, nitric oxide–mediated signaling cascade. Mol Biol Cell 20(18):4083–4090. doi:10.1091/mbc.E09-05-0362
39. Ho J, Man H, Marsden P (2012) Nitric oxide signaling in hypoxia. J Mol Med 90(3):217–231. doi:10.1007/s00109-012-0880-5
40. Cosby K, Partovi KS, Crawford JH, Patel RP, Reiter CD, Martyr S, Yang BK, Waclawiw MA, Zalos G, Xu X, Huang KT, Shields H, Kim-Shapiro DB, Schechter AN, Cannon RO, Gladwin MT (2003) Nitrite reduction to nitric oxide by deoxyhemoglobin vasodilates the human circulation. Nat Med 9(12):1498–1505, doi: http://www.nature.com/nm/journal/v9/n12/suppinfo/nm954_S1.html
41. Shiva S, Huang Z, Grubina R, Sun J, Ringwood LA, MacArthur PH, Xu X, Murphy E, Darley-Usmar VM, Gladwin MT (2007) Deoxymyoglobin is a nitrite reductase that generates nitric oxide and regulates mitochondrial respiration. Circ Res 100(5):654–661. doi:10.1161/01.RES.0000260171.52224.6b
42. Jensen FB, Rohde S (2010) Comparative analysis of nitrite uptake and hemoglobin-nitrite reactions in erythrocytes: sorting out uptake mechanisms and oxygenation dependencies. Am J Physiol Regul Integr Comp Physiol 298(4):R972–R982. doi:10.1152/ajpregu.00813.2009
43. Tiso M, Tejero J, Basu S, Azarov I, Wang X, Simplaceanu V, Frizzell S, Jayaraman T, Geary L, Shapiro C (2011) Human neuroglobin functions as a redox-regulated nitrite reductase. J Biol Chem 286(20):18277–18289
44. Li H, Hemann C, Abdelghany TM, El-Mahdy MA, Zweier JL (2012) Characterization of the mechanism and magnitude of cytoglobin-mediated nitrite reduction and nitric oxide generation under anaerobic conditions. J Biol Chem 287(43):36623–36633. doi:10.1074/jbc.M112.342378
45. Tiso M, Tejero J, Kenney C, Frizzell S, Gladwin MT (2012) Nitrite reductase activity of nonsymbiotic hemoglobins from *Arabidopsis thaliana*. Biochemistry 51(26):5285–5292
46. Hardison RC (1996) A brief history of hemoglobins: plant, animal, protist, and bacteria. Proc Natl Acad Sci U S A 93(12):5675
47. Godber BLJ, Doel JJ, Sapkota GP, Blake DR, Stevens CR, Eisenthal R, Harrison R (2000) Reduction of nitrite to nitric oxide catalyzed by xanthine oxidoreductase. J Biol Chem 275(11):7757–7763. doi:10.1074/jbc.275.11.7757
48. Millar TM, Stevens CR, Benjamin N, Eisenthal R, Harrison R, Blake DR (1998) Xanthine oxidoreductase catalyses the reduction of nitrates and nitrite to nitric oxide under hypoxic conditions. FEBS Lett 427(2):225–228. doi:10.1016/s0014-5793(98)00430-x
49. Li H, Samouilov A, Liu X, Zweier JL (2001) Characterization of the magnitude and kinetics of xanthine oxidase-catalyzed nitrite reduction. J Biol Chem 276(27):24482–24489. doi:10.1074/jbc.M011648200
50. Casey DB, Badejo AM, Dhaliwal JS, Murthy SN, Hyman AL, Nossaman BD, Kadowitz PJ (2009) Pulmonary vasodilator responses to sodium nitrite are mediated by an allopurinol-sensitive mechanism in the rat. Am J Physiol Heart Circ Physiol 296(2):H524–H533. doi:10.1152/ajpheart.00543.2008
51. Webb A, Bond R, McLean P, Uppal R, Benjamin N, Ahluwalia A (2004) Reduction of nitrite to nitric oxide during ischemia protects against myocardial ischemia-reperfusion damage. Proc Natl Acad Sci U S A 101(37):13683–13688. doi:10.1073/pnas.0402927101
52. Tosha T, Shiro Y (2013) Crystal structures of nitric oxide reductases provide key insights into functional conversion of respiratory enzymes. IUBMB Life 65(3):217–226. doi:10.1002/iub.1135
53. Castello PR, David PS, McClure T, Crook Z, Poyton RO (2006) Mitochondrial cytochrome oxidase produces nitric oxide under hypoxic conditions: Implications for oxygen sensing and hypoxic signaling in eukaryotes. Cell Metab 3(4):277–287, doi: http://dx.doi.org/10.1016/j.cmet.2006.02.011

54. Shiva S (2013) Nitrite: a physiological store of nitric oxide and modulator of mitochondrial function. Redox Biol 1(1):40–44, doi: http://dx.doi.org/10.1016/j.redox.2012.11.005
55. Anderson IC, Levine JS (1986) Relative rates of nitric oxide and nitrous oxide production by nitrifiers, denitrifiers, and nitrate respirers. Appl Environ Microbiol 51(5):938–945
56. Cohen MF, Lamattina L, Yamasaki H (2010) Nitric oxide signaling by plant-associated bacteria. In: Hayat S, Mori M, Pichtel J, Ahmad A (eds) Nitric oxide in plant physiology. Wiley-VCH, Weinheim, pp 161–172
57. Ducluzeau A-L, van Lis R, Duval S, Schoepp-Cothenet B, Russell MJ, Nitschke W (2009) Was nitric oxide the first deep electron sink? Trends Biochem Sci 34(1):9–15. doi:10.1016/j.tibs.2008.10.005
58. Igamberdiev AU, Hill RD (2004) Nitrate, NO and haemoglobin in plant adaptation to hypoxia: an alternative to classic fermentation pathways. J Exp Bot 55(408):2473–2482
59. Dean JV, Harper JE (1986) Nitric oxide and nitrous oxide production by soybean and winged bean during the in vivo nitrate reductase assay. Plant Physiol 82(3):718–723
60. Yamasaki H, Sakihama Y, Takahashi S (1999) An alternative pathway for nitric oxide production in plants: new features of an old enzyme. Trends Plant Sci 4(4):128–129
61. Gilberthorpe NJ, Poole RK (2008) Nitric oxide homeostasis in *Salmonella typhimurium*: roles of respiratory nitrate reductase and flavohemoglobin. J Biol Chem 283(17):11146–11154
62. Vine CE, Purewal SK, Cole JA (2011) NsrR-dependent method for detecting nitric oxide accumulation in the Escherichia coli cytoplasm and enzymes involved in NO production. FEMS Microbiol Lett 325(2):108–114. doi:10.1111/j.1574-6968.2011.02385.x
63. Tripathi P, Tripathi P, Kashyap L, Singh V (2007) The role of nitric oxide in inflammatory reactions.FEMSImmunolMedMicrobiol51(3):443–452.doi:10.1111/j.1574-695X.2007.00329.x
64. Modolo LV, Augusto O, Almeida IMG, Magalhaes JR, Salgado I (2005) Nitrite as the major source of nitric oxide production by *Arabidopsis thaliana* in response to *Pseudomonas syringae*. FEBS Lett 579(17):3814–3820. doi:10.1016/j.febslet.2005.05.078
65. Oliveira HC, Saviani EE, Oliveira JFP, Salgado I (2010) Nitrate reductase-dependent nitric oxide synthesis in the defense response of *Arabidopsis thaliana* against *Pseudomonas syringae*. Trop Plant Pathol 35:104–107
66. Yamamoto A, Katou S, Yoshioka H, Doke N, Kawakita K (2003) Nitrate reductase, a nitric oxide-producing enzyme: induction by pathogen signals. J Gen Plant Pathol 69(4):218–229. doi:10.1007/s10327-003-0039-x
67. Yamamoto-Katou A, Katou S, Yoshioka H, Doke N, Kawakita K (2006) Nitrate reductase is responsible for elicitin-induced nitric oxide production in *Nicotiana benthamiana*. Plant Cell Physiol 47(6):726–735. doi:10.1093/pcp/pcj044
68. Lundberg JO, Weitzberg E, Shiva S, Gladwin MT (2011) The nitrate–nitrite–nitric oxide pathway in mammals. In: Byan NS, Loscalzo L (eds) Nitrite and nitrate in human health and disease. Springer, New York, pp 21–48
69. Flores-Santana W, Switzer C, Ridnour L, Basudhar D, Mancardi D, Donzelli S, Thomas D, Miranda K, Fukuto J, Wink D (2009) Comparing the chemical biology of NO and HNO. Arch Pharm Res 32(8):1139–1153. doi:10.1007/s12272-009-1805-x
70. Fukuto JM, Carrington SJ (2011) HNO signaling mechanisms. Antioxid Redox Signal 14:1649–1657
71. Switzer CH, Flores-Santana W, Mancardi D, Donzelli S, Basudhar D, Ridnour LA, Miranda KM, Fukuto JM, Paolocci N, Wink DA (2009) The emergence of nitroxyl (HNO) as a pharmacological agent. Biochim Biophys Acta 1787(7):835–840. doi:10.1016/j.bbabio.2009.04.015
72. Myshkin AE, Konyaeva VS, Gumargalieva KZ, Moiseev YV (1991) Mechanism of nitrosation of ascorbic acid by nitrite in neutral aqueous media. Russ Chem Bull 40(10):1961–1965. doi:10.1007/bf00963487
73. Kirsch M, Buscher A-M, Aker S, Schulz R, de Groot H (2009) New insights into the *S*-nitrosothiol-ascorbate reaction. The formation of nitroxyl. Org Biomol Chem 7(9):1954–1962

74. Marnett LJ, Riggins JN, West JD (2003) Endogenous generation of reactive oxidants and electrophiles and their reactions with DNA and protein. J Clin Invest 111(5):583–594
75. Sakihama Y, Tamaki R, Shimoji H, Ichiba T, Fukushi Y, Tahara S, Yamasaki H (2003) Enzymatic nitration of phytophenolics: evidence for peroxynitrite-independent nitration of plant secondary metabolites. FEBS Lett 553(3):377–380
76. Sampson JB, Ye YZ, Rosen H, Beckman JS (1998) Myeloperoxidase and horseradish peroxidase catalyze tyrosine nitration in proteins from nitrite and hydrogen peroxide. Arch Biochem Biophys 356(2):207–213
77. Boyland E, Nice E, Williams K (1971) The catalysis of nitrosation by thiocyanate from saliva. Food Cosmet Toxicol 9(5):639–643
78. Roediger WEW (2008) Review article: nitric oxide from dysbiotic bacterial respiration of nitrate in the pathogenesis and as a target for therapy of ulcerative colitis. Aliment Pharmacol Ther 27(7):531–541. doi:10.1111/j.1365-2036.2008.03612.x
79. McKnight GM, Duncan CW, Leifert C, Golden MH (1999) Dietary nitrate in man: friend or foe? Br J Nutr 81:349–358
80. Sobko T, Reinders CI, Jansson E, Norin E, Midtvedt T, Lundberg JO (2005) Gastrointestinal bacteria generate nitric oxide from nitrate and nitrite. Nitric Oxide 13(4):272–278. doi:10.1016/j.niox.2005.08.002
81. Sobko T, Elfström K, Navér L, Lundberg JO, Norman M (2009) Intestinal hydrogen and nitric oxide gases in preterm infants—effects of antibiotic therapy. Neonatology 95:68–73
82. Sobko T, Huang L, Midtvedt T, Norin E, Gustafsson LE, Norman M, Jansson E, Lundberg JO (2006) Generation of NO by probiotic bacteria in the gastrointestinal tract. Free Radic Biol Med 41(6):985–991. doi:10.1016/j.freeradbiomed.2006.06.020
83. Kapil V, Haydar S, Pearl V, Lundberg JO, Weitzberg E, Ahluwalia A (2013) Physiological role for nitrate-reducing oral bacteria in blood pressure control. Free Radic Biol Med 55(1): 93–100
84. Kapil V, Milsom AB, Okorie M, Maleki-Toyserkani S, Akram F, Rehman F, Arghandawi S, Pearl V, Benjamin N, Loukogeorgakis S (2010) Inorganic nitrate supplementation lowers blood pressure in humans role for nitrite-derived NO. Hypertension 56(2):274–281
85. Martin RM, Gunnell D, Davey Smith G (2005) Breastfeeding in infancy and blood pressure in later life: systematic review and meta-analysis. Am J Epidemiol 161(1):15–26
86. Pintucci JP, Corno S, Garotta M (2010) Biofilms and infections of the upper respiratory tract. Eur Rev Med Pharmacol Sci 14:683–690
87. Bjarnsholt T, Jensen P, Fiandaca MJ, Pedersen J, Hansen CR, Andersen CB, Pressler T, Givskov M, Høiby N (2009) *Pseudomonas aeruginosa* biofilms in the respiratory tract of cystic fibrosis patients. Pediatr Pulmonol 44(6):547–558. doi:10.1002/ppul.21011
88. Harmsen M, Yang L, Pamp SJ, Tolker-Nielsen T (2010) An update on *Pseudomonas aeruginosa* biofilm formation, tolerance, and dispersal. FEMS Immunol Med Microbiol 59(3):253–268. doi:10.1111/j.1574-695X.2010.00690.x
89. Pinar E, Oncel S, Karagoz U, Sener G, Calli C, Tatar B (2008) Demonstration of bacterial biofilms in chronic otitis media. Mediterr J Otol 4:64–68
90. Thomas DD, Ridnour LA, Isenberg JS, Flores-Santana W, Switzer CH, Donzelli S, Hussain P, Vecoli C, Paolocci N, Ambs S, Colton CA, Harris CC, Roberts DD, Wink DA (2008) The chemical biology of nitric oxide: implications in cellular signaling. Free Radic Biol Med 45(1):18–31. doi:10.1016/j.freeradbiomed.2008.03.020
91. Hotchkiss J (1988) Nitrate, nitrite balance, and *de novo* synthesis of nitrate. Am J Clin Nutr 47(1):161–162
92. Capra F (1975) The Tao of physics: an exploration of the parallels between modern physics and eastern mysticism. Shambhala, Berkeley
93. Navarro González R, McKay CP, Nna Mvondo D (2001) A possible nitrogen crisis for Archaean life due to reduced nitrogen fixation by lightning. Nature 64:61–64
94. Calvert JW, Lefer DJ (2010) Clinical translation of nitrite therapy for cardiovascular diseases. Nitric Oxide 22(2):91–97

95. Ohta N, Tsukahara H, Ohshima Y, Nishii M, Ogawa Y, Sekine K, Kasuga K, Mayumi M (2004) Nitric oxide metabolites and adrenomedullin in human breast milk. Early Hum Dev 78(1):61–65
96. Iizuka T, Sasaki M, Oishi K, Uemura S, Koike M, Shinozaki M (1999) Non-enzymatic nitric oxide generation in the stomachs of breastfed neonates. Acta Paediatr 88(10):1053–1055
97. Fulhan J, Collier S, Duggan C (2003) Update on pediatric nutrition: breastfeeding, infant nutrition, and growth. Curr Opin Pediatr 15(3):323–332
98. Shu XO, Linet MS, Steinbuch M, Wen WQ, Buckley JD, Neglia JP, Potter JD, Reaman GH, Robison LL (1999) Breast-feeding and risk of childhood acute leukemia. J Natl Cancer Inst 91(20):1765–1772
99. Rich-Edwards JW, Stampfer MJ, Manson JAE, Rosner B, Hu FB, Michels KB, Willett WC (2004) Breastfeeding during infancy and the risk of cardiovascular disease in adulthood. Epidemiology 15(5):550–556
100. Hehemann JH, Correc G, Barbeyron T, Helbert W, Czjzek M, Michel G (2010) Transfer of carbohydrate-active enzymes from marine bacteria to Japanese gut microbiota. Nature 464(7290):908–912
101. Yamasaki H, Sakihama Y (2000) Simultaneous production of nitric oxide and peroxynitrite by plant nitrate reductase: in vitro evidence for the NR-dependent formation of active nitrogen species. FEBS Lett 468:89–92

Chapter 4
Nitric Oxide and Endothelial Dysfunction

Eisei Noiri and Kousuke Minami

Abbreviations

ADMA	Asymmetrical dimethyl arginine
ADP	Adenosine diphosphate
AMP	Adenosine monophosphate
ARE/EpRE	Antioxidant/electrophile-responsive element
ATP	Adenosine triphosphate
BH4	Tetrahydrobiopterin
CKD	Chronic kidney disease
COX-2	Cyclooxygenase-2
Cys	Cysteine
DDAH1	Dimethylarginine dimethylaminohydrolase-1
eNOS	Endothelial NO synthase
ESRD	End-stage renal disease
GMP	Guanosine monophosphate
GTP	Guanosine triphosphate
L-FABP	L-type fatty acid-binding protein
L-NAME	N^{α}-nitro-L-arginine methyl
L-NMMA	N^{α}-monomethyl-L-arginine
5-MTHF	5-Methyltetrahydrofolate
5-MTHFR	5, 10-Methylenetetrahydrofolate reductase
NADPH	Nicotinamide adenine dinucleotide phosphate

E. Noiri, M.D., Ph.D. (✉)
Nephrology and Endocrinology, Hemodialysis and Apheresis,
The University of Tokyo Hospital, Tokyo, Japan
e-mail: noiri-tky@umin.ac.jp

K. Minami, Ph.D.
The International Center for Materials Nanoarchitectonics (MANA),
National Institute for Materials Science (NIMS), Ibaraki, Japan

H. Tsukahara and K. Kaneko (eds.), *Studies on Pediatric Disorders*, Oxidative Stress in Applied Basic Research and Clinical Practice, DOI 10.1007/978-1-4939-0679-6_4,

8-nitro-cGMP	8-Nitroguanosine 3′ 5′-cyclic monophosphate
NO	Nitric oxide
NOS	Nitric oxide synthase
NOX	Nicotinamide adenine dinucleotide phosphate oxidase
OA-NO_2	Nitro fatty acids
15d-PGJ_2	15-Deoxy-$\Delta^{12,14}$-prostaglandin J_2
RNOS	Reactive nitrogen oxide species
SNO-hemoglobin	S-nitrosohemoglobin

4.1 Endothelial Dysfunction, eNOS, and Postulated Mechanisms

Endothelial dysfunction is the pathologic vascular tone caused by impaired platelet activity, leukocyte adhesion, and angiogenesis. Pioneering work on nitric oxide synthase (NOS) demonstrated the crucial role of endothelial NOS (eNOS) to regulate vascular tone [1]. It was known that NOS inhibition by arginine analogues such as N^G-nitro-L-arginine methyl (L-NAME)-induced hypertension. Mice deficient in eNOS do not demonstrate acetylcholine-induced vascular relaxation and were modestly hypertensive (wild-type mice 90 ± 12 mmHg, $n = 15$, eNOS mutant mice 109 ± 11 mmHg, $n = 18$, under halothane anesthesia). When other subtypes of NOS (inducible or neuronal) were deleted together with eNOS, creating doubly NOS-deficient mice, blood pressure was increased to almost the same levels. This observation was the same even in the triply (inducible, neuronal, and endothelial) NOS-deficient mice [2]. The bioavailability of nitric oxide (NO) is affected by certain risk factors, but the question remains as to how these risk factors affect blood pressure through NO. For example, increased endogenous levels of eNOS inhibitors such as asymmetrical dimethyl arginine (ADMA) propagate endothelial dysfunction in diabetes mellitus and chronic kidney diseases. ADMA is metabolized by dimethylarginine dimethylaminohydrolase-1 (DDAH1). Indeed, DDAH1-deficient mice showed a hypertensive phenotype [3]. Similarly, smooth muscle cell-specific deletion of guanylyl cyclase also successfully produced hypertension in the murine model [4].

In addition, the activation of eNOS is dependent on local ADP levels. Endoplasmic reticulum stress in endothelial cells is mediated in part by the regulation of the sarco/endoplasmic reticulum calcium ATPase and calcium homeostasis that affects eNOS coupling (uncoupled eNOS can produce superoxide). Induction of endoplasmic reticulum stress is also dependent on the AMP kinase level. The balance between AMP kinase, Akt, A kinase, C kinase phosphorylate, Thr495, Ser677, and Ser1177 residue thus affect the enzymatic activity of eNOS. Interestingly, phosphorylation at Ser633 and Ser1177 activate eNOS but that at Thr495 is inhibitory [5]. The activation of eNOS is necessary for the scenario of angiogenesis to switch from stationary to locomotive phenotypes, namely *podokinesis* [6–9].

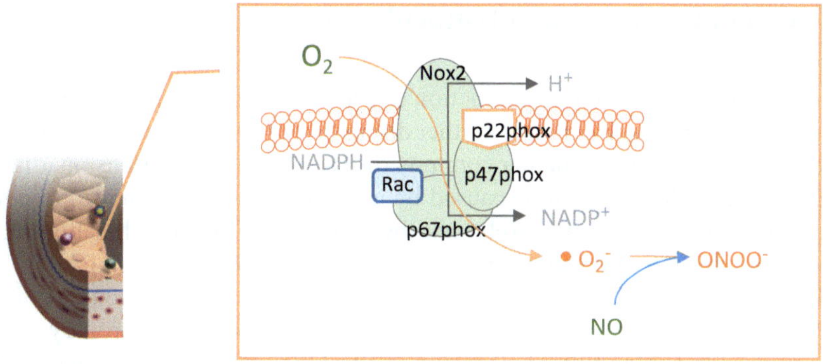

Fig. 4.1 Vascular NADPH oxidase complex. NADPH oxidases are a family of enzyme complexes. Their primary function is the catalytic action of transferring electrons from NADPH to molecular oxygen, resulting in the generation of superoxide. The Nox subunit binds to p22phox in the plasma membrane, and both cooperatively stabilize each other. In hypertension, the disarrangement of these subunits possibly increases the oxidative stress level, reduces the bioavailabilities of NO, and simultaneously results in the formation of more powerful oxidants—peroxynitrite ($ONOO^-$) and hydroxyl radicals ($OH^·$)

The study of human endothelial dysfunction was recently examined in relation to gene polymorphism. Induction of oxidative stress by homocysteine impairs endothelial function, thus decreasing physiologic NO levels. We have investigated C/T polymorphism at the nucleotide position 677 of 5,10-methylenetetrahydrofolate reductase [5-MTHFR; this enzyme converts 5,10-methylenetetrahydrofolate to 5-methyltetrahydrofolate (5-MTHF)], the key enzyme in the remethylation pathway [10], in 170 men with end-stage renal disease (ESRD) and 666 healthy men. The allele frequency of the 677 T variant was significantly increased in ESRD compared with that seen in controls ($p<0.0001$); odds ratio 2.11 (95 % CI: 1.49–2.99) [11]. Antoniades et al. examined tissue and plasma homocysteine and 5-MTHF levels in saphenous veins and internal mammary arteries obtained during bypass surgery from patients with polymorphisms of 5-MTHFR [12]. They observed that higher tissue 5-MTHF levels correlated with endothelium-dependent vasodilation and lower superoxide production in the isolated vascular tissue. Similarly, NAD(P)H oxidase is one of the most important sources of reactive oxygen species closely acting with NO (Fig. 4.1). NADPH oxidase consist of cytosolic components (p47phox, p67phox), a G-protein (Rac1 or 2), and membrane-associated cytochrome b558 comprised p22phox and gp91phox. In nonphagocytic cells, the gp91phox subunit is often substituted by another isoform, such as NOX1, 2, 4, and 5 [13, 14]. NADPH has been demonstrated to be upregulated by angiotensin II in afferent arterioles of the glomerulus [15]. We have identified a risk haplotype (242CC-640AA, $p=0.035$) for nondiabetic ESRD in the cytochrome b558 gene encoding the p22phox component of NAD(P)H oxidase [16]. This genetic study compared 467 patients with ESRD and 490 healthy individuals, and found that 242C/T has a protective effect for ESRD only in nondiabetic subjects ($p=0.0095$), and showed by haplotype estimation and linkage disequilibrium analysis that 242C-640A was more frequent in nondiabetic ESRD subjects than in controls.

4.2 eNOS Glu298Asp Polymorphism

The potential for endothelial dysfunction due to eNOS polymorphism has attracted many research groups. But before further discussion of eNOS polymorphism, it will be necessary to establish whether common diseases such as hypertension, diabetes, CKD, and heart diseases can be caused by a single gene polymorphism or equivalent genetic factor. It is more likely that they are caused by a combination of gene activations subject to predisposing factors and that eNOS is one such factor.

G894T is the most investigated eNOS polymorphism locus in exon 7 coding for Glu298Asp, reportedly having a remarkable association with coronary spasms, acute myocardial infarction [17], hypertension [18], carotid atherosclerosis [19], and preeclampsia [20]. Using the candidate gene approach, we have analyzed 185 patients with ESRD and compared our findings with those from 304 unrelated healthy individuals and found that the Glu298Asp polymorphism of eNOS predisposed to diabetic ESRD (the odds ratio 2.02, 95 % CI: 1.37–3.07, $p = 0.001$) [21]. We further clarified the functional difference in NO generation depending on eNOS with either Glu or Asp at the position of 298 by using site-directed mutagenesis and both fluorometric nitrite assay and NO electrode detection methods. Tsukahara et al., for the first time, reported L-arginine-induced increases in NO production with the use of an NO-selective electrochemical detection method [22].

The observed accumulation of Glu298Asp in diabetic ESRD may partially explain the poor prognosis of patients with this condition and their significantly lower 5-year survival rate compared to that of patients with nondiabetic ESRD [23]. It is possible that the prevalence of atherosclerotic comorbidities and therefore the lesser NO production in 298Asp individuals will increase the amount of proinflammatory leukocyte adhesion to endothelial cells and decrease vasodilatation capabilities, thereby propagating the vicious cycle.

Differences in vascular reactivity to an $\alpha 1$-adrenoceptor agonist, phenylephrine, was investigated by Philip et al.; individuals with 894GT and 894TT genotype showed virtually the same increase of mean blood pressure with the lower dose of phenylephrine compared with responses in 894GG individuals [24]. A more detailed study for vasospastic angina was performed by Naber et al.; coronary vasomotor tone was investigated during diagnostic coronary catheterization to clarify unclear episodes of chest pain [25]. They used intracoronary Doppler ultrasonography to measure coronary vascular resistance and velocity and found that individuals with the 894TT genotype showed higher coronary vascular resistance and lower average peak velocity compared to those with 894GG and 894GT. Cherney and colleagues further accomplished functional characterization of this polymorphism in humans [26]. They challenged subjects with intravenous L-arginine and measured changes in renal hemodynamics. L-arginine increased effective plasma flow and glomerular filtration rate in 894GG individuals, but this increase was not as much as that seen in 894GT and 894TT subjects. Given that L-arginine is the substrate of NO generation, there is a question of how further biochemical properties of L-arginine can affect NO bioavailability.

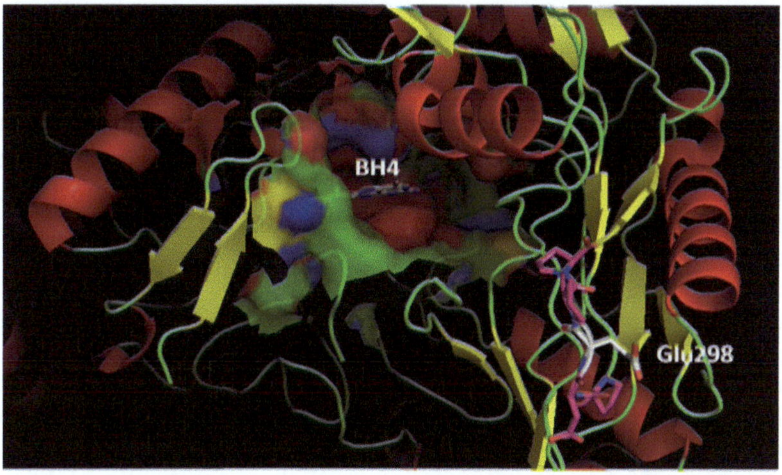

Fig. 4.2 Ribbon diagram of the structure of eNOS viewing from Glu298 to BH4. Glu298 is located in between the protein beta-sheet that is the site of a loop structure flanked by prolines from both sides (Pro-Asp-Glu-Pro-Pro). The change from Glu to Asp at the 298 site causes a small but distinct steric hindrance. *Magenta* represents proline. The active site of BH4 is marked in the pocket at the *center* of this image

The efficacy of L-arginine infusion may provide a better subcellular distribution of L-arginine in close proximity to the cationic amino acid transporter-1 channel, which is also present in caveolae. In addition, because the accumulation of ADMA interferes with L-arginine's function to induce endothelial dysfunction, the relative depletion of L-arginine might be reversed in high-risk individuals through the infusion of L-arginine. The increased chance of binding between L-arginine and the eNOS coupling complex protects a cofactor in NO synthesis, BH4 from consumption by oxidative and nitrosative stress [27] partially induced by uncoupled eNOS. The guanidino nitrogen group in L-arginine possesses radical-scavenging activity, thus protecting NO from the reaction with oxidative stress. The non-substrate action of the guanidino nitrogen group in L-arginine to α2-adrenoceptor and imidazoline receptors, similarly to agmatine, would lead to the activation of eNOS and NO production through a second messenger pathway [28] seen in an L-arginine infusion study.

Finally, the protein structural analysis of eNOS is difficult because of its large molecular weight of 133 kDa. The analysis of the Glu298Asp polymorphism is a virtually unexplored area. The position of Glu298Asp is remotely displaced from the active sites (the main reason it is unexplored) and is instead located in between the protein beta-sheet that is the site of the loop structure flanked by prolines from both sides. (Pro-Asp-Glu-Pro-Pro) (Fig. 4.2). The change from Glu to Asp at the 298 site does not change the electric charge but does cause a small but distinct steric hindrance. This is because proline is notoriously less flexible compared with other amino acids and therefore this substitution will induce a critical change in the folding and binding pockets of eNOS. Future structural analysis of this site is awaited.

4.3 Early Manifestation of Dysfunctional Endothelial Cells

Using a low dose of L-NMMA or ADMA without affecting systemic blood pressure, Goligorsky and coworkers clarified that chronic nonpressor doses of NOS inhibitors induce chronic eNOS uncoupling and the reduction of mitochondrial volume [29] and result in the upregulation of collagen XVIII and its C-terminal fragment, endostatin; microvascular rarefaction readily detectable in the renal medulla; and ultrastructural changes in microvascular endothelia; but no appreciable degree of hypoxia, as judged by the lack of pimonidazole retention [30]. They found markedly depressed expression of the key enzyme of the Krebs cycle, aconitase 2, and reduced oxidative phosphorylation with enhanced glycolysis, as judged by the accumulation of lactate in the culture medium and by the inability of cells to survive in the glucose-to-galactose substituted medium. This is further supported by the fact that α-ketoglutarate, a downstream substrate bypassing the mitochondrial enzymatic block at the level of isocitrate formation, partially restored cell viability in the glucose-free medium [29]. The observed normoxic glycolysis in dysfunctional endothelial cells was initiated under the circumstances of eNOS uncoupling and ADMA/L-NMMA-induced mitochondrial oxidative stress. Therefore, they claimed the analogy of the observed preferential glycolytic metabolism with the Warburg effect, a hallmark of tumors.

4.4 Non-NOS-Mediated Mechanism to Endothelial Dysfunction

The target molecule of NO is the soluble form of guanylyl cyclase, an enzyme that forms cyclic GMP from GTP. The heme moieties on this enzyme can bind NO to induce a 400 times increase of enzymatic activity [31, 32]. Similar to the heme moieties in guanylyl cyclase, the heme proteins of hemoglobin can bind NO. In practice, hemoglobin has been utilized for disarming NO bioactivities during experiments in the life sciences. The reaction of hemoglobin and NO is oxidation, in which NO reacts with oxy-hemoglobin to yield met-hemoglobin and plasma nitrate. NO reacts with Fe(II)-hemoglobin to form NO-Fe(II)-hemoglobin in this process. The dominant percentage of oxy-hemoglobin in the circulation enables the rapid reaction of NO at close to the diffusion limit. As suggested by Castillo et al. [33], the estimated scavenging of mammalian NO via the NO/oxy-Hb reaction is slower than that expected by the detected level of NO in biological systems. Because NO is rapidly metabolized, an alternative mechanism for its biologic activity must be involved.

Stamler and colleagues clearly demonstrated a third relevant reaction of NO with hemoglobin [34], the S-nitrosylation of a cysteine residue conserved in β-globins, which yields SNO-hemoglobin. SNO-hemoglobin retains NO-like bioactivity and is capable of transferring NO to low-molecular-weight thiol-containing molecules. Moreover, S-nitrosothiols do not react with Fe(II)-hemoglobin, precluding the

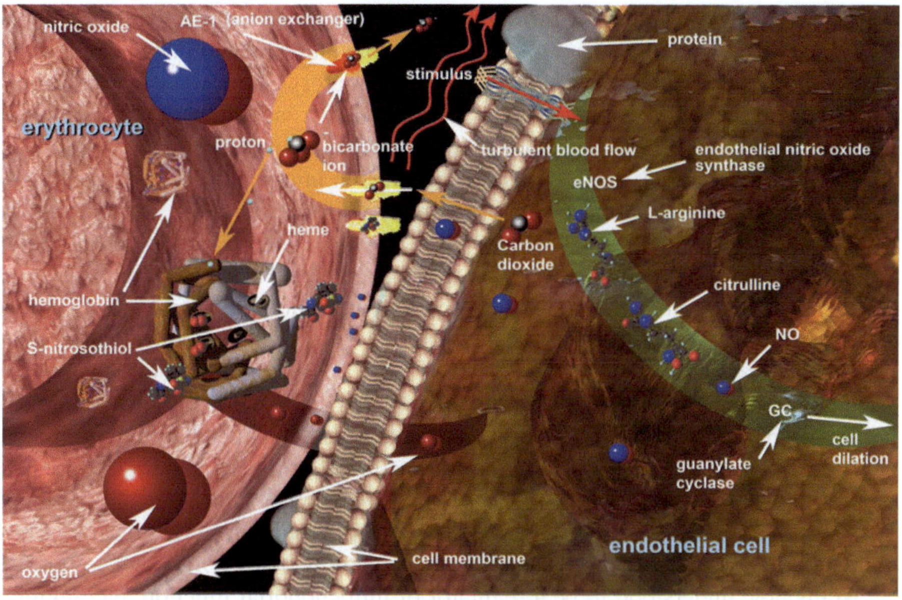

Fig. 4.3 Erythrocyte and NO-mediated endothelial function. SNO-hemoglobin (hemoglobin with S-nitrosothiol) retains NO-like bioactivity and is capable of transferring NO to low-molecular-weight thiol-containing molecules, further releasing NO to the endothelial cells. Of note is that S-nitrosothiols do not react with Fe(II)-hemoglobin, precluding the consumption of bioactive NO by erythrocytes, but they do facilitate vascular smooth muscle relaxation via the soluble form of guanylyl cyclase (GC). The endothelial cell itself can generate NO with eNOS, as is very well known

consumption of bioactive NO by erythrocytes, but they do facilitate vascular smooth muscle relaxation (Fig. 4.3). In addition, the S-nitrosylation of hemoglobin occurs preferentially in higher oxygen-tension organs such as the lung, and the release of NO from the resultant SNO-hemoglobin occurs at lower oxygen-tension peripheral sites such as the capillaries. In peripheral areas, NO release is triggered by allosteric structural transitions in hemoglobin when O_2 is released from hemoglobin in a low pO_2 region; the shift from a relaxed- (R-) to a tense- (T-) structure of hemoglobin triggers the NO release. Therefore, the X-SNO form enables NO equivalents to be released preferentially at the lowest pO_2 in peripheral dilating blood capillaries, directing blood flow to the most ischemic tissues and thus improving endothelial function.

Similarly, an NO-derived metabolite, plasma nitrate, awaited redefinition for its role in NO bioavailability. Scientists advocating nitrite-dependent NO bioactivity are focusing on the difference between arterial and venous nitrite levels [35]. But arterial–venous differences of nitrite were not similarly observed in comparisons using either S-nitrosylation of hemoglobin or albumin even after 80 ppm inhalation of NO [36]. Therefore, they claimed that capillary blood flow is rather regulated by nitrite-mediated NO-like activities. Nitrite can be reduced to bioactive NO along a physiological pH and oxygen gradient through enzymatic reduction by hemoglobin, myoglobin, components of the mitochondrial respiratory chain, and xanthine oxido-reductase, as well as nonenzymatic reduction via acidic disproportionation. Therefore, nitrite is considered to be an important endocrine reservoir of NO. NO is known to protect tissues from ischemia–reperfusion injury, but its therapeutic window is limited in terms of dose and duration of exposure.

Admittedly, nitrite has been shown to mediate potent cytoprotection after ischemia–reperfusion injury in the heart, liver, kidney, and brain when administered during ischemia or immediately before reperfusion. Nanomole doses (1.2–48 nmol) of nitrite reduced rodents' myocardial infarction volume and apoptosis more than 50 % compared with those not receiving nitrite [37, 38]. In a rat model of kidney ischemia–reperfusion injury, topically administered 30 nmol of nitrite doubled urine volume for 6 h of reperfusion time significantly reduced serum creatinine levels and protected renal histologic architecture. Similarly, nanomole doses of nitrite reduced cerebral infarct size by 75 % and enhanced neurological functional recovery in a rat model of stroke [39]. The lower pH found in tissues during ischemia–reperfusion enhances the reduction of nitrite to liberate NO.

NO is now understood to be a regulator of mitochondrial function. NO inhibits cytochrome c oxidase reversibly with nanomolar concentration [40], regulates reactive oxygen species formation [41], initiates biogenesis [42, 43], and limits apoptotic cytochrome c release [44, 45]. Nitrite directly modifies the electron transfer chain complex I activity in association with S-nitrosation of this complex [46], which dampens electron transfer and effectively reduces reperfusion reactive oxygen species generation and ameliorates oxidative inactivation of complexes II–IV and aconitase, thus preventing mitochondrial permeability transition pore opening and cytochrome c release.

4.5 Cue to Improve Renal Endothelial Dysfunction Via Controlling Oxidative Stress

One of the ways that NO is involved in endothelial function is by mediating the balance between oxidative and nitrosative stresses. The endogenous electrophiles generated downstream of NO will increase nitrosative stress. Endothelial dysfunction is not induced solely on ischemia–reperfusion injury or drug-induced injury but potentially also by electrophilic stress that is the primary factor in oxidative, environmental, and xenobiotic stresses. These stresses consist of the expression of both oxidative and anti-oxidative genes. The expression of stress-responsive genes involved in cellular protection is activated by a transduced signal through a stress-sensing protein.

A recent discovery among multiple stress response systems in vertebrates is the system regulated by the Nrf2-Keap1 pathway. Under stable conditions Nrf2 is continuously ubiquitinated by the Cul3-Keap1 ubiquitin E3 ligase complex and is rapidly degraded in proteasomes [47, 48]. Under the pathological conditions of exposure to electrophilic and oxidative stresses, the reactive cysteine residues of Keap1 become modified, leading to a decline in the E3 ligase activity, thus stabilizing Nrf2 and inducing a robust induction of a battery of cytoprotective genes, such as the primary targets for NAD(P)H-quinone oxidoreductase 1 and glutathione *S*-transferase; and the subsidiary targets for glutathione peroxidase 2, glutamate cysteine lyase catalytic and regulatory subunits, and heme oxygenase-1. Nrf2 activates the expression of these genes through a *cis*-acting element called the antioxidant/electrophile-responsive element (ARE/EpRE) [49, 50]. The known Nrf2 inducers are diethyl maleate, tert-butyl-hydroquinone, sulforaphane, and 1-cyano-3, 12-dioxooleana-1, 9-dien-28-oic acid derivatives.

Two prevalent inflammatory signaling cascades, the cyclooxygenase (COX)-2 pathway and the NO synthesis pathway, generate Nrf2-activating molecules. COX-2 catalyzes arachidonic acid and produces various bioactive prostaglandins. One of the COX-2 pathway products is 15-deoxy-$\Delta^{12,14}$-prostaglandin J_2 (15d-PGJ_2). Of note is that 15d-PGJ_2 binds directly to the cysteine residues of Keap1 [51]. 15d-PGJ2 is primarily produced by macrophages for resolution of inflammation; thus, abrogating 15d-PGJ_2 production with COX-2 inhibitors could cause persistent neutrophil infiltration after renal ischemia–reperfusion injury. NO can modify the reactive cysteine residues of Keap1 directly and/or indirectly through reactive nitrogen oxide species (RNOS). NO directly S-nitrosylates the Cys151 of Keap1 [52, 53]. By contrast, the indirect hypothesis is supported by the observation that NO generates RNOS that nitrosylate cGMP and produce 8-nitroguanosine 3′5′-cyclic monophosphate (8-nitro-cGMP) [54]. S-guanylation of Keap1 at Cys434 by 8-nitro-cGMP abrogates the Keap1-mediated inhibition of Nrf2 [55] (Fig. 4.4).

Alternatively, nitro fatty acids (OA-NO_2) are produced by RNOS through the nitration of unsaturated fatty acids. OA-NO_2 modifies Keap1 cysteines, primarily Cys273/288 [56, 57]. These observations support that NO-derived direct and indirect mechanisms and nitrosative stress act upon the Keap1-Nrf2 system, which stimulates ARE/EpRE and further initiates the diverse expression of anti-oxidative genes.

Evidence is accumulating that a broad range of anomalies associated with oxygen biology including hypoxia and oxidative stress are implicated in diabetic nephropathy. Interstitial fibrosis usually is an indicator of poor prognosis [58]. Interstitial hypoxia is one of the major predictive factors for interstitial fibrosis. Thus, early intervention is an important clinical concern in progressive glomerular diseases. Urinary L-type fatty acid binding protein (L-FABP) is a renal biomarker indicating low oxygen conditions through the hypoxia response element on its promoter region. Although the majority of the physiological observations have been conducted in renal ischemia–reperfusion injury models [59, 60], the recent clinical report on diabetic nephropathy demonstrated that the patients with more progressive disease showed higher urinary L-FABP levels [61–63]. Therefore, urinary L-FABP may be a better prognostic marker compared with the often fluctuating levels of urinary albumin.

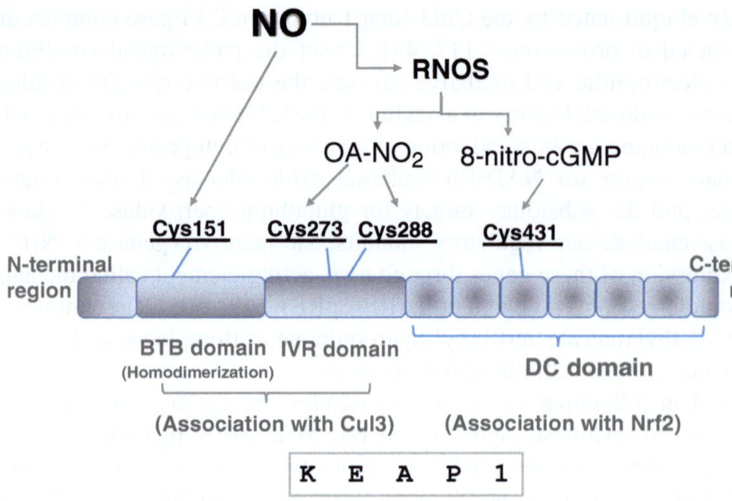

Fig. 4.4 Keap1 modification site mediated by NO and related products. Modification by NO and related products will affect molecular conformation (homodimerization) and interaction with Cul3 and Nrf2. *Cys* cysteine, *RNOS* reactive nitrogen oxide species, *8-nitro-cGMP* 8-nitroguanosine 3′5′-cyclic monophosphate, *OA-NO*$_2$ nitro fatty acids

Nrf2 activators or Keap1 inhibitors are potential therapeutic candidates to alleviate oxidative stress in endothelial function, leading to diabetic nephropathy and to improve bioavailability of NO in the renal microcirculation [64]. A potent inducer of Nrf2, bardoxolone methyl (2-cyano-3,12-dioxooleane-1,9 (11)-dien-28-oic methyl ester; CDDO-Me; or RTA 402) [65], which is the derivative of a natural oleanolic acid and originally was developed as an anticancer drug, demonstrated an unexpected improvement in renal function for patients during a clinical trial. The structure and activity profile of bardoxolone methyl resembles that of 15d-PGJ$_2$ addressed earlier. It is explained by Pergola et al. that bardoxolone methyl reduced serum creatinine levels with a corresponding improvement in estimated glomerular filtration rate in 2 phase 1 trials that included 81 oncology patients [66]. Further improvements were seen in a subset of patients with established CKD and effects were maintained for 5 months.

As a result, a phase II clinical trial, known as the BEAM (Bardoxolone Methyl Treatment: Renal Function in CKD/Type 2 Diabetes) study, has been undertaken in patients with advanced CKD and type 2 diabetes [67]. Bardoxolone improved renal function at 24 weeks with persistent favorable results at 52 weeks with only mild side effects, such as muscle spasm, weight loss, gastrointestinal effects, mild increases in alanine aminotransferase levels, and hypomagnesemia. Although this was a promising phase II clinical trial, a subsequent Phase III BEACON (Bardoxolone methyl EvAluation in patients with Chronic kidney disease and type 2 diabetes: the Occurrence of renal eveNts) trial in patients with stage 4 CKD and type 2 diabetes had to be terminated early because of serious adverse outcomes in the primary endpoint; the time to first event of the composite endpoint consisting of ESRD (need for renal transplantation or long-term dialysis) or cardiovascular death.

This study enrolled patients with type 2 diabetes and stage 4 chronic kidney disease with high potential risk to develop cardiovascular events. A cohort of patients with milder diseases such as those with G3bA2 in the chronic kidney disease guideline [68] may be a more suitable population to consider for the next clinical studies. The combination of albuminuria and urinary L-FABP measurements or equivalent testing will serve to discriminate active cases and could clarify the beneficial cohort for bardoxolone methyl therapy. Further studies are necessary to delineate the mechanism of adverse effects seen in the clinical trials.

4.6 Conclusion

NO is known to be indispensable in maintaining vascular tone. And it is known that NO deficiency plays a role in cardiovascular and cerebrovascular diseases, atherosclerosis, and chronic kidney diseases. But does the malfunction of NO come first to initiate endothelial dysfunction and then to develop these pathological conditions, or are these chronic disease conditions damaging endothelial cells and thereby draining endogenous NO?

Whichever comes first, the topics that are described herein are key mechanisms of NO and endothelial dysfunction that should be addressed in therapeutic interventions. There are some clues to assisting in this process. One such example was thought to be bardoxolone methyl therapy, but its second clinical study has failed, and more studies will be required.

There has been no drug throughout the history of modern medicine that has successfully reduced oxidative stress or improved redox status. One day, perhaps soon, such a drug will be a reality. For now, prevention is the best scenario in clinical practice. For those patients already suffering from NO-related diseases, strategies to improve endothelial function or maintain functional NO levels will be an important aspect of clinical intervention.

References

1. Huang PL, Huang Z, Mashimo H, Bloch KD, Moskowitz MA, Bevan JA et al (1995) Hypertension in mice lacking the gene for endothelial nitric oxide synthase. Nature 377:239–242
2. Morishita T, Tsutsui M, Shimokawa H, Sabanai K, Tasaki H, Suda O et al (2005) Nephrogenic diabetes insipidus in mice lacking all nitric oxide synthase isoforms. Proc Natl Acad Sci U S A 102:10616–10621
3. Hu X, Xu X, Zhu G, Atzler D, Kimoto M, Chen J et al (2009) Vascular endothelial-specific dimethylarginine dimethylaminohydrolase-1-deficient mice reveal that vascular endothelium plays an important role in removing asymmetric dimethylarginine. Circulation 120:2222–2229

4. Boerrigter G, Burnett JC Jr (2009) Soluble guanylate cyclase: not a dull enzyme. Circulation 119:2752–2754
5. Mount PF, Kemp BE, Power DA (2007) Regulation of endothelial and myocardial NO synthesis by multi-site eNOS phosphorylation. J Mol Cell Cardiol 42:271–279
6. Noiri E, Hu Y, Bahou WF, Keese CR, Giaever I, Goligorsky MS (1997) Permissive role of nitric oxide in endothelin-induced migration of endothelial cells. J Biol Chem 272: 1747–1752
7. Noiri E, Lee E, Testa J, Quigley J, Colflesh D, Keese CR et al (1998) Podokinesis in endothelial cell migration: role of nitric oxide. Am J Physiol Cell Physiol 274:C236–C244
8. Noiri E, Peresleni T, Srivastava N, Weber P, Bahou WF, Peunova N et al (1996) Nitric oxide is necessary for a switch from stationary to locomoting phenotype in epithelial cells. Am J Physiol Cell Physiol 270:C794–C802
9. Goligorsky MS, Budzikowski AS, Tsukahara H, Noiri E (1999) Co-operation between endothelin and nitric oxide in promoting endothelial cell migration and angiogenesis. Clin Exp Pharmacol Physiol 26:269–271
10. Frosst P, Blom HJ, Milos R, Goyette P, Sheppard CA, Matthews RG et al (1995) A candidate genetic risk factor for vascular disease: a common mutation in methylenetetrahydrofolate reductase. Nat Genet 10:111–113
11. Noiri E, Taguchi J, Nakao A, Fujita T (2000) MTHFR gene polymorphism as an exacerbation factor of diabetic nephropathy in type 2 diabetes. Analysis in Japanese male hemodialysis patients. Diabetes Care 23:260
12. Antoniades C, Shirodaria C, Leeson P, Baarholm OA, Van-Assche T, Cunnington C et al (2009) MTHFR 677 C>T Polymorphism reveals functional importance for 5-methyltetrahydrofolate, not homocysteine, in regulation of vascular redox state and endothelial function in human atherosclerosis. Circulation 119:2507–2515
13. Taniyama Y, Griendling KK (2003) Reactive oxygen species in the vasculature: molecular and cellular mechanisms. Hypertension 42:1075–1081
14. Drummond GR, Selemidis S, Griendling KK, Sobey CG (2011) Combating oxidative stress in vascular disease: NADPH oxidases as therapeutic targets. Nat Rev Drug Discov 10:453–471
15. Wang D, Chen Y, Chabrashvili T, Aslam S, Borrego Conde LJ, Umans JG et al (2003) Role of oxidative stress in endothelial dysfunction and enhanced responses to angiotensin II of afferent arterioles from rabbits infused with angiotensin II. J Am Soc Nephrol 14:2783–2789
16. Doi K, Noiri E, Nakao A, Fujita T, Kobayashi S, Tokunaga K (2005) Haplotype analysis of NAD(P)H oxidase p22 phox polymorphisms in end-stage renal disease. J Hum Genet 50:641–647
17. Hibi K, Ishigami T, Tamura K, Mizushima S, Nyui N, Fujita T et al (1998) Endothelial nitric oxide synthase gene polymorphism and acute myocardial infarction. Hypertension 32: 521–526
18. Miyamoto Y, Saito Y, Kajiyama N, Yoshimura M, Shimasaki Y, Nakayama M et al (1998) Endothelial nitric oxide synthase gene is positively associated with essential hypertension. Hypertension 32:3–8
19. Lembo G, De Luca N, Battagli C, Iovino G, Aretini A, Musicco M et al (2001) A common variant of endothelial nitric oxide synthase (Glu298Asp) is an independent risk factor for carotid atherosclerosis. Stroke 32:735–740
20. Yoshimura T, Yoshimura M, Tabata A, Shimasaki Y, Nakayama M, Miyamoto Y et al (2000) Association of the missense Glu298Asp variant of the endothelial nitric oxide synthase gene with severe preeclampsia. J Soc Gynecol Investig 7:238–241
21. Noiri E, Satoh H, Taguchi J, Brodsky SV, Nakao A, Ogawa Y et al (2002) Association of eNOS Glu298Asp polymorphism with end-stage renal disease. Hypertension 40:535–540
22. Tsukahara H, Gordienko DV, Goligorsky MS (1993) Continuous monitoring of nitric oxide release from human umbilical vein endothelial cells. Biochem Biophys Res Commun 193:722–729
23. McMillan MA, Briggs JD, Junor BJ (1990) Outcome of renal replacement treatment in patients with diabetes mellitus. BMJ 301:540–544

24. Philip I, Plantefeve G, Vuillaumier-Barrot S, Vicaut E, LeMarie C, Henrion D et al (1999) G894T polymorphism in the endothelial nitric oxide synthase gene is associated with an enhanced vascular responsiveness to phenylephrine. Circulation 99:3096–3098
25. Naber CK, Baumgart D, Altmann C, Siffert W, Erbel R, Heusch G (2001) eNOS 894T allele and coronary blood flow at rest and during adenosine-induced hyperemia. Am J Physiol Heart Circ Physiol 281:H1908–H1912
26. Cherney DZ, Scholey JW, Zhou J, Zimpelmann J, Kennedy C, Burns KD et al (2009) Endothelial nitric oxide synthase gene polymorphisms and the renal hemodynamic response to L-arginine. Kidney Int 75:327–332
27. Noiri E, Nakao A, Uchida K, Tsukahara H, Ohno M, Fujita T et al (2001) Oxidative and nitrosative stress in acute renal ischemia. Am J Physiol Renal Physiol 281:F948–F957
28. Joshi MS, Ferguson TB Jr, Johnson FK, Johnson RA, Parthasarathy S, Lancaster JR Jr (2007) Receptor-mediated activation of nitric oxide synthesis by arginine in endothelial cells. Proc Natl Acad Sci U S A 104:9982–9987
29. Addabbo F, Ratliff B, Park HC, Kuo MC, Ungvari Z, Csiszar A et al (2009) The Krebs cycle and mitochondrial mass are early victims of endothelial dysfunction: proteomic approach. Am J Pathol 174:34–43
30. Stoessel A, Paliege A, Theilig F, Addabbo F, Ratliff B, Waschke J et al (2008) Indolent course of tubulointerstitial disease in a mouse model of subpressor, low-dose nitric oxide synthase inhibition. Am J Physiol Renal Physiol 295:F717–F725
31. Lee YC, Martin E, Murad F (2000) Human recombinant soluble guanylyl cyclase: expression, purification, and regulation. Proc Natl Acad Sci U S A 97:10763–10768
32. Humbert P, Niroomand F, Fischer G, Mayer B, Koesling D, Hinsch KD et al (1990) Purification of soluble guanylyl cyclase from bovine lung by a new immunoaffinity chromatographic method. Eur J Biochem 190:273–278
33. Castillo L, Beaumier L, Ajami AM, Young VR (1996) Whole body nitric oxide synthesis in healthy men determined from [15N] arginine-to-[15N]citrulline labeling. Proc Natl Acad Sci U S A 93:11460–11465
34. Jia L, Bonaventura C, Bonaventura J, Stamler JS (1996) S-nitrosohaemoglobin: a dynamic activity of blood involved in vascular control. Nature 380:221–226
35. Gladwin MT, Shelhamer JH, Schechter AN, Pease-Fye ME, Waclawiw MA, Panza JA et al (2000) Role of circulating nitrite and S-nitrosohemoglobin in the regulation of regional blood flow in humans. Proc Natl Acad Sci U S A 97:11482–11487
36. Cannon RO III, Schechter AN, Panza JA, Ognibene FP, Pease-Fye ME, Waclawiw MA et al (2001) Effects of inhaled nitric oxide on regional blood flow are consistent with intravascular nitric oxide delivery. J Clin Invest 108:279–287
37. Duranski MR, Greer JJ, Dejam A, Jaganmohan S, Hogg N, Langston W et al (2005) Cytoprotective effects of nitrite during in vivo ischemia-reperfusion of the heart and liver. J Clin Invest 115:1232–1240
38. Webb A, Bond R, McLean P, Uppal R, Benjamin N, Ahluwalia A (2004) Reduction of nitrite to nitric oxide during ischemia protects against myocardial ischemia-reperfusion damage. Proc Natl Acad Sci U S A 101:13683–13688
39. Jung KH, Chu K, Ko SY, Lee ST, Sinn DI, Park DK et al (2006) Early intravenous infusion of sodium nitrite protects brain against in vivo ischemia-reperfusion injury. Stroke 37: 2744–2750
40. Brown GC, Cooper CE (1994) Nanomolar concentrations of nitric oxide reversibly inhibit synaptosomal respiration by competing with oxygen at cytochrome oxidase. FEBS Lett 356:295–298
41. Brookes P, Darley-Usmar VM (2002) Hypothesis: the mitochondrial NO(*) signaling pathway, and the transduction of nitrosative to oxidative cell signals: an alternative function for cytochrome C oxidase. Free Radic Biol Med 32:370–374
42. Nisoli E, Clementi E, Paolucci C, Cozzi V, Tonello C, Sciorati C et al (2003) Mitochondrial biogenesis in mammals: the role of endogenous nitric oxide. Science 299:896–899

43. Nisoli E, Falcone S, Tonello C, Cozzi V, Palomba L, Fiorani M et al (2004) Mitochondrial biogenesis by NO yields functionally active mitochondria in mammals. Proc Natl Acad Sci U S A 101:16507–16512
44. Brookes PS, Salinas EP, Darley-Usmar K, Eiserich JP, Freeman BA, Darley-Usmar VM et al (2000) Concentration-dependent effects of nitric oxide on mitochondrial permeability transition and cytochrome c release. J Biol Chem 275:20474–20479
45. Kim YM, Kim TH, Seol DW, Talanian RV, Billiar TR (1998) Nitric oxide suppression of apoptosis occurs in association with an inhibition of Bcl-2 cleavage and cytochrome c release. J Biol Chem 273:31437–31441
46. Shiva S, Sack MN, Greer JJ, Duranski M, Ringwood LA, Burwell L et al (2007) Nitrite augments tolerance to ischemia/reperfusion injury via the modulation of mitochondrial electron transfer. J Exp Med 204:2089–2102
47. Kobayashi A, Kang MI, Okawa H, Ohtsuji M, Zenke Y, Chiba T et al (2004) Oxidative stress sensor Keap1 functions as an adaptor for Cul3-Based E3 ligase to regulate proteasomal degradation of Nrf2. Mol Cell Biol 24:7130–7139
48. Zhang DD, Lo SC, Cross JV, Templeton DJ, Hannink M (2004) Keap1 is a redox-regulated substrate adaptor protein for a Cul3-dependent ubiquitin ligase complex. Mol Cell Biol 24:10941–10953
49. Friling RS, Bensimon A, Tichauer Y, Daniel V (1990) Xenobiotic-inducible expression of murine glutathione S-transferase Ya subunit gene is controlled by an electrophile-responsive element. Proc Natl Acad Sci U S A 87:6258–6262
50. Rushmore TH, Morton MR, Pickett CB (1991) The antioxidant responsive element. Activation by oxidative stress and identification of the DNA consensus sequence required for functional activity. J Biol Chem 266:11632–11639
51. Itoh K, Mochizuki M, Ishii Y, Ishii T, Shibata T, Kawamoto Y et al (2004) Transcription factor Nrf2 regulates inflammation by mediating the effect of 15-deoxy-Delta(12,14)-prostaglandin j(2). Mol Cell Biol 24:36–45
52. McMahon M, Lamont DJ, Beattie KA, Hayes JD (2010) Keap1 perceives stress via three sensors for the endogenous signaling molecules nitric oxide, zinc, and alkenals. Proc Natl Acad Sci U S A 107:18838–18843
53. Um HC, Jang JH, Kim DH, Lee C, Surh YJ (2011) Nitric oxide activates Nrf2 through S-nitrosylation of Keap1 in PC12 cells. Nitric Oxide 25:161–168
54. Sawa T, Zaki MH, Okamoto T, Akuta T, Tokutomi Y, Kim-Mitsuyama S et al (2007) Protein S-guanylation by the biological signal 8-nitroguanosine 3′,5′-cyclic monophosphate. Nat Chem Biol 3:727–735
55. Fujii S, Sawa T, Ihara H, Tong KI, Ida T, Okamoto T et al (2010) The critical role of nitric oxide signaling, via protein S-guanylation and nitrated cyclic GMP, in the antioxidant adaptive response. J Biol Chem 285:23970–23984
56. Kansanen E, Bonacci G, Schopfer FJ, Kuosmanen SM, Tong KI, Leinonen H et al (2011) Electrophilic nitro-fatty acids activate NRF2 by a KEAP1 cysteine 151-independent mechanism. J Biol Chem 286:14019–14027
57. Tsujita T, Li L, Nakajima H, Iwamoto N, Nakajima-Takagi Y, Ohashi K et al (2011) Nitro-fatty acids and cyclopentenone prostaglandins share strategies to activate the Keap1-Nrf2 system: a study using green fluorescent protein transgenic zebrafish. Genes Cells 16:46–57
58. Olson JL (2007) Renal disease caused by hypertension. In: Chap 21, Heptinstall's pathology of the kidney, 6th edn. Lippincott Williams & Wilkins, Philadelphia, p 960
59. Noiri E, Doi K, Negishi K, Tanaka T, Hamasaki Y, Fujita T et al (2009) Urinary fatty acid-binding protein 1: an early predictive biomarker of kidney injury. Am J Physiol Renal Physiol 296:F669–F679
60. Yamamoto T, Noiri E, Ono Y, Doi K, Negishi K, Kamijo A et al (2007) Renal L-type fatty acid—binding protein in acute ischemic injury. J Am Soc Nephrol 18:2894–2902
61. Araki S, Haneda M, Koya D, Sugaya T, Isshiki K, Kume S et al (2013) Predictive effects of urinary liver-type fatty acid-binding protein for deteriorating renal function and incidence of

cardiovascular disease in type 2 diabetic patients without advanced nephropathy. Diabetes Care 36:1248–1253
62. Nakamura T, Sugaya T, Kawagoe Y, Ueda Y, Osada S, Koide H (2005) Effect of pitavastatin on urinary liver-type fatty acid-binding protein levels in patients with early diabetic nephropathy. Diabetes Care 28:2728–2732
63. Nielsen SE, Sugaya T, Hovind P, Baba T, Parving HH, Rossing P (2010) Urinary liver-type fatty acid-binding protein predicts progression to nephropathy in type 1 diabetic patients. Diabetes Care 33:1320–1324
64. Brownlee M (2001) Biochemistry and molecular cell biology of diabetic complications. Nature 414:813–820
65. Dinkova-Kostova AT, Liby KT, Stephenson KK, Holtzclaw WD, Gao X, Suh N et al (2005) Extremely potent triterpenoid inducers of the phase 2 response: correlations of protection against oxidant and inflammatory stress. Proc Natl Acad Sci U S A 102:4584–4589
66. Pergola PE, Krauth M, Huff JW, Ferguson DA, Ruiz S, Meyer CJ et al (2011) Effect of bardoxolone methyl on kidney function in patients with T2D and stage 3b–4 CKD. Am J Nephrol 33:469–476
67. Pergola PE, Raskin P, Toto RD, Meyer CJ, Huff JW, Grossman EB et al (2011) Bardoxolone methyl and kidney function in CKD with type 2 diabetes. N Engl J Med 365:327–336
68. Stevens PE, Levin A (2013) Evaluation and management of chronic kidney disease: synopsis of the kidney disease: improving global outcomes 2012 clinical practice guideline. Ann Intern Med 158:825–830

Chapter 5
Multifunctional Roles of Nitric Oxide (NO) in Neurons

Tayo Katano and Seiji Ito

Abbreviations

AD	Alzheimer's disease
AMPAR	α-Amino-3-hydroxy-5-methyl-4-isoxazole propionic acid receptor
Aβ	Amyloid-β
CaMKII	Ca^{2+}/CaM-dependent protein kinase II
Cdk-5	Cyclin-dependent kinase-5
CFA	Complete Freund's adjuvant
cGK	cGMP-dependent protein kinase
COX	Cyclooxygenase
CP-AMPAR	Ca^{2+} permeable AMPAR
DG	Diacylglycerol
DRG	Dorsal root ganglia
Drp-1	Dynamin-related protein 1
ER	Endoplasmic reticulum
GRIP	Glutamate receptor-interacting protein
Gs/Gq	Class of guanine nucleotide-binding protein
GSNO	S-nitrosoglutathione
MLC	Myosin light chain
NMDAR	*N*-methyl-D-aspartate receptor
nNOS	Neuronal nitric oxide synthase
NO	Nitric oxide
NSAIDs	Nonsteroidal inflammatory drugs
NSF	*N*-ethylmaleimide-sensitive factor
PACAP	Pituitary adenylate cyclase-activating peptide
PD	Parkinson's disease

T. Katano • S. Ito (✉)
Department of Medical Chemistry, Kansai Medical University, Hirakata 573-1010, Japan
e-mail: katanot@hirakata.kmu.ac.jp; ito@hirakata.kmu.ac.jp

H. Tsukahara and K. Kaneko (eds.), *Studies on Pediatric Disorders*, Oxidative Stress in Applied Basic Research and Clinical Practice, DOI 10.1007/978-1-4939-0679-6_5,

PDI	Protein disulfide isomerase
PI3	Phosphatidylinositol 3-phosphate
PICK1	Protein interacting with C kinase 1
PKA	Protein kinase A
PKC	Protein kinase C
PKG	Protein kinase G
PLC	Phospholipase C
PPI	Protein phosphatase I
PSD	Postsynaptic density
RyRs	Ryanodine receptors
SAP97	Synapse-associated protein 97
sGC	Soluble guanylyl cyclase
TARPs	Transmembrane AMPAR regulatory proteins
VASP	Vasodilator-stimulated phosphoprotein

5.1 Introduction

Nitric oxide (NO) is a gaseous signaling mediator in our body, that is discovered as endothelium-derived relaxing factor in 1980 [1]. Since 1988 when NO was first shown to be released in a Ca^{2+}-dependent manner following *N*-methyl-D-aspartate receptor (NMDAR) activation and to act as an intercellular messenger in the brain [2], the physiological and pathophysiological roles of NO in the nervous system, such as signal transmission, neurotoxicity/neuroprotection, neurodegeneration/neuroregeneration, learning/memory, and chronic pain, have been extensively studied [3, 4]. Synaptic plasticity is fundamental to many neurobiological functions including pain and memory. Analysis of the molecular mechanisms underlying the generation and maintenance of central sensitization based on chronic pain and of long-term potentiation based on learning/memory indicates that there are striking similarities between those contributing to memory and pain [5].

Reactive gaseous NO exerts versatile effects on proteins via phosphorylation by cGMP-dependent protein kinase (cGK), nitrosylation at cysteine residues (S-nitrosylation), and nitration at tyrosine residues (Fig. 5.1). In this chapter, we introduce the recent progress in studies on pathophysiological role of NO in relation to its action mechanisms, especially pain.

5.2 Structure and Activation Mechanism of nNOS

NO is generated by three isoforms of NO synthase (NOS); neuronal NOS (nNOS, type I), inducible NOS (iNOS, type II), and endothelial NOS (eNOS, type III). All NOS isoforms consist of two domains, an oxygenase (N-terminal) domain and a reductase (C-terminal) one that are separated by a calmodulin (CaM)-binding motif (Fig. 5.2i). Binding of CaM to this motif facilitates electron flow from NADPH to

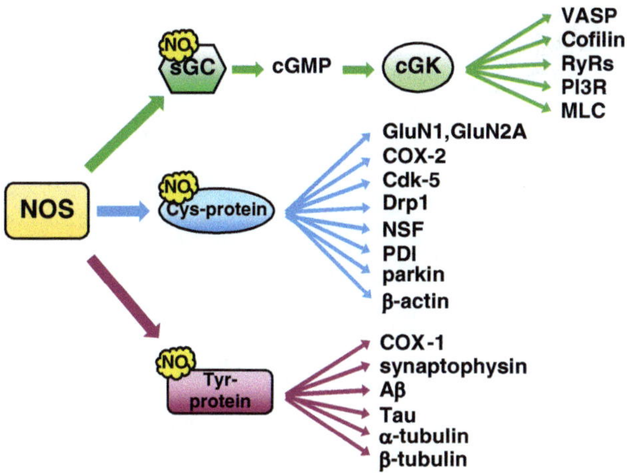

Fig. 5.1 Actions and target proteins of NO in neurons. NO generated by NOS exerts actions on proteins via three pathways, i.e., activation of sGC-cGK, nitrosylation of Cys residues, and nitration of Tyr residues. Since target proteins of NO are multiple, NO is able to be involved in many intracellular events in neurons

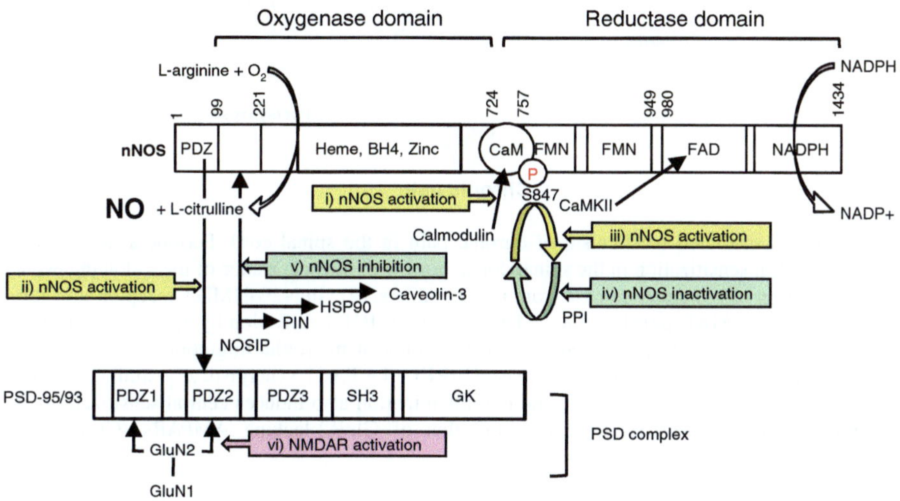

Fig. 5.2 Structure and regulatory factors of nNOS. NOS family has two common domains, oxygenase and reductase. The PDZ domain at the N-terminus is unique to nNOS and interacts with PDZ2 of PSD-95 in the PSD complex. nNOS activity is regulated by several factors, CaM and other modulatory proteins, the binding to PSD-95, and phosphorylation of Ser847 by CaMKII and its dephosphorylation by PPI

arginine via cofactors from NADPH, FAD, and FMN to heme and stimulates NO synthesis from L-arginine and oxygen by the oxygenase domain (Fig. 5.2) [6]. Ca^{2+}/CaM acts as an allosteric activator of nNOS, and its activity depends on CaM activation by the intracellular Ca^{2+} concentration ([Ca^{2+}]i). Since the gradient of [Ca^{2+}]i is produced by Ca^{2+} influx through the NMDAR, the translocation of nNOS close to the NMDAR can activate its activity efficiently (Fig. 5.3) [7, 8]. The N-terminal PDZ domain (1–99 amino acids) unique to nNOS interacts with PDZ2 of PSD-95 (Fig. 5.2ii), and nNOS can be anchored to the postsynaptic density (PSD) in the spine (Fig. 5.3).

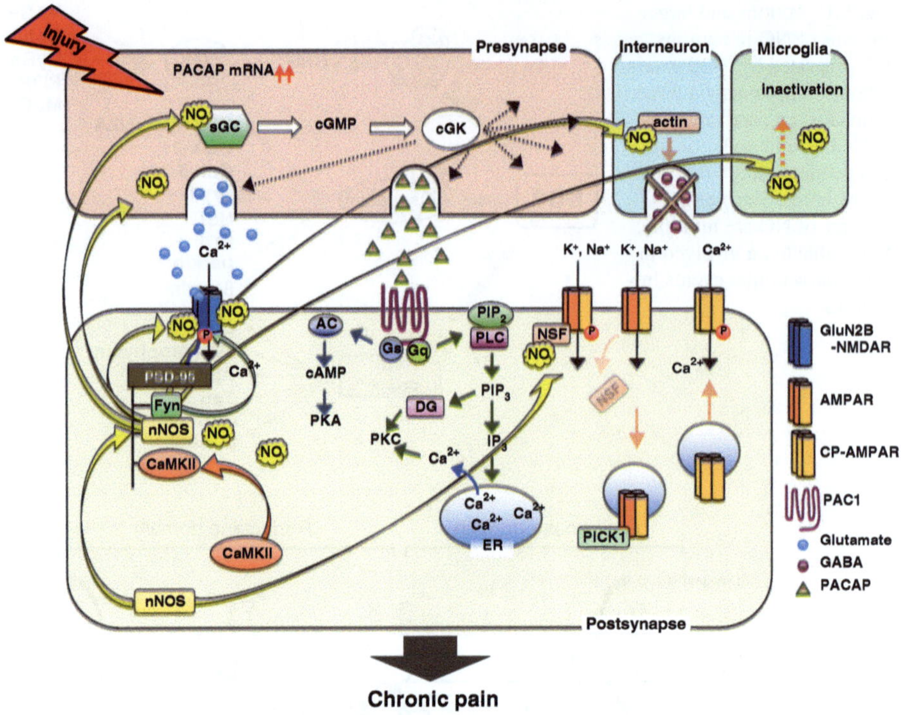

Fig. 5.3 NO-mediated mechanism of chronic pain in the spinal cord. Peripheral nerve injury induces central sensitization in the spinal dorsal horn with the occurrence of the following events: (1) stimulation of glutamate release from nerve endings, (2) GluN2B-NMDA receptor activation, (3) nNOS and CaMKII translocation and activation, (4) PAC1 activation by up-regulated PACAP, (5) inhibition of inhibitory interneurons, (6) inhibition of microglia migration, (7) activation of protein kinase A (PKA) and protein kinase C (PKC) by PAC1 synergistically stimulates nNOS translation with GluN2B activation, and (8) inflammation also induces central sensitization, and content of GluR2-AMPARs decreases and that of GluR2-lacking AMPARs (CP-AMPAR) increases in an NSF-dependent manner

Ca^{2+}/CaM-dependent protein kinase II (CaMKII)α enhances nNOS activity by phosphorylating nNOS at its Ser847 in an auto-inhibitory loop in which the kinase interacts with the FAD region in the reductase domain for activation of nNOS (Fig. 5.2iii) and protein phosphatase I (PPI) inactivates nNOS by dephosphorylating at Ser847 (Fig. 5.2iv).

5.3 Action Mechanisms of Nitric Oxide (NO) in Neurons

In the nervous system, information is unidirectionally transmitted via synapses between neurons. NO has at least three pathways for modulation of neuronal functions: (1) phosphorylation of serine or threonine residues by cGK mediated by soluble guanylyl cyclase (sGC) [9], (2) S-nitrosylation of cysteine residues, and (3) nitration of tyrosine and tryptophan residues in multiple target proteins (Fig. 5.1) [10–12].

5.3.1 NO-cGMP-cGK Pathway

sGC contains an NO-binding domain that stimulates the production of the second messenger cGMP upon activation by NO, and its principal intracellular target is cGK, also called protein kinase G (PKG). The isoform of cGK, the α isoform of cGKI (cGK-Iα), is abundantly expressed in the cerebellum and dorsal root ganglia (DRG). In the DRG, nNOS is co-expressed with cGK-I during embryonic development and after peripheral nerve axotomy. The cGK pathway has been implicated in the development of primary afferent fibers, preventing apoptosis of DRG neurons and facilitating nerve elongation [13, 14].

5.3.2 S-nitrosylation: Reversible Post-Translational Modification by NO

S-nitrosylation of proteins is a kind of post-translational modification by NO at cysteine residues. S-nitrosylation of plasma proteins and glyceraldehyde-3-phosphate dehydrogenase by exogenous NO donors was first reported in 1992 [15, 16]. Jaffrey et al. identified NMDAR subunits as endogenous S-nitrosylated proteins in mouse brain [12]. From 2001 onward, many S-nitrosylated proteins have been identified in the nervous system, and the significance of these individual proteins has been clarified (Fig. 5.1). The inhibition of S-nitrosylation of GluN2A at Cys399 using mutagenesis technique reduced the inhibition of the GluN2A-NMDAR-induced response by NO, implying that inhibition of NMDAR function by NO may serve as a negative-feedback mechanism for excess NMDAR-associated channel activity leading to neurotoxicity [11]. Surprisingly, cyclooxygenase (COX)-2, a key enzyme for the synthesis of prostaglandins interacts nNOS via its PDZ domain and activates it by S-nitrosylation [17]. Dominant-negative constructs of nNOS-PDZ and COX-2 attenuate NMDAR-induced neurotoxicity.

5.3.3 Tyr-Nitration: Reversible Post-Translational Modification by NO

Tyr-nitration is another kind of post-translational modification by NO. Whereas, COX-2 is S-nitrosylated in the cerebellum [17] and cardiac muscle [18], COX-1 is Tyr-nitrated in vascular smooth muscle cells and in human atherosclerotic tissues (Fig. 5.1) [19]. In contrast to activation of COX-2 by S-nitrosylation [17], COX-1 activity is inhibited by Tyr-nitration, resulting in reduced PGE_2 synthesis [19]. Nitration of amyloid-β (Aβ), a critical factor of Alzheimer's disease (AD) accelerates aggregation. S-nitrosylation and Tyr-nitration are a common consequence of overproduction of NO by excessive activation of NMDAR under various pathophysiological conditions (Fig. 5.1).

5.4 The Role of PSD and NO in Synaptic Transmission

5.4.1 Properties and NO-Related Regulation of Glutamate Receptors

Glutamate is the main excitatory neurotransmitter at the vast majority of excitatory synapses and mediates fast neurotransmission via NMDARs and non-NMDARs. Non-NMDARs are further divided into α-Amino-3-hydroxy-5-methyl-4-isoxazole propionic acid (AMPA) and kainate subtypes. NMDAR is an ionotropic glutamate receptor that is voltage- and ligand-gated and shows high permeability to Ca^{2+} [20, 21]. These properties enable it to detect coincident synaptic input and postsynaptic depolarization and to mediate the effect of glutamate through activation of intracellular signaling cascades. The Ca^{2+} influx through NMDARs is essential for synaptic transmission and experience-dependent synaptic plasticity such as development, learning and memory, and chronic pain. Functional NMDARs are heteromeric assemblies of the essential subunit GluN1, and modulatory subunits GluN2A–GluN2D. The C-terminal domain of GluN2 and GluN1 subunits contain a PDZ motif (–ESDV and –STVV, respectively) that binds to the PDZ1 or PDZ2 of PSD-95 [22–24] (Fig. 5.2). Also, nNOS is coupled to the PDZ2 domain of PSD-95 as a functional molecule in the PSD complex [25–28]. Excessive stimulation of NMDARs by a stroke insult activates nNOS, causing its translocation from the cytoplasm to the synaptic membranes where it interacts with NMDARs via PSD-95 [27, 28]. The glutamate-induced neurotoxicity in cortical neurons is significantly inhibited by a small-molecule inhibitor of the nNOS–PSD-95 interaction, and it is significantly reduced in nNOS knockout mouse [28]. Thus, the interaction of nNOS with NMDARs via PSD-95 is crucial for neuronal injury after a stroke as well as for neuronal plastic changes; the control of nNOS activation is effective against NMDAR-mediated neuronal death.

On the other hand, GluN1 and GluN2 are S-nitrosylated [11, 12]. S-nitrosylation of GluN1 induced by hypoxia and ischemia/reperfusion, as well as the phosphorylation by c-Src is diminished by inhibition of NOS. The NMDAR antagonist MK-801 also attenuates the S-nitrosylation of GluN1 and activation of c-Src [29, 30]. These reports indicate that glutamate receptors are regulated by NO.

5.4.2 The Complex of PSD Proteins

AMPA and NMDARs are concentrated in the PSD, a specialized structure localized in the inner part of the postsynaptic membrane, where they are clustered with receptors such as PAC1, a receptor for pituitary adenylate cyclase-activating peptide (PACAP), adaptor proteins such as PSD-95 and *N*-ethylmaleimide-sensitive factor (NSF), and signal transducers including nNOS and CaMKII (Fig. 5.3) [31].

PSD-95 is a principal scaffold protein in the PSD, containing an SH3, a GK, and three PDZ domains (Fig. 5.2); more than 40 proteins have been identified as interacting via these domains of PSD-95. nNOS is anchored to PSD-95 (Fig. 5.2ii) and exists close to GluN2-NMDARs (Fig. 5.3). Conversely, NO regulates targeting of PSD-95 to synapses via mutually competitive modifications by palmitoylation; the modification at Cys3 and Cys5 is targeting signal of PSD-95 and S-nitrosylation [32, 33]. Thus, it appears that NO shows mutual regulation of target proteins. NO generated upon activation of NMDARs dynamically regulates the protein–protein interaction in the PSD and effectively modulates functions of PSD proteins via phosphorylation by cGK, Tyr-nitration, and S-nitrosylation under both physiological and pathological conditions (Fig. 5.3) [34, 35]. Furthermore, the free gas NO is able to regulate multiple functions not only at the synapse, but also in terms of cell-to-cell communication such as by inhibitory interneurons and microglia in the spinal cord (Fig. 5.3). We recently demonstrated that NO inhibits dopamine release via S-nitrosylation of β-actin and inhibits ATP-induced migration of microglia [36, 37].

5.5 Involvement of NO in Intractable Pain in the Spinal Cord

5.5.1 The Roles of NO in Synaptic Plasticity in the Nervous System

Sensory signals produced by thermal, mechanical, and chemical stimuli in the skin are transmitted to and processed in the dorsal horn of the spinal cord, and a part of them are recognized as pain in the brain. Intractable pain following inflammation or peripheral nerve injury is associated with increased excitability of nociceptive neurons, i.e., central sensitization, in the spinal cord. The process of plastic changes in neurons is caused by morphological and functional changes [38, 39]. cGK-I is required for the guidance and connectivity of sensory axons via the phosphorylation of target proteins such as cofilin [40–42]. Inhibition of NOS leads to a significant decrease in cofilin phosphorylation in the spinal cord and in nociceptive behavior induced by zymosan, which activates the toll-like receptor 2 and induces inflammatory pain [43]. On the other hand, semaphorin 3A-induced collapse of growth cones is overcome by cGMP/cGK-I signaling.

Trafficking of AMPARs is a mechanism of synaptic plasticity and is regulated by various interacting proteins: NSF, transmembrane AMPAR regulatory proteins (TARPs), protein interacting with C kinase 1 (PICK1), synapse-associated protein 97 (SAP97), and glutamate receptor-interacting protein (GRIP). Among these proteins regulated by NO, NSF, and stargazin stimulate AMPAR trafficking by S-nitrosylation. S-nitrosylation of NSF enhances its binding to GluA2 (Fig. 5.3), and this binding is reduced in nNOS-deficient mice [44].

5.5.2 *Regulation of AMPAR Trafficking by NO in Inflammatory Pain*

Many studies have demonstrated that the NO/cGMP signaling pathway is present in neurons of the spinal cord and contributes to the development of hyperalgesia in models of acute and chronic pain [45–49]. nNOS is expressed in a specified population of inhibitory interneurons, and cGK-Iα is expressed in a subpopulation of small- and medium-sized neurons in the DRG and superficial lamina of the dorsal horn of the spinal cord [13, 48]. nNOS activity is increased in the spinal dorsal horn after inflammation [43, 50, 51]. The amount of cGK-Iα protein is also increased following inflammation elicited by a formalin injection, and this formalin-induced up-regulation in the spinal dorsal horn is blocked by inhibitors of NMDARs and nNOS [48, 49].

The modulation of inflammatory pain by NO is altered during the course of the pathological process. COX-2, a target of nonsteroidal inflammatory drugs (NSAIDs), is activated by S-nitrosylation. Whereas inflammatory pain is significantly relieved by indomethacin, an NSAID, in the early phase after a Complete Freund's adjuvant (CFA) injection, it is resistant to the drug in the late phase (Fig. 5.3) [52]. The content of GluA2 subunits is decreased along with that of NSF in the spinal dorsal horn; and, conversely, that of GluA1 subunits is significantly increased in the late phase, suggesting that Ca^{2+} permeable AMPAR (CP-AMPAR) is increased by the subunit switch from GluA2 to GluA1 in AMPARs [52]. These results suggest that NO is involved in COX activation and GluA2-NSF association via S-nitrosylation in the early phase of inflammatory pain and that denitrosylation of NSF promotes GluA2-NSF dissociation and the subunit switch of AMPARs in the late phase.

5.5.3 *Involvement of NMDAR-nNOS Activation in Intractable Pain*

NO production by nNOS is involved in the maintenance of neuropathic pain produced by nerve injury, and inhibition of this production alleviates it [53]. Further, the disruption of nNOS–PSD95 blocks cGMP production and thermal hyperalgesia and mechanical allodynia [54]. Whereas CP-101,606, the GluN2B-selective antagonist, ameliorated neuropathic pain, and decreased nNOS activity as assessed by NADPH diaphorase histochemistry. GluN2B subunit is essential for the maintenance of neuropathic pain [29]; and, further, we demonstrated that the phosphorylation of Tyr1472 at the C-terminal tail of GluN2B subunits is fundamental to the maintenance of neuropathic pain in the spinal cord that lasts over 1 week. Tyr1472 is located in the internalization motif YEKL, and the phosphorylation of Tyr1472 may have dual functions, i.e., synaptic localization of GluN2B-NMDAR in the PSD [55] and its signal transduction, i.e., nNOS activation.

Tyr1472 phosphorylation is not always crucial for the interaction of nNOS with GluN2B and that Ca^{2+} influx through NMDARs containing mutated GluN2B is sufficient for nNOS activation [56–58]. Whereas nNOS is significantly translocated from the cytosol to the membrane only above 500 μM NMDA alone, this concentration could be markedly reduced to 10 μM in the presence of 5 nM PACAP. PACAP is up-regulated in the DRG of neuropathic pain model mice and synergistically stimulates NO formation with glutamate or NMDA. Since many molecules are involved in chronic pain, PACAP may provide an alternative to nNOS activation [7] (Fig. 5.3). Like that of nNOS, CaMKII activity is controlled by $[Ca^{2+}]i$; and the translocation of the kinase to the PSD is enhanced after NMDAR activation (Fig. 5.3).

cGK-Iα-expressing neurons are involved in nociceptive processing by central nerve terminals of laminas I and II in the spinal cord, but not in that by spinal interneurons [13, 59]. By contrast, nNOS is present in only 1–2 % of lumbar DRG neurons; but nNOS-containing fibers and small interneurons are present in all layers of the spinal cord, especially in lamina II. Therefore, NO has been supposed to act as a retrograde messenger, and stimulates glutamate release from primary afferent terminals through the NO/cGMP/cGK pathway [13, 45, 60]; moreover, NO donors attenuate dopamine release via S-nitrosylation of β-actin [61]. The NO donor sodium nitroprusside inhibits ongoing impulse activity in 49 % of all spinal neurons in the laminas I and II and activates only 28 % of them [62]. These opposing effects of NO mediated by cGK and S-nitrosylation suggest that the contribution of these two pathways of NO action is determined to a considerable extent by the presence of cGK in the cells.

5.6 The Role of NO in Neurodegenerative Diseases

The cause of neurodegenerative diseases such as AD and Parkinson's disease (PD) is attributed to neuronal cell damage in specific brain regions. These neurodegenerative processes are produced in part by NO-mediated neuronal injury and synaptic damage.

5.6.1 *Alzheimer's Disease (AD)*

AD is the most common neurodegenerative disease, and the brains of AD patients contain aberrant accumulation of aggregated Aβ, a critical factor in the development of AD. Aggregated Aβ is supposed to be a molecular link between oxidative stress and AD-associated brain dysfunction by attracting inflammatory mediators that in turn up-regulate iNOS and generate NO radicals. Nitration of Aβ at its Tyr10 enhances Aβ aggregation and plaque formation [63], and is found in the core of Aβ plaques of amyloid precursor protein/presenilin 1 transgenic mice, and in the brains of AD patients [63]. Also Tyr-nitration of synaptophysin may contribute to Aβ-induced cholinergic dysfunction [64]. Also, Aβ-induced S-nitrosylation of the mitochondrial division protein Dynamin-related protein 1 (Drp-1) at its Cys644 leads to excessive mitochondrial fission, synaptic loss, and neuronal damage by dimerization and activation of the GTPase activity of Drp-1 [65]. Cyclin-dependent kinase-5 (Cdk-5) is thought to be important for normal brain functions and neurodegenerative disease pathogenesis. Cdk5 is activated by S-nitrosylation at its Cys83 and Cys157 and contributes to Aβ-induced dendritic spine loss. Interestingly, S-nitrosylated Cdk5 can transfer its NO group by transnitrosylation to Drp1, which may be involved in Aβ-induced dendritic spine loss and neuronal apoptosis [66]. Moreover, protein-disulphide isomerase (PDI) is S-nitrosylated following NMDAR activation by the physiological NO donor S-nitrosocysteine and in the brain of AD patients. The chaperone activity of PDI is attenuated by S-nitrosylation, resulting in an accumulation of misfolded proteins that are not eliminated by the ubiquitin proteasome system [67].

5.6.2 *Parkinson's Disease (PD)*

PD is caused by progressive loss of dopamine neurons. Parkin is an E3 ubiquitin ligase that is important for the survival of dopamine neurons in PD. Parkin is S-nitrosylated by the NO donor S-nitrosoglutathione (GSNO) and in the brain of patients with PD and diffuse Lewy body disease. S-nitrosylation inhibits parkin activity and may promote the degenerative process of PD by impairing the ubiquitination of parkin substrates [68].

Taken together, the pathogenesis of AD and PD includes not only specific targets such as Aβ, but also common molecules involving neurotoxicity targeted by NO synthesized following NMDAR activation.

5.7 Concluding Remarks

The reactive gaseous NO plays versatile roles in the nervous system via three pathways, phosphorylation by cGK and modifications at Cys and Tyr residues. Since the target molecules and modifications by NO may differ concentration-dependently in the same cell under pathophysiological conditions, NO is able to modify cellular functions diversely or bidirectionally regardless of the specificity [35]. The target protein of NO and cGK in the nervous systems remains to be clarified. In recent years, the techniques for imaging [69] and identification of proteins after post-translational modifications such as phosphorylation and S-nitrosylation by mass spectrometry have been constantly advancing and expanding. These outcomes may lead to identification of target molecules and the development drugs for neuronal degenerative disorders and chronic pain.

References

1. Furchgott RF, Zawadzki JV (1980) The obligatory role of endothelial cells in the relaxation of arterial smooth muscle by acetylcholine. Nature 288:373–376
2. Garthwaite J, Charles SL, Chess-Williams R (1988) Endothelium-derived relaxing factor release on activation of NMDA receptors suggests role as intercellular messenger in the brain. Nature 336:385–388
3. Calabrese V, Mancuso C, Calvani M, Rizzarelli E, Butterfield DA, Stella AM (2007) Nitric oxide in the central nervous system: neuroprotection versus neurotoxicity. Nat Rev Neurosci 8:766–775
4. Contestabile A, Ciani E (2004) Role of nitric oxide in the regulation of neuronal proliferation, survival and differentiation. Neurochem Int 45:903–914
5. Ji RR, Kohno T, Moore KA, Woolf CJ (2003) Central sensitization and LTP: do pain and memory share similar mechanisms? Trends Neurosci 26:696–705
6. Matsuda H, Iyanagi T (1999) Calmodulin activates intramolecular electron transfer between the two flavins of neuronal nitric oxide synthase flavin domain. Biochim Biophys Acta 1473: 345–355
7. Ohnishi T, Okuda-Ashitaka E, Matsumura S, Katano T, Nishizawa M, Ito S (2008) Characterization of signaling pathway for the translocation of neuronal nitric oxide synthase to the plasma membrane by PACAP. J Neurochem 105:2271–2285
8. Lee SJ, Escobedo-Lozoya Y, Szatmari EM, Yasuda R (2009) Activation of CaMKII in single dendritic spines during long-term potentiation. Nature 458:299–304
9. Bredt DS, Snyder SH (1989) Nitric oxide mediates glutamate-linked enhancement of cGMP levels in the cerebellum. Proc Natl Acad Sci U S A 86:9030–9033
10. Stamler JS, Lamas S, Fang FC (2001) Nitrosylation. The prototypic redox-based signaling mechanism. Cell 106:675–683
11. Choi YB, Tenneti L, Le DA, Ortiz J, Bai G, Chen HS et al (2000) Molecular basis of NMDA receptor-coupled ion channel modulation by S-nitrosylation. Nat Neurosci 3:15–21
12. Jaffrey SR, Erdjument-Bromage H, Ferris CD, Tempst P, Snyder SH (2001) Protein S-nitrosylation: a physiological signal for neuronal nitric oxide. Nat Cell Biol 3:193–197
13. Qian Y, Chao DS, Santillano DR, Cornwell TL, Nairn AC, Greengard P et al (1996) cGMP-dependent protein kinase in dorsal root ganglion: relationship with nitric oxide synthase and nociceptive neurons. J Neurosci 16:3130–3138
14. Thippeswamy T, Morris R (2002) The roles of nitric oxide in dorsal root ganglion neurons. Ann N Y Acad Sci 962:103–110
15. Stamler JS, Simon DI, Osborne JA, Mullins ME, Jaraki O, Michel T et al (1992) S-nitrosylation of proteins with nitric oxide: synthesis and characterization of biologically active compounds. Proc Natl Acad Sci U S A 89:444–448
16. Molina y Vedia L, McDonald B, Reep B, Brüne B, Di Silvio M, Billiar TR et al (1992) Nitric oxide-induced S-nitrosylation of glyceraldehyde-3-phosphate dehydrogenase inhibits enzymatic activity and increases endogenous ADP-ribosylation. J Biol Chem 267:24929–24932
17. Tian J, Kim SF, Hester L, Snyder SH (2008) S-nitrosylation/activation of COX-2 mediates NMDA neurotoxicity. Proc Natl Acad Sci U S A 105:10537–10540
18. Atar S, Ye Y, Lin Y, Freeberg SY, Nishi SP, Rosanio S et al (2006) Atorvastatin-induced cardioprotection is mediated by increasing inducible nitric oxide synthase and consequent S-nitrosylation of cyclooxygenase-2. Am J Physiol Heart Circ Physiol 290:H1960–H1968
19. Deeb RS, Resnick MJ, Mittar D, McCaffrey T, Hajjar DP, Upmacis RK (2002) Tyrosine nitration in prostaglandin H(2) synthase. J Lipid Res 43:1718–1726
20. MacDermott AB, Mayer ML, Westbrook GL, Smith SJ, Barker JL (1986) NMDA-receptor activation increases cytoplasmic calcium concentration in cultured spinal cord neurones. Nature 321:519–522
21. Lim IA, Merrill MA, Chen Y, Hell JW (2003) Disruption of the NMDA receptor-PSD-95 interaction in hippocampal neurons with no obvious physiological short-term effect. Neuropharmacology 45:738–754

22. Kornau HC, Schenker LT, Kennedy MB, Seeburg PH (1995) Domain interaction between NMDA receptor subunits and the postsynaptic density protein PSD-95. Science 269: 1737–1740
23. Niethammer M, Kim E, Sheng M (1996) Interaction between the C terminus of NMDA receptor subunits and multiple members of the PSD-95 family of membrane-associated guanylate kinases. J Neurosci 16:2157–2163
24. Lau CG, Zukin RS (2007) NMDA receptor trafficking in synaptic plasticity and neuropsychiatric disorders. Nat Rev Neurosci 8:413–426
25. Brenman JE, Chao DS, Gee SH, McGee AW, Craven SE, Santillano DR et al (1996) Interaction of nitric oxide synthase with the postsynaptic density protein PSD-95 and alpha1-syntrophin mediated by PDZ domains. Cell 84:757–767
26. Doucet MV, Harkin A, Dev KK (2012) The PSD-95/nNOS complex: new drugs for depression? Pharmacol Ther 133:218–229
27. Fan J, Vasuta OC, Zhang LY, Wang L, George A, Raymond LA (2010) *N*-methyl-D-aspartate receptor subunit- and neuronal-type dependence of excitotoxic signaling through post-synaptic density 95. J Neurochem 115:1045–1056
28. Zhou L, Li F, Xu HB, Luo CX, Wu HY, Zhu MM et al (2010) Treatment of cerebral ischemia by disrupting ischemia-induced interaction of nNOS with PSD-95. Nat Med 16:1439–1443
29. Abe T, Matsumura S, Katano T, Mabuchi T, Takagi K, Xu L et al (2005) Fyn kinase-mediated phosphorylation of NMDA receptor NR2B subunit at Tyr1472 is essential for maintenance of neuropathic pain. Eur J Neurosci 22:1445–1454
30. Tang LJ, Li C, Hu SQ, Wu YP, Zong YY, Sun CC et al (2012) S-nitrosylation of c-Src via NMDAR-nNOS module promotes c-Src activation and NR2A phosphorylation in cerebral ischemia/reperfusion. Mol Cell Biochem 365:363–377
31. Husi H, Grant SG (2001) Proteomics of the nervous system. Trends Neurosci 24:259–266
32. El-Husseini AE, Craven SE, Chetkovich DM, Firestein BL, Schnell E, Aoki C et al (2000) Dual palmitoylation of PSD-95 mediates its vesiculotubular sorting, postsynaptic targeting, and ion channel clustering. J Cell Biol 148:159–172
33. Ho GP, Selvakumar B, Mukai J, Hester LD, Wang Y, Gogos JA et al (2011) S-nitrosylation and S-palmitoylation reciprocally regulate synaptic targeting of PSD-95. Neuron 71:131–141
34. Ahern GP, Klyachko VA, Jackson MB (2002) cGMP and S-nitrosylation: two routes for modulation of neuronal excitability by NO. Trends Neurosci 25:510–517
35. Akhtar MW, Sunico CR, Nakamura T, Lipton SA (2012) Redox regulation of protein function via cysteine S-nitrosylation and its relevance to neurodegenerative diseases. Int J Cell Biol 2012:463756
36. Lu J, Katano T, Uta D, Furue H, Ito S (2011) Rapid S-nitrosylation of actin by NO-generating donors and in inflammatory pain model mice. Mol Pain 7:101
37. Kunori S, Matsumura S, Okuda-Ashitaka E, Katano T, Audoly LP, Urade Y et al (2011) A novel role of prostaglandin E2 in neuropathic pain: blockade of microglial migration in the spinal cord. Glia 59:208–218
38. Ito S, Okuda-Ashitaka E, Minami T (2001) Central and peripheral roles of prostaglandins in pain and their interactions with novel neuropeptides nociceptin and nocistatin. Neurosci Res 41:299–332
39. Mao J, Mayer DJ (2001) Spinal cord neuroplasticity following repeated opioid exposure and its relation to pathological pain. Ann N Y Acad Sci 933:175–184
40. Schmidt H, Werner M, Heppenstall PA, Henning M, Moré MI, Kühbandner S et al (2002) cGMP-mediated signaling via cGKIalpha is required for the guidance and connectivity of sensory axons. J Cell Biol 159:489–498
41. Zhao Z, Wang Z, Gu Y, Feil R, Hofmann F, Ma L (2009) Regulate axon branching by the cyclic GMP pathway via inhibition of glycogen synthase kinase 3 in dorsal root ganglion sensory neurons. J Neurosci 29:1350–1360
42. Yamazaki M, Chiba K, Mohri T (2005) Fundamental role of nitric oxide in neuritogenesis of PC12h cells. Br J Pharmacol 146:662–669

43. Zulauf L, Coste O, Marian C, Möser C, Brenneis C, Niederberger E (2009) Cofilin phosphorylation is involved in nitric oxide/cGMP-mediated nociception. Biochem Biophys Res Commun 390:1408–1413
44. Huang Y, Man HY, Sekine-Aizawa Y, Han Y, Juluri K, Luo H et al (2005) S-nitrosylation of *N*-ethylmaleimide sensitive factor mediates surface expression of AMPA receptors. Neuron 46:533–540
45. Meller ST, Gebhart GF (1993) Nitric oxide (NO) and nociceptive processing in the spinal cord. Pain 52:127–136
46. Sluka KA, Willis WD (1998) Increased spinal release of excitatory amino acids following intradermal injection of capsaicin is reduced by a protein kinase G inhibitor. Brain Res 798:281–286
47. Minami T, Nishihara I, Ito S, Sakamoto K, Hyodo M, Hayaishi O (1995) Nitric oxide mediates allodynia induced by intrathecal administration of prostaglandin E2 or prostaglandin F2 alpha in conscious mice. Pain 61:285–290
48. Tao YX, Johns RA (2000) Activation of cGMP-dependent protein kinase Ialpha is required for *N*-methyl-D-aspartate- or nitric oxide-produced spinal thermal hyperalgesia. Eur J Pharmacol 392:141–145
49. Schmidtko A, Ruth P, Geisslinger G, Tegeder I (2003) Inhibition of cyclic guanosine 5′-monophosphate-dependent protein kinase I (PKG-I) in lumbar spinal cord reduces formalin-induced hyperalgesia and PKG upregulation. Nitric Oxide 8:89–94
50. Lam HH, Hanley DF, Trapp BD, Saito S, Raja S, Dawson TM et al (1996) Induction of spinal cord neuronal nitric oxide synthase (NOS) after formalin injection in the rat hind paw. Neurosci Lett 210:201–204
51. Chu YC, Guan Y, Skinner J, Raja SN, Johns RA, Tao YX (2005) Effect of genetic knockout or pharmacologic inhibition of neuronal nitric oxide synthase on complete Freund's adjuvant-induced persistent pain. Pain 119:113–123
52. Katano T, Furue H, Okuda-Ashitaka E, Tagaya M, Watanabe M, Yoshimura M et al (2008) *N*-ethylmaleimide-sensitive fusion protein (NSF) is involved in central sensitization in the spinal cord through GluR2 subunit composition switch after inflammation. Eur J Neurosci 27: 3161–3170
53. Mabuchi T, Matsumura S, Okuda-Ashitaka E, Kitano T, Kojima H, Nagano T et al (2003) Attenuation of neuropathic pain by the nociceptin/orphanin FQ antagonist JTC-801 is mediated by inhibition of nitric oxide production. Eur J Neurosci 17:1384–1392
54. Florio SK, Loh C, Huang SM, Iwamaye AE, Kitto KF, Fowler KW et al (2009) Disruption of nNOS-PSD95 protein–protein interaction inhibits acute thermal hyperalgesia and chronic mechanical allodynia in rodents. Br J Pharmacol 158:494–506
55. Roche KW, Standley S, McCallum J, Dune Ly C, Ehlers MD, Wenthold RJ (2001) Molecular determinants of NMDA receptor internalization. Nat Neurosci 4:794–802
56. Matsumura S, Kunori S, Mabuchi T, Katano T, Nakazawa T, Abe T et al (2010) Impairment of CaMKII activation and attenuation of neuropathic pain in mice lacking NR2B phosphorylated at Tyr1472. Eur J Neurosci 32:798–810
57. Katano T, Nakazawa T, Nakatsuka T, Watanabe M, Yamamoto T, Ito S (2011) Involvement of spinal phosphorylation cascade of Tyr1472-NR2B, Thr286-CaMKII, and Ser831-GluR1 in neuropathic pain. Neuropharmacology 60:609–616
58. Nakazawa T, Komai S, Watabe AM, Kiyama Y, Fukaya M, Arima-Yoshida F et al (2006) NR2B tyrosine phosphorylation modulates fear learning as well as amygdaloid synaptic plasticity. EMBO J 25:2867–2877
59. Schlossmann J, Hofmann F (2005) cGMP-dependent protein kinases in drug discovery. Drug Discov Today 10:627–634
60. Xu L, Okuda-Ashitaka E, Matsumura S, Mabuchi T, Okamoto S, Sakimura K et al (2007) Signal pathways coupled to activation of neuronal nitric oxide synthase in the spinal cord by nociceptin/orphanin FQ. Neuropharmacology 52:1318–1325
61. Lu J, Katano T, Okuda-Ashitaka E, Oishi Y, Urade Y, Ito S (2009) Involvement of S-nitrosylation of actin in inhibition of neurotransmitter release by nitric oxide. Mol Pain 5:58

62. Pehl U, Schmid HA (1997) Electrophysiological responses of neurons in the rat spinal cord to nitric oxide. Neuroscience 77:563–573
63. Kummer MP, Hermes M, Delekarte A, Hammerschmidt T, Kumar S, Terwel D et al (2011) Nitration of tyrosine 10 critically enhances amyloid β aggregation and plaque formation. Neuron 71:833–844
64. Tran MH, Yamada K, Nakajima A, Mizuno M, He J, Kamei H et al (2003) Tyrosine nitration of a synaptic protein synaptophysin contributes to amyloid beta-peptide-induced cholinergic dysfunction. Mol Psychiatry 8:407–412
65. Cho DH, Nakamura T, Fang J, Cieplak P, Godzik A, Gu Z et al (2009) S-nitrosylation of Drp1 mediates beta-amyloid-related mitochondrial fission and neuronal injury. Science 324: 102–105
66. Qu J, Nakamura T, Cao G, Holland EA, McKercher SR, Lipton SA (2011) S-nitrosylation activates Cdk5 and contributes to synaptic spine loss induced by beta-amyloid peptide. Proc Natl Acad Sci U S A 108:14330–14335
67. Uehara T, Nakamura T, Yao D, Shi ZQ, Gu Z, Ma Y et al (2006) S-nitrosylated protein-disulphide isomerase links protein misfolding to neurodegeneration. Nature 441:513–517
68. Chung KK, Thomas B, Li X, Pletnikova O, Troncoso JC, Marsh L et al (2004) S-nitrosylation of parkin regulates ubiquitination and compromises parkin's protective function. Science 304:1328–1331
69. Jo A, Do H, Jhon GJ, Suh M, Lee Y (2011) Electrochemical nanosensor for real-time direct imaging of nitric oxide in living brain. Anal Chem 83:8314–8319

Part II
Basic Science Topics

Chapter 6
Oxidative Stress Biomarkers: Current Status and Future Perspective

Hirokazu Tsukahara

Abbreviations

AD	Atopic dermatitis
ADMA	Asymmetric dimethylarginine
AG	Aminoguanidine
ASL	Argininosuccinate lyase
ASS	Argininosuccinate synthase
BAP	Biological antioxidative potential
BOM	Bilirubin oxidative metabolites
CAAT	Cationic amino acid transporter
CAT	Catalase
CML	Carboxymethyllysine
CO	Carbon monoxide
Cr	Creatinine
CSF	Cerebrospinal fluid
DDAH	Dimethylarginine dimethylaminohydrolase
ELISA	Enzyme-linked immunosorbent assay
GPX	Glutathione peroxidase
GSH	Glutathione
HNE	4-Hydroxy-2-nonenal
H_2O_2	Hydrogen peroxide
L-NAME	N^G-nitro-L-arginine methyl ester
LO	Lipid alkoxyl radical
LOO	Lipid peroxyl radical

H. Tsukahara, M.D., Ph.D. (✉)
Department of Pediatrics, Okayama University Graduate School of Medicine, Dentistry and Pharmaceutical Sciences, Okayama 700-8558, Japan
e-mail: hirokazu010661@yahoo.co.jp

H. Tsukahara and K. Kaneko (eds.), *Studies on Pediatric Disorders*, Oxidative Stress in Applied Basic Research and Clinical Practice, DOI 10.1007/978-1-4939-0679-6_6,

LOOH	Lipid hydroperoxide
MDA	Malondialdehyde
NO	Nitric oxide
NO_2	Nitrogen dioxide
NOS	Nitric oxide synthase
$^1O_2$	Singlet oxygen
O_2^-	Superoxide anion
OH	Hydroxyl radical
8-OHdG	8-Hydroxy-2′-deoxyguanosine
ONOO	Peroxynitrite radical
$ONOO^-$	Peroxynitrite anion
PRMT	Protein arginine methyltransferase
ROS	Reactive oxygen species
SOD	Superoxide dismutase
TBARS	Thiobarbituric acid reactive substances
TP	Total hydroperoxides

6.1 Introduction

Reactive oxygen species (ROS) are generated as by-products of cellular metabolism and ionizing radiation, usually indicating the following four species: superoxide anion (O_2^-), hydrogen peroxide (H_2O_2), hydroxyl radical (OH), and singlet oxygen ($^1O_2$). Reactivity of either O_2^- or H_2O_2 by itself with other molecules is rather low, but in the presence of transition metals, each is converted to OH via the Fenton or Haber–Weiss reaction. Other biologically important free radicals and their equivalents are lipid hydroperoxide (LOOH), lipid peroxyl radical (LOO), and lipid alkoxyl radical (LO), which are associated with membrane lipids. Still others are nitric oxide (NO), nitrogen dioxide (NO_2), peroxynitrite radical (ONOO), and peroxynitrite anion ($ONOO^-$), which are reactive nitrogen species. Excess NO can injure tissues, and does so mainly by its rapid reaction with O_2^- to give $ONOO^-$ [1, 2].

The ROS originate primarily from mitochondria and cellular enzymes such as nicotinamide adenine dinucleotide phosphate-oxidases. Small amounts of ROS are cellular requirements because they are involved in cell signaling as messenger molecules of the autocrine and paracrine systems. They are also involved in host defenses against invading pathogens. At the system level, ROS contribute to complex functions such as blood pressure regulation, cognitive function, and immune function [3, 4].

Under normal physiological conditions, enzymes and antioxidants maintain the balance between production and elimination of ROS. Biologically active enzymes and antioxidants are categorized into three groups: enzymes such as superoxide dismutase (SOD), catalase (CAT), glutathione peroxidase (GPX), glutathione reductase, glutathione-*S*-transferase, thioredoxin reductase, and heme oxygenase; proteins such as albumin, ferritin, transferrin, lactoferrin, ceruloplasmin, and thioredoxin; and low-molecular-weight molecules such as bilirubin, ascorbate, tocopherols, carotenoids, ubiquinol/ubiquinone, cysteine, and glutathione (GSH) [3].

"Oxidative stress" is an imbalance between ROS generation and antioxidative defense mechanisms with any excess of the former, engendering macromolecular damage and dysfunction. Recently, that definition has been expanded to include more subtle perturbations in redox signaling mechanisms that control various cellular functions such as enzyme activation/inhibition, membrane signal transduction, transcription factor binding/gene expression, proliferation/apoptosis, and precursor cell ontogeny [2].

6.2 Nitric Oxide System, Endothelial Dysfunction, and Oxidative Stress

NO is synthesized in every cell type by nitric oxide synthases (NOSs) from L-arginine and molecular oxygen. The NOSs exist in three isoforms, which differ in the way their activity is controlled. They are neuronal (NOS1), inducible (NOS2), and endothelial (NOS3). The NOS1 and NOS3, which are present constitutively in cells of various types, are activated by transient increases in intracellular calcium. The third isoform, NOS2, is induced in response to inflammatory and immunological stimuli in myriad cells, including vascular endothelial cells, smooth muscle cells, and activated immune cells. The output of NO from NOS2 is about 1,000 times that of other constitutive isoforms [3].

The vascular endothelium, rather than being a mere barrier between intravascular and interstitial compartments, is a widely distributed organ that is responsible for the regulation of hemodynamics, angiogenic vascular remodeling, and metabolic, synthetic, anti-inflammatory, and antithrombogenic processes. Understanding the interrelationship of NO system blockade, endothelial dysfunction, and oxidative stress is expected to enable the delineation of a rational therapeutic strategy in conditions that are associated with oxidative damage [4].

Indeed, NO formation and oxidative stress are mutually related [3]. Previously, we established a young rat model of chronic NO deficiency and endothelial dysfunction. We examined the effects of endogenous NO blockade on oxidative stress status and renal function in young rats [5, 6]. Two NOS inhibitors were used: N^G-nitro-L-arginine methyl ester (L-NAME) as a nonselective inhibitor, and aminoguanidine (AG) as a selective inhibitor of NOS2 (Fig. 6.1). Oral administration of L-NAME, but not AG, for 4 weeks induced systemic hypertension, significant reduction in urinary nitrite/nitrate (a marker of endogenous NO formation), and a significant increase in urinary 8-hydroxy-2′-deoxyguanosine (8-OHdG), also known as 8-oxo-7,8-dihydro-2-deoxyguanosine (a marker of oxidative stress), compared with non-treated animals. On combining all the data, there was a significant negative correlation between urinary nitrite/nitrate and 8-OHdG. The L-NAME-treated rats also developed proteinuria and tubular enzymuria. The effects of L-NAME on blood pressure and urinary parameters were restored by a large dose of L-arginine.

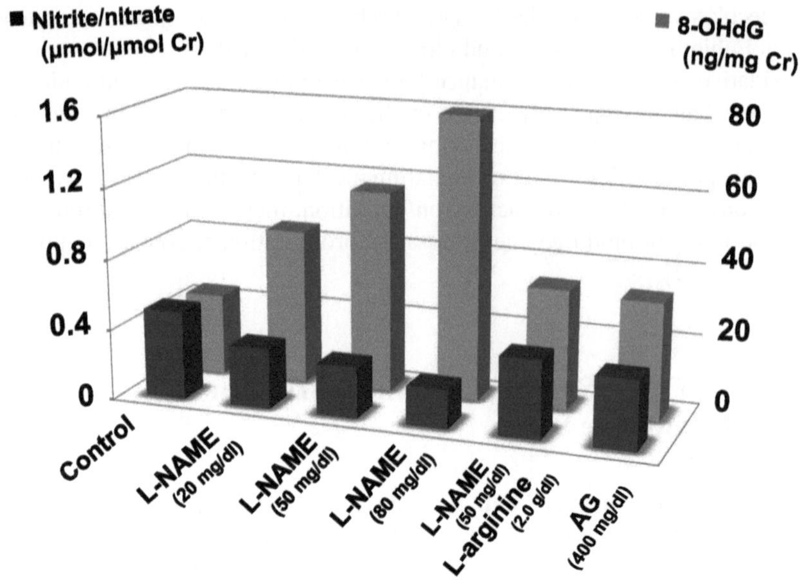

Fig. 6.1 Effects of chronic nitric oxide blockade on oxidative stress status in young rats. *Abbreviations*: *AG* aminoguanidine, *Cr* creatinine, *L-NAME* N^G-nitro-L-arginine methyl ester, *8-OHdG* 8-hydroxy-2′-deoxyguanosine. Presented data are mean values of the markers. Oral administration of L-NAME (20, 50, and 80 mg/dL of drinking water), but not AG (400 mg/dL), for 4 weeks induced systemic hypertension and a significant reduction in urinary excretion of nitrite/nitrate. Rats treated with L-NAME also showed a significant increase in urinary 8-OHdG excretion compared with the control animals. The above effects were dependent on the dosage of L-NAME. The effects of L-NAME (50 mg/dL) on blood pressure and urinary nitrite/nitrate and 8-OHdG were restored by a large dose of L-arginine (2.0 g/dL), a precursor for nitric oxide synthesis

These observations underscore the importance of continuous generation of NO by constitutive NOS (especially NOS3) for the control of vascular tone, renal function, and antioxidative capacity in young animals.

Moreover, rats receiving chronic nonselective NOS inhibitor treatment were reported to exhibit various parenchymal lesions [7]. Enhanced oxidative stress is likely to participate in the development of such organ damage in animals with chronic NO deficiency and endothelial dysfunction.

6.3 Asymmetric Dimethylarginine, an Endogenous Nitric Oxide Synthase Inhibitor

Asymmetric dimethylarginine (ADMA), which serves a regulatory role in the L-arginine/NOS/NO pathway by inhibiting NOS [3, 8], is formed when arginine residues in proteins are methylated by the action of protein arginine methyltransferase (PRMT) activity and by the subsequent protein turnover. ADMA is mainly

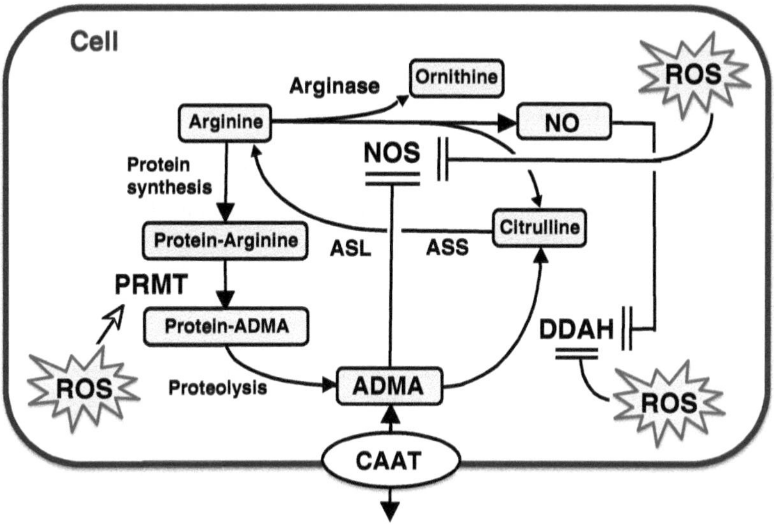

Fig. 6.2 L-Arginine/nitric oxide synthase/nitric oxide pathway versus protein arginine methyltransferase/asymmetric dimethylarginine/dimethylarginine dimethylaminohydrolase pathway. *Abbreviations*: *ADMA* asymmetric dimethylarginine, *ASL* argininosuccinate lyase, *ASS* argininosuccinate synthase, *CAAT* cationic amino acid transporter, *DDAH* dimethylarginine dimethylaminohydrolase, *NO* nitric oxide, *NOS* nitric oxide synthase, *PRMT* protein arginine methyltransferase, *ROS* reactive oxygen species. Note that the activity of PRMT is enhanced by ROS and that the activity of DDAH is suppressed by ROS (or high levels of NO). Also note that the activity of endothelial NOS is suppressed by ROS. Thus, oxidative stress enhancement and endothelial NOS inhibition play a pivotal role in cardiovascular pathologies by managing PRMT/DDAH activities and NO synthesis and leading to a common result—"endothelial dysfunction"

metabolized by dimethylarginine dimethylaminohydrolase (DDAH). That enzyme hydrolyzes ADMA to L-citrulline and dimethylamine. Free ADMA can be transported by cationic amino acid transporter (CAAT) to move in or out of the cells. ADMA can be transported to various organs such as the liver, kidneys, lungs, intestine, and blood vessels (Fig. 6.2). Although DDAH-1 is present in many tissues that express NOS1, DDAH-2 is present mainly in vascular tissues that express NOS3.

ADMA inhibits NOS3 and decreases NO production in the endothelium of vessel walls. Consequently, when ADMA levels are elevated, endothelial dysfunction and oxidative stress enhancement are likely to result. Elevated ADMA levels are detected in cases of various vascular diseases including hypertension, atherosclerosis, diabetes mellitus, renal insufficiency, hepatic failure, hypercholesterolemia, hypertriglyceridemia, and hyperhomocysteinemia [8]. Dysregulation of the balance between the L-arginine/NOS/NO pathway and the PRMT/ADMA/DDAH pathway is likely to be involved in the accumulation of ADMA in tissues and blood, thereby contributing to cardiovascular pathologies.

A reliable enzyme-linked immunosorbent assay (ELISA) method for ADMA measurement was developed by Schulze et al. [9]. At our laboratory, serum concentrations of ADMA were measured using this ELISA method (DLD Diagnostika

GmbH, Germany). Using that method, it has been found that the ADMA concentrations in umbilical blood from newborns were significantly higher (about twofold) (mean ± SD: 1.71 ± 0.47 μmol/L, $n = 33$) than in blood from healthy children (0.71 ± 0.11 μmol/L, $n = 19$) or healthy adults (0.52 ± 0.12 μmol/L, $n = 10$). Nitrite/nitrate concentrations (Nitrate/Nitrite colorimetric assay; Cayman Chemical, USA) were significantly lower (about half) in umbilical blood from newborns (23.3 ± 8.3 μmol/L) than in blood from children (43.4 ± 22.9 μmol/L) and adults (40.8 ± 27.8 μmol/L). Consequently, the ratios of ADMA to nitrite/nitrate were much higher (about fourfold) in umbilical blood from newborns (0.085 ± 0.045) than in blood from children (0.020 ± 0.009) or adults (0.019 ± 0.012) [10].

Physiologically high ADMA concentrations in umbilical blood might be attributable to increased synthesis, decreased metabolism by DDAH, decreased clearance by the placenta or fetal kidney, or some combination of those factors. The high ADMA levels per se or relative to NO are likely to represent an important mechanism that maintains blood pressure during the birth process.

6.4 Biomarkers for Oxidative Stress

Any biomolecule can be damaged by ROS (including $ONOO^-$). The primary cellular target of oxidative stress depends on the cell type, the characteristics of the imposed stress, the generation site, the ROS proximity to a specific target, and the stress severity. Usually, because the half-lives of ROS are short, special techniques must be used to detect ROS in vivo. "Oxidative stress biomarkers" can not only determine the extent of oxidative injury but also indicate the oxidant source. They are important for predicting the consequences of oxidation and for providing a basis for designing appropriate interventions to alleviate or prevent injury [3].

Briefly, oxidative stress biomarkers are separable into two categories: (a) formation of modified molecules by ROS; and (b) consumption or induction of enzymes or antioxidants. Measurement of these biomarkers in body fluids enables repeated monitoring of the oxidativc stress status in vivo, which cannot be done with invasive tests.

The first category (a) includes molecules that are generated in a reaction with ROS. Molecules are subjected to scission, cross-linking, or covalent modification in these reactions. Accordingly, these molecules are increased when ROS are generated. Some are removed rapidly or repaired rapidly, but others remain for a long time in intracellular or extracellular compartments. Major targets of ROS in the molecular components of the cells are membrane lipids, proteins, nucleic acids, and carbohydrates. These markers are measurable using stable adducts that are produced in vivo as a result of oxidative processes. Clinically applicable biomarkers include 4-hydroxy-2-nonenal (HNE), malondialdehyde (MDA), acrolein, F2-isoprostane (markers of lipid oxidation), 8-OHdG, 8-nitroguanine (markers of oxidative DNA damage), carboxymethyllysine (CML), pentosidine (markers of glycoxidation), 2-pyrrolidone, 3-nitrotyrosine (markers of protein oxidation), nitrite/nitrate (a marker of nitro-oxidation), and bilirubin oxidative metabolites (BOM) (a marker of heme oxygenase activity) (Table 6.1; Fig. 6.3).

The second category (b) includes antioxidative enzymes and molecules associated with ROS metabolism. In most cases, these molecules are destroyed or modified. They exhibit decreased activity or quantity after exposure to ROS. Conversely, they often show an overshooting response for a matter of hours, days, or weeks. Various antioxidative enzymes, proteins, and low-molecular-weight molecules are listed in Table 6.1.

Here, one developing topic can be mentioned. Assessment of oxidative stress markers in exhaled breath has proven useful for managing airway inflammatory diseases [11, 12]. These "lung biomarkers" might be helpful in making diagnoses, defining specific phenotypes of diseases, monitoring exacerbations, and evaluating the effects of drugs in airway diseases. A list of these biomarkers is presented in the next section.

6.5 Oxidative Stress Biomarkers in Pediatric Medicine

Many diseases of children and adolescents appear to be linked to oxidative damage attributable to ROS in their pathogenesis and progression (Table 6.2; Fig. 6.4) [13–166]. Oxidative stress might also contribute to tissue damage induced by certain drugs (such as analgesics, anticancer drugs, and immunosuppressive drugs) [64, 85, 92].

Most studies have determined oxidative stress biomarkers in blood samples (such as serum, plasma, erythrocytes, granulocytes, and lymphocytes) or urine samples. In some, however, the parameters were measured using different body fluids (such as cerebrospinal fluid (CSF) [32, 38, 64, 79, 83, 92, 110, 165], bronchoalveolar lavage fluid [48, 101], nasal lavage fluid [102], middle-ear fluid [132], joint fluid, and breast milk [138]), tissues [58, 62, 71, 72], or exhaled breath [13, 14, 34, 39, 99, 125, 137, 150, 156, 159], either alone or in combination with samples of blood or urine.

Analyzing exhaled breath involves the direct measurement of gases such as NO and carbon monoxide (CO) in addition to inflammatory indicators in exhaled breath condensate such as nitrite/nitrate, H_2O_2, HNE, MDA, F2-isoprostane, "thiobarbituric acid reactive substances (TBARS)," ADMA, GSH, arachidonic acid metabolites, and inflammatory cytokines (Fig. 6.5).

Previous studies measured oxidative stress biomarkers using high-performance liquid chromatography or gas chromatography/mass spectrometry. Recently, specific monoclonal antibodies have been developed. ELISA systems have been constructed for various oxidative stress biomarkers [3].

Table 6.1 Oxidative stress biomarkers of two categories

(a) Formation of modified molecules by reactive oxygen species	(b) Antioxidative enzymes and molecules
Lipid peroxidation:	Enzymes:
4-Hydroxy-2-nonenal-lysine	Superoxide dismutase
Malondialdehyde-lysine	Catalase
Acrolein-lysine	Glutathione peroxidase
Hexanoyl-lysine	Glutathione reductase
1,4-Dihydroxynonane-mercapturic acid	Glutathione-*S*-transferase
F2-isoprostane	Thioredoxin reductase
Lipid peroxide	Heme oxygenase
Total hydroperoxides	Arylesterase
Oxidative DNA damage:	Paraoxonase
8-Hydroxy-2′-deoxyguanosine	Proteins:
8-Nitroguanine	Albumin
Glyoxidation:	Ferritin
Carboxymethyllysine	Transferrin
Pentosidine	Lactoferrin
Argpyrimidine	Ceruloplasmin
Methylglyoxal	Thioredoxin
Advanced glycation endproducts	Peroxidoxin
Nitro-oxidation:	L-type fatty acid binding protein
3-Nitrotyrosine	Low-molecular-weight molecules:
Nitrite/nitrate	Bilirubin
Others:	Ascorbate
Hydrogen peroxide	Tocopherols
o,o′-Dityrosine	Carotenoids
ortho-Tyrosine	Ubiquinol/ubiquinone (coenzyme Q_{10})
2-Pyrrolidone	Retinol
Ischemia-modified albumin	Folate
Protein carbonyls	Cysteine
Advanced oxidation protein products	Glutathione
Bilirubin oxidative metabolites	Polyphenols
Dehydroascorbate	Nitrite/nitrate
Oxidized glutathione	Urate
Oxidized low-density lipoprotein	Selenium
Conjugated diene	Zinc
"Thiobarbituric acid reactive substances"	Carbon monoxide
	Others:
	"Total antioxidative capacity"
	"Total radical trapping antioxidative potential"
	"Biological antioxidative potential"

Nitric oxide behaves either as a prooxidant or as an antioxidant according to its environment. Therefore, its stable metabolites, nitrite/nitrate, are listed in both categories (a) and (b)

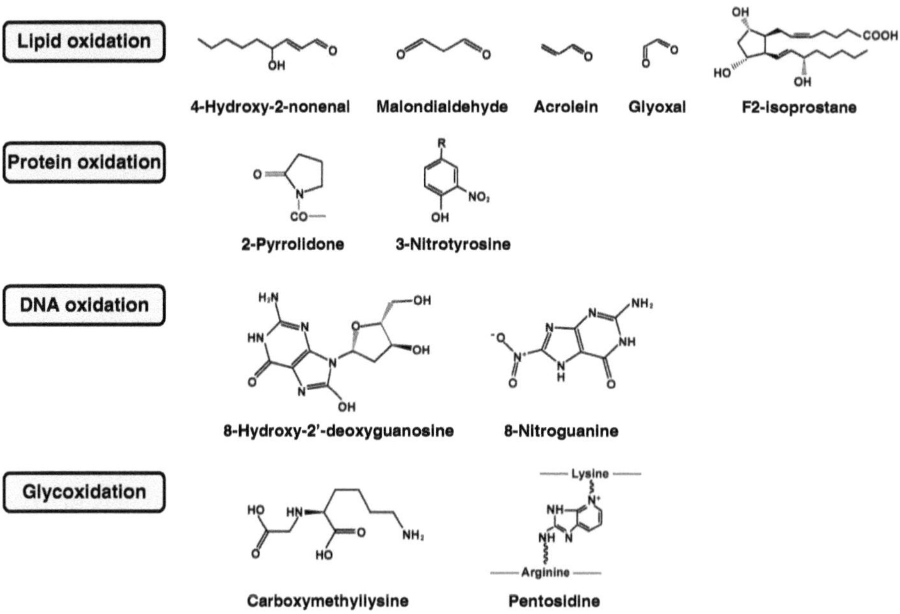

Fig. 6.3 Chemical structures of representative biomarkers for oxidative stress. Reproduced with slight modification with permission from © 2013 Medical Review Co., Ltd., Japan

Rapid diagnosis and prompt treatment are regarded as extremely important for the management of oxidative stress-related diseases in children. Determinations by which oxidative stress is quantified rapidly include total hydroperoxides (TP), which function as a measure of overall oxidative injury because hydroperoxides are the intermediate oxidative product of lipids, peptides, and amino acids, and biological antioxidative potential (BAP), which functions as a measure of total antioxidative activity (Free Radical Analytic System; Diacron International, Italy). Using the percentage ratio of TH to BAP ("oxidative stress index"), any shift of oxidative/antioxidative balance toward the oxidative direction can be analyzed. This machine is compact: only 20 μL of serum or plasma and 10 min of processing time are necessary for analyses. The measurement of TP and BAP (plus "oxidative stress index") is a simple and informative outpatient or bedside measurement in clinical medicine [33, 68, 83, 107, 110, 113, 128, 141, 151, 158].

For bedside point-of-care testing to measure urinary 8-OHdG, an automated analyzer (ICR-001; Techno Medica Co. Ltd., Japan) has been developed. The measurement of 8-OHdG and creatinine (Cr) is performed, respectively, using competitive immunochromatography and Jaffe method. The sample-to-answer time is only 5 min for the simultaneous determination of urinary 8-OHdG and Cr concentrations [153].

Oxidative damage often takes place selectively. For example, lipid peroxidation and oxidative DNA damage are not always accompanied by overproduction of

Table 6.2 Pediatric diseases associated with enhanced oxidative stress

Allergic/inflammatory:	Neonatal:
Adenoidal hypertrophy[a] [13]	Asphyxia[a] [56]
Allergic rhinitis[a] [82, 125, 143, 150, 157]	Hyperbilirubinemia[a] [26]
Atopic dermatitis[a] [137, 144]	Hypoxic ischemic encephalopathy[a] [37, 83]
Bronchial asthma[a] [14, 34, 39, 48, 87, 99, 125, 150]	Maternal chorioamnionitis[a] [51]
Burn[a] [50]	Maternal preeclampsia[a] [78]
Chronic arthritis	Neonatal sepsis[a] [56]
Henoch–Schönlein purpura[a] [17, 127]	Periventricular leukomalacia[a] [33]
Kawasaki disease[a] [35, 88, 107, 151]	Premature birth
Systemic lupus erythematosus	Respiratory distress syndrome[a] [56]
Vasculitis syndrome	Retinopathy
Cardiovascular:	Small-for-gestational age and large-for-gestational age children[a] [63]
Cardiac transplantation	Neurologic/muscular:
Cardiopulmonary bypass[a] [57]	Ataxia telangiectasia[a] [43]
Congenital cardiac defects[a] [152]	Attention deficit hyperactivity disorder[a] [129]
Congenital portosystemic venous shunt[a] [94]	Autism spectrum disorders[a] [27, 44, 104, 105, 129, 130]
Primary hypertension[a] [154]	Cerebral palsy[a] [163]
Endocrinologic:	Chronic fatigue syndrome/myalgic encephalomyelitis[a] [76]
Hyperthyroidism	Congenital muscular dystrophy[a] [93, 149]
Iodine-deficient goiter	Developmental brain disorders[a] [62]
Thyroid carcinoma[a] [16]	Epilepsy (treatment with valproic acid)[a] [41, 113]
Thyroiditis[a] [16]	Febrile seizures[a] [70]
Environmental/toxicologic:	Friedreich ataxia[a] [106]
Air pollution[a] [30, 39, 61, 84]	Hereditary spastic paraparesis[a] [77]
Carcinogenic metal (cadmium, chromium, arsenic) exposure[a] [138]	Inflammatory myopathy
Chronic fluoride exposure[a] [46]	Migraine[a] [25, 97, 161]
Exercise[a] [34, 89]	Mitochondrial encephalopathy (complex I deficiency)[a] [22]
Ozone exposure	Psychosis[a] [155]
Passive smoking[a] [117, 123]	Rett syndrome[a] [45, 112]
Gastrointestinal/hepatologic:	Selenium-deficient skeletal muscle disorder
Autoimmune hepatitis	Traumatic brain injury[a] [38, 145]
Celiac disease[a] [36]	Nutritional:
Chronic constipation	Constitutional leanness[a] [42]
Inflammatory bowel disease	Hypercholesterolemia[a] [31, 80]
Live failure[a] [140]	Hyperlipidemia
Living-related liver transplantation[a] [128, 158]	Kwashiorkor[a] [98, 115]
Nonalcoholic fatty liver disease[a] [160, 162]	Multimetabolic syndrome
Postoperative biliary atresia[a] [72]	Obesity[a] [42, 75, 81, 103, 141]
Viral hepatitis	Rickets[a] [146]

(continued)

Table 6.2 (continued)

Genetic:	Pharmacologic/therapeutic:
Alagille syndrome[a] [114]	Analgesics
Down syndrome[a] [15, 142]	Anticancer drugs[a] [64, 85, 92]
Zellweger syndrome	Immunosuppressive drugs
Hematologic/neoplastic:	Laparoscopic surgery[a] [55]
Acute leukemia[a] [23, 40, 64]	Total body irradiation
Bone marrow transplantation[a] [136]	Renal:
Erythropoietic protoporphyria	Congenital hydronephrosis[a] [134]
Fanconi anemia	Glomerulonephritis
Pheochromocytoma[a] [108]	Hemolytic uremic syndrome[a] [121]
Sickle cell anemia[a] [20, 28, 133]	Idiopathic renal hypouricemia[a] [68]
Solid tumors[a] [23, 131]	Nephrotic syndrome[a] [69, 126, 153]
Thalassemia major[a] [28, 54]	Renal insufficiency/failure[a] [24, 29, 67, 73, 135, 166]
Infectious:	Urinary tract infection
Acute bronchiolitis	Respiratory:
Acute infectious mononucleosis	Bronchiolitis obliterans[a] [101]
Acute otitis media/tonsillitis	Chronic pulmonary disease
Adenovirus infection	Cystic fibrosis[a] [21, 124]
Chronic nail candidiasis	Obstructive sleep apnea[a] [13, 156]
Chronic otitis media/tonsillitis[a] [109, 132]	
Cutaneous leishmaniasis	
Fasciola hepatica infection[a] [100]	
Helicobacter pylori infection[a] [164]	
HIV infection[a] [19, 116]	
Malaria[a] [147]	
Measles encephalitis	
Meningitis[a] [32, 79, 165]	
Pandemic influenza (H1N1)[a] [139]	
Respiratory syncytial virus bronchiolitis[a] [102]	
Sepsis[a] [111]	
Metabolic:	
Citrin deficiency[a] [60]	
Diabetes mellitus (type 1, type 2)[a] [49, 52, 53, 65, 66, 96]	
Glutathione synthetase deficiency	
Glycogen storage disease type 1a	
Homocystinuria[a] [120]	
Mucopolysaccharidosis type 2[a] [118]	
Nephropathic cystinosis[a] [91, 122]	
Niemann–Pick disease type C[a] [95]	
Organic aciduria[a] [71, 86, 90]	
Phenylketonuria[a] [47, 74, 119]	
Urea cycle enzyme defects[a] [59]	
Wilson disease[a] [58]	
X-linked adrenoleukodystrophy[a] [18, 110, 148]	

[a]These studies (published during 2008–2013) are cited in the reference list

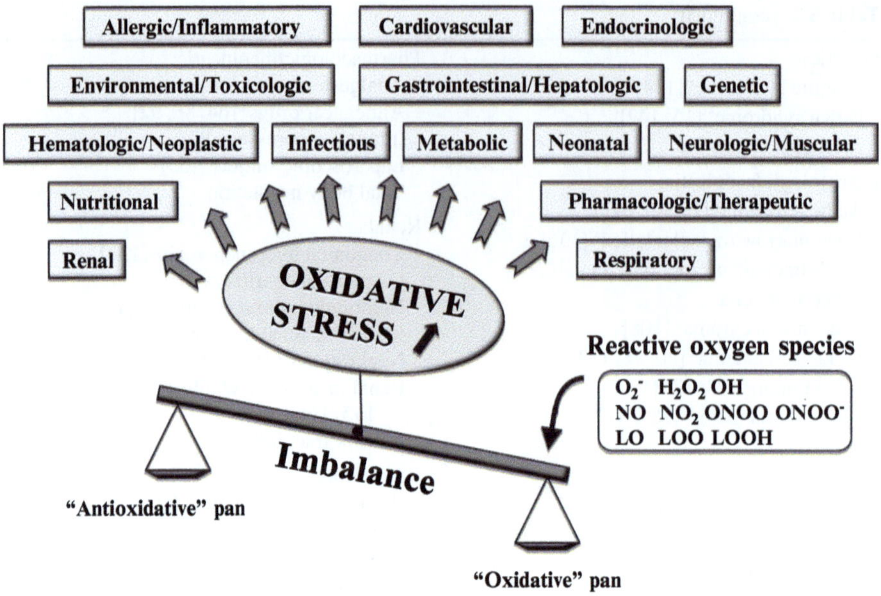

Fig. 6.4 Pediatric diseases associated with enhanced oxidative stress. *Abbreviations*: H_2O_2 hydrogen peroxide, *LO* lipid alkoxyl radical, *LOO* lipid peroxyl radical, *LOOH* lipid hydroperoxide, *NO* nitric oxide, NO_2 nitrogen dioxide, O_2^- superoxide anion, *OH* hydroxyl radical, *ONOO* peroxynitrite radical, $ONOO^-$ peroxynitrite anion. Note that it remains to be clarified yet whether excessive ROS formation is a primary cause or a downstream consequence of the pathological process in each disease

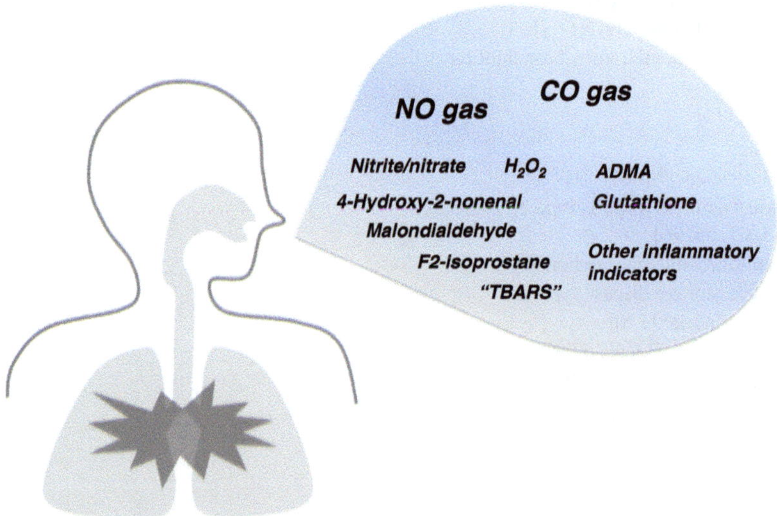

Fig. 6.5 Oxidative stress biomarkers in exhaled breath ("lung biomarkers"). *Abbreviations*: *ADMA* asymmetric dimethylarginine, *CO* carbon monoxide, H_2O_2 hydrogen peroxide, *NO* nitric oxide, *TBARS* thiobarbituric acid reactive substances

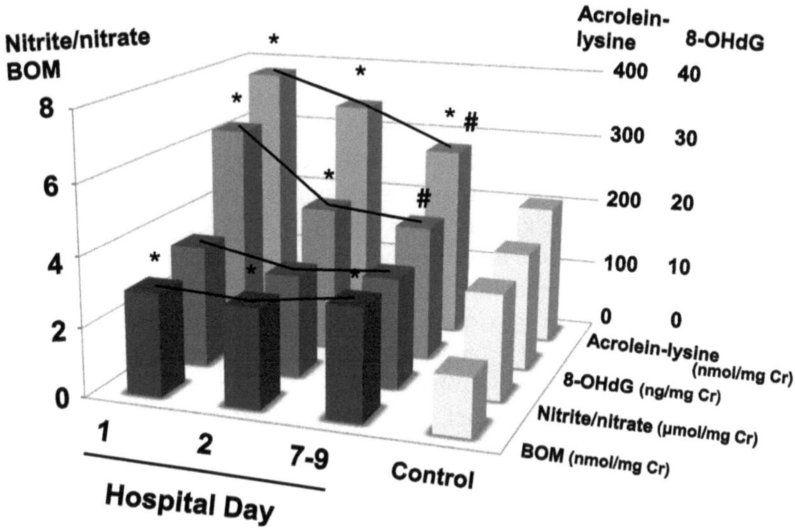

Fig. 6.6 Changes in urinary oxidative stress biomarker levels in children with acute exacerbation of atopic dermatitis. *Abbreviations*: *BOM* bilirubin oxidative metabolites, *Cr* creatinine, *8-OHdG* 8-hydroxy-2′-deoxyguanosine. Presented data are mean values of the markers. *$p<0.05$ versus controls; #$p<0.05$ versus Day 1 (the day of admission)

NO, especially in chronic disease conditions. We studied 13 children who were hospitalized for acute exacerbation of atopic dermatitis (AD) with purulent skin infection by *Staphylococcal aureus* (age: 1.5–10.0 years), along with 28 age-matched healthy controls [167] (Fig. 6.6). Urine samples obtained from the patients on admission, and on the second and seventh to ninth hospital days, and those from the controls were analyzed for acrolein-lysine, 8-OHdG, nitrite/nitrate, and BOM. Of these, urinary concentrations of acrolein-lysine, 8-OHdG, and BOM, but not nitrite/nitrate, were significantly higher in AD children on admission than in the controls. Response to treatment (with systemic antibiotics, topical antiseptics, and corticosteroids) was associated with significant decreases in levels of acrolein-lysine and 8-OHdG. Urinary BOM remained almost constant. They were considerably higher in AD children during hospitalization. These results do not indicate that endogenous NO formation in children with acute exacerbation of AD is enhanced in comparison with that in control subjects (although they do not disprove the enhanced production of peroxynitrite or "nitrative stress" in the patients either).

6.6 Reference Normal Values of Oxidative Stress Biomarkers

Investigating the role of oxidative stress in pediatric diseases requires information about the oxidative stress status of healthy young populations. The contribution of oxidative stress to various pediatric diseases can be evaluated and better approaches for each disease can be established only when normal levels of oxidative stress in children and adolescents are known.

Kauffman et al. [168] made the first report of a study evaluating oxidative stress in a large population of apparently healthy children using urine samples. They determined urinary levels of F2-isoprostane in 342 healthy children aged <7 years. Other studies examined small groups of healthy children and adolescents as age-matched controls, but not as study subjects.

Urine collection is simple, quick, comfortable, noninvasive, and particularly easy to do with children. Our data for the evaluation of oxidative stress levels in healthy young subjects were of 113 healthy Japanese people of a broad age range (1.5–21.0 years) [169]. Early morning void urine samples were obtained for analyses of biomarkers reflecting oxidative damage to lipids, DNA, and carbohydrates. Acrolein-lysine and 8-OHdG were determined using competitive ELISA kits (ACR-Lysine Adduct ELISA; NOF Corp., Japan and 8-OHdG Check; Institute for the Control of Aging, Japan). Nitrite/nitrate was measured using colorimetric assay (Nitrate/Nitrite colorimetric assay; Cayman Chemical). Pentosidine was measured using high-performance liquid chromatography with fluorometric detection. Our subjects were classified into the following five groups: 1–6 years ($n=33$), 6–11 years ($n=34$), 11–16 years ($n=20$), 16–21 years ($n=13$), and 21–30 years ($n=13$) (Fig. 6.7). The levels of urinary acrolein-lysine, 8-OHdG, nitrite/nitrate, and pentosidine were highest in the youngest age group (1–6 years). They decreased with age to reach constant levels by early adolescence. The physiological meaning and mechanisms for the high levels of urinary oxidative stress biomarkers in younger subjects remain to be clarified. They warrant further research.

Kaneko et al. [170] verified the reliability of the new device ICR-001 (Techno Medica Co. Ltd.) in measuring 8-OHdG concentration in healthy children by comparing the results with those obtained from ELISA (as described above) [169]. Both urinary 8-OHdG and Cr measured using ICR-001 correlate very well with those measured using ELISA and enzymatic assay, respectively.

6.7 Therapeutic Intervention

As described in earlier sections, clinical results of studies indicate that an imbalance between oxidative and antioxidative activities in favor of the former contributes to the pathogeneses of many diseases in the field of pediatric medicine. Therapeutic interventions that decrease exposure to ROS or which augment antioxidative defenses are expected to be beneficial as adjunctive therapies for oxidative stress-related diseases.

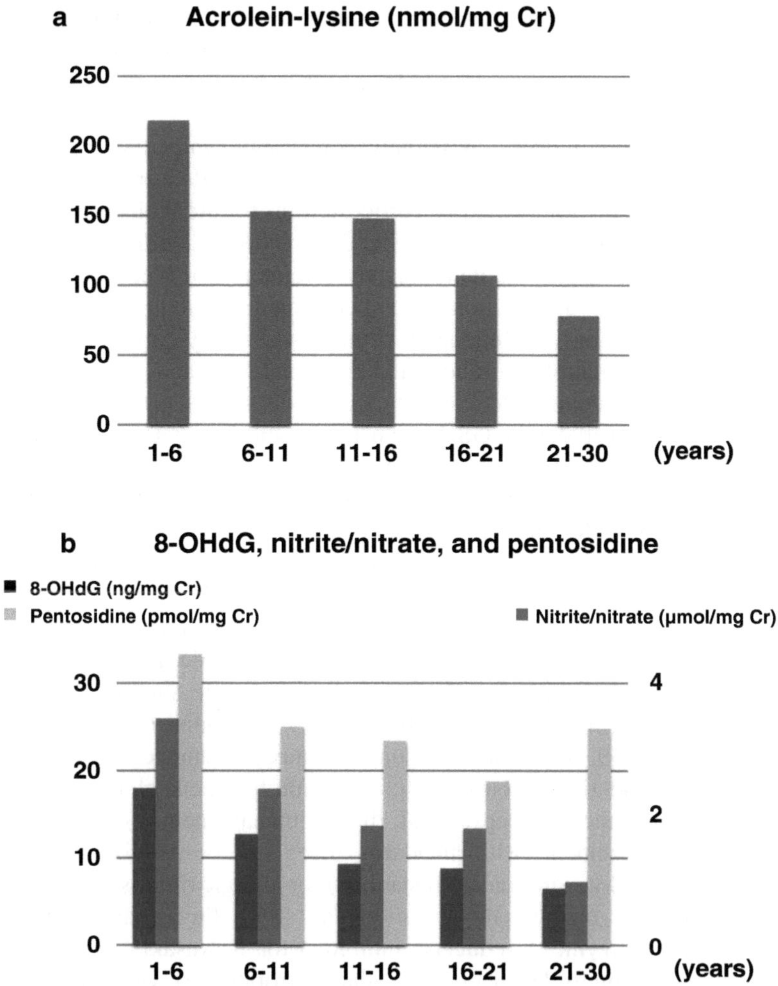

Fig. 6.7 Age-related changes of urinary levels of acrolein-lysine (**a**), 8-hydroxy-2′-deoxyguanosine, nitrite/nitrate, and pentosidine (**b**) in healthy children. *Abbreviations*: *Cr* creatinine, *8-OHdG* 8-hydroxy-2′-deoxyguanosine. Presented data are mean values of the markers. Note that younger subjects exhibit higher levels of urinary markers

Antioxidative strategies, including administration of pharmacological or dietary agents, are based on two main mechanisms: enhancement of ROS elimination and inhibition of ROS generation. The agents acting according to the former mechanism include antioxidative enzymes catalyzing ROS degradation and scavengers neutralizing ROS. Classic antioxidants include ascorbate, tocopherols, carotenoids, and acetylcysteine. The potency of these antioxidants might be limited because they work as scavengers of already-formed ROS in a stoichiometric manner. They might be regarded as a more "symptomatic" treatment rather than as a treatment for the underlying causes of oxidative stress-associated clinical problems.

Reportedly, angiotensin-converting enzyme inhibitors, angiotensin II type-1 receptor antagonists, and lipid-lowering agents suppress intracellular overproduction of ROS and function as "causal" antioxidants. Many of their beneficial effects are the result of this property [3].

The effects of certain drugs on oxidative stress status in pediatric patients have been examined in clinical studies. Drugs about which reports were published during 2000–2006 include the following [3]: corticosteroids for bronchial asthma and bacterial meningitis; L-arginine for endothelial dysfunction in cardiac transplantation; angiotensin-converting enzyme inhibitors and angiotensin II type-1 receptor antagonists for endothelial dysfunction in diabetes mellitus; melatonin for neonatal asphyxia and for epilepsy; folinic acid, betaine, and methylcobalamin for autism; tocopherols and ubiquinol/ubiquinone for Friedreich ataxia; selenium for skeletal muscle disorder in selenium deficiency; tocopherols and ascorbate for endothelial dysfunction in hyperlipidemia; GSH and α-lipoic acid for kwashiorkor; amifostine for total body irradiation and for anticancer drug use; tocopherols, carotenoids, and ascorbate for immunosuppressive drug use (cyclosporine A, tacrolimus); and ubiquinol/ubiquinone for anthracycline use.

Relative to this field, of the 154 reports that have been published recently (2008–2013) [13–166], 28 have enriched the list of studies revealing "antioxidative" therapeutic strategies for pediatric patients (Table 6.2): dietary management for cystic fibrosis-related diabetes [21]; methylcobalamin and folinic acid for autism [27]; intravenous immunoglobulin for Kawasaki disease (acute stage) [35, 107, 151]; therapeutic hypothermia for traumatic brain injury [38]; inhaled corticosteroids for bronchial asthma [39, 99]; betamethasone for ataxia telangiectasia [43]; tocopherols, ascorbate, and zinc for burns [50]; melatonin for neonatal asphyxia, respiratory distress syndrome, and sepsis [56]; dietary management for phenylketonuria [74]; dietary management (with mandarin juice) for obesity [75]; desloratadine (nonsedating second-generation antihistamine) for allergic rhinitis [82]; L-carnitine for propionic and methylmalonic acidemias [86]; fluvastatin (3-hydroxy-3-methylglutaryl coenzyme A reductase inhibitor) for Kawasaki disease (late after onset) [88]; melatonin for Duchenne muscular dystrophy [93]; vitamin/mineral supplement for autism [105]; idebenone and deferiprone for Friedreich ataxia [106]; edaravone for X-linked adrenoleukodystrophy [110]; enzyme replacement for mucopolysaccharidosis type II [118]; vitamin supplement for cystic fibrosis [124]; deferasirox (iron chelator) and hydroxyurea for sickle cell anemia [133]; plasma exchange for acute-on-chronic liver failure [140]; nimodipine (calcium channel blocker) for severe head trauma [145]; bone marrow transplantation for X-linked adrenoleukodystrophy [148]; antihypertensive therapy for hypertensive organ damage [154]; and tocopherols for nonalcoholic fatty liver disease [160].

The authors of the studies presented above document the considerable favorable effects of the respective strategies. Recent studies (2008–2013) have analyzed oxidative stress status in subjects using more numerous parameters. They are therefore more persuasive and illustrative than earlier studies (2000–2006) were. However, these results should be interpreted cautiously and should be corroborated with studies that are conducted in appropriately designed randomized controlled trials with more numerous patients.

6.8 Future Perspective

This chapter presents current knowledge related to noninvasive means of assessing oxidative stress status in the field of pediatric medicine. Challenges for the future include the elucidation of the molecular mechanisms that engender oxidative stress in pediatric diseases, the demonstration of the therapeutic benefits to patients, and the development of early detection and prevention strategies for oxidative stress-related diseases [3, 171] (Fig. 6.8).

Indeed, development of therapeutic interventions remains as an important challenge in this field. Antioxidative strategies would be successful if they were available at the site of pathological lesions. Therefore, the route of administration, bioavailability, and tissue distribution all stand as important considerations when planning future therapeutic interventions. Moreover, before antioxidative therapy becomes fully accepted in clinical practice, detailed longitudinal studies must be conducted to evaluate panels of oxidative stress biomarkers together with traditional clinical endpoints in patients who are undergoing treatment for acute and chronic diseases.

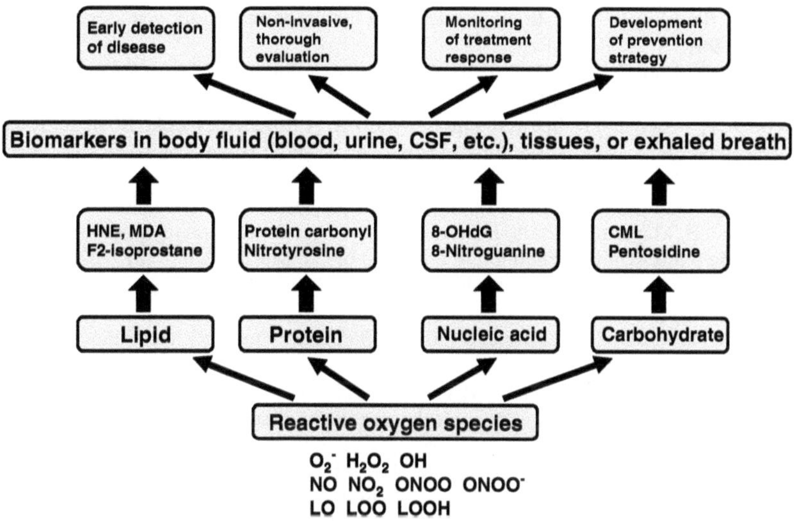

Fig. 6.8 Perspective about clinical use of oxidative stress biomarkers. *Abbreviations*: *CML* carboxymethyllysine, *CSF* cerebrospinal fluid, *HNE* 4-hydroxy-2-nonenal, H_2O_2 hydrogen peroxide, *LO* lipid alkoxyl radical, *LOO* lipid peroxyl radical, *LOOH* lipid hydroperoxide, *MDA* malondialdehyde, *NO* nitric oxide, NO_2 nitrogen dioxide, O_2^- superoxide anion, *OH* hydroxyl radical, *8-OHdG* 8-hydroxy-2′-deoxyguanosine, *ONOO* peroxynitrite radical, $ONOO^-$ peroxynitrite anion. Only representative markers are presented

Acknowledgments I thank Professor Tsuneo Morishima (Okayama University, Japan), Professor Mitsufumi Mayumi (University of Fukui), Professor Michael S. Goligorsky (New York Medical College, USA), and Professor Takanobu Ishida (State University of New York at Stony Brook, USA) for invaluable help with these studies. This work was supported by the Japanese Ministry of Health, Labour and Welfare.

References

1. Dröge W (2002) Free radicals in the physiological control of cell function. Physiol Rev 82:47–95
2. Auten RL, Davis JM (2009) Oxygen toxicity and reactive oxygen species: the devil is in the details. Pediatr Res 66:121–127
3. Tsukahara H (2007) Biomarkers for oxidative stress: clinical application in pediatric medicine. Curr Med Chem 14:339–351
4. Fenster BE, Tsao PS, Rockson SG (2003) Endothelial dysfunction: clinical strategies for treating oxidant stress. Am Heart J 146:218–226
5. Tsukahara H, Imura T, Tsuchida S, Nunose M, Hori C, Hiraoka M, Gejyo F, Sudo M (1996) Renal functional measurements in young rats with chronic inhibition of nitric oxide synthase. Acta Paediatr Jpn 38:614–618
6. Tsukahara H, Hiraoka M, Kobata R, Hata I, Ohshima Y, Jiang MZ, Noiri E, Mayumi M (2000) Increased oxidative stress in rats with chronic nitric oxide depletion: measurement of urinary 8-hydroxy-2′-deoxyguanosine excretion. Redox Rep 5:23–28
7. Zatz R, Baylis C (1998) Chronic nitric oxide inhibition model six years on. Hypertension 32:958–964
8. Teerlink T, Luo Z, Palm F, Wilcox CS (2009) Cellular ADMA: regulation and action. Pharmacol Res 60:448–460
9. Schulze F, Wesemann R, Schwedhelm E, Sydow K, Albsmeier J, Cooke JP, Böger RH (2004) Determination of asymmetric dimethylarginine (ADMA) using a novel ELISA assay. Clin Chem Lab Med 42:1377–1383
10. Tsukahara H, Ohta N, Tokuriki S, Nishijima K, Kotsuji F, Kawakami H, Ohta N, Sekine K, Nagasaka H, Mayumi M (2008) Determination of asymmetric dimethylarginine, an endogenous nitric oxide synthase inhibitor, in umbilical blood. Metabolism 57:215–220
11. Popov TA (2011) Human exhaled breath analysis. Ann Allergy Asthma Immunol 106:451–456
12. Dodig S, Richter D, Zrinski-Topić R (2011) Inflammatory markers in childhood asthma. Clin Chem Lab Med 49:587–599
13. Biltagi MA, Maguid MA, Ghafar MA, Farid E (2008) Correlation of 8-isoprostane, interleukin-6 and cardiac functions with clinical score in childhood obstructive sleep apnoea. Acta Paediatr 97:1397–1405
14. Dut R, Dizdar EA, Birben E, Sackesen C, Soyer OU, Besler T, Kalayci O (2008) Oxidative stress and its determinants in the airways of children with asthma. Allergy 63:1605–1609
15. Tiano L, Padella L, Carnevali P, Gabrielli O, Bruge F, Principi F, Littarru GP (2008) Coenzyme Q10 and oxidative imbalance in Down syndrome: biochemical and clinical aspects. Biofactors 32:161–167
16. Moncayo R, Kroiss A, Oberwinkler M, Karakolcu F, Starzinger M, Kapelari K, Talasz H, Moncayo H (2008) The role of selenium, vitamin C, and zinc in benign thyroid diseases and of selenium in malignant thyroid diseases: low selenium levels are found in subacute and silent thyroiditis and in papillary and follicular carcinoma. BMC Endocr Disord 8:2
17. Ece A, Kelekçi S, Kocamaz H, Hekimoğlu A, Balik H, Yolbaş I, Erel O (2008) Antioxidant enzyme activities, lipid peroxidation, and total antioxidant status in children with Henoch-Schönlein purpura. Clin Rheumatol 27:163–169

18. Uto T, Contreras MA, Gilg AG, Singh I (2008) Oxidative imbalance in nonstimulated X-adrenoleukodystrophy-derived lymphoblasts. Dev Neurosci 30:410–418
19. Srinivas A, Dias BF (2008) Antioxidants in HIV positive children. Indian J Pediatr 75:347–350
20. Ren H, Ghebremeskel K, Okpala I, Lee A, Ibegbulam O, Crawford M (2008) Patients with sickle cell disease have reduced blood antioxidant protection. Int J Vitam Nutr Res 78:139–147
21. Ntimbane T, Krishnamoorthy P, Huot C, Legault L, Jacob SV, Brunet S, Levy E, Guéraud F, Lands LC, Comte B (2008) Oxidative stress and cystic fibrosis-related diabetes: a pilot study in children. J Cyst Fibros 7:373–384
22. Khurana DS, Salganicoff L, Melvin JJ, Hobdell EF, Valencia I, Hardison HH, Marks HG, Grover WD, Legido A (2008) Epilepsy and respiratory chain defects in children with mitochondrial encephalopathies. Neuropediatrics 39:8–13
23. Mazor D, Abucoider A, Meyerstein N, Kapelushnik J (2008) Antioxidant status in pediatric acute lymphocytic leukemia (ALL) and solid tumors: the impact of oxidative stress. Pediatr Blood Cancer 51:613–615
24. Mishra OP, Pooniya V, Ali Z, Upadhyay RS, Prasad R (2008) Antioxidant status of children with acute renal failure. Pediatr Nephrol 23:2047–2051
25. Boćkowski L, Sobaniec W, Kułak W, Smigielska-Kuzia J (2008) Serum and intraerythrocyte antioxidant enzymes and lipid peroxides in children with migraine. Pharmacol Rep 60:542–548
26. Davutoglu M, Guler E, Olgar S, Kurutas EB, Karabiber H, Garipardic M, Ekerbicer HC (2008) Oxidative stress and antioxidant status in neonatal hyperbilirubinemia. Saudi Med J 29:1743–1748
27. James SJ, Melnyk S, Fuchs G, Reid T, Jernigan S, Pavliv O, Hubanks A, Gaylor DW (2009) Efficacy of methylcobalamin and folinic acid treatment on glutathione redox status in children with autism. Am J Clin Nutr 89:425–430
28. Claster S, Wood JC, Noetzli L, Carson SM, Hofstra TC, Khanna R, Coates TD (2009) Nutritional deficiencies in iron overloaded patients with hemoglobinopathies. Am J Hematol 84:344–348
29. Elshamaa MF, Sabry S, Nabih M, Elghoroury EA, El-Saaid GS, Ismail AA (2009) Oxidative stress markers and C-reactive protein in pediatric patients on hemodialysis. Ann Nutr Metab 55:309–316
30. Kelishadi R, Mirghaffari N, Poursafa P, Gidding SS (2009) Lifestyle and environmental factors associated with inflammation, oxidative stress and insulin resistance in children. Atherosclerosis 203:311–319
31. Pignatelli P, Loffredo L, Martino F, Catasca E, Carnevale R, Zanoni C, Del Ben M, Antonini R, Basili S, Violi F (2009) Myeloperoxidase overexpression in children with hypercholesterolemia. Atherosclerosis 205:239–243
32. Hamed SA, Hamed EA, Zakary MM (2009) Oxidative stress and S-100B protein in children with bacterial meningitis. BMC Neurol 9:51
33. Kakita H, Hussein MH, Yamada Y, Henmi H, Kato S, Kobayashi S, Ito T, Kato I, Fukuda S, Suzuki S, Togari H (2009) High postnatal oxidative stress in neonatal cystic periventricular leukomalacia. Brain Dev 31:641–648
34. Barreto M, Villa MP, Olita C, Martella S, Ciabattoni G, Montuschi P (2009) 8-Isoprostane in exhaled breath condensate and exercise-induced bronchoconstriction in asthmatic children and adolescents. Chest 135:66–73
35. Takatsuki S, Ito Y, Takeuchi D, Hoshida H, Nakayama T, Matsuura H, Saji T (2009) IVIG reduced vascular oxidative stress in patients with Kawasaki disease. Circ J 73:1315–1318
36. Stojiljković V, Todorović A, Pejić S, Kasapović J, Saicić ZS, Radlović N, Pajović SB (2009) Antioxidant status and lipid peroxidation in small intestinal mucosa of children with celiac disease. Clin Biochem 42:1431–1437
37. Fatemi A, Wilson MA, Johnston MV (2009) Hypoxic-ischemic encephalopathy in the term infant. Clin Perinatol 36:835–858

38. Bayir H, Adelson PD, Wisniewski SR, Shore P, Lai Y, Brown D, Janesko-Feldman KL, Kagan VE, Kochanek PM (2009) Therapeutic hypothermia preserves antioxidant defenses after severe traumatic brain injury in infants and children. Crit Care Med 37:689–695
39. Liu L, Poon R, Chen L, Frescura AM, Montuschi P, Ciabattoni G, Wheeler A, Dales R (2009) Acute effects of air pollution on pulmonary function, airway inflammation, and oxidative stress in asthmatic children. Environ Health Perspect 117:668–674
40. Yang Y, Tian Y, Yan C, Jin X, Tang J, Shen X (2009) Determinants of urinary 8-hydroxy-2′-deoxyguanosine in Chinese children with acute leukemia. Environ Toxicol 24:446–452
41. Yiş U, Seçkin E, Kurul SH, Kuralay F, Dirik E (2009) Effects of epilepsy and valproic acid on oxidant status in children with idiopathic epilepsy. Epilepsy Res 84:232–237
42. Giannini C, de Giorgis T, Scarinci A, Cataldo I, Marcovecchio ML, Chiarelli F, Mohn A (2009) Increased carotid intima-media thickness in pre-pubertal children with constitutional leanness and severe obesity: the speculative role of insulin sensitivity, oxidant status, and chronic inflammation. Eur J Endocrinol 161:73–80
43. Russo I, Cosentino C, Del Giudice E, Broccoletti T, Amorosi S, Cirillo E, Aloj G, Fusco A, Costanzo V, Pignata C (2009) In ataxia-teleangiectasia betamethasone response is inversely correlated to cerebellar atrophy and directly to antioxidative capacity. Eur J Neurol 16: 755–759
44. James SJ, Rose S, Melnyk S, Jernigan S, Blossom S, Pavliv O, Gaylor DW (2009) Cellular and mitochondrial glutathione redox imbalance in lymphoblastoid cells derived from children with autism. FASEB J 23:2374–2383
45. De Felice C, Ciccoli L, Leoncini S, Signorini C, Rossi M, Vannuccini L, Guazzi G, Latini G, Comporti M, Valacchi G, Hayek J (2009) Systemic oxidative stress in classic Rett syndrome. Free Radic Biol Med 47:440–448
46. Ailani V, Gupta RC, Gupta SK, Gupta K (2009) Oxidative stress in cases of chronic fluoride intoxication. Indian J Clin Biochem 24:426–429
47. Sitta A, Barschak AG, Deon M, Barden AT, Biancini GB, Vargas PR, de Souza CF, Netto C, Wajner M, Vargas CR (2009) Effect of short- and long-term exposition to high phenylalanine blood levels on oxidative damage in phenylketonuric patients. Int J Dev Neurosci 27:243–247
48. Fitzpatrick AM, Teague WG, Holguin F, Yeh M, Brown LA, Severe Asthma Research Program (2009) Airway glutathione homeostasis is altered in children with severe asthma: evidence for oxidant stress. J Allergy Clin Immunol 123:146–152
49. Heilman K, Zilmer M, Zilmer K, Tillmann V (2009) Lower bone mineral density in children with type 1 diabetes is associated with poor glycemic control and higher serum ICAM-1 and urinary isoprostane levels. J Bone Miner Metab 27:598–604
50. Barbosa E, Faintuch J, Machado Moreira EA, Gonçalves da Silva VR, Lopes Pereima MJ, Martins Fagundes RL, Filho DW (2009) Supplementation of vitamin E, vitamin C, and zinc attenuates oxidative stress in burned children: a randomized, double-blind, placebo-controlled pilot study. J Burn Care Res 30:859–866
51. Malaeb S, Dammann O (2009) Fetal inflammatory response and brain injury in the preterm newborn. J Child Neurol 24:1119–1126
52. Kostolanská J, Jakus V, Barák L (2009) HbA1c and serum levels of advanced glycation and oxidation protein products in poorly and well controlled children and adolescents with type 1 diabetes mellitus. J Pediatr Endocrinol Metab 22:433–442
53. Kostolanská J, Jakus V, Barák L (2009) Glycation and lipid peroxidation in children and adolescents with type 1 diabetes mellitus with and without diabetic complications. J Pediatr Endocrinol Metab 22:635–643
54. Cakmak A, Soker M, Koc A, Erel O (2009) Paraoxonase and arylesterase activity with oxidative status in children with thalassemia major. J Pediatr Hematol Oncol 31:583–587
55. Baysal Z, Togrul T, Aksoy N, Cengiz M, Celik H, Boleken ME, Kaya M, Yavuz G (2009) Evaluation of total oxidative and antioxidative status in pediatric patients undergoing laparoscopic surgery. J Pediatr Surg 44:1367–1370

56. Gitto E, Pellegrino S, Gitto P, Barberi I, Reiter RJ (2009) Oxidative stress of the newborn in the pre- and postnatal period and the clinical utility of melatonin. J Pineal Res 46:128–139
57. Caputo M, Mokhtari A, Rogers CA, Panayiotou N, Chen Q, Ghorbel MT, Angelini GD, Parry AJ (2009) The effects of normoxic versus hyperoxic cardiopulmonary bypass on oxidative stress and inflammatory response in cyanotic pediatric patients undergoing open cardiac surgery: a randomized controlled trial. J Thorac Cardiovasc Surg 138:206–214
58. Nagasaka H, Takayanagi M, Tsukahara H (2009) Children's toxicology from bench to bed–liver injury (3): oxidative stress and anti-oxidant systems in liver of patients with Wilson disease. J Toxicol Sci 34(Suppl 2):S229–S236
59. Nagasaka H, Tsukahara H, Yorifuji T, Miida T, Murayama K, Tsuruoka T, Takatani T, Kanazawa M, Kobayashi K, Okano Y, Takayanagi M (2009) Evaluation of endogenous nitric oxide synthesis in congenital urea cycle enzyme defects. Metabolism 58:278–282
60. Nagasaka H, Okano Y, Tsukahara H, Shigematsu Y, Momoi T, Yorifuji J, Miida T, Ohura T, Kobayashi K, Saheki T, Hirano K, Takayanagi M, Yorifuji T (2009) Sustaining hypercitrullinemia, hypercholesterolemia and augmented oxidative stress in Japanese children with aspartate/glutamate carrier isoform 2-citrin-deficiency even during the silent period. Mol Genet Metab 97:21–26
61. Svecova V, Rossner P Jr, Dostal M, Topinka J, Solansky I, Sram RJ (2009) Urinary 8-oxodeoxyguanosine levels in children exposed to air pollutants. Mutat Res 662:37–43
62. Hayashi M (2009) Oxidative stress in developmental brain disorders. Neuropathology 29:1–8
63. Chiavaroli V, Giannini C, D'Adamo E, de Giorgis T, Chiarelli F, Mohn A (2009) Insulin resistance and oxidative stress in children born small and large for gestational age. Pediatrics 124:695–702
64. Caron JE, Krull KR, Hockenberry M, Jain N, Kaemingk K, Moore IM (2009) Oxidative stress and executive function in children receiving chemotherapy for acute lymphoblastic leukemia. Pediatr Blood Cancer 53:551–556
65. Stringer DM, Sellers EA, Burr LL, Taylor CG (2009) Altered plasma adipokines and markers of oxidative stress suggest increased risk of cardiovascular disease in First Nation youth with obesity or type 2 diabetes mellitus. Pediatr Diabetes 10:269–277
66. Hernández-Marco R, Codoñer-Franch P, Pons Morales S, Del Castillo VC, Boix García L, Valls Bellés V (2009) Oxidant/antioxidant status and hyperfiltration in young patients with type 1 diabetes mellitus. Pediatr Nephrol 24:121–127
67. Cengiz N, Baskin E, Sezgin N, Agras P, Haberal M (2009) Oxidative stress in children on hemodialysis: value of autoantibodies against oxidized low-density lipoprotein. Pediatr Nephrol 24:387–393
68. Kaneko K, Taniguchi N, Tanabe Y, Nakano T, Hasui M, Nozu K (2009) Oxidative imbalance in idiopathic renal hypouricemia. Pediatr Nephrol 24:869–871
69. Bakr A, Abul Hassan S, Shoker M, Zaki M, Hassan R (2009) Oxidant stress in primary nephrotic syndrome: does it modulate the response to corticosteroids? Pediatr Nephrol 24:2375–2380
70. Güneş S, Dirik E, Yiş U, Seçkin E, Kuralay F, Köse S, Unalp A (2009) Oxidant status in children after febrile seizures. Pediatr Neurol 40:47–49
71. de Keyzer Y, Valayannopoulos V, Benoist JF, Batteux F, Lacaille F, Hubert L, Chrétien D, Chadefeaux-Vekemans B, Niaudet P, Touati G, Munnich A, de Lonlay P (2009) Multiple OXPHOS deficiency in the liver, kidney, heart, and skeletal muscle of patients with methylmalonic aciduria and propionic aciduria. Pediatr Res 66:91–95
72. Asakawa T, Tanaka Y, Asagiri K, Kobayashi H, Tanikawa K, Yagi M (2009) Oxidative stress profile in the post-operative patients with biliary atresia. Pediatr Surg Int 25:93–97
73. Zwolinska D, Grzeszczak W, Szczepanska M, Makulska I, Kilis-Pstrusinska K, Szprynger K (2009) Oxidative stress in children on peritoneal dialysis. Perit Dial Int 29:171–177
74. Schulpis KH, Papastamataki M, Stamou H, Papassotiriou I, Margeli A (2010) The effect of diet on total antioxidant status, ceruloplasmin, transferrin and ferritin serum levels in phenylketonuric children. Acta Paediatr 99:1565–1570

75. Codoñer-Franch P, López-Jaén AB, De La Mano-Hernández A, Sentandreu E, Simó-Jordá R, Valls-Bellés V (2010) Oxidative markers in children with severe obesity following low-calorie diets supplemented with mandarin juice. Acta Paediatr 99:1841–1846
76. Kennedy G, Khan F, Hill A, Underwood C, Belch JJ (2010) Biochemical and vascular aspects of pediatric chronic fatigue syndrome. Arch Pediatr Adolesc Med 164:817–823
77. Gücüyener K, Pinarli FG, Erbaş D, Hasanoğlu A, Serdaroğlu A, Topaloğlu H (2010) Is oxidative damage in operation in patients with hereditary spastic paraparesis? Brain Dev 32: 130–136
78. Jayet PY, Rimoldi SF, Stuber T, Salmòn CS, Hutter D, Rexhaj E, Thalmann S, Schwab M, Turini P, Sartori-Cucchia C, Nicod P, Villena M, Allemann Y, Scherrer U, Sartori C (2010) Pulmonary and systemic vascular dysfunction in young offspring of mothers with preeclampsia. Circulation 122:488–494
79. Miric D, Katanic R, Kisic B, Zoric L, Miric B, Mitic R, Dragojevic I (2010) Oxidative stress and myeloperoxidase activity during bacterial meningitis: effects of febrile episodes and the BBB permeability. Clin Biochem 43:246–252
80. Pirinccioglu AG, Gökalp D, Pirinccioglu M, Kizil G, Kizil M (2010) Malondialdehyde (MDA) and protein carbonyl (PCO) levels as biomarkers of oxidative stress in subjects with familial hypercholesterolemia. Clin Biochem 43:1220–1224
81. Araki S, Dobashi K, Yamamoto Y, Asayama K, Kusuhara K (2010) Increased plasma isoprostane is associated with visceral fat, high molecular weight adiponectin, and metabolic complications in obese children. Eur J Pediatr 169:965–970
82. Sadowska-Woda I, Bieszczad-Bedrejczuk E, Rachel M (2010) Influence of desloratadine on selected oxidative stress markers in patients between 3 and 10 years of age with allergic perennial rhinitis. Eur J Pharmacol 640:197–201
83. Hussein MH, Daoud GA, Kakita H, Kato S, Goto T, Kamei M, Goto K, Nobata M, Ozaki Y, Ito T, Fukuda S, Kato I, Suzuki S, Sobajima H, Hara F, Hashimoto T, Togari H (2010) High cerebrospinal fluid antioxidants and interleukin 8 are protective of hypoxic brain damage in newborns. Free Radic Res 44:422–429
84. Vujovic A, Kotur-Stevuljevic J, Kornic D, Spasic S, Spasojevic-Kalimanovska V, Bogavac-Stanojevic N, Stefanovic A, Deanovic M, Babka S, Aleksic B, Jelic-Ivanovic Z (2010) Oxidative stress and anti-oxidative defense in schoolchildren residing in a petrochemical industry environment. Indian Pediatr 47:233–239
85. Monsuez JJ, Charniot JC, Vignat N, Artigou JY (2010) Cardiac side-effects of cancer chemotherapy. Int J Cardiol 144:3–15
86. Ribas GS, Manfredini V, de Mari JF, Wayhs CY, Vanzin CS, Biancini GB, Sitta A, Deon M, Wajner M, Vargas CR (2010) Reduction of lipid and protein damage in patients with disorders of propionate metabolism under treatment: a possible protective role of L-carnitine supplementation. Int J Dev Neurosci 28:127–132
87. Heidenfelder B, Johnson M, Hudgens E, Inmon J, Hamilton RG, Neas L, Gallagher JE (2010) Increased plasma reactive oxidant levels and their relationship to blood cells, total IgE, and allergen-specific IgE levels in asthmatic children. J Asthma 47:106–111
88. Hamaoka A, Hamaoka K, Yahata T, Fujii M, Ozawa S, Toiyama K, Nishida M, Itoi T (2010) Effects of HMG-CoA reductase inhibitors on continuous post-inflammatory vascular remodeling late after Kawasaki disease. J Cardiol 56:245–253
89. Nasca MM, Zhang R, Super DM, Hazen SL, Hall HR (2010) Increased oxidative stress in healthy children following an exercise program: a pilot study. J Dev Behav Pediatr 31:386–392
90. Pedersen CB, Zolkipli Z, Vang S, Palmfeldt J, Kjeldsen M, Stenbroen V, Schmidt SP, Wanders RJ, Ruiter JP, Wibrand F, Tein I, Gregersen N (2010) Antioxidant dysfunction: potential risk for neurotoxicity in ethylmalonic aciduria. J Inherit Metab Dis 33:211–222
91. Sansanwal P, Li L, Hsieh SC, Sarwal MM (2010) Insights into novel cellular injury mechanisms by gene expression profiling in nephropathic cystinosis. J Inherit Metab Dis 33: 775–786

92. Stenzel SL, Krull KR, Hockenberry M, Jain N, Kaemingk K, Miketova P, Moore IM (2010) Oxidative stress and neurobehavioral problems in pediatric acute lymphoblastic leukemia patients undergoing chemotherapy. J Pediatr Hematol Oncol 32:113–118
93. Chahbouni M, Escames G, Venegas C, Sevilla B, García JA, López LC, Muñoz-Hoyos A, Molina-Carballo A, Acuña-Castroviejo D (2010) Melatonin treatment normalizes plasma pro-inflammatory cytokines and nitrosative/oxidative stress in patients suffering from Duchenne muscular dystrophy. J Pineal Res 48:282–289
94. Nagasaka H, Okano Y, Aizawa M, Miida T, Yorifuji T, Tajima G, Sakura N, Takatani T, Sanayama Y, Sugamoto K, Mayumi M, Kobayashi K, Hirano K, Takayanagi M, Tsukahara H (2010) Altered metabolisms of mediators controlling vascular function and enhanced oxidative stress in asymptomatic children with congenital portosystemic venous shunt. Metabolism 59:107–113
95. Fu R, Yanjanin NM, Bianconi S, Pavan WJ, Porter FD (2010) Oxidative stress in Niemann-Pick disease, type C. Mol Genet Metab 101:214–218
96. Codoñer-Franch P, Pons-Morales S, Boix-García L, Valls-Bellés V (2010) Oxidant/antioxidant status in obese children compared to pediatric patients with type 1 diabetes mellitus. Pediatr Diabetes 11:251–257
97. Erol I, Alehan F, Aldemir D, Ogus E (2010) Increased vulnerability to oxidative stress in pediatric migraine patients. Pediatr Neurol 43:21–24
98. Bandsma RH, Mendel M, Spoelstra MN, Reijngoud DJ, Boer T, Stellaard F, Brabin B, Schellekens R, Senga E, Heikens GT (2010) Mechanisms behind decreased endogenous glucose production in malnourished children. Pediatr Res 68:423–428
99. Zhang J, Yao X, Yu R, Bai J, Sun Y, Huang M, Adcock IM, Barnes PJ (2010) Exhaled carbon monoxide in asthmatics: a meta-analysis. Respir Res 11:50
100. Karsen H, Sunnetcioglu M, Ceylan RM, Bayraktar M, Taskin A, Aksoy N, Erten R (2011) Evaluation of oxidative status in patients with *Fasciola hepatica* infection. Afr Health Sci 11(Suppl 1):S14–S18
101. Mallol J, Aguirre V, Espinosa V (2011) Increased oxidative stress in children with post infectious *Bronchiolitis Obliterans*. Allergol Immunopathol (Madr) 39:253–258
102. Hosakote YM, Jantzi PD, Esham DL, Spratt H, Kurosky A, Casola A, Garofalo RP (2011) Viral-mediated inhibition of antioxidant enzymes contributes to the pathogenesis of severe respiratory syncytial virus bronchiolitis. Am J Respir Crit Care Med 183:1550–1560
103. Codoñer-Franch P, Tavárez-Alonso S, Murria-Estal R, Megías-Vericat J, Tortajada-Girbés M, Alonso-Iglesias E (2011) Nitric oxide production is increased in severely obese children and related to markers of oxidative stress and inflammation. Atherosclerosis 215:475–480
104. Al-Yafee YA, Al-Ayadhi LY, Haq SH, El-Ansary AK (2011) Novel metabolic biomarkers related to sulfur-dependent detoxification pathways in autistic patients of Saudi Arabia. BMC Neurol 11:139
105. Adams JB, Audhya T, McDonough-Means S, Rubin RA, Quig D, Geis E, Gehn E, Loresto M, Mitchell J, Atwood S, Barnhouse S, Lee W (2011) Effect of a vitamin/mineral supplement on children and adults with autism. BMC Pediatr 11:111
106. Velasco-Sánchez D, Aracil A, Montero R, Mas A, Jiménez L, O'Callaghan M, Tondo M, Capdevila A, Blanch J, Artuch R, Pineda M (2011) Combined therapy with idebenone and deferiprone in patients with Friedreich's ataxia. Cerebellum 10:1–8
107. Yahata T, Suzuki C, Hamaoka A, Fujii M, Hamaoka K (2011) Dynamics of reactive oxygen metabolites and biological antioxidant potential in the acute stage of Kawasaki disease. Circ J 75:2453–2459
108. Pacak K (2011) Pheochromocytoma: a catecholamine and oxidative stress disorder. Endocr Regul 45:65–90
109. Koc S, Aksoy N, Bilinc H, Duygu F, Uysal IÖ, Ekinci A (2011) Paraoxonase and arylesterase activity and total oxidative/anti-oxidative status in patients with chronic adenotonsillitis. Int J Pediatr Otorhinolaryngol 75:1364–1367
110. Kawashima H, Nishimata S, Ishii C, Yamanaka G, Kashiwagi Y, Takekuma K, Hoshika A, Watanabe Y (2011) New treatment of free-radical scavenger in adrenoleukodystrophy. J Clin Pharm Ther 36:412–415

111. von Dessauer B, Bongain J, Molina V, Quilodrán J, Castillo R, Rodrigo R (2011) Oxidative stress as a novel target in pediatric sepsis management. J Crit Care 26:103–130
112. D'Esposito M, Filosa S, Pecorelli A, Valacchi G, Hayek J (2011) F2-dihomo-isoprostanes as potential early biomarkers of lipid oxidative damage in Rett syndrome. J Lipid Res 52:2287–2297
113. Grosso S, Longini M, Rodriguez A, Proietti F, Piccini B, Balestri P, Buonocore G (2011) Oxidative stress in children affected by epileptic encephalopathies. J Neurol Sci 300: 103–106
114. Radi E, Formichi P, Di Maio G, Battisti C, Federico A (2011) Oxidative stress-induced apoptosis in two patients with Alagille syndrome. J Neurol Sci 308:49–56
115. Bandsma RH, Spoelstra MN, Mari A, Mendel M, van Rheenen PF, Senga E, van Dijk T, Heikens GT (2011) Impaired glucose absorption in children with severe malnutrition. J Pediatr 158:282–287
116. Oliveira KF, Cunha DF, Weffort VR (2011) Analysis of serum and supplemented vitamin C and oxidative stress in HIV-infected children and adolescents. J Pediatr (Rio J) 87:517–522
117. Yıldırım F, Sermetow K, Aycicek A, Kocyigit A, Erel O (2011) Increased oxidative stress in preschool children exposed to passive smoking. J Pediatr (Rio J) 87:523–528
118. Filippon L, Vanzin CS, Biancini GB, Pereira IN, Manfredini V, Sitta A, Peralba Mdo C, Schwartz IV, Giugliani R, Vargas CR (2011) Oxidative stress in patients with mucopolysaccharidosis type II before and during enzyme replacement therapy. Mol Genet Metab 103:121–127
119. Sanayama Y, Nagasaka H, Takayanagi M, Ohura T, Sakamoto O, Ito T, Ishige-Wada M, Usui H, Yoshino M, Ohtake A, Yorifuji T, Tsukahara H, Hirayama S, Miida T, Fukui M, Okano Y (2011) Experimental evidence that phenylalanine is strongly associated to oxidative stress in adolescents and adults with phenylketonuria. Mol Genet Metab 103:220–225
120. Vanzin CS, Biancini GB, Sitta A, Wayhs CA, Pereira IN, Rockenbach F, Garcia SC, Wyse AT, Schwartz IV, Wajner M, Vargas CR (2011) Experimental evidence of oxidative stress in plasma of homocystinuric patients: a possible role for homocysteine. Mol Genet Metab 104:112–117
121. Ferraris V, Acquier A, Ferraris JR, Vallejo G, Paz C, Mendez CF (2011) Oxidative stress status during the acute phase of haemolytic uraemic syndrome. Nephrol Dial Transplant 26:858–864
122. Vaisbich MH, de Faria P, Guimaraes L, Shimizu H, Seguro AC (2011) Oxidative stress in cystinosis patients. Nephron Extra 1:73–77
123. Wilson KM, Finkelstein JN, Blumkin AK, Best D, Klein JD (2011) Micronutrient levels in children exposed to secondhand tobacco smoke. Nicotine Tob Res 13:800–808
124. Sadowska-Woda I, Rachel M, Pazdan J, Bieszczad-Bedrejczuk E, Pawliszak K (2011) Nutritional supplement attenuates selected oxidative stress markers in pediatric patients with cystic fibrosis. Nutr Res 31:509–518
125. Bakkeheim E, Mowinckel P, Carlsen KH, Burney P, Carlsen KC (2011) Altered oxidative state in schoolchildren with asthma and allergic rhinitis. Pediatr Allergy Immunol 22:178–185
126. Mishra OP, Gupta AK, Prasad R, Ali Z, Upadhyay RS, Mishra SP, Tiwary NK, Schaefer FS (2011) Antioxidant status of children with idiopathic nephrotic syndrome. Pediatr Nephrol 26:251–256
127. Keskin N, Civilibal M, Elevli M, Koldas M, Duru NS, Ozturk H (2011) Elevated plasma advanced oxidation protein products in children with Henoch-Schonlein purpura. Pediatr Nephrol 26:1989–1993
128. Hussein MH, Hashimoto T, Daoud GA, Kakita H, Kato S, Goto T, Hibi M, Kato T, Okumura N, Tomishige H, Hara F, Ito T, Fukuda S, Kato I, Suzuki T, Suzuki S, Togari H (2011) Oxidative stress after living related liver transplantation subsides with time in pediatric patients. Pediatr Surg Int 27:17–22
129. Kawatani M, Tsukahara H, Mayumi M (2011) Evaluation of oxidative stress status in children with pervasive developmental disorder and attention deficit hyperactivity disorder using urinary-specific biomarkers. Redox Rep 16:45–46

130. Damodaran LP, Arumugam G (2011) Urinary oxidative stress markers in children with autism. Redox Rep 16:216–222
131. Stachowicz-Stencel T, Synakiewicz A, Owczarzak A, Sliwinska A, Lysiak-Szydlowska W, Balcerska A (2012) Association between intestinal and antioxidant barriers in children with cancer. Acta Biochim Pol 59:237–242
132. Testa D, Guerra G, Marcuccio G, Landolfo PG, Motta G (2012) Oxidative stress in chronic otitis media with effusion. Acta Otolaryngol 132:834–837
133. Belini Junior E, da Silva DG, Torres Lde S, de Almeida EA, Cancado RD, Chiattone C, Bonini-Domingos CR (2012) Oxidative stress and antioxidant capacity in sickle cell anaemia patients receiving different treatments and medications for different periods of time. Ann Hematol 91:479–489
134. Li Z, Liu X, Liu S, Gu C, Tian F, Wen J (2012) Urinary heme oxygenase-1 in children with congenital hydronephrosis due to ureteropelvic junction obstruction. Biomarkers 17: 471–476
135. Hamed EA, El-Abaseri TB, Mohamed AO, Ahmed AR, El-Metwally TH (2012) Hypoxia and oxidative stress markers in pediatric patients undergoing hemodialysis: cross section study. BMC Nephrol 13:136
136. Sabuncuoğlu S, Kuşkonmaz B, Uckun Çetinkaya D, Ozgüneş H (2012) Evaluation of oxidative and antioxidative parameters in pediatric hematopoietic SCT patients. Bone Marrow Transplant 47:651–656
137. Peroni DG, Bodini A, Corradi M, Coghi A, Boner AL, Piacentini GL (2012) Markers of oxidative stress are increased in exhaled breath condensates of children with atopic dermatitis. Br J Dermatol 166:839–843
138. Kippler M, Hossain MB, Lindh C, Moore SE, Kabir I, Vahter M, Broberg K (2012) Early life low-level cadmium exposure is positively associated with increased oxidative stress. Environ Res 112:164–170
139. Kelekçi S, Evliyaoğlu O, Sen V, Yolbaş I, Uluca U, Tan I, Gürkan MF (2012) The relationships between clinical outcome and the levels of total antioxidant capacity (TAC) and coenzyme Q (CoQ 10) in children with pandemic influenza (H1N1) and seasonal flu. Eur Rev Med Pharmacol Sci 16:1033–1038
140. Liu H, Han T, Tian J, Zhu ZY, Liu Y, Li Y, Xiao SX, Li Y, Feng YY (2012) Monitoring oxidative stress in acute-on-chronic liver failure by advanced oxidation protein products. Hepatol Res 42:171–180
141. Faienza MF, Francavilla R, Goffredo R, Ventura A, Marzano F, Panzarino G, Marinelli G, Cavallo L, Di Bitonto G (2012) Oxidative stress in obesity and metabolic syndrome in children and adolescents. Horm Res Paediatr 78:158–164
142. Sulthana SM, Kumar SN, Sridhar MG, Bhat BV, Rao KR (2012) Levels of non enzymatic antioxidants in Down syndrome. Indian J Pediatr 79:1473–1476
143. Emin O, Hasan A, Aysegul D, Rusen D (2012) Total antioxidant status and oxidative stress and their relationship to total IgE levels and eosinophil counts in children with allergic rhinitis. J Investig Allergol Clin Immunol 22:188–192
144. Toyran M, Kaymak M, Vezir E, Harmanci K, Kaya A, Giniş T, Köse G, Kocabaş CN (2012) Trace element levels in children with atopic dermatitis. J Investig Allergol Clin Immunol 22:341–344
145. Aslan A, Gurelik M, Cemek M, Buyukokuroglu M, Goksel HM, Eser O (2012) Nimodipine can diminish oxidative stress in patients with severe head trauma. J Neurosurg Sci 56: 247–253
146. Doğan M, Cesur Y, Zehra Doğan Ş, Kaba S, Bulan K, Cemek M (2012) Oxidant/antioxidant system markers and trace element levels in children with nutritional rickets. J Pediatr Endocrinol Metab 25:1129–1139
147. Narsaria N, Mohanty C, Das BK, Mishra SP, Prasad R (2012) Oxidative stress in children with severe malaria. J Trop Pediatr 58:147–150
148. Rockenbach FJ, Deon M, Marchese DP, Manfredini V, Mescka C, Ribas GS, Habekost CT, Castro CG Jr, Jardim LB, Vargas CR (2012) The effect of bone marrow transplantation on oxidative stress in X-linked adrenoleukodystrophy. Mol Genet Metab 106:231–236

149. Renjini R, Gayathri N, Nalini A, Srinivas Bharath MM (2012) Oxidative damage in muscular dystrophy correlates with the severity of the pathology: role of glutathione metabolism. Neurochem Res 37:885–898
150. Celik M, Tuncer A, Soyer OU, Saçkesen C, Tanju Besler H, Kalayci O (2012) Oxidative stress in the airways of children with asthma and allergic rhinitis. Pediatr Allergy Immunol 23:556–561
151. Kaneko K, Takahashi M, Yoshimura K, Kitao T, Yamanouchi S, Kimata T, Tsuji S (2012) Intravenous immunoglobulin counteracts oxidative stress in Kawasaki disease. Pediatr Cardiol 33:1086–1088
152. Pirinccioglu AG, Alyan O, Kizil G, Kangin M, Beyazit N (2012) Evaluation of oxidative stress in children with congenital heart defects. Pediatr Int 54:94–98
153. Kaneko K, Kimata T, Takahashi M, Shimo T, Tanaka S, Tsuji S (2012) Change in urinary 8-hydroxydeoxyguanosine in idiopathic nephrotic syndrome. Pediatr Nephrol 27:155–156
154. Sladowska-Kozłowska J, Litwin M, Niemirska A, Płudowski P, Wierzbicka A, Skorupa E, Wawer ZT, Janas R (2012) Oxidative stress in hypertensive children before and after 1 year of antihypertensive therapy. Pediatr Nephrol 27:1943–1951
155. Fraguas D, Gonzalez-Pinto A, Micó JA, Reig S, Parellada M, Martínez-Cengotitabengoa M, Castro-Fornieles J, Rapado-Castro M, Baeza I, Janssen J, Desco M, Leza JC, Arango C (2012) Decreased glutathione levels predict loss of brain volume in children and adolescents with first-episode psychosis in a two-year longitudinal study. Schizophr Res 137:58–65
156. Malakasioti G, Alexopoulos E, Befani C, Tanou K, Varlami V, Ziogas D, Liakos P, Gourgoulianis K, Kaditis AG (2012) Oxidative stress and inflammatory markers in the exhaled breath condensate of children with OSA. Sleep Breath 16:703–708
157. Ozkaya E, Akduman H, Erenberk U, Demir A, Dundaroz MR (2013) Plasma paraoxonase activity and oxidative stress and their relationship to disease severity in children with allergic rhinitis. Am J Rhinol Allergy 27:13–17
158. Hussein MH, Hashimoto T, Suzuki T, Daoud GA, Goto T, Nakajima Y, Kato T, Hibi M, Tomishige H, Hara F, Kato S, Kakita H, Kamei M, Ito T, Kato I, Sugioka A, Togari H (2013) Children undergoing liver transplantation for treatment of inherited metabolic diseases are prone to higher oxidative stress, complement activity and transforming growth factor-β1. Ann Transplant 18:63–68
159. Carraro S, Giordano G, Piacentini G, Kantar A, Moser S, Cesca L, Berardi M, Di Gangi IM, Baraldi E (2013) Asymmetric dimethylarginine (ADMA) in exhaled breath condensate and serum of asthmatic children. Chest 144:405–410
160. D'Adamo E, Marcovecchio ML, Giannini C, de Giorgis T, Chiavaroli V, Chiarelli F, Mohn A (2013) Improved oxidative stress and cardio-metabolic status in obese prepubertal children with liver steatosis treated with lifestyle combined with vitamin E. Free Radic Res 47:146–153
161. Vurucu S, Karaoglu A, Paksu MS, Yesilyurt O, Oz O, Unay B, Akin R (2013) Relationship between oxidative stress and chronic daily headache in children. Hum Exp Toxicol 32: 113–119
162. Pirgon O, Bilgin H, Kurku H, Cekmez F, Dündar BN (2013) Association between insulin resistance and oxidative stress parameters in obese adolescents with non-alcoholic fatty liver disease. J Clin Res Pediatr Endocrinol 5:33–39
163. Schoendorfer NC, Vitetta L, Sharp N, DiGeronimo M, Wilson G, Coombes JS, Boyd R, Davies PS (2013) Micronutrient, antioxidant, and oxidative stress status in children with severe cerebral palsy. J Parenter Enteral Nutr 37:97–101
164. Soundaravally R, Pukazhvandthen P, Zachariah B, Hamide A (2013) Plasma ferritin and indices of oxidative stress in *Helicobacter pylori* infection among schoolchildren. J Pediatr Gastroenterol Nutr 56:519–522
165. Srivastava R, Lohokare R, Prasad R (2013) Oxidative stress in children with bacterial meningitis. J Trop Pediatr 59:305–308
166. Kotur-Stevuljević J, Peco-Antić A, Spasić S, Stefanović A, Paripović D, Kostić M, Vasić D, Vujović A, Jelić-Ivanović Z, Spasojević-Kalimanovska V, Kornic-Ristovski D (2013)

Hyperlipidemia, oxidative stress, and intima media thickness in children with chronic kidney disease. Pediatr Nephrol 28:295–303
167. Tsukahara H, Shibata R, Ohshima Y, Todoroki Y, Sato S, Ohta N, Hiraoka M, Yoshida A, Nishima S, Mayumi M (2003) Oxidative stress and altered antioxidant defenses in children with acute exacerbation of atopic dermatitis. Life Sci 72:2509–2516
168. Kauffman LD, Sokol RJ, Jones RH, Awad JA, Rewers MJ, Norris JM (2003) Urinary F2-isoprostanes in young healthy children at risk for type 1 diabetes mellitus. Free Radic Biol Med 35:551–557
169. Tamura S, Tsukahara H, Ueno M, Maeda M, Kawakami H, Sekine K, Mayumi M (2006) Evaluation of a urinary multi-parameter biomarker set for oxidative stress in children, adolescents and young adults. Free Radic Res 40:1198–1205
170. Kaneko K, Kimata T, Tsuji S, Ohashi A, Imai Y, Sudo H, Kitamura N (2012) Measurement of urinary 8-oxo-7,8-dihydro-2-deoxyguanosine in a novel point-of-care testing device to assess oxidative stress in children. Clin Chim Acta 413:1822–1826
171. Ogino K, Wang DH (2007) Biomarkers of oxidative/nitrosative stress: an approach to disease prevention. Acta Med Okayama 61:181–189

Chapter 7
Urinary L-Type Fatty Acid-Binding Protein as a New Renal Biomarker

Tsuyoshi Oikawa and Takeshi Sugaya

Abbreviations

ACE	Angiotensin converting enzyme
ADPKD	Autosomal dominant polycystic kidney disease
AER	Albumin excretion rate
AKI	Acute kidney injury
AKIN	Acute Kidney Injury Network
APACHE	Acute physiology and chronic health evaluation
ARB	Angiotensin II receptor blocker
ARF	Acute renal failure
AUC	Area under the curve
BSA	Bovine serum albumin
CCD	Charge coupled devices
CI-AKI	Contrast-induced acute kidney injury
CKD	Chronic kidney disease
CPB	Cardiopulmonary bypass
DN	Diabetic nephropathy
ELISA	Enzyme-linked immunosorbent assay
FABP	Fatty acid-binding protein
FFA	Free fatty acid
FGS	Focal glomerulosclerosis

T. Oikawa, Ph.D.
CMIC Holdings Co. Ltd., Tokyo 113-0034, Japan

T. Sugaya, Ph.D. (✉)
CMIC Holdings Co. Ltd., Tokyo 113-0034, Japan

Department of Nephrology and Hypertension, Internal Medicine,
St. Marianna University School of Medicine, Kanagawa 216-8511, Japan
e-mail: takeshi-sugaya@marianna-u.ac.jp; takeshi-sugaya@cmic.co.jp

H. Tsukahara and K. Kaneko (eds.), *Studies on Pediatric Disorders*, Oxidative Stress in Applied Basic Research and Clinical Practice, DOI 10.1007/978-1-4939-0679-6_7,

HIF	Hypoxia-inducible factor
HNF	Hepatocyte nuclear factor
ICU	Intensive care unit
IL-18	Interleukin-18
IVD	In vitro diagnostics
KDIGO	Kidney Disease Improving Global Outcomes
KIM-1	Kidney injury molecule-1
LDL	Low-density lipoprotein
L-FABP	Liver-type fatty acid-binding protein
MCNS	Minimal change nephrotic syndrome
MG	Microglobulin
NAG	*N*-acetyl-β-D-glucosaminidase
NGAL	Neutrophil gelatinase-associated lipocalin
8-OHdG	8-Hydroxydeoxyguanosine
pAKI	Pediatric acute kidney injury
PMX-F	Polymyxin B-immobilized fiber
PPAR	Peroxisome proliferator-activated receptor
ROC	Receiver operating characteristic
UAE	Urinary albumin excretion

7.1 Introduction

Chronic kidney disease (CKD) affects 20 million adults in the USA, and another 20–40 million are at risk for developing the disease during their lifetime. The mainstays of screening and testing for renal disease are serum creatinine and urinary protein tests. While the standards of care today, the tests have a number of shortcomings. Neither test can accurately diagnose the type of renal injury present. Serum creatinine is a poor marker of early kidney disease. Also, both are indirect measures of glomerular filtration using distantly produced substances, rather than a reflection of the state of the renal tissue [1]. What is needed, are new biomarkers produced in the kidney in response to disease or other chronic or acute assault.

7.2 L-FABP and the Tubular Response to Injury

Fatty acid-binding proteins (FABPs) are a group of intracellular proteins of 14 kDa, belonging to the lipocalin family. Several isoforms of FABPs have been identified including liver, intestinal, heart (muscle), adipocyte, epidermal, ileal, brain, myelin, and testis types [2, 3]. In the human kidney, two types of FABP have been identified: liver-type FABP (L-FABP) that is expressed in the proximal tubule, and heart-type FABP, produced in the distal tubule. L-FABP has a key role in the binding and trafficking of fatty acids and some other hydrophobic molecules, as shown in Fig. 7.1.

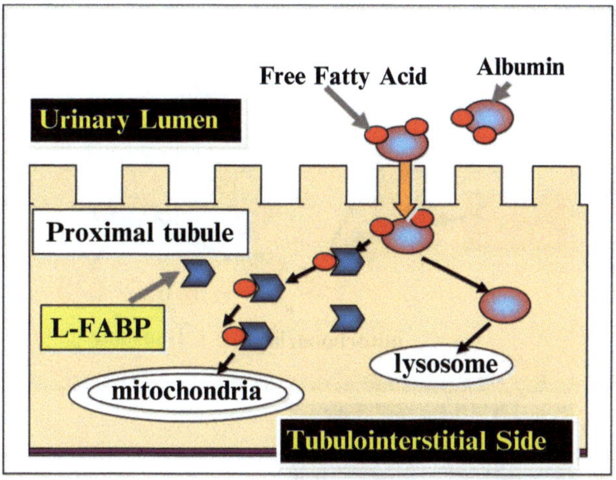

Fig. 7.1 Physiological role of L-FABP. Free fatty acids (FFAs) are bound to serum albumin, filtered through glomeruli and reabsorbed into the proximal tubule along with albumin. FFAs cause up-regulation of L-FABP gene expression. L-FABP plays a role in the intracellular transport of FFAs to mitochondria and/or peroxisomes for metabolism

L-FABP helps transport free fatty acids (FFAs) to organelles such as mitochondria and lysosomes for β-oxidation [4]. It has also been suggested that L-FABP exerts an influence on the activity of peroxisome proliferator-activated receptor (PPAR) by transporting ligands to the nucleus thereby influencing transcriptional regulation of genes involved in the proliferation and differentiation of cells and in the metabolism of lipids and carbohydrates [5].

A number of studies have shown that FABPs are associated with tissue ischemia and damage in their organ of production. For example, L-FABP expression and secretion into urine is increased during tubulointerstitial damage in the kidney. Kamijo and colleagues used transgenic mice containing the human *l-fabp* gene to study the effects of FFAs on human L-FABP expression. They found that human L-FABP expression in the transgenic mouse kidney was more upregulated, and urinary human L-FABP excretion increased to a greater extent, upon injection of FFA replete bovine serum albumin (BSA) vs. BSA alone [6]. A number of studies in kidney proximal tubular cells in animals or in culture have shown that increased uptakc of fatty acids causes apoptosis and tubulointerstitial cell damage [7–10].

Figure 7.2 shows a proposed mechanism for L-FABP's relationship to tubulointerstitial damage. FFAs that are overloaded in tubulointerstitial cells during massive proteinuria are oxidized in the mitochondria and lysosomes, thus overloading the cells with lipid hydroperoxides. These molecules with reactive oxygen species damage the cell membranes. L-FABP binds these oxidized molecules and escapes through damaged cell membranes into the urinary lumen. Increased damage and assault lead to increased synthesis of L-FABP in the cell, and increased excretion of L-FABP into the urine.

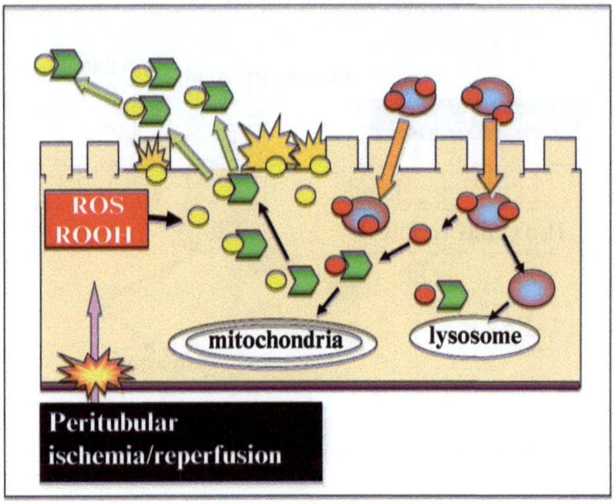

Fig. 7.2 Rationale for diagnostic use of L-FABP. Lipid hydroperoxides (i.e., ROOH/ROS) are accumulated in proximal tubules during renal ischemia/reperfusion and other assaults to proximal tubules. L-FABP is excreted from the proximal tubules into urine by binding three cytotoxic lipids, and escaping through damaged cell membranes

Many laboratories have been searching for new, endogenous biomarkers for kidney health to complement current tests of kidney function. The goal of these efforts is to gain a window on the health of the organ itself by measuring a protein produced in the kidney, that reflects the organ's relative health and/or injury, as for example, troponin C is used as an indicator of injury to heart muscle [1]. We believe that L-FABP is a strong candidate to become a new and sensitive endogenous biomarker of kidney health and disease, due to its synthesis and secretion by the kidney during disease and other assaults. L-FABP will not likely replace conventional functional testing, but will complement it. Our research laboratory has developed L-FABP enzyme-linked immunosorbent assay (ELISA) kit for diagnostic use in Japan [11]. The urinary L-FABP assay provides physicians with a more sensitive, precise, and earlier measure for diagnosis and prognosis, than possible with current methods.

7.3 Clinical Results

To date, urinary L-FABP testing has undergone recent extensive clinical research and evaluation mainly in Japan, where this assay was developed. The following reviews the highlights of clinical evaluation and use to date, as summarized in Table 7.1.

Table 7.1 Summary of recent clinical studies examining the utility of urinary L-FABP testing

Disease type	Population	n^a	Key outcomes	Reference
Normal	Annual physical exam subjects	908	Average urinary L-FABP concentration in normal patients was 3.6±0.2 μg/g creatinine. Urinary L-FABP was significantly higher in patients with hypertension, diabetes, and chronic hepatitis.	[12]
Diabetic nephropathy	Type 2 diabetes	356	Urinary L-FABP increased significantly with advancing stages of diabetic nephropathy.	[18]
		78	Urinary L-FABP levels associated with the progression of diabetic nephropathy, and pitavastatin might be effective in ameliorating tubulointerstitial damage in early diabetic nephropathy.	[11]
		108	Pioglitazone appeared to be effective in reducing urinary albumin excretion and the urinary L-FABP level.	[19]
		50	L-FABP levels were significantly higher in smokers than in non-smokers.	[42]
		60	Telmisartan decreased proteinuria and the urinary L-FABP level in dose-dependent manner that is independent of its blood pressure-lowering effect in diabetic nephropathy.	[43]
		36	Urinary markers (8-OHdG and L-FABP) were significantly lower in group A (red wine) than in groups B (white wine) and C (control) at 3 and 6 months. Red wine was renoprotective whereas white wine had no such effect in type 2 diabetes mellitus patients with nephropathy.	[15]
		170	Urinary L-FABP was significantly increased in diabetes patients compared with healthy controls. Urinary L-FABP could be a novel biomarker for chronic intrarenal ischemia.	[16]
		552	Urinary L-FABP accurately reflected the severity of diabetic nephropathy in type 2 diabetes, and its level was high in the patients with normoalbuminuria. Higher urinary L-FABP was a risk factor for progression of diabetic nephropathy.	[20]
		618	Urinary L-FABP might be a predictive marker for renal and cardiovascular prognosis in type 2 diabetic patients without advanced nephropathy.	[17]

(continued)

Table 7.1 (continued)

Disease type	Population	n^a	Key outcomes	Reference
		45	The addition of azelnidipine treatment to therapy with angiotensin II receptor blockers (ARBs) had dose-dependent antioxidant and renoprotective effects beyond blood pressure-lowering effects in hypertensive diabetic nephropathy patients.	[44]
		27	Urinary L-FABP levels might reflect the stress induced by FFAs to the proximal tubules, leading to severe tubulointerstitial damage.	[13]
		136	Systolic and diastolic blood pressures, urinary albumin excretion, urinary L-FABP and 8-OHdG excretion were significantly reduced 6 and 12 months compared with baseline on any of the ARBs treated groups. The renoprotective effect of telmisartan appeared to be more potent than that of losartan, candesartan, or olmesartan in early-stage diabetic nephropathy patients.	[45]
	Type 1 diabetes	205	Urinary L-FABP levels were associated with early and progressive tubulointerstitial damage and albuminuria. Angiotensin converting enzyme (ACE) inhibition reduced the tubular and glomerular damage and dysfunction.	[46]
		277 (204)	High levels of urinary L-FABP predicted the initiation and progression to diabetic nephropathy and all-cause mortality, independent of urinary albumin excretion rate and other established risk factors.	[47]
		2,454	L-FABP was an independent predictor of progression of DN irrespective of disease stage. L-FABP used alone or together with albumin excretion rate might not improve the risk prediction of DN progression in patients with type 1 diabetes.	[48]
Non-diabetic nephropathy	CKD (serum creatinine conc. less than 2.5 mg/dL)	120	Urinary L-FABP reflected the clinical prognosis of chronic renal disease.	[21]
	CKD (serum creatinine conc. 1.2–3 mg/dL in men and 0.9–2.3 mg/dL in women	48	Urinary excretion of L-FABP increased with the deterioration of kidney function. Urinary L-FABP was a useful clinical marker in the monitoring of CKD.	[5]

ADPKD	40	Increased urinary L-FABP levels might be associated with the development of ADPKD, and candesartan cilexetil had a beneficial effect on reducing their levels.	[22]
Focal glomerulosclerosis (FGS) and minor glomerular abnormality	61	Urinary L-FABP might be a useful diagnostic indicator for differentiation between FGS and minor glomerular abnormalities. LDL apheresis might be effective in ameliorating tubulointerstitial lesions associated with FGS.	[23]
IgA nephropathy	50	Urinary L-FABP level could be used to discriminate between IgA nephropathy and thin basement membrane nephropathy in patients with microscopic hematuria.	[49]
Normotensive patients with IgA nephropathy	44	A combination therapy of ARB plus ACE inhibitor had a greater beneficial effect on renal injury compared with monotherapy using ARB or ACE inhibitor in normotensive patients with IgA nephropathy.	[50]
Study 1) Non-diabetic CKD Study 2) CKD and liver disease	1) 48 2) 215	Urinary excretion of L-FABP increased with the deterioration of renal function. Serum L-FABP did not influence on urinary L-FABP.	[14]
Non-diabetic mild CKD	50	Pitavastatin reduced the urinary protein excretion, while the urinary L-FABP levels fell. Pitavastatin ameliorated tubulointerstitial damage in CKD patients independent of the lipid-lowering effect.	[51]
Chronic renal failure and anemia	40	The urinary L-FABP levels were decreased as compared with the baseline level in group A (treated with erythropoietin) but not in group B (not treated with erythropoietin). Erythropoietin supplementation might ameliorate renal tubular damage, in part, due to a reduction of oxidative stress in chronic renal failure patients with anemia.	[52]
Hypertensive CKD	60	Azelnidipine was renoprotective in hypertensive patients with mild CKD and this action was, at least in part, due to the antioxidative effect.	[53]
	30	Telmisartan reduced proteinuria, urinary L-FABP excretion, and urinary collagen IV levels in hypertensive CKD patients.	[54]
	40	Benidipine was more effective than amlodipine for protecting renal function and potentially for ameliorating atherosclerosis in hypertensive patients with mild CKD.	[55]
	30	Telmisartan resulted in a greater reduction of urinary markers than treated enalapril. This effect occurred by a mechanism independent of blood pressure reduction.	[56]

(continued)

Table 7.1 (continued)

Disease type	Population	n^a	Key outcomes	Reference
Acute kidney injury	Contrast-induced acute kidney injury (CI-AKI)	96	Urinary L-FABP levels could serve clinically as a predictive marker for CI-AKI.	[29]
		220	Urinary L-FABP level was useful for predicting the onset of CI-AKI before contrast medium exposure.	[30]
	Septic shock	110	Urinary L-FABP level was significantly increased in patients with septic shock and that PMX-F treatment was effective in reducing this level.	[34]
	Kidney transplant	12	Only urinary L-FABP correlated to the increase of reciprocal unit of peritubular capillary blood flow.	[36]
	Preterm neonates	40	Urinary L-FABP level showed significant positive correlation with those of urinary NAG and 8-OHdG. L-FABP was expressed in the neonatal kidney. It might also point to potential effects of proximal tubular damage and oxidative stress on urinary excretion of L-FABP.	[37]
	Pediatric patients (cardiopulmonary bypass surgery)	40	Urinary L-FABP levels represented a sensitive and predictive early biomarker of AKI after cardiac surgery.	[25]
	AKI patients admitted to the ICU	25	L-FABP was a useful biomarker for early detection of AKI. Podocyte injury observed by elevation of urinary podocalyxin was induced during the recovery phase of AKI.	[26]
	Critically ill patients (in ICU)	339	The combination of L-FABP and NGAL could serve as new biomarkers of mortality prediction in critical care.	[41]
	After cardiac surgery	85	Urinary L-FABP was a useful biomarker for early detection of AKI and was a good early predictor of the onset of AKI.	[24]
	Prerenal acute kidney injury	337	Urinary L-FABP, NGAL, IL-18, NAG, and albumin in patients with prerenal AKI showed modest by significantly higher concentrations than in patients with non-AKI.	[27]
	Critically ill patients	145	Urinary L-FABP could be an adjunctive and independent biomarker for both the detection of AKI as well as the prediction of prognosis in heterogeneous ICU patients.	[28]

[a]Total subjects, including normal volunteers

7.3.1 Establishing Normal Values

To establish normal values, Ishimitsu and colleagues measured L-FABP in the first morning urine of 908 people undergoing annual physical examinations [12]. Among 150 healthy subjects, urinary L-FABP was found to average 3.6±0.2 μg/g creatinine. L-FABP values were significantly higher in patients with hypertension, diabetes mellitus, and chronic hepatitis. In regression analysis, L-FABP correlated better with α1-microglobulin (MG) than albumin, supporting the notion that L-FABP is a marker of renal tubular injury, rather than of glomerular leakage.

7.3.2 Chronic Kidney Disease

Urinary L-FABP is associated with the degree of tubulointerstitial injury in human kidney biopsy tissue [6] and is a marker reflecting tubular stress burden, and has been proven useful when predicting a kidney prognosis in clinical studies targeting non-diabetic chronic renal disease patients [5, 6]. Urinary L-FABP has greater sensitivity than conventional urinary protein markers, and can determine patients whose renal disease will progress. Also, because urinary L-FABP increases as renal disease progresses and decreases with its remission, it is useful for the monitoring of renal disease.

Eight patients with diabetic nephropathy diagnosed by renal biopsy and 12 patients with minimal change nephrotic syndrome (MCNS) were studied to examine the correlation between histological tubulointerstitial injury in CKD and urinary markers. Urine was collected from the above subjects at the time of hospitalization just before renal biopsy for 24 h, and measured values for urinary protein, urinary albumin, urinary *N*-acetyl-β-D-glucosaminidase (NAG), and urinary L-FABP were compared with pathological histology evaluations across both groups. No significant difference was found in both groups for urinary protein, urinary albumin, and urinary NAG, but significantly high values were shown for urinary L-FABP alone in the diabetic nephropathy cases, matching the result of histological tubulointerstitial injury evaluations (Fig. 7.3). Because urinary L-FABP is able to detect early changes of tubulointerstitial injury, which is determined only occasionally with conventional urinary markers, the implication is that the appearance of urinary L-FABP is possibly tubular injury-specific in earlier stage diabetic nephropathy [13].

Because L-FABP is also expressed in the liver and intestinal tract, the possibility of blood L-FABP levels rising in cases of liver or intestinal disease is conceivable. There is a report of clinical analysis in such instances of whether urinary L-FABP is affected by blood L-FABP [14]. The results of blood and urinary L-FABP measurements for 71 liver disease patients with normal kidney function, 73 CKD patients with normal liver function, and 71 healthy subjects are shown in Fig. 7.4. In liver disease patients with normal kidney function, blood L-FABP showed significantly

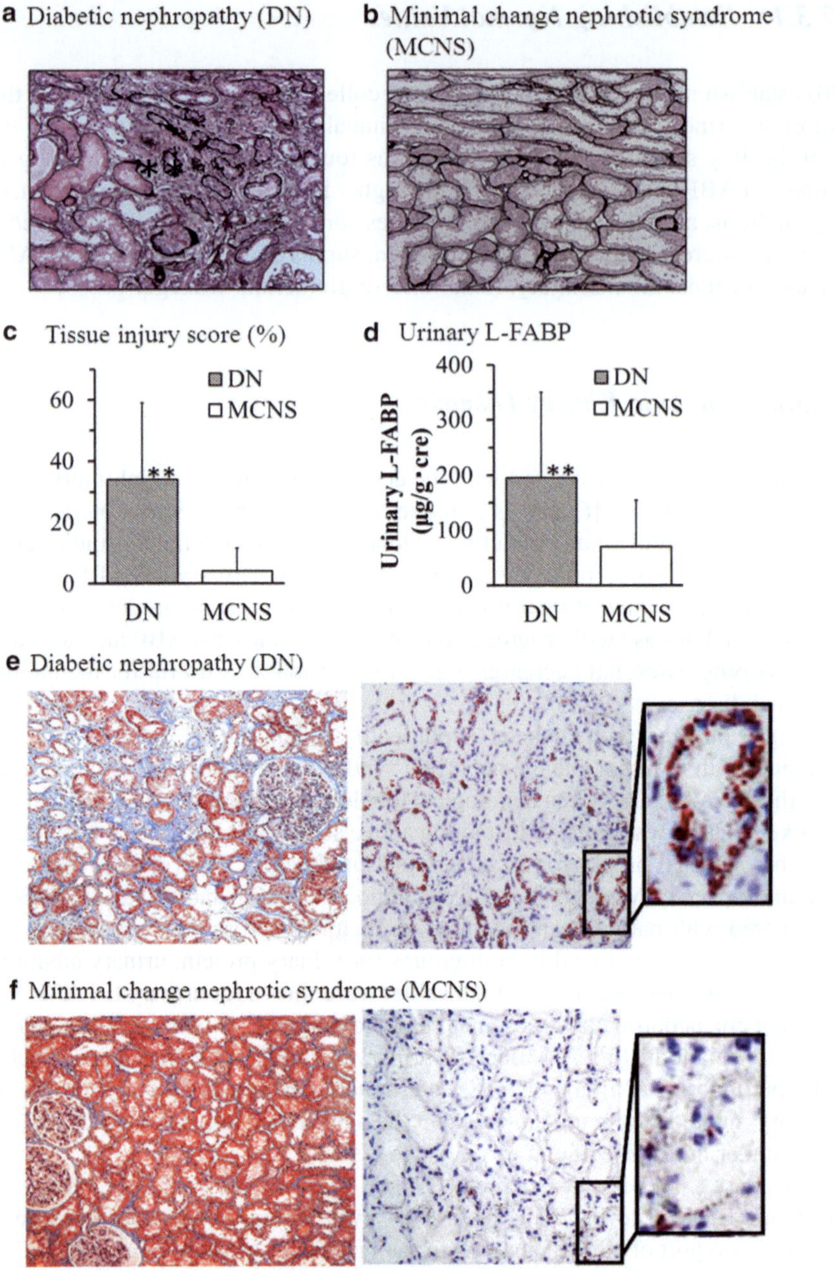

Fig. 7.3 Histological evaluation of renal tubulointerstitial lesions and urinary L-FABP. Pathological tissue images (×100) of (**a**) diabetic nephropathy (DN) and (**b**) minimal change nephrotic syndrome (MCNS). (**c**) Tubulointerstitial injury score (%) for both disease groups and (**d**) urinary L-FABP (mean ± standard deviation, $^{**}p<0.01$ vs. MCNS) (no significant differences seen in other urinary markers in both disease groups). In diabetic nephropathy, stronger tubulointerstitial injury compared to MCNS was found, and adipose accumulation in renal tubules was confirmed. (**e**) Diabetic nephropathy (DN). (**f**) Minimal change nephritic syndrome (MCNS). Masson trichrome stain (*left*). Fat stain (*right*)

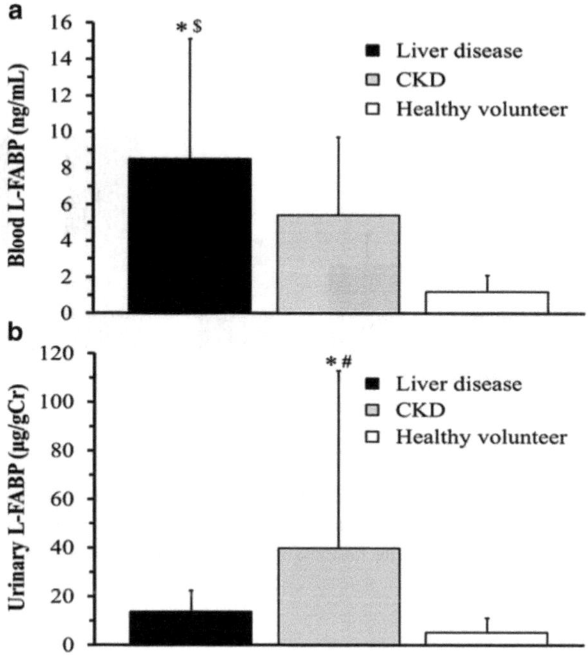

Fig. 7.4 Comparison of blood L-FABP and urinary L-FABP (kidney disease specificity of urinary L-FABP). Comparison of (**a**) blood L-FABP and (**b**) urinary L-FABP in liver disease patients (71 cases, *closed column*), CKD patients (73 cases, *dotted column*), and healthy subjects (71 cases, *open column*) (mean ± standard deviation. $^{*}p<0.05$ compared to healthy volunteers, $^{\$}p<0.05$ compared to patients with CKD, $^{\#}p<0.05$ compared to patients with liver disease

high values more than seven times greater than that of healthy subjects, but it was confirmed that no significant difference in urinary L-FABP compared to healthy subjects exists. Meanwhile, in CKD patients with normal liver function, urinary L-FABP showed significantly high values compared to both healthy subjects and liver disease patients, while blood L-FABP showed a nonsignificant tendency to increase compared with healthy subjects. This is thought to be caused by a clearance decrease in CKD patients. In the previous study, the implications have been that even in diseases that show high values for blood L-FABP, urinary L-FABP is not affected if kidney function is normal, with no differences according to gender or age. Urinary L-FABP responds sensitively to tubular injury by means of a mechanism that is different from urinary NAG and urinary α1-MG, and it might be a kidney disease-specific index that is excreted intrinsically from kidney tissue.

7.3.3 Diabetic Nephropathy (DN)

A number of investigators have studied urinary L-FABP in type 2 diabetes, with the following results [15–17]. L-FABP levels were found to be significantly associated with the stage of diabetic nephropathy [18]. Pioglitazone ameliorates both

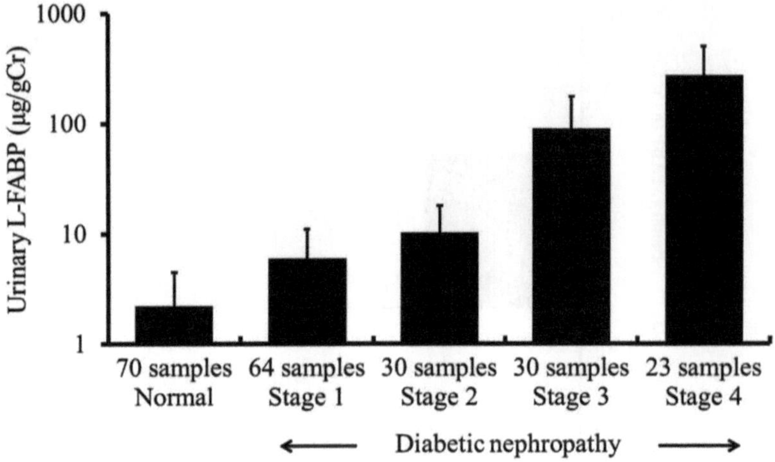

Fig. 7.5 Relationship between urinary L-FABP levels and progression of diabetic nephropathy. The level of urinary L-FABP increased significantly according to the severity of diabetic nephropathy. Note that urinary L-FABP in the patients with normoalbuminuria was significantly higher than in normal control subjects

glomerular and tubulointerstitial damage in type 2 diabetics, as reflected by the reduction in urinary albumin excretion (UAE) and urinary L-FABP, respectively [19].

Kamijo and colleagues measured urinary L-FABP levels of 147 diabetic nephropathy patients. At the same time, urinary L-FABP levels in healthy subjects were also measured. The urinary L-FABP levels in the diabetic nephropathy patients increase as the stages become aggravated, and in comparison with healthy subjects significantly high values are shown from early stages of nephropathy (Fig. 7.5). The result demonstrates urinary L-FABP is useful for early diagnosis of diabetic nephropathy [20]. With respect to diabetic nephropathy, it was confirmed after a cross-sectional analysis of type 1 diabetes at the Steno Diabetes Center of Denmark that urinary L-FABP significantly elevates at a stage earlier than microalbuminuria. Furthermore, in an inspection of the prognostic diagnosis performance of urinary L-FABP before nephropathy progression targeting patients who had undergone prognostic observation for roughly 30 years from disease onset, it became clear that, even in early-stage nephropathy, the patient group with high urinary L-FABP values had significantly high values during the 20+ year follow-up period for both microalbuminuria and risk of death. Also, in a 4-year prospective clinical trial targeting ambulatory patients with type 2 diabetes in Japan, it became clear that, of the patients who were CKD stage I or stage II, compared to microalbuminuria patients, positive patients for whom the normal upper limit was the cut-off for urinary L-FABP had a significantly high risk of diabetic nephropathy stage progression over the following 4 years. It has been found that urinary L-FABP is important as a risk factor in the progression of type 2 diabetes-induced nephropathy [20].

7.3.4 Non-diabetic Nephropathy (Non-DN)

Studies in non-diabetes have shown similar positive results for urinary L-FABP as both a diagnostic and prognostic marker of chronic renal disease.

While urinary protein is widely used to predict the risk of renal disease progression and the risk of dialysis, urinary L-FABP may in fact be a better prognostic marker [21]. Urinary L-FABP was more sensitive than urinary protein in predicting the progression of CKD. However, urinary protein showed greater specificity than did urinary L-FABP. Over time, the progression of CKD tended to correlate with changes in urinary L-FABP, but not urinary protein [5]. In a later report of the same study, the authors showed that serum L-FABP was not correlated with urinary L-FABP [14].

In autosomal dominant polycystic kidney disease (ADPKD), elevated urinary L-FABP is associated with ADPKD, and that candesartan mitigates this effect [22].

In focal glomerulosclerosis (FGS), L-FABP may be used as an aid to differentially diagnose patients with FGS, and as a prognostic marker for the likely success of drug treatment in those patients. The drug-resistant group was later treated with a course of low-density lipoprotein (LDL) apheresis, and urinary L-FABP re-measured. L-FABP was found to be significantly lower after LDL apheresis, perhaps due to the treatment ameliorating the tubulointerstitial lesions associated with FGS [23].

7.3.5 Acute Kidney Injury (AKI)

In cardiovascular postoperative patients, the patients were divided into an AKI group and non-AKI group according to Acute Kidney Injury Network (AKIN) criteria. The serum creatinine and urinary L-FABP were measured diachronically pre- and post-operation. In the conventional AKI diagnostic procedure that assumes serum creatinine as an indicator, the onset of AKI was established 24 h post-operation. On the other hand, urinary L-FABP showed a pronounced rise in the AKI group immediately post-operation. It is clear that urinary L-FABP can predict earlier onset of AKI (Fig. 7.6) [24]. Similarly, in cardiac surgery using cardiopulmonary bypass (CPB) for the correction or palliation of congenital cardiac defects, urinary L-FABP levels represent a sensitive and predictive early biomarker of AKI after cardiac surgery [25]. In another report, urinary L-FABP was significantly higher in AKI patients [26, 27] and urinary L-FABP was an independent predictor for 90-day mortality [28].

7.3.6 Contrast Medium-Induced Nephropathy

Manabe and colleagues performed a prospective study of 220 consecutive patients with CKD who underwent elective catheterization. L-FABP levels were measured immediately before and 1 and 2 days after the procedure. Urinary L-FABP levels

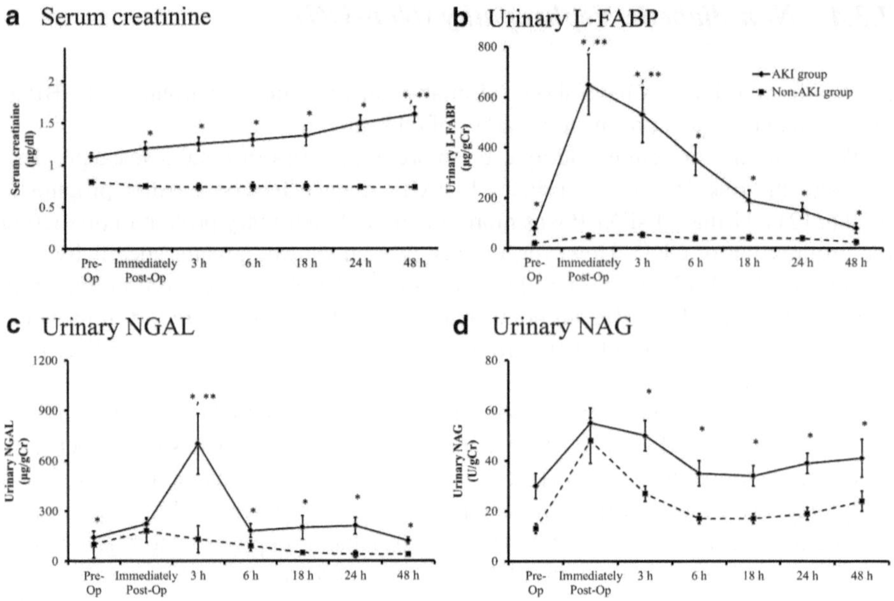

Fig. 7.6 Urinary L-FABP is useful as a biomarker that predicts the onset of acute kidney injury. Changes in various urinary parameters in the acute kidney injury (*diamond*, *solid line*) and non-AKI groups (*square*, *broken line*). (**a**) Serum creatinine. (**b**) Urinary L-FABP. (**c**) Urinary NGAL. (**d**) Urinary NAG $*p<0.05$, compared with the non-AKI group at the same time point; $**p<0.05$ compared with the respective preoperative level in the same group

were significantly higher in patients with contrast-induced acute kidney injury (CI-AKI) than those without CI-AKI before contrast medium exposure. Thus several studies have suggested that urinary L-FABP is useful for predicting the onset of CI-AKI before contrast medium exposure [29, 30].

7.3.7 Septic Shock

The patients with severe sepsis may be treated with hemoperfusion through a polymyxin B-immobilized fiber (PMX-F) adsorbent column. Numerous reports show this to be safe and effective [31–33]. Nakamura and colleagues measured urinary L-FABP in 40 patients with septic shock, 20 acute renal failure (ARF) patients without septic shock, and 30 healthy volunteers. Urinary L-FABP levels in ARF patients without septic shock (120 μg/g creatinine) were significantly lower than those in both surviving and non-surviving septic shock patients. Among the survivors, PMX-F treatment significantly reduced urinary L-FABP levels from 1,420 to 240 μg/g creatinine. However, PMX-F treatment had little effect on urinary L-FABP in the non-survivors (2,880 vs. 2,460 μg/g creatinine). The authors suggested that urinary L-FABP levels might be used to reflect the clinical prognosis of patients with septic shock, and also to monitor the effectiveness of treatment [34, 35].

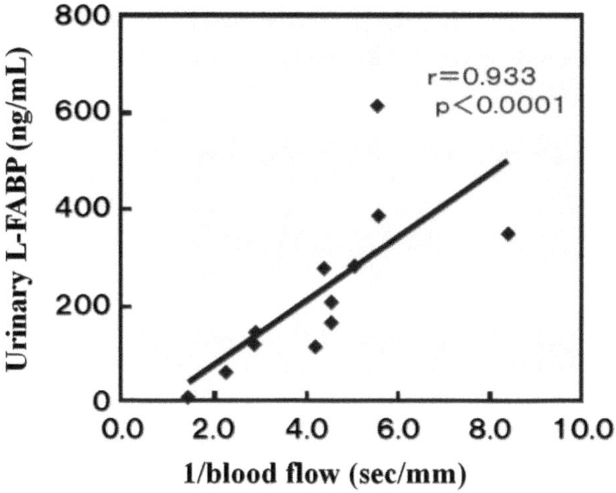

Fig. 7.7 Correlation between peritubular capillary blood flow and urinary L-FABP. Urinary L-FABP is compared with 1/blood flow. Where renal ischemia is more severe, 1/blood flow is larger. Urinary L-FABP is correlated with the decrease of peritubular blood flow

7.3.8 Microcirculation Disorder Caused by Transplant

Human *fabp* gene expression has been confirmed mainly in the liver, kidneys, and large intestine, and in those transcriptional regulation regions, coupling regions for transcription factors relating to ischemia and fat metabolism such as hypoxia-inducible factor 1 (HIF-1), hepatocyte nuclear factors (HNF-4, HNF-1), and PPAR exist. Therefore L-FABP can be inferred to correlate with ischemia. Yamamoto and colleagues studied the peritubular capillary blood flow immediately after kidney reperfusion during a human live kidney transplant was measured using noninvasive charge coupled devices (CCD) video recording, and compared with L-FABP value in simultaneously collected initial urine. Urinary L-FABP had a very high correlation ($r=0.933$, $p<0.0001$, Fig. 7.7), with tubular circumference blood flow, and it became clear that renal tubule disorder could be subtly detected using a mechanism different from existing urinary analytic indicators (NAG, α1-MG, β2-MG). It has been proven that in a normal human kidney, L-FABP located in the cytoplasm of the proximal tubule is swiftly excreted into the lumen in response to ischemia and oxidative stress (Fig. 7.8) [36].

7.3.9 Pediatric Acute Kidney Injury

Pediatric acute kidney injury (pAKI) is a heterogeneous disorder with various clinical presentations and an unpredictable outcome. In industrialized countries, the incidence of AKI in hospitalized children is rising, and the etiology of pAKI is dramatically changing from isolated acute renal disease to multiple organ failure.

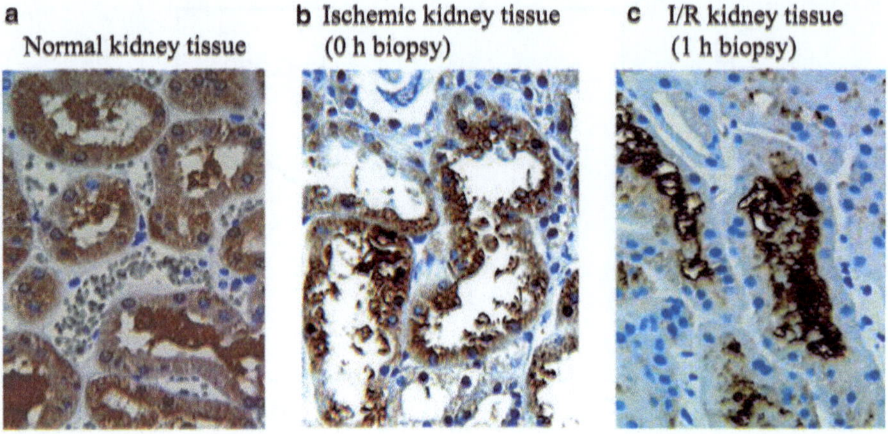

Fig. 7.8 Immunohistochemical distribution of L-FABP. L-FABP was stained predominantly in the cytoplasmic region of the proximal tubule in intact human kidney (normal kidney tissue) (**a**), ischemic kidney tissue (**b**) and ischemia reperfusion (I/R) kidney tissue (**c**)

AKI is associated with prolonged hospitalization, reduced quality of life, increased risk of progressive kidney disease, and increased mortality rate.

Tsukahara and colleagues measured urinary L-FABP in 40 preterm neonates born with gestational ages of between 23 and 35 weeks. They found that urinary L-FABP levels were approximately ten times higher than adult values during the 30-day period after birth, and did not drop significantly during that period. The levels were negatively correlated with both gestational age at birth and birth weight. Urinary L-FABP was positively correlated with NAG (a marker for proximal tubule cell injury) and 8-hydroxydeoxyguanosine (8-OHdG; a marker for systemic oxidative stress) levels. Proximal tubular damage and oxidative stress may have a causative role in the high L-FABP levels observed [37].

Ivanisevic and colleagues report that urinary L-FABP can be used for early identification of AKI after pediatric cardiac surgery [38]. Twenty-seven children (median age 360 days) without preexisting CKD or other major comorbidities undergoing CPB procedures were enrolled. AKI was defined as a 50 % increase in serum creatinine concentrations within 48 h after surgery. AKI developed in 11 patients (41 % of the cohort), three needed renal replacement therapy (peritoneal dialysis); there were two deaths. There were significant differences between patients with and without AKI in L-FABP levels at 2, 6, 24, and 48 h after surgery, length of hospital stay, and CPB time; there were no differences in gender, patient age, and body weight. Patients with transient AKI (lasting $<$ 24 h) were excluded. Urinary L-FABP excretion was significantly higher in the AKI group than in the non-AKI group at all time points. Peak urinary L-FABP was measured at 2 h. The area under the curve (AUC)–receiver operating characteristic (ROC) were 0.867 for the 2 h and 0.867 for the 6 h postoperatively. Correlation coefficient between L-FABP and length of hospitalization after surgery was statistically significant. Finally, they conclude that urinary

L-FABP can be used to diagnose AKI earlier than rise in serum creatinine in children undergoing CPB.

However, Schiffl and Lang proposed the unsolved problems of biomarkers in pediatric patients that most pediatric studies come from single-center studies and homogeneous patient populations (post-surgery AKI) [39]. Further work is needed before this or other novel biomarkers (alone or in combination) can be implemented in clinical practice. Large-scale observational studies are needed to test these biomarkers against hard clinical endpoints, independent of serial measurements of serum creatinine concentration. Prospective randomized interventional trials using exclusively high biomarker levels to define AKI should demonstrate improved clinical outcomes.

7.4 Current Development Stage

In, 2012, the Clinical Practice Guideline for Acute Kidney Injury was published by the Kidney Disease Improving Global Outcomes (KDIGO) foundation, which was established with the aim of developing international guidelines for AKI. In this guideline, a necessity of biomarkers for the diagnosis of AKI earlier has also been pointed out. Five items are introduced as the so-called five biomarkers of AKI in the above-mentioned KDIGO guidelines; neutrophil gelatinase-associated lipocalin (NGAL), cystatin C, interleukin-18 (IL-18), kidney injury molecule-1 (KIM-1), and L-FABP [40]. For example, there is a study to evaluate the use of new AKI biomarkers including urinary L-FABP, NGAL, and IL-18 with a heterogeneous cohort in a mixed medical–surgical intensive care unit (ICU). In the prediction of mortality rate within 2 weeks of admission, the AUC-ROC values of urinary L-FABP, NGAL, and acute physiology and chronic health evaluation (APACHE) II scores were significantly higher than those of NAG and albumin (Table 7.2) [41]. Based on the above evidence, the usefulness of urinary L-FABP has also been approved for reimbursement for in vitro diagnostics (IVD) from the Japanese Ministry of Health, Labour and Welfare.

Table 7.2 The AUC-ROC values of AKI biomarkers and 14-day mortality

	AUC-ROC (95 % confidence interval)
L-FABP	0.896 (0.835–0.937)
NGAL	0.827 (0.688–0.912)
IL-18	0.826 (0.679–0.914)
NAG	0.664 (0.495–0.800)
Albumin	0.717 (0.606–0.806)
Serum creatinine	0.733 (0.614–0.826)
APACHE II	0.897 (0.761–0.960)

AUC area under the curve, *ROC* receiver operating characteristic

7.5 Conclusions

The earlier diagnosis of kidney disease and the surrogate markers for therapeutic are all needed in the clinical laboratory. Primary care physicians and nephrologists need better tools to identify patients at risk for kidney disease and to monitor therapeutic effect. While current tests are based on estimated renal function, there is an unmet medical need for biomarkers that reflect directly kidney health. L-FABP is a clinically useful biomarker easily measured in urine. As a description above, urinary L-FABP has been shown to correlate with renal disease state, diagnosis, prognosis, and therapeutic effect in the studies of both CKD and AKI. Similarly, urinary L-FABP can be used to diagnose AKI earlier than rise in serum creatinine in children undergoing CPB. Prospective randomized interventional trials using urinary L-FABP levels to define AKI should demonstrate improved clinical outcomes. Further research to elucidate the usefulness of L-FABP must be performed in experimental and clinical settings.

Acknowledgments The authors thank Dr. Kenjiro Kimura and Dr. Atsuko Kamijo-Ikemori (Department of Nephrology and Hypertension, Internal Medicine, St. Marianna University School of Medicine, Kanagawa, Japan), Dr. Yasuhiko Tomino and Dr. Yusuke Suzuki (Division of Nephrology, Department of Internal Medicine, Juntendo University Faculty of Medicine, Tokyo, Japan) for scientific encouragement and helpful discussions.

References

1. Hewitt MH, Dear J, Star RA (2004) Discovery of protein biomarkers for renal diseases. J Am Soc Nephrol 15:1677–1689
2. Veerkamp JH, Peeters RA, Maatman RG (1991) Structural and functional features of different types of cytoplasmic fatty acid binding proteins. Biochim Biophys Acta 1081:1–24
3. Chmurznska A (2006) The multigene family of fatty acid-binding proteins (FABPs): function, structure and polymorphism. J Appl Genet 47:39–48
4. Nakamura K, Ito K, Kato U, Sugaya T, Kubo Y, Tsuji A (2008) L-type fatty acid binding protein transgenic mouse as a novel tool to explore cytotoxicity to renal proximal tubules. Drug Metab Pharmacokinet 23:271–278
5. Kamijo A, Sugaya T, Hikawa A, Yamanouchi M, Hirata Y, Ishimitsu T, Numabe A, Takagi M, Hayakawa H, Tabei F, Sugimoto T, Mise N, Kimura K (2005) Clinical evaluation of urinary excretion of liver-type fatty acid-binding protein as a marker for the monitoring of chronic kidney disease: a multicenter trial. J Lab Clin Med 145:125–133
6. Kamijo A, Sugaya T, Hikawa A, Okada M, Okumura F, Yamanouchi M, Honda A, Okabe M, Fujino T, Hirata Y, Omata M, Kaneko R, Fujii H, Fukamizu A, Kimura K (2004) Urinary excretion of fatty acid-binding protein reflects stress overload on the proximal tubules. Am J Pathol 165:1243–1255
7. Kamijo A, Kimura K, Sugaya T, Yamanouchi M, Hase K, Kaneko T, Hirata Y, Goto A, Fujita T, Omata M (2002) Urinary free fatty acids bound to albumin aggravate tubulointerstitial damage. Kidney Int 62:1628–1637
8. Arici M, Chana R, Lewington A, Brown J, Brunskill NK (2003) Stimulation of proximal tubular cell apoptosis by albumin-bound fatty acids mediated by peroxisome proliferators activated receptor gamma. J Am Soc Nephrol 14:17–27

9. Thomas ME, Schreiner GF (1993) Contribution of proteinuria to progressive renal injury: consequences of tubular uptake of fatty acid bearing albumin. Am J Nephrol 13:385–389
10. Thomas ME, Morrison AR, Schreiner GF (1995) Metabolic effects of fatty acid bearing albumin on a proximal tubule cell line. Am J Physiol 268:F1177–F1184
11. Nakamura T, Sugaya T, Kawagoe Y, Ueda Y, Osada S, Koide H (2005) Effect of pitavastatin on urinary liver-type fatty acid-protein levels in patients with early diabetic nephropathy. Diabetes Care 28:2728–2732
12. Ishimitsu T, Ohta S, Saito M, Teranishi M, Inada H, Yoshii M, Minami J, Ono H, Hikawa A, Shibata N, Sugaya T, Kamijo A, Kimura K, Ohrui M, Matsuoka H (2005) Urinary excretion of liver fatty acid-binding protein in health-check participants. Clin Exp Nephrol 9:34–39
13. Sasaki H, Kamijo-Ikemori A, Sugaya T, Yamashita K, Yokoyama T, Koike J, Sato T, Yasuda T, Kimura K (2009) Urinary fatty acids and liver-type fatty acid binding protein in diabetic nephropathy. Nephron Clin Pract 112:c148–c156
14. Kamijo A, Sugaya T, Hikawa A, Yamanouchi M, Hirata Y, Ishimitsu T, Numabe A, Hayakawa H, Takagi M, Tabei F, Sugimoto T, Mise N, Omata M, Kimura K (2006) Urinary liver-type fatty acid binding protein as a useful biomarker in chronic kidney disease. Mol Cell Biochem 284:175–182
15. Nakamura T, Fujiwara N, Sugaya T, Ueda Y, Koide H (2009) Effect of red wine on urinary protein, 8-hydroxydeoxyguanosine, and liver-type fatty acid-binding protein excretion in patients with diabetic nephropathy. Metabolism 58:1185–1190
16. von Eynatten M, Baumann M, Heemann U, Zdunek D, Hess G, Nawroth PP, Bierhaus A, Humpert PM (2010) Urinary L-FABP and anaemia: distinct roles of urinary markers in type 2 diabetes. Eur J Clin Invest 40:95–102
17. Araki SI, Haneda M, Koya D, Sugaya T, Isshiki K, Kume S, Kashiwagi A, Uzu T, Maegawa H (2013) Predictive effects of urinary liver-type fatty acid-binding protein for deteriorating renal function and incidence of cardiovascular disease in type 2 diabetic patients without advanced nephropathy. Diabetes Care 36:1248–1253
18. Suzuki K, Murata H, Babazono T, Iwamoto Y (2005) Clinical significance of urinary liver-type fatty acid-binding protein in patients with diabetic nephropathy. Diabetes Care 28:2038–2039
19. Nakamura T, Sugaya T, Kawagoe Y, Ueda Y, Osada S, Koide H (2006) Effect of pioglitazone on urinary liver-type fatty acid-binding protein concentrations in diabetes patients with microalbuminuria. Diabetes Metab Res Rev 22:385–389
20. Kamijo-Ikemori A, Sugaya T, Yasuda T, Kawata T, Ota A, Tatsunami S, Kaise R, Ishimitsu T, Tanaka Y, Kimura K (2011) Clinical significance of urinary liver-type fatty acid-binding protein in diabetic nephropathy of type 2 diabetic patients. Diabetes Care 34:691–696
21. Kamijo A, Kimura K, Sugaya T, Yamanouchi M, Hikawa A, Hirano N, Hirata Y, Goto A, Omata M (2004) Urinary fatty acid-binding protein as a new clinical marker of the progression of chronic renal disease. J Lab Clin Med 143:23–30
22. Nakamura T, Sugaya T, Kawagoe Y, Ueda Y, Osada S, Koide H (2005) Candesartan reduces urinary fatty acid-binding protein excretion in patients with autosomal dominant polycystic kidney disease. Am J Med Sci 330:161–165
23. Nakamura T, Sugaya T, Kawagoe Y, Ueda Y, Osada S, Koide H (2006) Urinary liver-type fatty acid-binding protein levels for differential diagnosis of idiopathic focal glomerulosclerosis and minor glomerular abnormalities and effect of low density lipoprotein apheresis. Clin Nephrol 65:1–6
24. Matsui K, Kamijo-Ikemori A, Sugaya T, Yasuda T, Kimura K (2011) Usefulness of urinary biomarkers in early detection of acute kidney injury after cardiac surgery in adults. Circ J 76:213–220
25. Portilla D, Dent C, Sugaya T, Nagothu KK, Kundi I, Moore P, Noiri E, Devarajan P (2008) Liver fatty acid-binding protein as a biomarker of acute kidney injury after cardiac surgery. Kidney Int 73:465–472
26. Matsui K, Kamijo-Ikemori A, Hara M, Sugaya T, Kodama T, Fujitani S, Taira Y, Yasuda T, Kimura K (2011) Clinical significance of tubular and podocyte biomarkers in acute kidney injury. Clin Exp Nephrol 15:220–225

27. Doi K, Katagiri D, Negishi K, Hasegawa S, Hamasaki Y, Fujita T, Matsubara T, Ishii T, Yahagi N, Sugaya T, Noiri E (2012) Mild elevation of urinary biomarkers in prerenal acute kidney injury. Kidney Int 82:1114–1120
28. Cho E, Yang HN, Jo SK, Cho WY, Kim HK (2013) The role of urinary liver-type fatty acid-binding protein in critically ill patients. J Korean Med Sci 28:100–105
29. Nakamura T, Sugaya T, Node K, Ueda Y, Koide H (2006) Urinary excretion of liver-type fatty acid-binding protein in contrast medium-induced nephropathy. Am J Kidney Dis 47:439–444
30. Manabe K, Kamihata H, Motohiro M, Senoo T, Yoshida S, Iwasaka T (2012) Urinary liver-type fatty acid-binding protein level as a predictive biomarker of contrast-induced acute kidney injury. Eur J Clin Invest 42:557–563
31. Uriu K, Osajima A, Hiroshige K, Watanabe H, Aibara K, Inada Y, Segawa K, Anai H, Takagi I, Ito A, Kamuchi M, Kaizu K (2002) Endotoxin removal by direct hemoperfusion with an adsorbent column using polymyxin B-immobilized fiber ameliorates systemic circulatory disturbance in patients with septic shock. Am J Kidney Dis 39:937–947
32. Shoji H (2003) Extracorporeal endotoxin removal for the treatment of sepsis: endotoxin adsorption cartridge (Toraymyxin). Ther Apher Dial 7:108–114
33. Ono S, Tsujimoto H, Matsumoto A, Ikuta S, Kinoshita M, Mochizuki H (2004) Modulation of human leukocyte antigen DR on monocytes and CD 16 on granulocytes in patients with septic shock using hemoperfusion with polymyxin B-immobilized fiber. Am J Surg 188:150–156
34. Nakamura T, Sugaya T, Koide H (2009) Urinary liver-type fatty acid-binding protein in septic shock: effect of polymyxin B-immobilized fiber hemoperfusion. Shock 31:454–459
35. Doi K, Noiri E, Sugaya T (2010) Urinary L-type fatty acid-binding protein as a new renal biomarker in critical care. Curr Opin Crit Care 16:545–549
36. Yamamoto T, Noiri E, Ono Y, Doi K, Negishi K, Kamijo A, Kimura K, Fujita T, Kinukawa T, Taniguchi H, Nakamura K, Goto M, Shinozaki N, Ohshima S, Sugaya T (2007) Renal L-type fatty acid-binding protein in acute ischemic injury. J Am Soc Nephrol 18:2894–2902
37. Tsukahara H, Sugaya T, Hayakawa K, Mori Y, Hiraoka M, Hata A, Mayumi M (2005) Quantification of L-type fatty acid binding protein in the urine of preterm neonates. Early Hum Dev 81:643–646
38. Ivanisevic I, Peco-Antric A, Vullcevic I, Hercog D, Milovanovic V, Kotur-Stevuljevic J, Stefanovic A, Kocev N (2013) L-FABP can be an early marker of acute kidney injury in children. Pediatr Nephrol 28:963–969
39. Schiffl H, Lang SM (2013) Urinary biomarkers and acute kidney injury in children: the long road to clinical application. Pediatr Nephrol 28:837–842
40. Moore E, Bellomo R, Nichol A (2010) Biomarkers of acute kidney injury in anesthesia, intensive care and major surgery: from the bench to clinical research to clinical practice. Minerva Anestesiol 76:425–440
41. Doi K, Negishi K, Ishizu T, Katagiri D, Fujita T, Matsubara T, Yahagi N, Sugaya T, Noiri E (2011) Evaluation of new acute kidney injury biomarkers in a mixed intensive care unit. Crit Care Med 39:2464–2469
42. Nakamura T, Sugaya T, Koide H (2006) Cigarette smoking affects urinary liver-type fatty acid-binding protein concentration in patients with early diabetic nephropathy. Diabetes Care 29:1717
43. Nakamura T, Sugaya T, Koide H (2007) Angiotensin II receptor antagonist reduces urinary liver-type fatty acid-binding protein levels in patients with diabetic nephropathy and chronic renal failure. Diabetologia 50:490–492
44. Nakamura T, Inoue T, Fujiwara N, Kawagoe Y, Sugaya T, Ueda Y, Koide H, Node K (2008) Additional renoprotective effects of azelinidipine combined with angiotensin receptor blockers in patients with diabetic nephropathy. Clin Nephrol 70:385–392
45. Nakamura T, Fujiwara N, Sato E, Ueda Y, Sugaya T, Koide H (2010) Renoprotective effects of various angiotensin II receptor blockers in patients with early-stage diabetic nephropathy. Kidney Blood Press Res 33:213–220

46. Nielsen SE, Sugaya T, Tarnow L, Lajer M, Schjoedt KJ, Astrup AS, Baba T, Parving HH, Rossing P (2009) Tubular and glomerular injury in diabetes and the impact of ACE inhibition. Diabetes Care 32:1684–1688
47. Nielsen SE, Sugaya T, Hovind P, Baba T, Parving HH, Rossing P (2010) Urinary liver-type fatty acid-binding protein predicts progression to nephropathy in type 1 diabetic patients. Diabetes Care 33:1320–1324
48. Panduru NM, Forsblom C, Saraheimo M, Thorn L, Bierhaus A, Humpert PM, Groop PH, on behalf of the FinnDiane Study Group (2013) Urinary liver-type fatty acid-binding protein and progression of diabetic nephropathy in type 1 diabetes. Diabetes Care 36(7):2077–2083
49. Nakamura T, Sugaya T, Ebihara I, Koide H (2005) Urinary liver-type fatty acid-binding protein: discrimination between IgA nephropathy and thin basement membrane nephropathy. Am J Nephrol 25:447–450
50. Nakamura T, Inoue T, Sugaya T, Kawagoe Y, Suzuki T, Ueda Y, Koide H, Node K (2007) Beneficial effects of olmesartan and temocapril on urinary liver-type fatty acid-binding protein levels in normotensive patients with immunoglobin A nephropathy. Am J Hypertens 20:1195–1201
51. Nakamura T, Sugaya T, Kawagoe Y, Suzuki T, Inoue T, Node K (2006) Effect of pitavastatin on urinary liver-type fatty-acid-binding protein in patients with nondiabetic mild chronic kidney disease. Am J Nephrol 26:82–86
52. Nakamura T, Sugaya T, Kawagoe Y, Suzuki T, Ueda Y, Koide H (2006) Effect of erythropoietin on urinary liver-type fatty-acid-binding protein in patients with chronic renal failure and anemia. Am J Nephrol 26:276–280
53. Nakamura T, Sugaya T, Kawagoe Y, Suzuki T, Ueda Y, Koide H, Inoue T, Node K (2007) Azelnidipine reduces urinary protein excretion and urinary liver-type fatty acid binding protein in patients with hypertensive chronic kidney disease. Am J Med Sci 333:321–326
54. Nakamura T, Inoue T, Sugaya T, Kawagoe Y, Suzuki T, Ueda Y, Node K (2008) Renoprotective effect of telmisartan in patients with chronic kidney disease. Clin Exp Hypertens 30:662–672
55. Nakamura T, Sato E, Fujiwara N, Kawagoe Y, Ueda Y, Sugaya T, Yamagishi S, Yamada S, Koide H (2010) Comparative effects of benidipine and amlodipine on proteinuria, urinary 8-OHdG, urinary L-FABP, and inflammatory and atherosclerosis markers in early-stage chronic kidney disease. Am J Med Sci 339:157–163
56. Nakamura T, Fujiwara N, Kawagoe Y, Sugaya T, Ueda Y, Koide H (2010) Effects of telmisartan and enalapril on renoprotection in patients with mild to moderate chronic kidney disease. Eur J Clin Invest 40:790–796

Chapter 8
Rapid Diagnostic Tests for Oxidative Stress Status

Kazunari Kaneko

Abbreviations

8-OHdG	8-hydroxy-2′-deoxyguanosine
8-oxodG	8-oxo-7,8-dihydro-2′-deoxyguanosine
BAP	Biological antioxidant potential
d-ROMs	Derivatives of reactive oxygen metabolites
ELISA	Enzyme-linked immunosorbent assay
FRAS4	Free radical analytical system 4
INS	Idiopathic nephrotic syndrome
IVIG	Intravenous immunoglobulin therapy
KD	Kawasaki disease
MELAS	Myopathy encephalopathy, lactic acidosis, and stroke-like episodes
OS	Oxidative stress
OSI	Oxidative stress index
POCT	Point-of-care testing
ROS	Reactive oxygen species

8.1 Introduction

Oxidative stress (OS) is a disturbance in the balance of reactive oxygen species (ROS) and antioxidants that leads to damage of lipids, proteins, and nucleic acids. This disturbance can result from either low levels of antioxidants or increased production of ROS. In general, ROS are generated as byproducts of cellular

K. Kaneko, M.D., Ph.D. (✉)
Department of Pediatrics, Kansai Medical University,
2-5-1 Shin-machi, Hirakata-shi, Osaka 573 1010, Japan
e-mail: kanekok@hirakata.kmu.ac.jp

H. Tsukahara and K. Kaneko (eds.), *Studies on Pediatric Disorders*, Oxidative Stress in Applied Basic Research and Clinical Practice, DOI 10.1007/978-1-4939-0679-6_8,

metabolism and ionizing radiation, and usually consist of the following four species: superoxide anion (O_2^-), hydrogen peroxide (H_2O_2), hydroxyl radical (OH), and singlet oxygen ($^1O_2$) [1]. All types of biomolecules can be damaged by ROS, and OS have now been shown to be associated with diverse human diseases or pathological conditions, such as aging, atherosclerosis, hypertension, renal failure, immune alterations, neurodegeneration, reperfusion injury, radiation damage, carcinogenesis, and many other degenerative disorders [2]. Therefore, information on the OS profile is useful in allowing physicians to treat their patient's diseases by ameliorating long-term OS.

By the end of 2012, approximately 10,000 scientific papers had been reported under the general topic of OS biomarkers based on a search of the PubMed database. In these papers, a wide range of markers were used to assess oxidative damage to nucleic acids, proteins, and lipids, as well as the activities of antioxidants and antioxidant status.

The object of the present work is to show the importance of assessing OS in children and to provide an overview of the recent advances in rapid diagnostic testing of OS using automatic analyzers.

8.2 Association of OS with Pediatric Diseases

Table 8.1 presents a list of pediatric diseases that have been reported to be associated with OS [1] (the corresponding references can be readily obtained on Medline or PubMed of the National Library of Medicine Gateway website http://www.nlm.nih.gov).

As can be seen, the pathogenesis and progression of numerous diseases of children and adolescents appear to be linked to ROS-induced OS. OS might also contribute to tissue damage induced by certain drugs (e.g., analgesics, anticancer agents, and immunosuppressive drugs). Oxidative tissue injury by pathological conditions might have more serious consequences in young people than in older people because of the need for subsequent tissue growth to match somatic growth and because life span is longer in children than in adults. Primary and secondary alleviation of OS might therefore be important in children. In fact, the use of antioxidants has led to new therapeutic approaches for diseases that involve enhanced OS [3–7].

8.3 Biomarkers for OS

Biomarkers of OS can be divided into two categories, i.e., products induced by oxidative damage and antioxidants, as shown in Table 8.2 [1]. In most previous studies, OS biomarkers were measured in blood (i.e., serum, plasma, erythrocytes, granulocytes, or lymphocytes) or urine samples. However, several studies have measured OS biomarkers using other body fluids (e.g., cerebrospinal fluid, bronchoalveolar

Table 8.1 Diseases and conditions associated with oxidative stress in children

Allergic/immunologic: Atopic dermatitis, bronchial asthma, chronic arthritis, Henoch-Schönlein purpura, Kawasaki disease, systemic lupus erythematosus, vasculitis syndrome
Cardiovascular: Cardiac surgery, cardiopulmonary bypass, essential hypertension
Endocrinologic/metabolic: Diabetes mellitus, glutathione synthetase deficiency, hyperthyroidism, iodine-deficient goiter, mitochondrial disorder, phenylketonuria, X-linked adrenoleukodystrophy
Environmental/toxicologic: Carcinogenic metal (chromium, arsenic) exposure, ozone exposure, passive smoking, urban residence
Gastrointestinal/hepatologic: Autoimmune hepatitis, chronic constipation, inflammatory bowel disease, nonalcoholic fatty liver, viral hepatitis, Wilson disease
Genetic: Cockayne syndrome, Down syndrome, Zellweger syndrome
Hematologic: Acute leukemia, β-thalassemia, erythropoietic protoporphyria, Fanconi anemia, sickle cell anemia
Infectious: Acute bronchiolitis, infectious mononucleosis, acute otitis media, acute tonsillitis, adenovirus infection, chronic otitis media, chronic tonsillitis, cutaneous leishmanias, HIV infection, measles encephalitis, meningitis, septic shock
Neonatal: Asphyxia, maternal preeclampsia, respiratory distress syndrome, premature birth, retinopathy
Neurologic/muscular: Ataxia telangiectasia, attention deficit hyperactivity disorder, autism spectrum disorders, cerebral organic acid disorder, cerebral palsy, congenital muscular dystrophy, epilepsy, Friedreich ataxia, inflammatory myopathy, selenium-deficient skeletal muscle disorder, spinal muscular atrophy
Nutritional: Hyperlipidemia, kwashiorkor, obesity
Pharmacologic/therapeutic: Analgesics, anticancer drugs, immunosuppressive drugs, total body irradiation
Renal: Glomerulonephritis, nephrotic syndrome, renal failure, urinary tract infection

Reproduced from [1] with permission from © 2007 Bentham Science Publishers Ltd

Table 8.2 Biomarkers for oxidative stress

I. Formation of modified molecules by reactive oxygen species	
Lipid peroxidation	Malondialdehyde-lysine, 4-hydroxy-2-nonenal-lysine, acrolein-lysine, F2-isoprostane
Oxidative DNA damage	8-oxo-7,8-dihydro-2′-deoxyguanosine (8-oxodG), also known as 8-hydroxy-2′-deoxyguanosine (8-OHdG)
Glyco-oxidation	Carboxymethyl-lysine, pentosidine, argpyrimidine, methylglyoxal
Nitro-oxidation	Nitrotyrosine, nitrite/nitrate
Others	*o*,*o*′-Dityrosine, orthotyrosine, bilirubin oxidative metabolites, dehydroascorbate, oxidized glutathione, thiobarbituric acid reactive substances
II. Antioxidant enzymes and molecules	
Enzymes	Superoxide dismutase, catalase, glutathione peroxidase, glutathione reductase, glutathione-*S*-transferase, thioredoxin reductase, heme oxygenase
Proteins	Albumin, ferritin, transferrin, lactoferrin, ceruloplasmin, thioredoxin, L-type fatty acid binding protein
Low molecular weight molecules	Bilirubin, tocopherols, carotenoids, ubiquinol/ubiquinone, ascorbate, glutathione, cysteine, urate, nitrite/nitrate, selenium
Others	Total antioxidant reactivity, total radical trapping antioxidant potential

Reproduced from [1] with permission from © 2007 Bentham Science Publishers Ltd
Only biomarkers that are determined in samples of blood or urine are shown

lavage, joint fluid, nasal lavage fluid, middle-ear fluid, tissues, or exhaled breath), either alone or in combination with samples of blood or urine [1, 2]. As the half-lives of ROS are generally short, it is difficult to measure the amount of ROS directly. Therefore, an assay for the formation of molecules modified by ROS is a widely accepted measure for assessment of oxidative injury. ROS-induced damages affect a diverse array of molecules and include their scission, cross-linking, or covalent modification. The amount of these molecules increases when ROS are generated. Major targets of ROS in the molecular components of the cells are membrane lipids, proteins, nucleic acids, and carbohydrates. These markers are often measurable using stable adducts that are produced *in vivo* as a result of oxidative processes. Clinically applicable biomarkers include malondialdehyde-lysine, 4-hydroxy-2-nonenal-lysine, acrolein-lysine, F2-isoprostane (markers of lipid peroxidation), 8-oxo-7,8-dihydro-2′-deoxyguanosine (8-oxodG, also known as 8-hydroxy-2′-deoxyguanosine [8-OHdG], a marker of oxidative DNA damage), carboxymethyl-lysine, pentosidine (markers of glycol-oxidation), nitrotyrosine, nitrite/nitrate (markers of nitro-oxidation), and bilirubin oxidative metabolites (a marker of heme oxygenase activity) [1, 2].

Another group of biomarkers for OS consists of antioxidant enzymes and molecules that are associated with ROS metabolism. In most cases, these molecules are destroyed or modified and exhibit decreased activity or quantity after exposure to ROS. Conversely, they often show an overshooting response for a matter of hours, days, or weeks after ROS exposure. A list of the wide variety of these antioxidant enzymes, proteins, and low-molecular-weight molecules is shown in Table 8.2.

In previous studies, OS biomarkers have been measured using analytic chemistry techniques, including high performance liquid chromatography and gas chromatography–mass spectrometry. Recently, specific monoclonal antibodies have been developed and enzyme-linked immunosorbent assay (ELISA) systems constructed for various OS biomarkers [8]. Although ELISA allows us to skip the steps of pretreatments and does not require expensive equipment, it is still too time consuming for clinical use. Therefore, to enhance its clinical value, there is need of an easy, rapid, sensitive, and specific assay, i.e., point-of-care testing (POCT) to assess OS levels.

8.4 Rapid Diagnostic Tests for OS

Recently, several automatic analyzers for the rapid diagnosis of OS have become available. These include the Free Radical Analytical System 4 (FRAS4; Diacron International, Grosseto, Italy) and ICR-001 (Techno Medica Co., Ltd. Yokohama, Japan). Because these systems are easy to use and provide rapid results, there are great hopes that they can be used for POCT, i.e., testing available at bedside or for

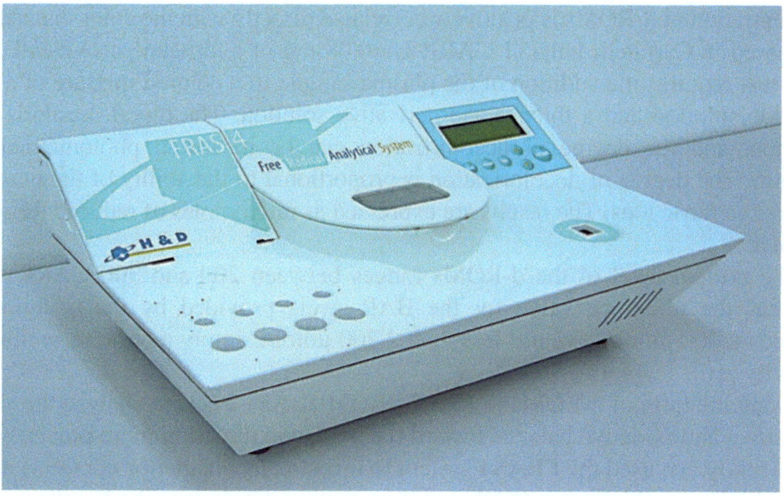

Fig. 8.1 Photographs of the FRAS4. The FRAS4 system was used under the following conditions: temperature 15–32 °C and humidity 20–80 %. The system is portable, being as small as 400 mm in width, 270 mm in depth, and 150 mm in height, and weighing as little as 4.0 kg. *FRAS4* free radical analytical system 4

pediatric outpatient clinics. Here, the details of FRAS 4 and ICR-001 are delineated with their clinical applications.

8.4.1 Assessment of the OS Profile Using the FRAS4

The analyzer known as FRAS4 (Diacron International, Grosseto, Italy) enables successive measurement of the derivatives of reactive oxygen metabolites (d-ROMs) and the biological antioxidant potential (BAP) by a spectrophotometric method [9, 10]. d-ROMs are primarily total hydroperoxides that can serve as surrogate biomarkers for overall ROS because hydroperoxides are the intermediate oxidative products of lipids, peptides, and amino acids while BAP can be a surrogate biomarker of the total power of antioxidants. This device can be used in routine clinical settings, such as an outpatient clinic (use conditions: temperature 15–32 °C; humidity 20–80 %) and is portable (width 400 mm, depth 270 mm, height 150 mm, weight 4.0 kg) as shown in Fig. 8.1. The required sample volume of blood is only 20 μL and the sample-to-answer time is only 15 min for successive detection of blood d-ROMs and BAP.

A brief description of methodology is as follows: d-ROMs, primarily hydroperoxides in the presence of iron released from plasma proteins by an acidic buffer generate alkoxyl and peroxyl radicals, in accordance with Fenton's reaction. Such radicals are in turn able to oxidize an alkyl-substituted aromatic amine, producing a pink-colored derivative that can be photometrically quantified at 505 nm. The concentration of d-ROMs is positively correlated directly with the color intensity and expressed in Carratelli units (1 CARR U=0.08 mg of hydrogen peroxide/dL). The BAP test requires the addition of the plasma sample to a colored mixture of a ferric chloride solution and a thiocyanate derivative solution. The blood decolorizes the solution, and the degree of decolorization can be measured photometrically at 505 nm. The degree of decolorization is proportional to the ability of the plasma to reduce the ferric ions. The results are expressed as micromoles of reduced ferric ions per liter.

The normal level of the d-ROMs ranges between 200 and 300 CARR U [9], whereas the reference value for the BAP levels provided by the manufacturer exceeds 2,200 μmol/L: values less than 2,200 μmol/L imply a reduced antioxidant capacity [10].

Using the ratio of d-ROMs to BAP (d-ROMs/BAP), one can analyze the shift of oxidative /antioxidative balance toward the oxidative direction more precisely.

Recently, we used the FRAS4 system to investigate whether or not OS is associated with Kawasaki disease (KD), and if so, then to determine whether the mechanism of intravenous immunoglobulin therapy (IVIG) is associated with scavenging ROS [11]. KD is an acute systemic vasculitis in children that primarily affects small and medium-sized arteries and is of unknown etiology [12]. The results obtained by FRAS4 showed a significantly higher OS as assessed by the d-ROMs/BAP ratio in patients with KD compared to those with febrile illnesses. Furthermore, children with KD had a significantly decreased d-ROMs/BAP ratio after IVIG. Interestingly enough, the d-ROMs/BAP ratio decreased in 12 of 14 patients with KD who defervesced after IVIG, implying that their cases were milder, whereas all but one patient who did not defervesce showed an increase in this ratio (Fig. 8.2). Thus, these results suggest that KD can be added to the possible list of OS-associated conditions, although the etiology remains a mystery. An assessment of OS in the acute phase of KD may be useful for early recognition of the necessity to perform additional IVIG treatment, because it is considered that IVIG therapy relieves OS by augmenting BAP.

We also conducted an investigation into whether hypoalbuminemia contributes to a decreased BAP in idiopathic nephrotic syndrome (INS) using FRAS4 [13]. A prior study had revealed that blood albumin plays an important role in the host defense mechanism by functioning as an antioxidant. In this study [13], 53 heparinized blood samples were obtained from 8 patients with INS (median age 13.5 years). Eighteen samples from 6 patients with Henoch–Schönlein purpura (median age 7 years) served as controls. The results showed that the blood levels of BAP were significantly lower in patients with nephrotic relapse than in patients with nephrotic remission or Henoch–Schönlein purpura. Furthermore, BAP was well correlated with the blood albumin levels in patients with INS. These findings suggest that decreased antioxidant potentials due to hypoalbuminemia may contribute at least in part to the aberrant immunity associated with immune system compromise in INS.

Others have also reported a derangement of the ratio of d-ROMs to BAP by the FRAS4. These include the presence of OS in neutropenic fever in children with leukemia and malignant lymphoma in adults. Nishikawa et al. reported that levels of d-ROMs can be early markers for predicting the severity of neutropenic fever in

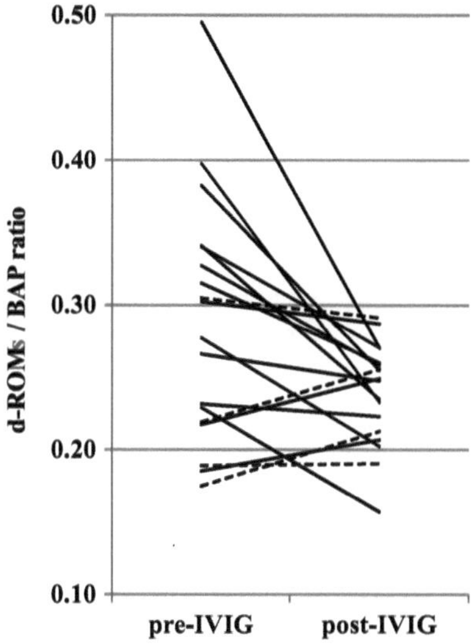

Fig. 8.2 Changes in the ratio of d-ROMs to BAP before and after intravenous IVIG in children with KD. The d-ROMs/BAP ratio decreased for 12 of 14 patients with KD who defervesced after IVIG (responders: *solid lines*), whereas all but one patient who did not defervesce (four nonresponders: *dashed line*) showed an increase in this ratio. *d-ROMs* derivatives of reactive oxygen metabolites, *BAP* biologic antioxidant potential, *IVIG* intravenous immunoglobulin therapy, *KD* Kawasaki disease (reproduced from [11] with permission from © 2012 Springer)

patients with hemato-oncological disorders since they were significantly lower in neutropenic fever patients with systemic inflammatory response syndrome than those without [14]. Nojima et al. found that the oxidative stress index (OSI), which is arbitrarily defined parameter based on the ratio of d-ROMs to BAP, was significantly higher in the sera of non-Hodgkin lymphoma patients than in those of healthy volunteers. Furthermore, when the patients were grouped into aggressive forms and non-malignant forms, the OSI was significantly higher in the former groups, than in those with the latter forms. From these findings, they suggest that the OSI might be a useful clinical marker for the evaluation of malignant grades and clinical stages of non-Hodgkin lymphoma [15]. In addition, the presence of OS in patients with myopathy, encephalopathy, lactic acidosis, and stroke-like episodes (MELAS) or in children with metabolic syndrome have been recently confirmed by the ratio of d-ROMs and BAP [16, 17].

8.4.2 Measurement of Urinary 8-oxodG Using a Novel POCT Device

8-oxodG, also known as 8-OHdG, is a product of an oxidatively modified guanine base excreted into urine and currently used as one of the sensitive biomarkers for oxidative cellular damage [18]. Therefore, it has been postulated that it is a useful

Fig. 8.3 Photographs of the automatic 8-oxodG analytical system, ICR-001. The ICR-001 system was used at a temperature of 18–30 °C and humidity of 30–80 %, and it had a width of 150 mm, depth of 160 mm, height of 170 mm, and weight of 2.8 kg. The sample-to-answer time was only 5 min for simultaneous detection of urinary creatinine and 8-oxodG concentrations. Test strips for 8-oxodG and creatinine are separately embedded in the sensor card. 8-oxodG; 8-oxo-7,8-dihydro-2′-deoxyguanosine

biomarker for evaluating the cardiovascular risk factors and inflammatory status in hypertensive patients [19, 20], for identifying persons at risk of developing cancer [21], for early prediction of lifestyle-related disease risks [22, 23] and for assessing the environmental conditions, including sunlight exposure, that are expected to have a serious long-term impact on the health status in adults [24, 25]. Despite growing interest in urinary 8-oxodG as a sensitive OS biomarker in sick children as well as in adults [1, 26], its use is hindered in clinical settings because the current assays depend on a complicated methodology to measure 8-oxodG, such as an ELISA [22, 24, 27–31]. Furthermore, limited information exists regarding age- and sex-related reference intervals [31] in healthy children.

A recently developed automatic analyzer (model no. ICR-001; Techno Medica Co., Ltd., Yokohama, Japan) to measure urinary 8-oxodG by competitive immunochromatography has been developed for POCT and has been commercially available in Japan since 2010. This device consists of a detector, an external computer and single-use sensor cards (Fig. 8.3). It can be used in routine clinical settings, such as an outpatient clinic (use conditions: temperature 18–30 °C, humidity 30–80 %) and is portable (width 150 mm, depth 160 mm, height 170 mm, weight 2.8 kg). The sample-to-answer time is only 5 min for simultaneous detection of urinary creatinine and 8-oxodG concentrations.

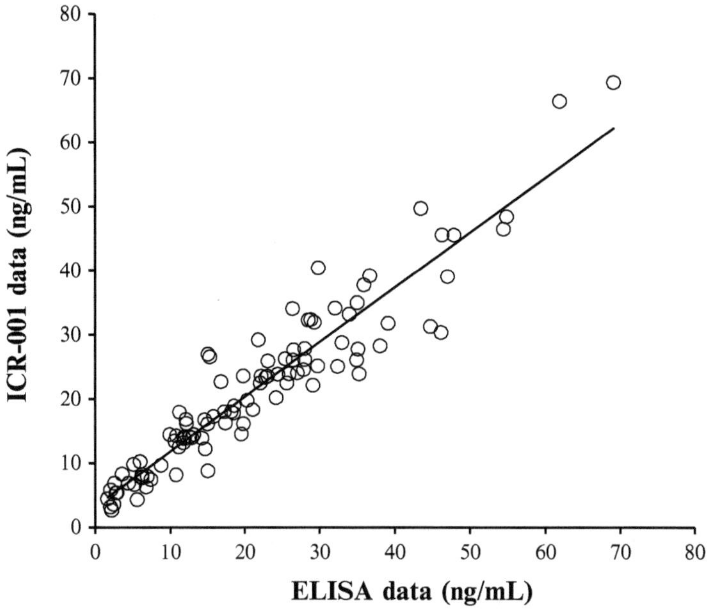

Fig. 8.4 Relationship between urinary 8-oxodG values obtained by the ICR-001 and ELISA. Linear regression analysis produced the following formula: 8-oxodG by ICR-001 = 0.85 × (8-oxodG by ELISA) + 3.23 ($r_s = 0.945$, $p < 0.00001$). (Reproduced from [32] with permission from © 2012 Elsevier B.V.)

We verified the reliability of this new device in measuring urinary 8-oxodG concentration by comparing the results of this device with those by the ELISA, and also determined sex- and age-related changes in healthy children and adults [32]. First, the results demonstrated that the values of urinary 8-oxodG and creatinine obtained by the ICR-001 were well correlated with those measured by ELISA and enzymatic method, respectively (Fig. 8.4), and the urinary levels of 8-oxodG corrected by creatinine of healthy adult Japanese were comparable to those of the published ones determined by the ELISA [22, 24]. Accordingly, the value of urinary 8-oxodG corrected by creatinine determined by ICR-001 was confirmed to have comparable reliability to the standard method. In addition, we determined the age-related changes in healthy children, which demonstrated that the values of urinary 8-oxodG corrected by creatinine were highest in the youngest subjects and decreased through adolescence to reach a plateau, showing a significant inverse correlation with age as shown in Table 8.3. This finding was consistent with previously published data in children [1, 31] that were determined using an ELISA. Furthermore, the resulting age-related reference intervals appeared to be accurate enough for clinical application, as we recently reported that children with heart diseases receiving cardiac catheterization showed an approximately fourfold increase in urinary 8-oxodG after radiation exposure and that 8 of 14 children showed extremely high levels above the 90th percentile of the reference intervals

Table 8.3 Age-related changes in urinary 8-oxodG as determined by ICR-001 (ng/mg creatinine)

	Neonates	1–4 years	5–9 years	10–15 years	Adults
n	24	95	19	19	100
Minimum	4.20	7.30	5.70	4.30	5.32
2.5 percentile	6.04	9.11	5.75	4.75	5.87
5 percentile	7.69	9.78	5.79	5.20	6.42
10 percentile	13.60	11.6	6.70	6.40	7.10
First quartile	15.33	13.35	8.75	6.95	8.46
Median	20.25	16.90	11.40	9.90	9.93
Third quartile	26.90	23.05	20.25	14.10	11.92
90 percentile	28.50	31.40	25.00	27.00	15.70
95 percentile	37.69	35.12	65.20	43.12	18.12
97.5 percentile	40.03	45.85	77.80	45.91	21.98
Maximum	42.90	111.20	90.40	48.70	25.63

Reproduced from [32] with permission from © 2012 Elsevier B.V

after radiation exposure. The mechanisms underlying the higher levels of urinary 8-oxodG in younger subjects remain unresolved. However, as a possible mechanism, it has been pointed out that children experiencing rapid growth and sustained immune activation are prone to be exposed to high levels of ROS and are therefore more vulnerable to oxidation of DNA, which could result in higher urinary excretion of 8-oxodG [1].

Acknowledgments This study was partly funded by grants from the Mami Mizutani Foundation and the Human Resource Chair for Neonatologists funded by Izumi-Ohtsu City, Osaka, Japan and by the Chair for Children with Developmental Disorders funded by Nabari City, Mie, Japan.

References

1. Tsukahara H (2007) Biomarkers for oxidative stress: clinical application in pediatric medicine. Curr Med Chem 14:339–351
2. Cutler RG, Plummer J, Chowdhury K, Heward C (2005) Oxidative stress profiling: part II. Theory, technology, and practice. Ann N Y Acad Sci 1055:136–158
3. Fulia F, Gitto E, Cuzzocrea S, Reiter RJ, Dugo L, Gitto P et al (2001) Increased levels of malondialdehyde and nitrite/nitrate in the blood of asphyxiated newborns: reduction by melatonin. J Pineal Res 31:343–349
4. Tsukahara H, Haruta T, Todoroki Y, Hiraoka M, Noiri E, Maeda M et al (2002) Oxidant and antioxidant activities in childhood meningitis. Life Sci 71:2797–2806
5. Corradi M, Folesani G, Andreoli R, Manini P, Bodini A, Piacentini G et al (2003) Aldehydes and glutathione in exhaled breath condensate of children with asthma exacerbation. Am J Respir Crit Care Med 167:395–9
6. Gupta M, Gupta YK, Agarwal S, Aneja S, Kalaivani M, Kohli K (2004) Effects of add-on melatonin administration on antioxidant enzymes in children with epilepsy taking carbamazepine monotherapy: a randomized, double-blind, placebo-controlled trial. Epilepsia 45:1636–1639

7. Ceriello A (2003) New insights on oxidative stress and diabetic complications may lead to a "causal" antioxidant therapy. Diabetes Care 26:1589–1596
8. Noiri E, Tsukahara H (2005) Parameters for measurement of oxidative stress in diabetes mellitus: applicability of enzyme-linked immunosorbent assay for clinical evaluation. J Investig Med 53:167–175
9. Cesarone MR, Belcaro G, Carratelli M, Cornelli U, De Sanctis MT, Incandela L et al (1999) A simple test to monitor oxidative stress. Int Angiol 18:127–130
10. Imatoh T, Kamimura S, Tanihara S (2013) Moderate oxidative stress and high antioxidative activity are associated with steatosis in Japanese males. Clin Transl Sci 6:45–49
11. Kaneko K, Takahashi M, Yoshimura K, Kitao T, Yamanouchi S, Kimata T et al (2012) Intravenous immunoglobulin counteracts oxidative stress in Kawasaki disease. Pediatr Cardiol 33:1086–1088
12. Burns JC, Glode MP (2004) Kawasaki syndrome. Lancet 364:533–544
13. Kaneko K, Kimata T, Tsuji S, Shimo T, Takahashi M, Tanaka S (2012) Serum albumin level accurately reflects antioxidant potentials in idiopathic nephrotic syndrome. Clin Exp Nephrol 16:411–414
14. Nishikawa T, Okamoto Y, Kodama Y, Tanabe T, Shinkoda Y, Kawano Y (2010) Serum derivative of reactive oxygen metabolites (d-ROMs) in pediatric hemato-oncological patients with neutropenic fever. Pediatr Blood Cancer 55:91–94
15. Nojima J, Motoki Y, Tsuneoka H, Kuratsune H, Matsui T, Yamamoto M et al (2011) 'Oxidation stress index' as a possible clinical marker for the evaluation of non-Hodgkin lymphoma. Br J Haematol 155:528–530
16. Faienza MF, Francavilla R, Goffredo R, Ventura A, Marzano F, Panzarino G et al (2012) Oxidative stress in obesity and metabolic syndrome in children and adolescents. Horm Res Paediatr 78:158–164
17. Ikawa M, Arakawa K, Hamano T, Nagata M, Nakamoto Y, Kuriyama M et al (2012) Evaluation of systemic redox states in patients carrying the MELAS A3243G mutation in mitochondrial DNA. Eur Neurol 67:232–237
18. European Standards Committee on Urinary (DNA) Lesion Analysis, Evans MD, Olinski R, Loft S, Cooke MS (2010) Toward consensus in the analysis of urinary 8-oxo-7,8-dihydro-2′-deoxyguanosine as a noninvasive biomarker of oxidative stress. FASEB J 24:1249–1260
19. Rosello-Lleti E, de Burgos FG, Morillas P, Cortes R, Martinez-Dolz L, Almenar L et al (2012) Impact of cardiovascular risk factors and inflammatory status on urinary 8-OHdG in essential hypertension. Am J Hypertens 25:236–242
20. Soeki T, Kitani M, Kusunose K, Yagi S, Taketani Y, Koshiba K et al (2012) Renoprotective and antioxidant effects of cilnidipine in hypertensive patients. Hypertens Res 35:1058–1062
21. Halliwell B (2000) Why and how should we measure oxidative DNA damage in nutritional studies? How far have we come? Am J Clin Nutr 72:1082–1087
22. Sakano N, Wang DH, Takahashi N, Wang B, Sauriasari R, Kanbara S et al (2009) Oxidative stress biomarkers and lifestyles in Japanese healthy people. J Clin Biochem Nutr 44:185–195
23. Ogino K, Takahashi N, Takigawa T, Obase Y, Wang DH (2011) Association of serum arginase I with oxidative stress in a healthy population. Free Radic Res 45:147–155
24. Abder-Rahman HA, Nusair S (2007) 8-Hydroxy-2′-deoxyguanosine (8-OHdG) as a short-term predictor of regional and occupational health problems. J UOEH 29:247–258
25. Kato M, Iida M, Goto Y, Kondo T, Yajima I (2011) Sunlight exposure-mediated DNA damage in young adults. Cancer Epidemiol Biomarkers Prev 20:1622–1628
26. Granot E, Kohen R (2004) Oxidative stress in childhood–in health and disease states. Clin Nutr 23:3–11
27. Orhan H, van Holland B, Krab B, Moeken J, Vermeulen NP, Hollander P et al (2004) Evaluation of a multi-parameter biomarker set for oxidative damage in man: Increased urinary excretion of lipid, protein and DNA oxidation products after one hour of exercise. Free Radic Res 38:1269–1279
28. Lenarczyk M, Cohen EP, Fish BL, Irving AA, Sharma M, Driscoll CD et al (2009) Chronic oxidative stress as a mechanism for radiation nephropathy. Radiat Res 171:164–172

29. Matsubasa T, Uchino T, Karashima S, Kondo Y, Maruyama K, Tanimura M et al (2002) Oxidative stress in very low birth weight infants as measured by urinary 8-OHdG. Free Radic Res 36:189–193
30. Shoji H, Shimizu T, Shinohara K, Oguchi S, Shiga S, Yamashiro Y (2004) Suppressive effects of breast milk on oxidative DNA damage in very low birthweight infants. Arch Dis Child Fetal Neonatal Ed 89:F136–F138
31. Tamura S, Tsukahara H, Ueno M, Maeda M, Kawakami H, Sekine K et al (2006) Evaluation of a urinary multi-parameter biomarker set for oxidative stress in children, adolescents and young adults. Free Radic Res 40:1198–1205
32. Kaneko K, Kimata T, Tsuji S, Ohashi A, Imai Y, Sudo H et al (2012) Measurement of urinary 8-oxo-7,8-dihydro-2′-deoxyguanosine in a novel point-of-care testing device to assess oxidative stress in children. Clin Chim Acta 413:1822–1826

Chapter 9
Analytical Procedures for Nitrative/Nitrosative Stress

Masayuki Kubo and Keiki Ogino

Abbreviations

BEC	*S*-(2-boronoethyl)-L-cysteine
ECD	Electrochemical detection
ELISA	Enzyme-linked immunosorbent assay
GC–MS	Gas chromatograph–mass spectrometry
GC–MS/MS	Gas chromatograph–tandem mass spectrometry
H_2O_2	Hydrogen peroxide
HPLC	High pressure liquid chromatography
HPLC-ECD	High pressure liquid chromatography coupled with electrochemical detection
LC–MS	Liquid chromatograph–mass spectrometry
LC-MS/MS	Liquid chromatograph–tandem mass spectrometry
N_2O_3	Dinitrogen trioxide
NO	Nitric oxide
NO_2	Nitrogen dioxide
NO_2^-	Nitrite
NO_3^-	Nitrate
nor-NOHA	N^{ω}-hydroxy-nor-arginine
NOS	Nitric oxide synthase
O_2^-	Superoxide
$ONOO^-$	Peroxynitrite
RNS	Reactive nitrogen species
ROS	Reactive oxygen species

M. Kubo, Ph.D. • K. Ogino, M.D., Ph.D. (✉)
Department of Public Health, Okayama University Graduate School of Medicine, Dentistry and Pharmaceutical Sciences, 2-5-1 Shikata-cho, Kita-ku, Okayama 700-8558, Japan
e-mail: kogino@md.okayama-u.ac.jp

H. Tsukahara and K. Kaneko (eds.), *Studies on Pediatric Disorders*, Oxidative Stress in Applied Basic Research and Clinical Practice, DOI 10.1007/978-1-4939-0679-6_9,

9.1 Introduction

Nitric oxide (NO) is a short-lived free radical gas that plays a pivotal role in maintaining homeostasis in the cardiovascular, nervous, and immune systems [1]. When NO, reactive nitrogen species (RNS), and reactive oxygen species generation (ROS) overwhelms the capacity of the antioxidant defense system to scavenge such molecules, the imbalance can generate nitrative/nitrosative damage to cellular constituents (DNA, protein, lipid, and sugar), which is nitrative/nitrosative stress [2]. Furthermore, nitrative/nitrosative stress might contribute to disease initiation and progression, including inflammatory diseases, ischemic heart diseases, diabetes, and atherosclerosis [3–5].

Given that RNS are highly reactive and have extremely short half-lives, it is impractical to directly measure such molecules in tissue or body fluids. Therefore, cellular constituents modified by RNS in biological samples have been measured to detect appropriate biomarkers for diseases that involve RNS. For example, 3-nitrotyrosine is produced via RNS-mediated tyrosine nitration, such as through peroxynitrite ($ONOO^-$), and it is a biomarker of RNS formation in vivo [6–8].

On the other hand, NO is primarily produced by NO synthase (NOS) using L-arginine as a substrate [9]. Because L-arginine is also used by arginase, which is an enzyme involved in urea cycling, arginase might inhibit NO production by reducing the intracellular L-arginine available to NOS [10, 11]. We have shown that arginase up-regulation modulates NO metabolism through L-arginine consumption [12–15]. Thus, arginase expression might be related to NO and subsequent RNS production.

In the following chapter, we consider nitrative/nitrosative stress and analytical procedures used to detect such stress, especially through 3-nitrotyrosine and arginase.

9.2 Nitrative/Nitrosative Stress

NO is produced as a substrate by L-arginine using NOS. There are three NOS isoforms: constitutive NOS [neuronal NOS (nNOS, NOS1), endothelial NOS (eNOS, NOS3)], and inducible NOS (iNOS, NOS2).

Constitutive NOS (nNOS and eNOS) produces low NO concentrations (pM-nM) and contributes to homeostatic function, such as in the cardiovascular, nervous, and immune systems [1, 9]. In general, NO is physiologically converted to nitrite (NO_2^-) and nitrate (NO_3^-), major oxidative products of NO metabolism, by different oxidation reactions [16].

On the other hand, iNOS produces higher levels (μM) of NO, which is a primary initiator of RNS, after exposure to proinflammatory and immune stimuli under pathological conditions [1, 9]. For example, high levels of iNOS-produced NO react with superoxide (O_2^-) to form $ONOO^-$.

$ONOO^-$ is a powerful oxidizing species and produces oxidized proteins, DNA, and lipids; most notably, it generates the nitration of free-tyrosine or tyrosine residues in proteins [6, 7]. Tyrosine nitration has been observed in many human diseases. Thus, increased formation of NO and RNS, such as $ONOO^-$, nitrogen dioxide (NO_2), and dinitrogen trioxide (N_2O_3), could lead to nitrative/nitrosative stress, which is responsible for the pathogenesis and progression of various diseases [17]. Therefore, measuring RNS-modified constituents as biomarkers in biological samples is likely to be useful for discerning the role of NO and RNS in such diseases.

9.3 3-Nitrotyrosine as a Marker of Nitrative/Nitrosative Stress

9.3.1 Mechanism of Tyrosine Nitration

3-Nitrotyrosine is a biomarker for nitrative/nitrosative stress. 3-Nitrotyrosine is generated by nitrating the amino acid tyrosine and protein-bound tyrosine. Initially, the formation of 3-nitrotyrosine was considered as a specific footprint of $ONOO^-$, which is produced by a reaction between NO and O_2^- [6, 7]. Interestingly, the depletion of L-arginine or tetrahydrobiopterin, which is a NOS cofactor, promotes O_2^- generation from NOS in a process termed "NOS uncoupling." Simultaneous O_2^- production from uncoupled NOS and NO production from coupled NOS can generate increased $ONOO^-$ levels, which can form 3-nitrotyrosine [18–20].

Alternatively, tyrosine nitration can be mediated by NO_2 derived from catalytic oxidation of NO_2^- in the presence of hydrogen peroxide (H_2O_2) via peroxidase [8]. Myeloperoxidase in neutrophils or macrophages and eosinophil peroxidase in eosinophils are related to this peroxidase-dependent pathway during active inflammation [21, 22]. Therefore, 3-nitrotyrosine is an indicator of RNS formation rather than a specific marker for $ONOO^-$ (Fig. 9.1).

Accumulating evidence suggests a correlation between increased 3-nitrotyrosine levels and diseases that involve RNS, such as asthma, atherosclerosis, rheumatoid arthritis, multiple sclerosis, and inflammatory bowel disease [17].

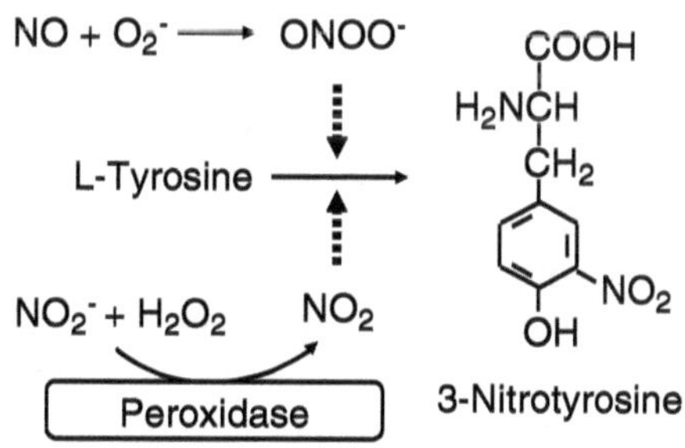

Fig. 9.1 Mechanism for 3-nitrotyrosine formation. Tyrosine nitration is generated by $ONOO^-$, which is produced by a reaction between NO and O_2^-, as well as by NO_2, which is derived from NO_2^- catalytic oxidation with H_2O_2 by peroxidase

On the other hand, certain studies have demonstrated that various components of whole blood samples and tissue homogenates or crude extracts from different organs can reduce 3-nitrotyrosine levels independent of apparent proteolytic activity [23–25]. These findings may support in vivo denitrase activity.

9.3.2 Analytical Procedures for 3-Nitrotyrosine

A wide spectrum of physicochemical and immunological methods have been used to detect and quantify 3-nitrotyrosine in biological samples, including plasma, serum, urine, and tissue, from humans and animals that are healthy or have a variety of disease states [26].

Mass spectrometry-based assays, including gas chromatograph–mass spectrometry (GC–MS), gas chromatograph–tandem mass spectrometry (GC–MS/MS), liquid chromatograph–mass spectrometry (LC–MS), liquid chromatograph–tandem mass spectrometry (LC–MS/MS), and high pressure liquid chromatography (HPLC)-based assays have been used as sensitive methods for the quantification of 3-nitrotyrosine [26–28]. Mass spectrometry approaches are fully quantitative, high sensitivity, and high specificity methods [26–28]. However, these approaches require expensive and specialized equipment, extensive sample preparation, and isotopically labeled internal standards [29, 30]. Thus, using such approaches might be technically difficult, time-consuming, and expensive.

Quantification of 3-nitrotyrosine using HPLC coupled with electrochemical detection (ECD) (HPLC-ECD) is a simple, relatively inexpensive, and sufficiently sensitive method. HPLC is a useful separation technique [26, 31]. Furthermore, ECD detection is more sensitive than ultraviolet and fluorescence. Because inert substrates are electrochemically invisible to an ECD, reactive components may be detected in a complex matrix without extensive sample purification [32]. Therefore, HPLC-ECD provides good separation and high sensitivity and maintains linearity across a wide range of 3-nitrityrosine concentrations. We have used an HPLC dual-mode ECD that comprises two serial electrochemical cells for reduction and oxidation detection models, and we have demonstrated increased 3-nitrotyrosine levels in the biological samples from the animal models of such diseases as asthma, atopic dermatitis, and sepsis [12, 33, 34].

Compared with mass spectrometry- or HPLC-based methods, the antibody-based method, enzyme-linked immunosorbent assay (ELISA), is technically easy, inexpensive, and time-efficient [26]. Thus, using such methods might be adequate for measuring many samples, such as for a population-based study [35]. However, this approach is likely semi-quantitative because it does not comprise rigorous assay validation, and it is difficult to assess reliability. Detection of immunoreactive proteins with nitration of tyrosine residues differs for disease or treatment samples. Divergent values have been reported using the ELISA method for plasma 3-nitrotyrosine that range between undetectable and remarkably higher compared with mass spectrometry- or HPLC-based methods. Therefore, parameters, including sample type, antibody, other components and time, should be considered for each experimental condition [26, 28, 36, 37].

Furthermore, biological samples should be prepared and analyzed with great care. Under acidic conditions, tyrosine could be nitrated by nitrite in biological samples, which may generate 3-nitrotyrosine artifacts. In the future, in addition to improving sensitivity and reliability, simple, inexpensive, and suitable analytical procedures for 3-nitrotyrosine quantification in multiple biological samples will be required to discern the contribution of RNS to disease initiation and progression and for early disease diagnosis.

9.3.3 Additional Markers for Nitrative/Nitrosative Stress

It has been reported that nitration/nitrasation of the cellular component occurs not only in tyrosine but also in tryptophan, cysteine, DNA, and lipid. Nitrotryptophan is formed by a reaction between RNS and free or protein-bound tryptophan [38, 39]. *S*-Nitrosothiols are generated by a reaction between the thiol of cysteine and nitrosating species, such as N_2O_3, after NO autooxidation [9]. Nitration of guanosine by RNS can produce 8-nitroguanosine [40, 41]. Nitrated fatty acids, such as nitrolinoleic acid and nitrooleic acid, have also been detected [42, 43]. Therefore, such products are markers of nitrative/nitrosative stress. Analyses for such products may also be useful in discerning the association between RNS and various diseases.

9.4 Arginase as a NOS Competitor

9.4.1 Arginase and NOS in L-Arginine Metabolism

L-Arginine, which is a NOS substrate, plays an important role in regulating the production of NO, and its availability depends on pathways for biosynthesis, cellular uptake, and catabolism by NOS and arginase (Fig. 9.2).

Arginase is the final enzyme in the urea cycle and catalyzes L-arginine hydrolysis to urea and L-ornithine, which is a precursor in polyamine and proline production [44–46]. There are two isoforms of arginase; arginase I is abundantly present in the

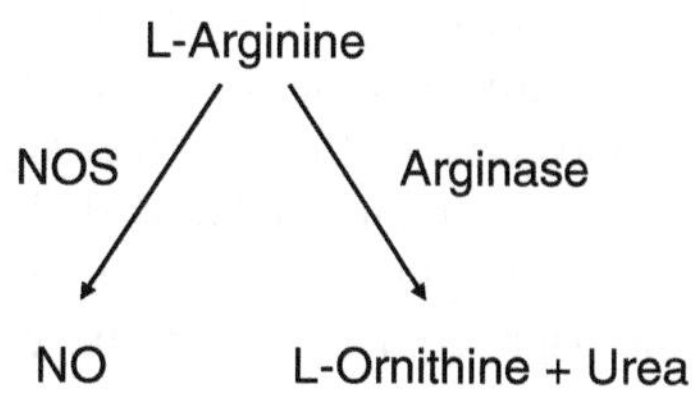

Fig. 9.2 L-Arginine metabolism by arginase and NOS. Arginase attenuates NO production by competing with NOS for their common substrate, L-arginine

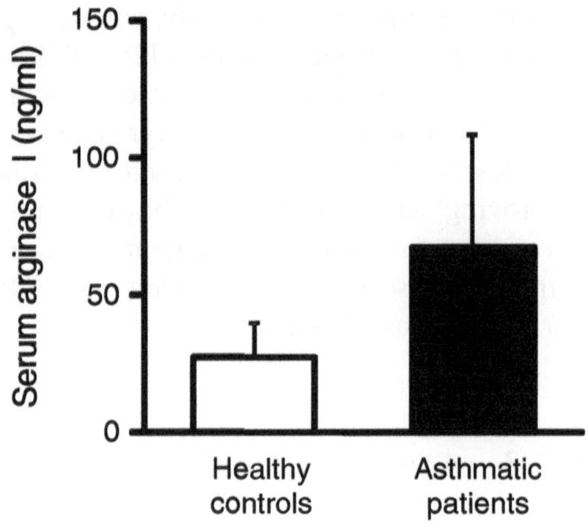

Fig. 9.3 High levels of serum arginase I in asthma. Serum levels of arginase I were significantly higher in asthmatic patients than in healthy controls. Data are expressed as mean ± SD

cytosol of the liver, while arginase II is in the mitochondria and primarily expressed extrahepatically, particularly in the kidney and prostate [44, 47]. While the affinity (Km) of L-arginine for arginase is in the low range (μM) compared with NOS affinity (mM), NOS and arginase compete for L-arginine because the V_{max} for arginase is 1,000-fold greater [44, 46]. By competing for L-arginine, a common substrate, it is likely that arginase attenuates L-arginine bioavailability for NOS, which limits NO production [10, 11].

The expression and activity of arginase I are increased in human disease states, such as asthma and inflammatory bowel disease [48, 49]. We also demonstrated that levels of serum arginase I were significantly higher in asthmatic patients compared with healthy controls (Fig. 9.3) [50]. Our previous studies on an animal model of diseases showed that NO production was decreased secondary to an increase in expression and activity of arginase I, and administration of the arginase inhibitor, N^{ω}-hydroxy-nor-arginine (nor-NOHA), ameliorated the attenuated NO production [12–15]. These findings suggest that increased activity and upregulation of arginase can reduce NO production by limiting the L-arginine supply for NOS, which is responsible for pathogenesis of such diseases. Therefore, arginase is an endogenous NOS antagonist, and it might be critical for regulating NO production.

We have shown that intranasal administration of arginase inhibitor, nor-NOHA, attenuated airway hyperresponsiveness and inflammation in a mite-induced NC/Nga mouse model of asthma [13]. Although there is a difference of the animal model of asthma (guinea-pig) and the allergen (ovalbumin), some studies demonstrated that treatment with nor-NOHA normalized airway hyperresponsiveness, and the effect was reversed by NOS inhibitor, N^{ω}-nitro-L-arginine methyl ester [51, 52]. These findings agree with our previous observations. Furthermore, there have been conflicting reports on the effect of other arginase inhibitor, *S*-(2-boronoethyl)-L-cysteine (BEC), in the ovalbumin-induced Balb/c mouse model of asthma [48, 53]. These different results might be due to the difference in the dose of nebulized BEC. Therefore, although arginase might be expected to be a preventive and therapeutic target, it will be necessary for additional studies to determine the optimal dose, route, and timing of administration of arginase inhibitor.

It is unknown whether arginase is associated with nitrative/nitrosative stress. It is likely that L-arginine consumption by arginase leads to uncoupled NOS, which produces nitrative/nitrosative stress via the formation of RNS, such as $ONOO^-$ [18–20, 54]. However, conversely, evidence suggests that administration of the arginase inhibitor, BEC, alters NO homeostasis and increases protein tyrosine nitration and S-nitrosylation in a mouse model for allergic airway disease [53]. Further studies are required to clarify the contribution of arginase to nitrative/nitrosative stress.

9.4.2 Analytical Procedures for Arginase

Arginase levels in biological samples, such as serum, plasma, and tissue, are analyzed by measuring activity or through an ELISA. Arginase activity is determined by calculating the level of urea generated via the arginase reaction [55]. The protein level of arginase can be assessed using an ELISA. Both approaches are advantageous because they are easy, inexpensive, and time-efficient.

9.5 Conclusion

Considering that nitrative/nitrosative stress is responsible for the pathogenesis and progression of various diseases, the quantification of RNS-modified constituents, such as 3-nitrotyrosine, might be beneficial in discerning the contribution of NO and RNS to such diseases. Furthermore, many studies have shown the upregulation of arginase in various diseases, which may reduce NOS-mediated NO production by limiting a common substrate, L-arginine. Additional improvements in sensitivity and reliability for analyzing such biomarkers, including 3-nitrotyrosine and arginase, will lead to early diagnosis of diseases and aid in monitoring development of disease and response to treatment.

Acknowledgments This work was supported by in part by Grant-in-Aid for Science Research No. 23390163 from the Ministry of Education, Culture, Sports, Science, and Technology of the Japanese Government.

References

1. Grisham MB, Jourd'Heuil D, Wink DA (1999) Nitric oxide. I. Physiological chemistry of nitric oxide and its metabolites:implications in inflammation. Am J Physiol 276:G315–G321
2. Sies H (1991) Oxidative stress: from basic research to clinical application. Am J Med 91: 31S–38S

3. Loscalzo J, Welch G (1995) Nitric oxide and its role in the cardiovascular system. Prog Cardiovasc Dis 38:87–104
4. Llorens S, Nava E (2003) Cardiovascular diseases and the nitric oxide pathway. Curr Vasc Pharmacol 1:335–346
5. Cutler RG, Plummer J, Chowdhury K, Heward C (2005) Oxidative stress profiling: part II. Theory, technology, and practice. Ann N Y Acad Sci 1055:136–158
6. Beckman JS, Koppenol WH (1996) Nitric oxide, superoxide, and peroxynitrite: the good, the bad, and ugly. Am J Physiol 271:C1424–C1437
7. Radi R, Peluffo G, Alvarez MN, Naviliat M, Cayota A (2001) Unraveling peroxynitrite formation in biological systems. Free Radic Biol Med 30:463–488
8. Brennan ML, Wu W, Fu X, Shen Z, Song W, Frost H, Vadseth C, Narine L, Lenkiewicz E, Borchers MT, Lusis AJ, Lee JJ, Lee NA, Abu-Soud HM, Ischiropoulos H, Hazen SL (2002) A tale of two controversies: defining both the role of peroxidases in nitrotyrosine formation in vivo using eosinophil peroxidase and myeloperoxidase-deficient mice, and the nature of peroxidase-generated reactive nitrogen species. J Biol Chem 277:17415–17427
9. Hanafy KA, Krumenacker JS, Murad F (2001) NO, nitrotyrosine, and cyclic GMP in signal transduction. Med Sci Monit 7:801–819
10. Boucher JL, Moali C, Tenu JP (1999) Nitric oxide biosynthesis, nitric oxide synthase inhibitors and arginase competition for L-arginine utilization. Cell Mol Life Sci 55:1015–1028
11. Zimmermann N, Rothenberg ME (2006) The arginine-arginase balance in asthma and lung inflammation. Eur J Pharmacol 533:253–262
12. Takemoto K, Ogino K, Shibamori M, Gondo T, Hitomi Y, Takigawa T, Wang DH, Takaki J, Ichimura H, Fujikura Y, Ishiyama H (2007) Transiently, paralleled upregulation of arginase and nitric oxide synthase and the effect of both enzymes on the pathology of asthma. Am J Physiol Lung Cell Mol Physiol 293:L1419–L1426
13. Takahashi N, Ogino K, Takemoto K, Hamanishi S, Wang DH, Takigawa T, Shibamori M, Ishiyama H, Fujikura Y (2010) Direct inhibition of arginase attenuated airway allergic reactions and inflammation in a Dermatophagoides farinae-induced NC/Nga mouse model. Am J Physiol Lung Cell Mol Physiol 299:L17–L24
14. Imagama T, Ogino K, Takemoto K, Kato Y, Kataoka H, Suzuki H, Ran Z, Setiawan H, Fujikura Y, Taguchi T (2012) Regulation of nitric oxide generation by up-regulated arginase I in rat spinal cord injury. J Clin Biochem Nutr 51:68–75
15. Akazawa Y, Kubo M, Zhang R, Matsumoto K, Yan F, Setiawan H, Takahashi H, Fujikura Y, Ogino K (2013) Inhibition of arginase ameliorates experimental ulcerative colitis in mice. Free Radic Res 47:137–145
16. Kelm M (1999) Nitric oxide metabolism and breakdown. Biochim Biophys Acta 1411: 273–289
17. Greenacre SA, Ischiropoulos H (2001) Tyrosine nitration: localisation, quantification, consequences for protein function and signal transduction. Free Radic Res 34:541–581
18. Vasquez-Vivar J, Kalyanaraman B, Martasek P, Hogg N, Masters BS, Karoui H, Tordo P, Pritchard KA Jr (1998) Superoxide generation by endothelial nitric oxide synthase: the influence of cofactors. Proc Natl Acad Sci U S A 95:9220–9225
19. Forstermann U, Li H (2011) Therapeutic effect of enhancing endothelial nitric oxide synthase (eNOS) expression and preventing eNOS uncoupling. Br J Pharmacol 164:213–223
20. Alkaitis MS, Crabtree MJ (2012) Recoupling the cardiac nitric oxide synthases: tetrahydrobiopterin synthesis and recycling. Curr Heart Fail Rep 9:200–210
21. van der Vliet A, Eiserich JP, Halliwell B, Cross CE (1997) Formation of reactive nitrogen species during peroxidase-catalyzed oxidation of nitrite. A potential additional mechanism of nitric oxide-dependent toxicity. J Biol Chem 272:7617–7625
22. Wu W, Chen Y, Hazen SL (1999) Eosinophil peroxidase nitrates protein tyrosyl residues. Implications for oxidative damage by nitrating intermediates in eosinophilic inflammatory disorders. J Biol Chem 274:25933–25944
23. Kamisaki Y, Wada K, Bian K, Balabanli B, Davis K, Martin E, Behbod F, Lee YC, Murad F (1998) An activity in rat tissues that modifies nitrotyrosine-containing proteins. Proc Natl Acad Sci U S A 95:11584–11589

24. Smallwood HS, Lourette NM, Boschek CB, Bigelow DJ, Smith RD, Pasa-Tolic L, Squier TC (2007) Identification of a denitrase activity against calmodulin in activated macrophages using high-field liquid chromatography–FTICR mass spectrometry. Biochemistry 46:10498–10505
25. Abello N, Kerstjens HA, Postma DS, Bischoff R (2009) Protein tyrosine nitration: selectivity, physicochemical and biological consequences, denitration, and proteomics methods for the identification of tyrosine-nitrated proteins. J Proteome Res 8:3222–3238
26. Duncan MW (2003) A review of approaches to the analysis of 3-nitrotyrosine. Amino Acids 25:351–361
27. Tsikas D, Caidahl K (2005) Recent methodological advances in the mass spectrometric analysis of free and protein-associated 3-nitrotyrosine in human plasma. J Chromatogr B Analyt Technol Biomed Life Sci 814:1–9
28. Tsikas D (2012) Analytical methods for 3-nitrotyrosine quantification in biological samples: the unique role of tandem mass spectrometry. Amino Acids 42:45–63
29. Crowley JR, Yarasheski K, Leeuwenburgh C, Turk J, Heinecke JW (1998) Isotope dilution mass spectrometric quantification of 3-nitrotyrosine in proteins and tissues is facilitated by reduction to 3-aminotyrosine. Anal Biochem 259:127–135
30. Leeuwenburgh C, Hardy MM, Hazen SL, Wagner P, Oh-ishi S, Steinbrecher UP, Heinecke JW (1997) Reactive nitrogen intermediates promote low density lipoprotein oxidation in human atherosclerotic intima. J Biol Chem 272:1433–1436
31. Nuriel T, Deeb RS, Hajjar DP, Gross SS (2008) Protein 3-nitrotyrosine in complex biological samples: quantification by high-pressure liquid chromatography/electrochemical detection and emergence of proteomic approaches for unbiased identification of modification sites. Methods Enzymol 441:1–17
32. Hensley K, Williamson KS, Floyd RA (2000) Measurement of 3-nitrotyrosine and 5-nitro-gamma-tocopherol by high-performance liquid chromatography with electrochemical detection. Free Radic Biol Med 28:520–528
33. Kubo M, Kambayashi Y, Takemoto K, Okuda J, Muto M, Ogino K (2005) Reactive nitrogen species formation in eosinophils and imbalance in nitric oxide metabolism are involved in atopic dermatitis-like skin lesions in NC/Nga mice. Free Radic Res 39:719–727
34. Hitomi YH, Okuda J, Nishino H, Kambayashi Y, Hibino Y, Takemoto K, Takigawa T, Ohno H, Taniguchi N, Ogino K (2007) Disposition of protein-bound 3-nitrotyrosine in rat plasma analysed by a novel protocol for HPLC-ECD. J Biochem 141:495–502
35. Sakano N, Takahashi N, Wang DH, Sauriasari R, Takemoto K, Kanbara S, Sato Y, Takigawa T, Takaki J, Ogino K (2009) Plasma 3-nitrotyrosine, urinary 8-isoprostane and 8-OHdG among healthy Japanese people. Free Radic Res 43:183–192
36. Tsikas D, Schwedhelm E, Frolich JC (2002) Methodological considerations on the detection of 3-nitrotyrosine in the cardiovascular system. Circ Res 90:E70
37. Ogino K, Wang DH (2007) Biomarkers of oxidative/nitrosative stress: an approach to disease prevention. Acta Med Okayama 61:181–189
38. Alvarez B, Radi R (2003) Peroxynitrite reactivity with amino acids and proteins. Amino Acids 25:295–311
39. Nuriel T, Hansler A, Gross SS (2011) Protein nitrotryptophan: formation, significance and identification. J Proteomics 74:2300–2312
40. Akaike T, Okamoto S, Sawa T, Yoshitake J, Tamura F, Ichimori K, Miyazaki K, Sasamoto K, Maeda H (2003) 8-nitroguanosine formation in viral pneumonia and its implication for pathogenesis. Proc Natl Acad Sci U S A 100:685–690
41. Murata M, Thanan R, Ma N, Kawanishi S (2012) Role of nitrative and oxidative DNA damage in inflammation-related carcinogenesis. J Biomed Biotechnol 2012:623019
42. Trostchansky A, Rubbo H (2008) Nitrated fatty acids: mechanisms of formation, chemical characterization, and biological properties. Free Radic Biol Med 44:1887–1896
43. Rubbo H, Radi R (2008) Protein and lipid nitration: role in redox signaling and injury. Biochim Biophys Acta 1780:1318–1324
44. Wu G, Morris SM Jr (1998) Arginine metabolism: nitric oxide and beyond. Biochem J 336(pt 1):1–17

45. Maarsingh H, Zaagsma J, Meurs H (2009) Arginase: a key enzyme in the pathophysiology of allergic asthma opening novel therapeutic perspectives. Br J Pharmacol 158:652–664
46. Benson RC, Hardy KA, Morris CR (2011) Arginase and arginine dysregulation in asthma. J Allergy 2011:736319
47. Iyer R, Jenkinson CP, Vockley JG, Kern RM, Grody WW, Cederbaum S (1998) The human arginases and arginase deficiency. J Inherit Metab Dis 21(suppl 1):86–100
48. North ML, Khanna N, Marsden PA, Grasemann H, Scott JA (2009) Functionally important role for arginase 1 in the airway hyperresponsiveness of asthma. Am J Physiol Lung Cell Mol Physiol 296:L911–L920
49. Horowitz S, Binion DG, Nelson VM, Kanaa Y, Javadi P, Lazarova Z, Andrekopoulos C, Kalyanaraman B, Otterson MF, Rafiee P (2007) Increased arginase activity and endothelial dysfunction in human inflammatory bowel disease. Am J Physiol Gastrointest Liver Physiol 292:G1323–G1336
50. Ogino K, Obase Y, Takahashi N, Shimizu H, Takigawa T, Wang DH, Ouchi K, Oka M (2011) High serum arginase I levels in asthma: its correlation with high-sensitivity C-reactive protein. J Asthma 48:1–7
51. Meurs H, McKay S, Maarsingh H, Hamer MA, Macic L, Molendijk N, Zaagsma J (2002) Increased arginase activity underlies allergen-induced deficiency of cNOS-derived nitric oxide and airway hyperresponsiveness. Br J Pharmacol 136:391–398
52. Maarsingh H, Bossenga BE, Bos IS, Volders HH, Zaagsma J, Meurs H (2009) L-arginine deficiency causes airway hyperresponsiveness after the late asthmatic reaction. Eur Respir J 34:191–199
53. Ckless K, Lampert A, Reiss J, Kasahara D, Poynter ME, Irvin CG, Lundblad LK, Norton R, van der Vliet A, Janssen-Heininger YM (2008) Inhibition of arginase activity enhances inflammation in mice with allergic airway disease, in association with increases in protein S-nitrosylation and tyrosine nitration. J Immunol 181:4255–4264
54. Durante W, Johnson FK, Johnson RA (2007) Arginase: a critical regulator of nitric oxide synthesis and vascular function. Clin Exp Pharmacol Physiol 34:906–911
55. Modolell M, Munder PG (1994) Macrophage mediated tumor cell destruction measured by an alkaline phosphatase assay. J Immunol Methods 174:203–208

Chapter 10
An Animal Model of Citrin Deficiency, the Citrin/Mitochondrial Glycerol 3-Phosphate Dehydrogenase Double-Knockout Mouse

Takeyori Saheki

Abbreviations

2OG	2-Oxoglutarate
AGC	Aspartate glutamate carrier
Ala	Alanine
Arg	Arginine
ASA	Argininosuccinate
ASS	Argininosuccinate synthetase
BMI	Body mass index
CD	Citrin deficiency
Cit	Citrulline
CP	Carbamoylphosphate
CTLN2	Adult-onset type II citrullinemia
Ctrn	Citrin
DHAP	Dihydroxyacetone phosphate
FTTDCD	Failure to thrive with dyslipidemia
G3P	Glycerol 3-phosphate
Gln	Glutamine
Glu	Glutamate
GP	Glycerophosphate
H^+	Proton
KO	Knockout
Lac	Lactate
Lys	Lysine
MA	Malate aspartate

T. Saheki (✉)
Institute of Resource Development and Analysis, Kumamoto University,
2-2-1 Honjo, Kumamoto 860-0811, Japan
e-mail: takesah@kumamoto-u.ac.jp

H. Tsukahara and K. Kaneko (eds.), *Studies on Pediatric Disorders*, Oxidative Stress in Applied Basic Research and Clinical Practice, DOI 10.1007/978-1-4939-0679-6_10,

Mal	Malate
MCT	Medium-chain triglyceride
mGPD	Mitochondrial glycerol 3-phosphate dehydrogenase
Na	Sodium
NASH	Nonalcoholic steatohepatitis
NICCD	Neonatal intrahepatic cholestasis caused by citrin deficiency
PFC	Protein, fat, and carbohydrate
Pyr	Pyruvate
Thr/Ser	Threonine/serine
wt	Wild-type

10.1 Introduction

Citrin deficiency (CD) is an autosomal recessive disease caused by mutations in *SLC25A13*, the gene encoding a mitochondrial aspartate glutamate carrier (AGC) named citrin. It is a novel disease concept involving adult-onset type II citrullinemia (CTLN2), neonatal intrahepatic cholestasis (NICCD), and failure to thrive with dyslipidemia (FTTDCD). CTLN2 is characterized by disturbances of consciousness with hyperammonemia and citrullinemia. However, some treatments for hyperammonemia caused by urea cycle enzyme deficiencies or conditions other than citrin deficiency may cause CD patients to deteriorate, often resulting in death, and all medical practitioners should keep this in mind. Such clinical features have been shown using a mouse model of CD.

10.2 Pathogenesis of CTLN2

Study of CTLN2 started in response to a finding showing high serum citrulline levels in patients with a kind of hepatocerebral diseases [1]. Saheki et al. [2] analyzed hepatic argininosuccinate synthetase (ASS)—thought to be a causative enzyme—from these and neonatal patients with citrullinemia, and found that there are two types in citrullinemia from a viewpoint of the enzymogical properties. One is classical citrullinemia or CTLN1 with a generalized ASS deficiency, found mainly in neonates and infants, and the other is CTLN2, in which the ASS defect is found only in liver, but not in cultured fibroblasts or kidney [3]. The pathogenesis of the former is easily understood because ASS deficiency is based on *ASS* gene abnormalities. However, the latter's pathogenesis has long been unclear. In 1999, Kobayashi et al. [4] localized the causative gene to chromosome 7q21.3 using homozygosity mapping and identified *SLC25A13* as the disease gene, which encodes a mitochondrial transporter named citrin.

10.3 Structure and Functions of Citrin

Citrin, encoded by *SLC25A13*, is a protein consisting of 675 amino acid residues with a molecular weight of approximately 74 kD. The carboxy-terminal portion demonstrates similarity with proteins of the mitochondrial solute-carrier family, with six transmembrane domains (Fig. 10.1). The amino-terminal portion contains four putative EF-hand domains that are conserved in calcium-binding proteins [4]. The structure of citrin is most similar to that of aralar, encoded by the *SLC25A12* gene identified by del Arco and Satrustegui in 1998 [5]. Citrin and aralar have been shown to be isoforms of AGC [6]. Citrin is mainly expressed in the liver, kidney, and heart; and aralar in the skeletal muscle, brain, kidney, and heart [4, 5, 7]. AGC exchanges glutamate (Glu) and a proton (H^+) for aspartate (Asp) across the inner mitochondrial membrane. In mitochondria with an active electron transport system which transports H^+ from mitochondria to cytosol, AGC unidirectionally transports Glu and H^+ from the cytosol to the mitochondrial matrix space and Asp from the mitochondria to the cytosol. AGC therefore plays a metabolic role in the supply of Asp from mitochondria for the synthesis of protein, nucleotides and urea, which occurs in the cytosol. Figure 10.2a illustrates urea synthesis from ammonia. More importantly, AGC plays a role in transporting NAD-reducing equivalents from the cytosol to the mitochondria as a component of the malate Asp (MA) shuttle (Fig. 10.2b) and in regulating the NADH/NAD^+ ratio in the cytosol; gluconeogenesis from lactate (Lac) inevitably needs AGC, namely citrin, in order to maintain stoichiometry of NADH in the cytosol (Fig. 10.2c), as pointed out by Krebs et al. [8].

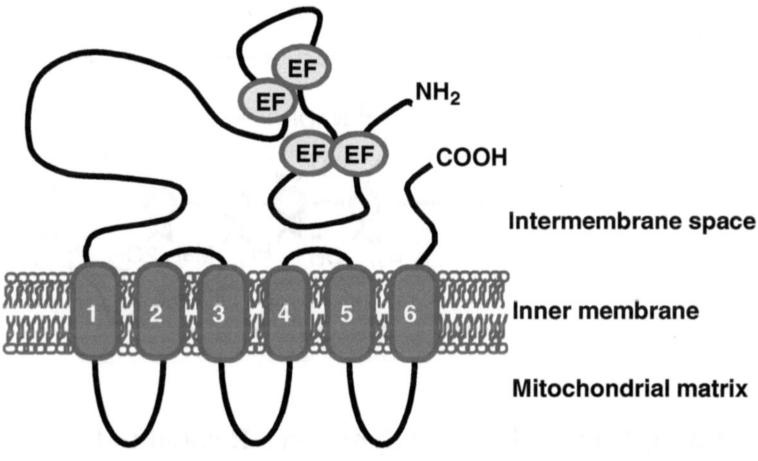

Fig. 10.1 Structure of citrin [12]. The structure of citrin is schematically drawn with six transmembrane domains (numbered 1–6) on the C-terminal half and EF-hand motifs to bind calcium on the N-terminal extension

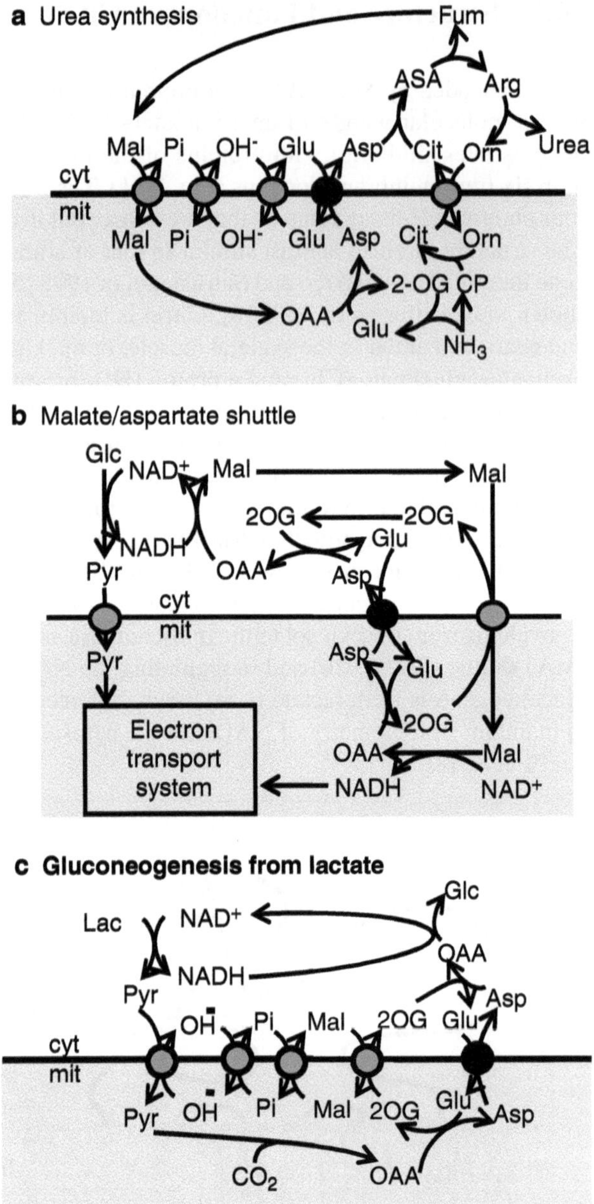

Fig. 10.2 Metabolic functions of citrin. (**a**) Urea synthesis from ammonia, (**b**) malate aspartate shuttle, (**c**) gluconeogenesis from lactate. *Grey* and *black circles* denote solute carriers located on the inner mitochondrial membrane and citrin, respectively. *Arg* arginine, *Asp* aspartate, *ASA* argininosuccinate, *Cit* citrulline, *CP* carbamoylphosphate, *cyt* cytosol, *Glu* glutamate, *Fum* fumarate, *Glc* glucose, *Mal* malate, *mit* mitochondria, *OAA* oxaloacetate, *2OG* 2-oxoglutarate, *Orn* ornithine, *Pi* inorganic phosphate, *Pyr* pyruvate

10.4 Multiple Disorders and Symptoms Related to CD

Following the discovery of *SLC25A13* as the causative gene for CTLN2, the same mutations in this gene were shown to be the cause of a type of neonatal hepatitis and/or cholestasis named NICCD, the symptoms of which are different from those

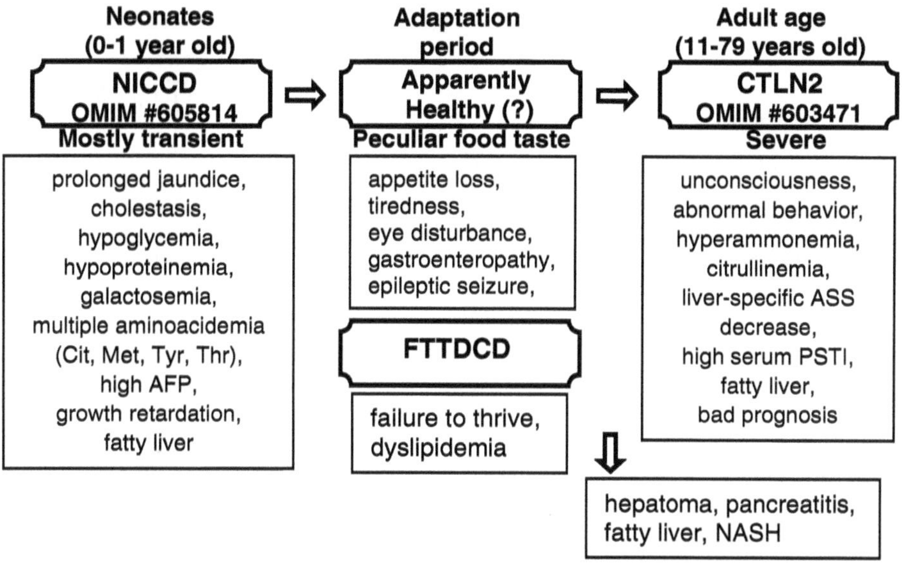

Fig. 10.3 Diverse symptoms of citrin deficiency during life time. *CTLN2* adult-onset type II citrullinemia, *FTTDCD* failure to thrive with dyslipidemia caused by citrin deficiency, *NICCD* neonatal intrahepatic cholestasis caused by citrin deficiency, *NASH* nonalcoholic steatohepatitis, *PSTI* pancreatic secretory trypsin inhibitor

of CTLN2 [9–11]. This discovery led us to propose CD as a disease entity [12]. Furthermore, CD was found to be related to diverse disorders or symptoms, such as hepatoma, pancreatitis, dyslipidemia, fatty liver, and NASH [13–20]. Figure 10.3 illustrates multiple disorders and symptoms related to CD during the lifetime of the patients. Even during apparently healthy adaptation periods, many CD patients have unusual food preferences, abnormal laboratory findings and signs of oxidative stress [18], and some have been diagnosed as suffering from FTTDCD [19]. It is also noteworthy that many CTLN2 patients have been diagnosed at the first stage as suffering from psychiatric diseases such as epilepsy, depression, and schizophrenia [15, 21]. Medical practitioners should add CD to the list of diseases to be differentiated during diagnosis of these.

10.5 Unique Characteristics of CD Patients

Unique features of CD patients are listed below; these are critical for considering the mechanisms of hyperammonemia in CD.

Since the time when CTLN2 was termed a hepatocerebral disease, it has been well known that CD patients consume a large amount of peanuts or soybeans, but the reasons for this were not known. A nutritional assessment by Saheki et al. [22] of 18 Japanese CD subjects in apparently healthy states (see Fig. 10.3) aged from 1 to 33 years provided some answers (Fig. 10.4). The proportion of protein, fat, and

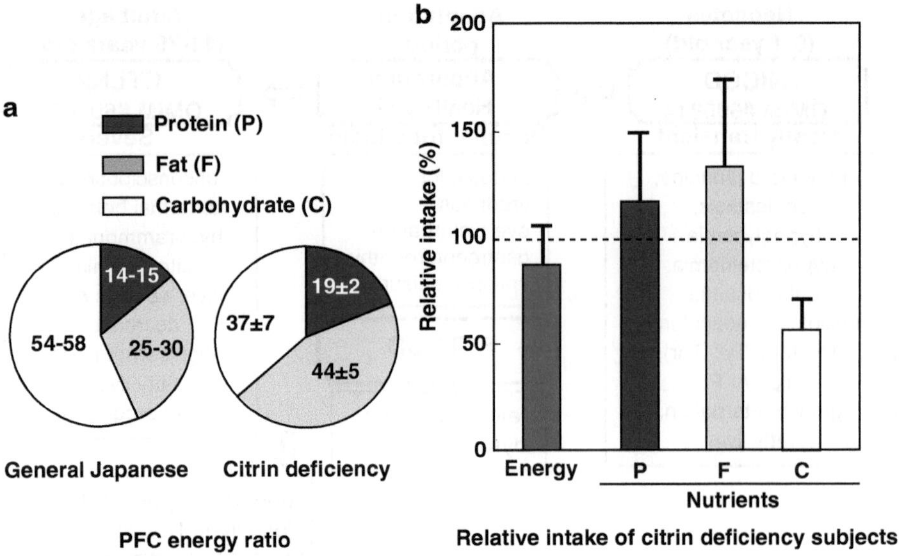

Fig. 10.4 Peculiar food taste: assessment of 18 citrin deficiency subjects [22]. (**a**) Comparison of proportions of protein, fat, and carbohydrate intake as energy (PFC ratios) between 18 citrin deficiency subjects and general Japanese population. (**b**) Average energy and nutrient intakes of citrin deficiency subjects matched by age and sex with the general Japanese population. The means ± SD values for total energy and nutrient intake of 18 citrin deficiency subjects were converted into percentages relative to age- and sex-matched controls

carbohydrate intake as energy (PFC ratio) (Fig. 10.4a), of the CD subjects was quite different from that of the general Japanese population, indicating a marked decrease in carbohydrate contributing to the total energy intake in CD subjects. Figure 10.4b shows the average relative intakes of total energy and nutrients of the CD subjects matched by age and sex with the general Japanese population. The average energy, protein, and fat intakes expressed as % of controls were 87 ± 20, 116 ± 34, and 134 ± 43, respectively, suggesting that the CD subjects did not differ substantially in their average total energy, protein, and fat intakes from the average population. In contrast, the average relative carbohydrate intake for the CD subjects was approximately half (56 ± 14 %) that of age- and sex-matched Japanese controls. The results strongly support an avoidance of carbohydrate-rich foods by CD subjects; they dislike sweet things and cooked rice, preferring protein- and fat-rich foods. Nakamura et al. [23] reached a similar conclusion with CTLN2 patients.

It is clear from case reports why CD subjects dislike carbohydrates (sugars). It is noteworthy that a CTLN2 case reported by Tamakawa et al. [24] showed consciousness disturbance due to hyperammonemia following infusion of a high glucose solution. Fukushima et al. [25] reported that recurrent disturbed consciousness from hyperammonemic attacks in the CTLN2 patients given a high carbohydrate diet (PFC ratio, 10/15/75) was gradually ameliorated, with a decrease in blood ammonia level, after a change to a carbohydrate-restricted diet (PFC ratio, 15/40/45) with arginine (Arg) administration. Dimmock et al. [26] reported that failure to thrive observed in an NICCD baby was ameliorated by a dietary change in PFC ratio from 11/42/47 to 15/50/35. An accidental increase in dietary carbohydrate content in a CTLN2 patient induced a marked increase in plasma triglyceride, as reported by Imamura et al. [27]. Many CTLN2 patients deteriorated and died shortly after infusion of Glyceol® (10 % glycerol and 5 % fructose) for the treatment of brain edema caused by hyperammonemia [28, 29]. Consciousness disturbance occurred after alcohol intake in a number of CTLN2 patients [13, 29]. These case reports and functions of citrin suggest a harmful effect of the substances that metabolically produce NADH in liver cytosol (see also Fig. 10.8, which is described in more detail in Sect. 10.8).

10.6 The First Mouse Model of CD, *Slc25a13* Knockout, Failed to Display Hallmarks of CTLN2

Animal models of disease are indispensable for analysis of pathophysiology and the development of therapeutics. The *Slc25a13* knockout (KO), or citrin (Ctrn)-KO mouse model was created using homologous recombination [30]. We have several examples of evidence for a marked loss of hepatic AGC activity in in vitro experiments together with loss of citrin mRNA and protein: (1) transport of Asp into the uncoupled mitochondria, following oxidation of mitochondrial NAD(P)H by the formation of oxaloacetate (OAA) and conversion to malate (Mal) in the presence of rotenone and 2-oxoglutarate (2-OG); (2) Asp formation from Glu and Mal in the mitochondria; (3) MA shuttle activity constructed with cytosolic components and mitochondria. Furthermore, liver perfusion experiments using Ctrn-KO mice showed that (1) no glucose production took place from Lac, although gluconeogenesis with Pyr was comparable to wild-type (wt) mice, (2) urea formation from ammonia with ornithine was inhibited and Lac/Pyr ratio was increased, and (3) the addition of asparagine to the perfusate increased the rate of urea formation, although it did not decrease the Lac/Pyr ratio. These results indicate that Asp is rate-limiting for ureogenesis in Ctrn-KO mice. Despite these in vitro data, there were initially almost no in vivo symptoms of CD, although we later found that C57BL/6-congenic Ctrn-KO mice suffer from hypoglycemia and fatty liver [31] (see Table 10.1).

Table 10.1 Comparison of symptoms between human citrin deficiency and mouse models

	Human citrin deficiency		Mouse models	
Symptoms	CTLN2	NICCD	Ctrn-KO	Double-KO
Hyperammonemia (fed conditions)	+	±	–	+
Citrullinemia	+	+	–	+
Hypoglycemia	–	+	–	+
Fatty liver	+	+	+(fasted)	+(fasted)
Failure to thrive or skinniness	+	+	–	+
Toxicity of sugars	+	(+)?	–	+
Psychiatric symptoms	+	–	?	?

We inquired into the reason why Ctrn-KO mice showed almost no hallmarks of CTLN2. There are two systems of NADH shuttle that transport cytosolic NADH-reducing equivalents to mitochondria: the MA shuttle, in which AGC is major component, and the glycerol phosphate (GP) shuttle, composed of cytosolic and mitochondrial glycerol 3-phosphate dehydrogenase (mGPD). Both systems are active in mouse liver; GP shuttle activity has been shown to be approximately 50 % that of MA shuttle [30]. On the other hand, GP shuttle activity should be very low in human liver, considering its mGPD activity is about one-twentieth that of mouse liver [32, Saheki et al., unpublished data]. Therefore, the NADH shuttle in human liver is almost completely dependent on the MA shuttle, while the GP shuttle may compensate for the defect of the MA shuttle in mouse liver. These observations helped us create Ctrn/mGPD double-KO mice in which both the MA and GP shuttles are defective [31]. The double-KO mice presented a number of symptoms very similar to those of human CD (Table 10.1).

10.7 Phenotypes of Ctrn/mGPD Double-KO Mice

Table 10.1 compares symptoms in human CD and the two mouse models. One of the most characteristic features of the double-KO mice is that hyperammonemia was observed underfed but not starved conditions (Fig. 10.5a) [31], which is comparable to the observation in CTLN2 patients [33]. We noted that hyperammonemia could be induced or enhanced by an enteral administration by a gastric tube of not only sucrose (5 or 10 g/kg body weight; Fig. 12.5b) [31], but also glycerol (1 or 2 g/kg), and ethanol (1.4 g/kg) [Saheki et al., unpublished data]. In addition, the double-KO mice avoid solutions of sucrose at high concentrations (≥10 %), ethanol (>3 % (v/v)), and glycerol (>5 %) [Saheki et al., unpublished data]. The other phenotypes of the double-KO mice comparable to those in human CD [13, 19, 34] are: a significantly higher level of plasma citrulline; hypoglycemia (also observed in congenic Ctrn-KO mice); fatty liver induced by overnight starvation (also observed in congenic Ctrn-KO mice); growth retardation during post-weaning period, followed by catch-up during adulthood.

10.8 Metabolic Perturbations Induced by Enteral Sucrose Administration in Ctrn/mGPD Double-KO Mice

To precisely clarify the pathophysiology of hyperammonemia induced by enteral sucrose (10 g/kg body weight) administration, we analyzed plasma amino acids and hepatic metabolites.

Plasma amino acid analysis revealed a slight but significant increase in citrulline (Cit) and ratio of threonine to serine (Thr/Ser ratio) in the starved double-KO mice [31], but no change was observed in plasma arginine (Arg) or Fischer ratio

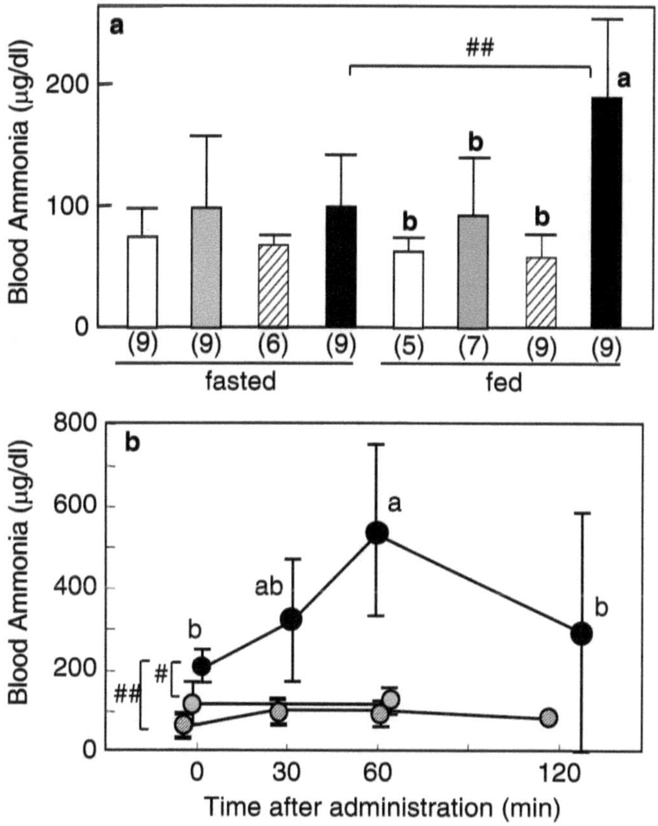

Fig. 10.5 Blood ammonia concentrations of fasted and fed wild-type (white column), Ctrn-KO (*grey*), mGPD-KO (*striped*), and Ctrn/mGPD double KO (*black*) mice in (**a**) and effect of sucrose administration on blood ammonia concentrations of Ctrn-KO (*grey circle*), mGPD-KO (*striped*), and double KO (*black*) mice in (**b**) [31]. Cardiac blood was taken under anesthesia from fed or fasted (for 16–17 h) mice in A. In B, sucrose (10 g/kg body weight) was enterally administered to mice with a gastric tube at zero time-point. Blood was taken from separate sets of mice for the initial time-point (0 time) and each subsequent time-point indicated. The number of animals used is shown in *parentheses* in (**a**) while 5–8 mice were used for each time point in (**b**). The means (±SD) were compared with a one-way ANOVA followed by the Tukey-Kramer test between four genotypes in (**a**) and between time points in (**b**). Values sharing identical alphabetical superscripts are not significantly different. The signs, # and ##, indicate a significant level of $p<0.05$ and $p<0.01$ (by Student's t test), respectively, between the two groups indicated

(ratio of plasma branched-chain amino acids to aromatic amino acids), which are markers of citrin deficiency or CTLN2 [13, 35]. It is likely that the plasma Arg level was not increased because the rise in plasma Cit was not high. Although the Fischer ratio was not different between the four mouse genotypes under the basal conditions, the ratio decreased 1 h after enteral sucrose administration, as did plasma alanine (Ala) in the Ctrn/mGPD double-KO mice [31], possibly reflecting the severe hyperammonemia and a decrease in the hepatic pyruvate (Pyr), respectively.

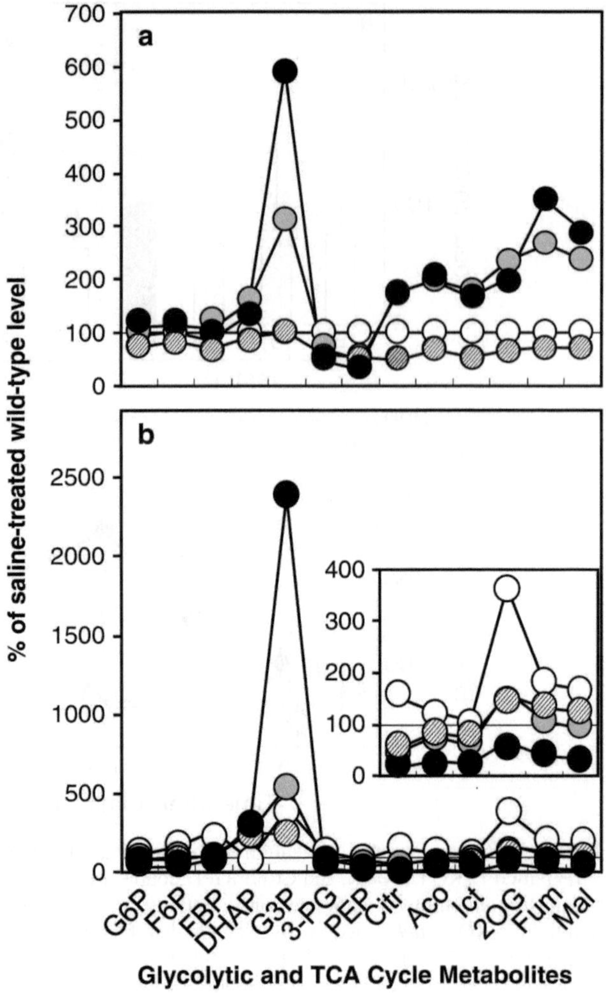

Fig. 10.6 Relative differences in hepatic glycolytic and TCA cycle metabolites among the mouse genotypes following enteral saline (**a**) or sucrose administration (**b**) [36]. The hepatic metabolites from each genotype are expressed as the percent of saline-treated wt mice. *White circles* denote wild-type; *grey* Ctrn-KO; *striped* mGPD-KO; and *black* double-KO mice. The *insert* in (**b**) shows a magnification of the data specifically for the TCA cycle intermediates. *G6P* glucose 6-phosphate, *F6P* fructose 6-phosphate, *FBP* fructose 1,6-bisphosphate, *DHAP* dihydroxyacetone phosphate, *G3P* glycerol 3-phosphate, *3PG* 3-phosphoglycerate, *PEP* phosphoenolpyruvate, *Citr* citrate, *Aco* cis-aconitate, *Ict* isocitrate, *2OG* 2-oxoglutarate, *Fum* fumarate, *Mal* malate

In fact, the hepatic Pyr concentration decreased (42 ± 21 to 21 ± 9 nmol/g liver), with a concomitant increase in the ratio of lactate (Lac) to Pyr (Lac/Pyr; 36 ± 14 to 75 ± 37) reflecting an increase in the cytosolic NADH/NAD^+ ratio [31].

In order to comprehensively analyze the metabolic perturbations induced by enteral sucrose administration, metabolomic analyses using gas chromatography/mass spectrometry and capillary electrophoresis/mass spectrometry have been adopted and the results have been confirmed by enzymatic analyses [36]. The following are the results of these analyses.

Figure 10.6a, b show the hepatic concentrations of glycolysis and TCA cycle intermediates of four mouse genotypes under basal conditions (administration of saline solution) and following administration of sucrose solution (10 g/kg) via a gastric tube [36]. The most marked difference in the metabolomics analyses was found in glycerol 3-phosphate (G3P) between the genotypes and between the basal

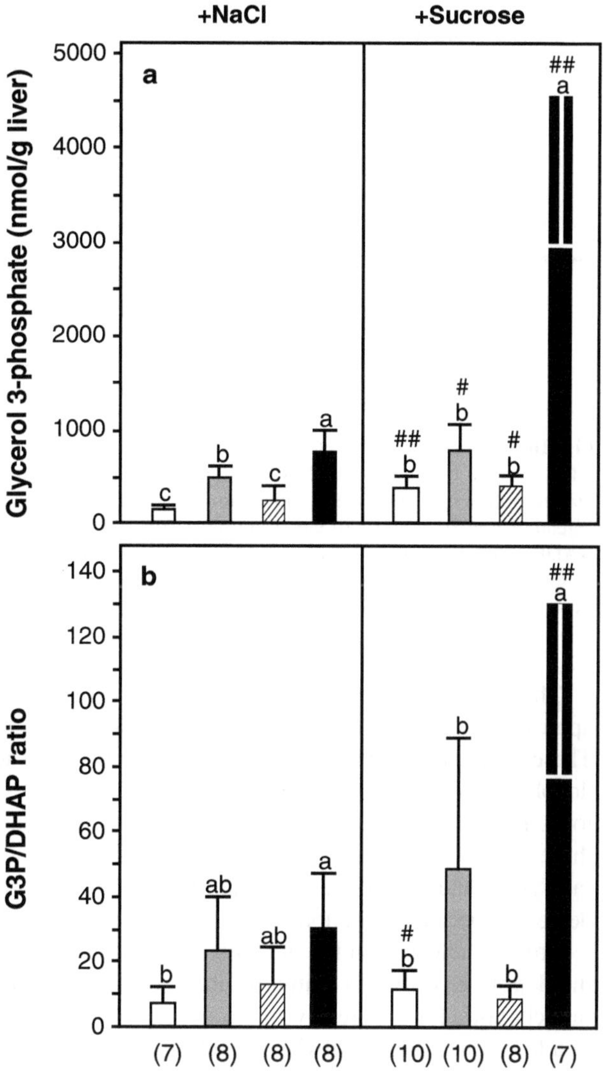

Fig. 10.7 Hepatic G3P (**a**) and the G3P/DHAP ratio (**b**) among the four genotypes following enteral saline and sucrose administration [36]. Saline (NaCl) or sucrose (10 g/kg body weight) was administered to the fed mice and after 1 h, liver was freeze-clamped and the extract was prepared. G3P and DHAP concentrations in the neutralized extract were determined by enzymatic methods [36]. *White column* denotes wt, *grey*, Ctrn-KO, *striped*, mGPD-KO, and *black*, double-KO mice. *Parenthesis* denotes number of mice. Data on each group of mice are expressed as mean ± SD. Differences among the mice were evaluated by ANOVA followed by the Tukey-Kramer test. Mean values with the same character are not different at $p < 0.05$. # and ## denote $p < 0.05$ and $p < 0.01$, respectively, using a Student's *t*-test to compare each sucrose-treated group versus saline treated

and sucrose-loaded conditions. As shown in Fig. 10.7a, enzymatic analysis revealed that the hepatic G3P level was several-fold higher in the double-KO and twofold higher in the Ctrn-KO mice than in the wt mice in the basal state, and greatly increased in the double-KO mice under sucrose load. Changes in the G3P/dihydroxyacetone phosphate (DHAP) ratio (Fig. 10.7b), one of the indicators of the cytosolic NADH/NAD$^+$ ratio, reflected the changes in G3P levels. This suggests that NADH formed during glycolysis inhibits glyceraldehyde 3-phosphate dehydrogenase, changing metabolic flow from glycolysis to G3P resulting in a large increase of the latter. The changes in the G3P/DHAP ratio were much greater than those in the Lac/Pyr ratio despite both parameters being indicative of the cytosolic NADH/NAD$^+$ ratio. The reasons for this are unclear.

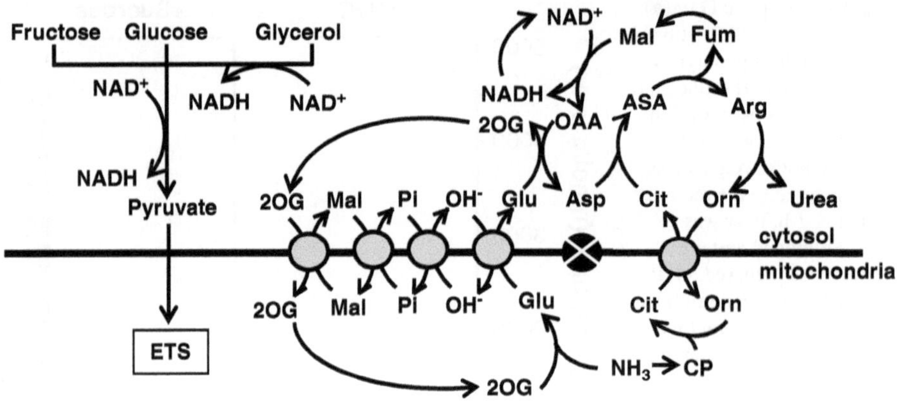

Fig. 10.8 Schematic representation of ureogenesis from ammonia in the absence of citrin and effect of sugar metabolism [22]. A *black circle* with a cross mark denotes defected citrin, and *grey circles*, solute carriers, on the inner mitochondrial membrane. Abbreviations are shown in the legend of Fig. 10.2, and *ETS* electron transport system. Without citrin, ureogenesis is possible, if NADH formed is oxidized back to NAD^+ for the operation of the metabolic pathway. Sugars generate NADH in the cytosol, which, if not oxidized properly, inhibits the ureogenesis, resulting in hyperammonemia

The TCA cycle intermediates in the Ctrn-KO and double-KO mice were approximately double those in the wt mice following saline administration (Fig. 10.6a). These decreased to basal level in Ctrn-KO mice, and to an even lower level in the double-KO mice, after sucrose load (Fig. 10.6b). The results can be interpreted as follows. The TCA cycle intermediates that increased during the postprandial period were those that accumulated due to inhibited gluconeogenesis, and should be converted to fatty acids and triglycerides. During the sucrose load, the TCA cycle intermediates decreased because of inhibited glycolysis following decreased supply of Pyr.

From a viewpoint of the function of citrin, the mechanism of hyperammonemia in the double-KO mice is primarily induced by a decreased supply of Asp from the mitochondria for the urea cycle. Formation of Asp in the cytosol is also inhibited due to a low concentration of OAA resulting from the high $NADH/NAD^+$ ratio and the equilibrium of malate dehydrogenase in the liver of the double-KO mice. In other words, ureogenesis in the absence of citrin takes place as shown in Fig. 10.8 (see also Fig. 10.2a, which illustrates ureogenesis in the presence of citrin): Glu leaves the mitochondria and Asp is formed by the transamination from Glu to OAA by the action of cytosolic Asp aminotransferase. Asp is one of the substrates of ASS, whereby urea is formed. In this metabolic pathway, OAA should be reformed from fumarate (Fum) liberated from argininosuccinate (ASA) via Mal, where NAD^+ is reduced to NADH. If the NADH can be oxidized back to NAD^+, ureogenesis can take place continuously. If cytosolic NADH formation is stimulated by the metabolism of carbohydrates or ethanol, ureogenesis is limited, resulting in hyperammonemia.

Figure 10.9 shows experimental data that reflect this [36]. The Asp concentration (Fig. 10.9a) was lower in Ctrn-KO and the double-KO mice than in wt and

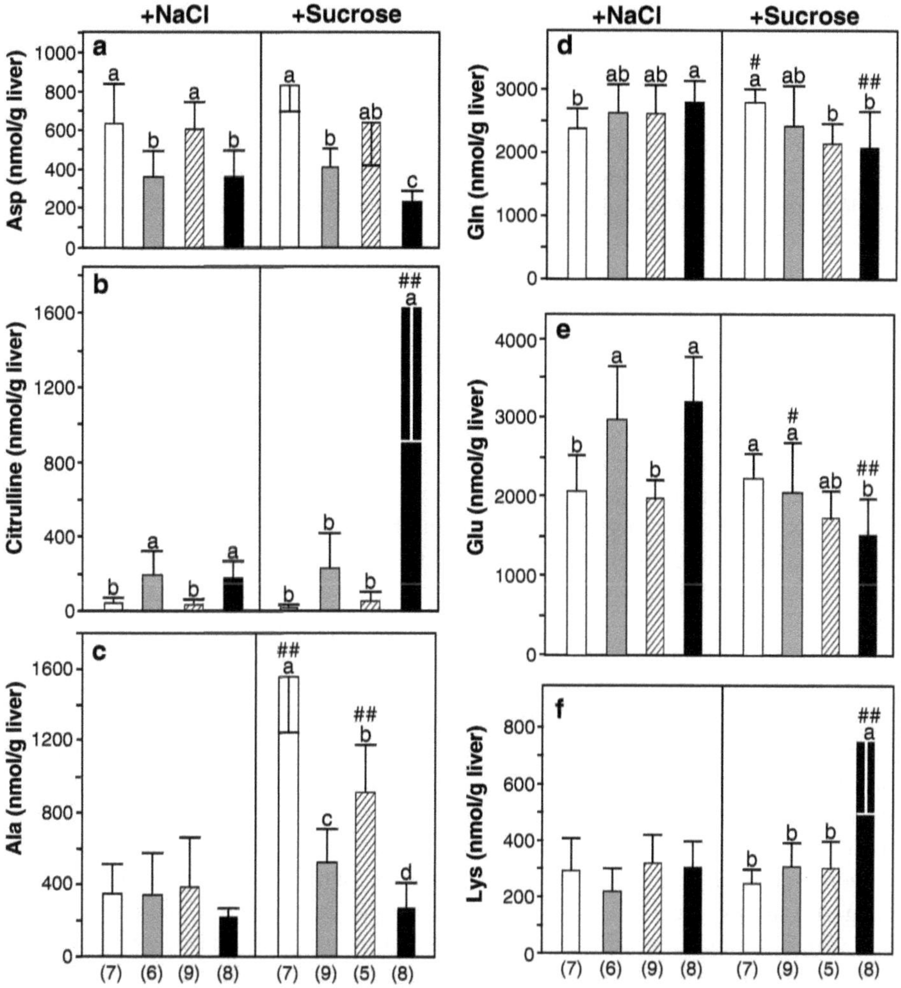

Fig. 10.9 Hepatic concentrations of aspartate (Asp; **a**), citrulline (**b**), alanine (Ala; **c**), glutamine (Gln; **d**), glutamate (Glu; **e**), and lysine (Lys; **f**) among the mouse genotypes following enteral saline (NaCl) or sucrose administration [36]. Saline (0.9 % NaCl) or sucrose (10 g/kg body weight) was enterally administered to the fed mice. One hour after the administration, liver was freeze-clamped and extract was prepared, as described in [36]. Hepatic amino acids in the neutralized extract were determined by liquid choromatography/tandem mass spectrometry (Acquity UPLC/TQD; Waters, Milford, USA) after solid phase extraction followed by derivatization, as described in [36]. *White bars* denote wt; *grey*, Ctrn-KO; *striped*, mGPD-KO; and *black*, double-KO mice. Data from each group of mice are expressed as mean ± SD. *Parentheses* denote the number of mice in each genotype examined. Differences among the genotypes within the same treatment group were evaluated by ANOVA followed by the Tukey-Kramer test, where mean values with the same character are not statistically different. Differences between two treatment groups within the same genotype were evaluated using a Student's *t*-test, where # denotes a statistical difference at $p<0.05$ and ## at $p<0.01$

mGPD-KO mice under basal conditions, and was not essentially changed under the sucrose load. This is probably because we could not assay the Asp concentrations in the cytosol and mitochondria separately under our experimental conditions. On the other hand, the pattern of Cit concentration in the liver of the four genotypes (Fig. 10.9b) was almost the same as that of hepatic G3P: there were higher levels of Cit in Ctrn-KO and double-KO mice than in other mice under basal conditions, while in response to enteral sucrose administration only the level in the double-KO mice increased markedly.

Concentrations of Ala (Fig. 10.9c), glutamine (Gln; Fig. 10.9d), Glu (Fig. 10.9e), and lysine (Lys; Fig. 10.9f) also exhibited significant differences between genotypes, or were altered significantly following enteral sucrose administration. The Ala concentration (Fig. 10.9c) was not different between the genotypes under basal conditions; however, enteral sucrose administration increased the concentration in wt and mGPD-KO mice, but not in Ctrn-KO and double-KO mice, indicating increases in Pyr in the former and no increase in the latter. Enteral sucrose administration increased Gln in wt, but decreased it in the double-KO mice (Fig. 10.9d). Glu, higher in Ctrn-KO and double-KO mice under basal conditions, was decreased significantly following sucrose administration in both genotypes (Fig. 10.9e). Lys was increased only in the double-KO mice under sucrose load (Fig. 10.9f).

To identify the enzyme reaction that is activated or inhibited by a certain metabolic change, the cross-over point method or theorem is useful. Relative changes in the urea cycle intermediates following the enteral sucrose administration in the double-KO mice are shown in Fig. 10.10a. A marked increase in Cit and decrease in ornithine and ASA produced two cross-over points, suggesting a reaction activated from ornithine to Cit and a reaction inhibited from Cit to ASA by sucrose administration. The inhibition of glycolysis and decreased TCA cycle intermediates suggest a decreased energy supply from these pathways, while decreases in Gln and Glu suggest a supply of energy. Degradation of these amino acids also increases ammonia, which stimulates synthesis of carbamoylphosphate (CP) and Cit, which in turn accumulates due to inhibition of the ASS reaction caused by shortage of Asp. While ATP supply is necessary for the synthesis of CP and ASA, it seems to be sufficient at least for CP synthesis.

The mechanism of the increase of hepatic Lys in the double-KO mice under sucrose load was elucidated by metabolomic analysis with capillary electrophoresis/mass spectrometry. Figure 10.10b shows a cross-over point between Lys and saccharopine in the Lys catabolic pathway, indicating the inhibition of L-Lys-2-OG reductase. Hyperlysinemia has been observed in several inborn metabolic abnormalities, including urea cycle disorders and pyruvate carboxylase deficiency. Kamoun et al. [37] postulated that the elevated plasma Lys levels in those disorders at least in part depend on the rate of mitochondrial Lys degradation due to the decreased availability of mitochondrial 2-OG in the liver.

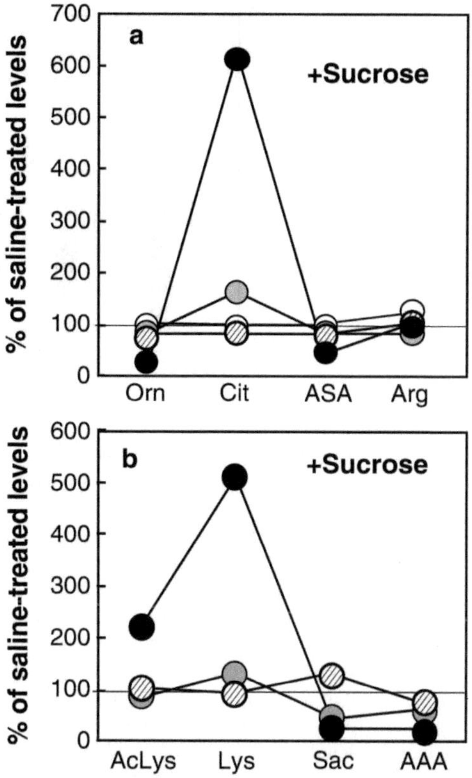

Fig. 10.10 Relative differences in hepatic urea cycle (**a**) and lysine catabolic (**b**) metabolites among the mouse genotypes following the enteral sucrose administration [36]. The hepatic metabolites following the enteral sucrose administration (10 g/kg body weight) are expressed as the percent of saline-treated levels within each of the mouse genotypes. *Orn* ornithine, *Cit* citrulline, *ASA* argininosuccinate, *Arg* arginine in (**a**); and *AcLys* N^6-acetyl-lysine, *Sac* saccharopine, *AAA* aminoadipic acid in (**b**)

10.9 Effect of Diet and Its Protein Content on Body Weight and Food Intake

One of the physical characteristics of most CD patients is a thin, lean body habitus (physique) beginning during the so-called adaptive or apparently healthy period; more than 90 % of CTLN2 patients have a body mass index (BMI) lower than 20, and approximately 40 % have a BMI less than 17 [12, 13]. At an earlier age, after resolution of NICCD, some patients developed a failure to thrive with dyslipidemia (FTTDCD) [19]. Mutoh et al. [38] reported a case of a 13-year-old girl whose BMI stayed at approximately 15 from an age of 6 years old until 13 years old, while the average BMI of age-matched control Japanese girls increased from 16 to 19 during the same time period. There has been no explanation so far for this apparent pathophysiologic aspect of CD, so we examined the effect of diets on the body weight of the four mouse genotypes [39].

Mice are usually maintained on a commercially available pelleted laboratory chow diet (CE-2; CLEA rodent diet, CLEA Japan, Inc., Tokyo, Japan) recommended for breeding rodents. To enable the assessment of dietary composition on

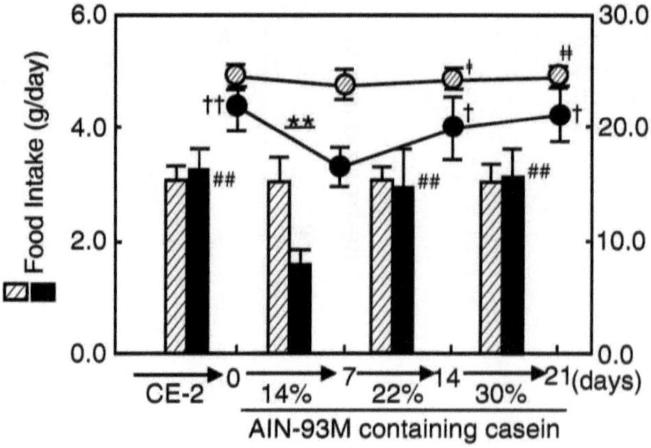

Fig. 10.11 Effects of diet change from CE-2 to AIN-93M, and supplementation of AIN-93M with casein, on body weight and food intake of mice [39]. Female mice were maintained on CE-2 prior to the experiment. On day 0, the diet was changed to AIN-93M containing 14 % casein (original AIN-93M formulation) and given for 7 days. The diet was subsequently changed again to the AIN-93M diet containing 22 % (by adding 8 % casein and reducing the corn starch by 8 %) for another 7 days, followed by the AIN-93M diet containing 30 % casein (16 % added casein with 16 % reduction of corn starch) for 7 days. The food intake and body weights of the mice were measured daily. Each value represents mean ± SD ($n=7$ for mGPD, and $n=11$ for Ctrn/mGPD-KO mice). Mean food intake values were calculated from the average food intake of each mouse over the 7 days. Although body weights of the mice were measured daily, mean body weight values were calculated using the body weight of each mouse on days 7, 14, and 21. *Striped* and *black columns* and *circles* denote food intakes and body weights of mGPD-KO littermate control mice and Ctrn/m GPD-KO mice, respectively. Differences in food intake for each diet (or supplementation) between genotypes was determined using an unpaired *t*-test (**$p<0.01$). Differences in food intake between diets within each genotype was determined using a paired *t*-test (##$p<0.01$). Differences in body weight for mGPD-KO mice (#$p<0.05$ and ##<0.01) and Ctrn-KO/mGPD-KO mice (†$p<0.05$ and ††$p<0.01$) comparing the AIN-93M diet versus other diets (or supplementation) was determined using a paired *t*-test

our mouse model of CD, the diet was changed from CE-2 to a purified rodent diet (AIN-93M) [39], which is recommended as sufficient for mature rodents by the American Institute for Nutrition. Monitoring the mice over a period of 7 days, the change in diet did not significantly affect the body weights of the Ctrn-KO or mGPD-KO mice, but slightly decreased the body weight of wt mice (−1.5 g or 6.6 %; $p<0.05$). In contrast, the Ctrn/mGPD double-KO mice showed a marked decrease in body weight after 7 days (−5.6 g or 26.7 %; $p<0.01$). The decrease in body weight of the double-KO mice was accompanied by a decrease in food intake (Fig. 10.11). The double-KO mice had an average food intake that was approximately 50 % of the average of the other genotypes. Limiting food (2 g CE-2/day) supply to mGPD-KO mice, the littermates of the double-KO mice, induced a decrease in body weight similar to that observed for the double-KO mice, indicating that lower food intake is the primary cause of the weight reduction in double-KO mice.

The major difference between the CE-2 and AIN-93M diets is the protein content of 25 % and 14 %, respectively. Addition of casein to the AIN-93M diet (+8 %; total protein content of 22 %), together with a reduction of cornstarch from 46.6 to 38.6 %, increased the food intake of the Ctrn/mGPD-KO mice to 3.0 ± 0.7 g/day and significantly increased their body weight within 7 days ($p < 0.01$; Fig. 10.11). These results indicate that although the double-KO mice dislike carbohydrate-rich food, they can increase food intake if it contains more protein. In fact, the increased food intake of the double-KO mice after modifying the protein content of AIN-93M diet also increased the total carbohydrate intake of the mice from 0.75 ± 0.11 to 1.14 ± 0.27 g/day ($p < 0.01$), based on the dietary intake and the calculated carbohydrate content of the original and modified AIN-93M diets (46.6 % and 38.6 %, respectively). This observation is noteworthy, as we were able to increase the overall carbohydrate tolerance of the mice by increasing the protein content of the AIN-93M diet, although we have shown that carbohydrates have a toxic effect on the double-KO mice [31, 36]. This has great clinical significance and is consistent with our clinical observations showing a dramatic improvement of a baby with growth retardation in response to a high-protein, low-carbohydrate diet [26].

While it remains to be elucidated how the protein-rich diet increases food intake of double-KO mice or CD patients, several kinds of amino acids seem to have effects similar to casein [Saheki et al., unpublished data]. The double-KO mice dislike a high concentration of sucrose solution, but they take more sucrose when casein hydrolysate, or tryptone, is added to the solution [Saheki et al., unpublished data]. The sucrose-induced hyperammonemia was ameliorated by the addition of casein hydrolysate or tryptone [39].

10.10 Effects of Single Amino Acids, Sodium Pyruvate or Fats on Food Intake, and Body Weight

Even the addition of a single amino acid, such as Ala or sodium glutamate (Na-Glu), to the AIN-93M increased the food intake and body weight of the double-KO mice, while it had almost no effect on mGPD-KO mice [39] (Fig. 10.12a, b).

We also tested sodium pyruvate (Na-Pyr). We previously showed that sodium pyruvate (Na-Pyr) activated the inhibited ureogenesis from ammonium chloride in the perfused liver of Ctrn-KO mice [40], and acted on elevated blood ammonia and hepatic metabolite derangements induced by oral sucrose administration [36]. Na-Pyr, again, increased the food intake and body weight of the double-KO mice [39] (Fig. 10.12c).

The proportion of fat in the PFC energy ratio of CD subjects is very high [22, 23]. However, among a variety of fats such as soybean oil, olive oil, fish oil, corn oil and lard, only medium-chain triglyceride (MCT) was effective in increasing the food intake and body weight of the double-KO mice [39] (Fig. 10.12d, e). This experiment also indicates that a decrease in carbohydrate proportion alone is not effective.

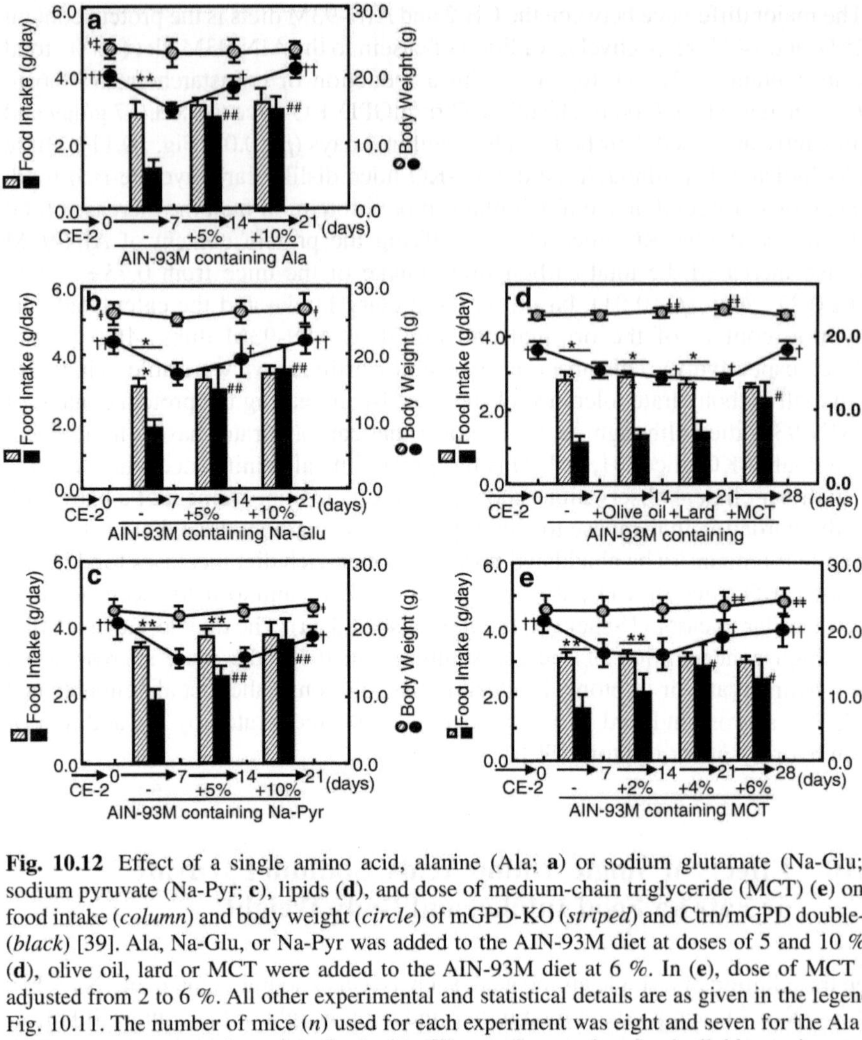

Fig. 10.12 Effect of a single amino acid, alanine (Ala; **a**) or sodium glutamate (Na-Glu; **b**), sodium pyruvate (Na-Pyr; **c**), lipids (**d**), and dose of medium-chain triglyceride (MCT) (**e**) on the food intake (*column*) and body weight (*circle*) of mGPD-KO (*striped*) and Ctrn/mGPD double-KO (*black*) [39]. Ala, Na-Glu, or Na-Pyr was added to the AIN-93M diet at doses of 5 and 10 %. In (**d**), olive oil, lard or MCT were added to the AIN-93M diet at 6 %. In (**e**), dose of MCT was adjusted from 2 to 6 %. All other experimental and statistical details are as given in the legend of Fig. 10.11. The number of mice (*n*) used for each experiment was eight and seven for the Ala and Na-Pyr experiments, eight and six for the Na-Glu experiments, four for the lipid experiments (**d**), and six and eight for the MCT experiments (**e**) of mGPD-KO and Ctrn/mGPD-KO mice, respectively

10.11 Effects of Na-Pyr, Amino Acids, and MCT on the Metabolic Perturbation Induced by Enteral Sucrose Administration

We tested the effects of Na-Pyr (20 mmol/kg), amino acids (Ala, 20 mmol/kg, or Na-Glu, 10 mmol/kg), and MCT (1 g/kg) on the metabolic perturbation induced by an enteral sucrose (5 g/kg body weight) administration [39]. The results are summarized as follows.

The increased hepatic G3P level of the double-KO mice induced by sucrose administration was effectively decreased to a level lower than the basal level by the simultaneous administration of Na-Pyr. Significant effects were also observed for Ala, Na-Glu, and MCT, although the effect was considerably less with the MCT (Fig. 10.13a).

The hepatic citrate concentration of the double-KO mice was decreased by sucrose administration, which was significantly ameliorated by the simultaneous administration of Na-Pyr, reaching the wt basal level. Neither of the other test supplements had a significant effect on the hepatic citrate level (Fig. 10.13b).

As already described, enteral sucrose administration dramatically increased hepatic Cit and Lys levels in the double-KO mice. Both Na-Pyr and Ala suppressed hepatic Cit near the wt basal level. Suppression of hepatic Cit was observed to a lesser extent for Na-Glu and MCT (Fig. 10.13c).

Suppression of hepatic Lys was only observed when enteral sucrose administration was supplemented with Na-Pyr, similar to normalization of hepatic citrate (Fig. 10.13d). On the other hand, the hepatic Glu level suppressed by enteral sucrose administration and the lower hepatic Asp level of the double-KO mice were only increased by the administration of Ala or Na-Glu, but not by Na-Pyr or MCT (Fig. 10.13e, f).

These results suggest that each of the three kinds of supplements, Na-Pyr, amino acids, and MCT, which increased food intake and body weight of the double-KO mice, acted in different ways, although their mechanisms of action are generally unknown. Therefore, combinations of these supplements may additively or synergistically act on the abnormal metabolism of the double-KO mice.

10.12 Therapeutic Procedures Based on Clinical Experiences and the Experimental Data Obtained from the Mouse Models

One of the most basic principles in treating a disease is that therapy should be based on the pathogenesis of the disease. In the case of CD, conventional therapies for hyperammonemia caused by conditions other than CD should be avoided, such as low-protein and high-carbohydrate diets, and infusions of high concentrations of glucose and Glyceol® (consisting of 10 % glycerol and 5 % fructose) for brain edema. These have been shown to enhance hyperammonemia in human CD and its corresponding mouse model. The metabolic functions of citrin and the experimental data obtained from mice recommend low-carbohydrate, high-protein and high-fat diets, which are favored by the CD subjects. Protein and/or amino acids ameliorate the toxicity of carbohydrates, although the mechanism for this is unknown. Branched-chain amino acids and Arg have been reported as effective for hyperammonemia in CTLN2 [27, 41]. The latter may have a protective effect against hypertriglyceridemia [27], of which mechanism remains to be solved. Na-Pyr has been shown to be effective from its physiological characteristic in metabolism; stimulating cytosolic

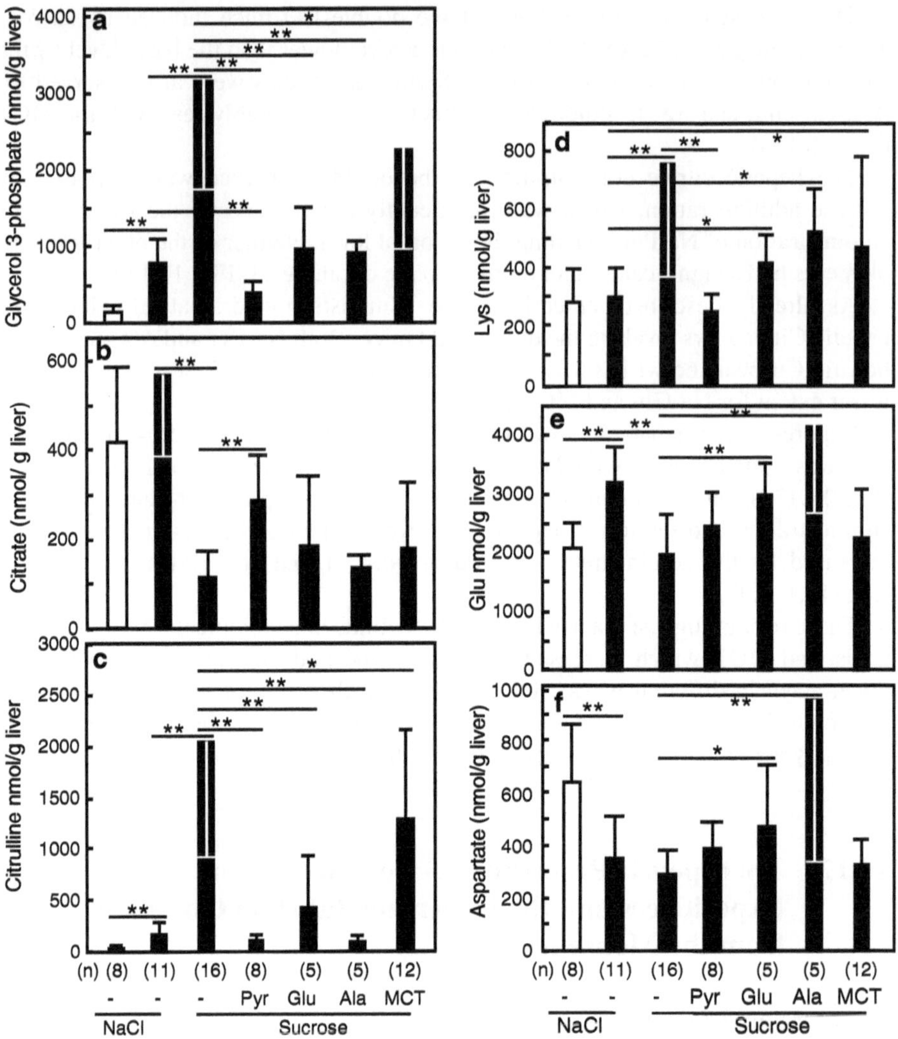

Fig. 10.13 Effect of supplementation on hepatic glycerol 3-phosphate (G3P; **a**), citrate (**b**), amino acids including citrulline (**c**), lysine (Lys; **d**), glutamate (Glu; **e**), and aspartate (Asp; **f**) of Ctrn/mGPD double-KO mice following enteral sucrose administration [36, 39]. Saline (NaCl) or sucrose solution (5 g/kg) was enterally administered alone, or supplemented with sodium pyruvate (Pyr; 20 mmol/kg), sodium glutamate (Glu; 10 mmol/kg), alanine (Ala; 20 mmol/kg), or MCT (1 g/kg). One hour after administration, livers were taken using a freeze-clamp method and extracted using 3 % perchloric acid. G3P and citrate concentrations in the neutralized extract were determined by enzymatic methods and amino acids, by liquid chromatography/mass spectrometry [36]. The number of mice used is shown in *brackets*. Differences between treatment groups were evaluated using the Student's *t*-test (*$p<0.05$; **$p<0.01$). *White* and *black columns* denote wt and the double-KO mice, respectively

NADH oxidation to NAD^+, and relieving the inhibited ureogenesis of Ctrn-KO mice, as described above [40]. Clinical trials of Na-Pyr have proven it effective [38, 42, 43]. On the other hand, among various fats, MCT alone showed effectiveness in mouse experiments [39]. Although fats composed of long-chain fatty acids were not effective in the experiments on food intake (Fig. 12.13a), these may prove to be effective under other experimental conditions. The specificity of MCT may derive from its metabolic properties; MCT can be digested without bile acid, transported via the portal vein, directly incorporated into liver, and metabolized without carnitine through unregulated mechanisms rather than via carnitine palmitoyltransferease-1 which is finely regulated. All of these suggest that MCT is a readily available energy source for the liver, and explain why CD subjects like cow's milk, coconut oil, and products rich in MCT. NICCD babies have been treated with galactose-free milk supplemented with MCT [44, 45]. The mechanism of action in increasing food intake of the double-KO mice, however, could not be solved through the analysis of hepatic metabolites. Finally, based on our experiments with the double-KO mice we recommend diets low in carbohydrates, and rich in protein and fat, Na-Pyr, amino acids, and MCT for the therapy of CD. Liver transplantation is the most effective approach to normalize diverse metabolic abnormalities in CD patients [16, 41], but this should be considered as a last resort.

10.13 Other Features of CD and Double-KO Mice to Be Solved

Our studies on CTLN2 started from enzymatic analysis of hepatic ASS of patients [2, 3]. The mechanism of the decrease in ASS is very important to clarify, because it makes the pathophysiology of citrin deficiency more complicated. Protein and/or amino acids may have a therapeutic effect under a defect of citrin alone, but ammonia is formed from these proteins and/or amino acids, which then becomes difficult to use in therapeutics, if a decrease in ASS is associated with a defect of citrin. Under such conditions, liver transplantation may currently be the only possible therapy. Therefore, clarification of the mechanism of ASS decrease would make the treatment of CTLN2 more effective and simpler. So far, we have found no pathogenically significant decrease in the urea cycle enzyme activities of the double-KO mice [31]. It is not clear why no decrease in ASS has been found in the double-KO mice. There are several such differences between human CD and its mouse model. It may also be important to note that the double-KO mice lack mGPD in their whole body, unlike human patients. We therefore need a model more closely mimicking human CD.

Concerning oxidative stress, the main theme of this book, Nagasaka et al. [18] reported several examples of evidence for oxidative stress in CD patients even for the adaptive or silent period. We have almost no data from the double-KO mice, but the Ctrn/mGPD double-KO mice would be a suitable model to explore oxidative stress.

References

1. Miyakoshi T, Takahashi T, Kato M, Watanabe M, Ito C (1968) Abnormal citrulline metabolism of inose-type hepatocerebral disease. Shinkeikagaku 7:88–91 (in Japanese)
2. Saheki T, Ueda A, Hosoya M, Kusumi K, Takada S, Tsuda M, Katsunuma T (1981) Qualitative and quantitative abnormalities of argininosuccinate synthetase in citrullinemia. Clin Chim Acta 109:325–335
3. Saheki T, Ueda A, Hosoya M, Sase M, Nakano K, Katsunuma T (1983) Enzymatic analysis of citrullinemia (12 cases) in Japan. Adv Exp Med Biol 153:63–76
4. Kobayashi K, Sinasac DS, Iijima M, Boright AP, Begum L, Lee JR, Yasuda T, Ikeda S, Hirano R, Terazono H, Crackower MA, Kondo I, Tsui LC, Scherer SW, Saheki T (1999) The gene mutated in adult-onset type II citrullinaemia encodes a putative mitochondrial carrier protein. Nat Genet 22:159–163
5. del Arco A, Satrústegui J (1998) Molecular cloning of Aralar, a new member of the mitochondrial carrier superfamily that binds calcium and is present in human muscle and brain. J Biol Chem 273:23327–23334
6. Palmieri L, Pardo B, Lasorsa FM, del Arco A, Kobayashi K, Iijima M, Runswick MJ, Walker JE, Saheki T, Satrústegui J, Palmieri F (2001) Citrin and aralar1 are Ca (2+)-stimulated aspartate/glutamate transporters in mitochondria. EMBO J 20:5060–5069
7. Begum L, Jalil MA, Kobayashi K, Iijima M, Li MX, Yasuda T, Horiuchi M, del Arco A, Satrústegui J, Saheki T (2002) Expression of three mitochondrial solute carriers, citrin, aralar1 and ornithine transporter, in relation to urea cycle in mice. Biochim Biophys Acta 1574: 283–292
8. Krebs HA, Gascoyne T, Notton BM (1967) Generation of extramitochondrial reducing power in gluconeogenesis. Biochem J 102:275–282
9. Tomomasa T, Kobayashi K, Kaneko H, Shimura H, Fukusato T, Tabata M, Inoue Y, Ohwada S, Kasahara M, Morishita Y, Kimura M, Saheki T, Morikawa A (2001) Possible clinical and histologic manifestations of adult-onset type II citrullinemia in early infancy. J Pediatr 138:741–743
10. Tazawa Y, Kobayashi K, Ohura T, Abukawa D, Nishinomiya F, Hosoda Y, Yamashita M, Nagata I, Kono Y, Yasuda T, Yamaguchi N, Saheki T (2001) Infantile cholestatic jaundice associated with adult-onset type II'citrullinemia. J Pediatr 138:735–740
11. Ohura T, Kobayashi K, Tazawa Y, Nishi I, Abukawa D, Sakamoto O, Iinuma K, Saheki T (2001) Neonatal presentation of adult-onset type II citrullinemia. Hum Genet 108:87–90
12. Saheki T, Kobayashi K (2002) Mitochondrial aspartate glutamate carrier (citrin) deficiency as the cause of adult-onset type II citrullinemia (CTLN2) and idiopathic neonatal hepatitis (NICCD). J Hum Genet 47(7):333–341
13. Kobayashi K, Saheki T, Song YZ (2005) Citrin deficiency. In: Pagon RA, Bird TD, Dolan CR, Stephens K, Adam MP (eds) GeneReviews™ [Internet]. University of Washington, Seattle, WA (updated 05 Jan 2012)
14. Tanaka T, Nagao M, Tsutsumi H (2002) Application of mutation analysis for the previously uncertain cases of adult-onset type II citrullinemia (CTLN2) and their clinical profiles. Tohoku J Exp Med 198:89–97
15. Chang KW, Chen HL, Chien YH, Chen TC, Yeh CT (2011) SLC25A13 gene mutations in Taiwanese patients with non-viral hepatocellular carcinoma. Mol Genet Metab 103:293–296
16. Ikeda S, Yazaki M, Takei Y, Ikegami T, Hashikura Y, Kawasaki S, Iwai M, Kobayashi K, Saheki T (2001) Type II (adult onset) citrullinaemia: clinical pictures and the therapeutic effect of liver transplantation. J Neurol Neurosurg Psychiatry 71:663–760
17. Terada R, Yamamoto K, Kobayashi K, Sakaguchi K, Iwasaki Y, Saheki T, Shiratori Y (2006) Adult-onset type II citrullinemia associated with idopathic hypertriglyceridemia as a preceding feature. J Gastroenterol Hepatol 21:1634–1635
18. Nagasaka H, Okano Y, Tsukahara H, Shigematsu Y, Momoi T, Yorifuji J, Miida T, Ohura T, Kobayashi K, Saheki T, Hirano K, Takayanagi M, Yorifuji T (2009) Sustaining

hypercitrullinemia, hypercholesterolemia and augmented oxidative stress in Japanese children with aspartate/glutamate carrier isoform 2-citrin-deficiency even during the silent period. Mol Genet Metab 97:21–26
19. Song YZ, Deng M, Chen FP, Wen F, Guo L, Cao SL, Gong J, Xu H, Jiang GY, Zhong L, Kobayashi K, Saheki T, Wang ZN (2011) Genotypic and phenotypic features of citrin deficiency: five-year experience in a Chinese pediatric center. Int J Mol Med 28:33–40
20. Komatsu M, Yazaki M, Tanaka N, Sano K, Hashimoto E, Takei Y, Song YZ, Tanaka E, Kiyosawa K, Saheki T, Aoyama T, Kobayashi K (2008) Citrin deficiency as a cause of chronic liver disorder mimicking non-alcoholic fatty liver disease. J Hepatol 49:810–820
21. Kawamoto S, Strong RW, Kerlin P, Lynch SV, Steadman C, Kobayashi K, Nakagawa S, Matsunami H, Akatsu T, Saheki T (1997) Orthotopic liver transplantation for adult-onset type II citrullinaemia. Clin Transplant 11:453–458
22. Saheki T, Kobayashi K, Terashi M, Ohura T, Yanagawa Y, Okano Y, Hattori T, Fujimoto H, Mutoh K, Kizaki Z, Inui A (2008) Reduced carbohydrate intake in citrin-deficient subjects. J Inherit Metab Dis 31:386–394
23. Nakamura M, Yazaki M, Kobayashi Y, Fukushima K, Ikeda S, Kobayashi K, Saheki T, Nakaya Y (2011) The characteristics of food intake in patients with type II citrullinemia. J Nutr Sci Vitaminol 57:239–245
24. Tamakawa S, Nakamura H, Katano T, Yoshizawa M, Ohtake K, Kubota T (1994) Hyperalimentation therapy produces a comatose state in a patient with citrullinemia. J Jpn Soc Intensive Care Med 1:37–41 (Japanese)
25. Fukushima K, Yazaki M, Nakamura M, Tanaka N, Kobayashi K, Saheki T, Takei H, Ikeda S (2010) Conventional diet therapy for hyperammonemia is risky in the treatment of hepatic encephalopathy associated with citrin deficiency. Intern Med 49:243–247
26. Dimmock D, Kobayashi K, Iijima M, Tabata A, Wong LJ, Saheki T, Lee B, Scaglia F (2007) Citrin deficiency: a novel cause of failure to thrive that responds to a high-protein, low-carbohydrate diet. Pediatrics 19:e773–e777
27. Imamura Y, Kobayashi K, Shibatou T, Aburada S, Tahara K, Kubozono O, Saheki T (2003) Effectiveness of carbohydrate-restricted diet and arginine granules therapy for adult-onset type II citrullinemia: a case report of siblings showing homozygous SLC25A13 mutation with and without the disease. Hepatol Res 26:68–72
28. Yazaki M, Takei Y, Kobayashi K, Saheki T, Ikeda S (2005) Risk of worsened encephalopathy after intravenous glycerol therapy in patients with adult-onset type II citrullinemia (CTLN2). Intern Med 44:188–195
29. Takahashi H, Kagawa T, Kobayashi K, Hirabayashi H, Yui M, Begum L, Mine T, Takagi S, Saheki T, Shinohara Y (2006) A case of adult-onset type II citrullinemia–deterioration of clinical course after infusion of hyperosmotic and high sugar solutions. Med Sci Monit 12: CS13–CS15
30. Sinasac DS, Moriyama M, Jalil MA, Begum L, Li MX, Iijima M, Horiuchi M, Robinson BH, Kobayashi K, Saheki T, Tsui LC (2004) Slc25a13-knockout mice harbor metabolic deficits but fail to display hallmarks of adult-onset type II citrullinemia. Mol Cell Biol 24:527–536
31. Saheki T, Iijima M, Li MX, Kobayashi K, Horiuchi M, Ushikai M, Okumura F, Meng XJ, Inoue I, Tajima A, Moriyama M, Eto K, Kadowaki T, Sinasac DS, Tsui LC, Tsuji M, Okano A, Kobayashi T (2007) Citrin/mitochondrial glycerol-3-phosphate dehydrogenase double knock-out mice recapitulate features of human citrin deficiency. J Biol Chem 282: 25041–25052
32. Sadava D, Depper M, Gilbert M, Bernard B, McCabe ER (1987) Development of enzymes of glycerol metabolism in human fetal liver. Biol Neonate 52:26–32
33. Yajima Y, Hirasawa T, Saheki T (1982) Diurnal fluctuation of blood ammonia levels in adult-type citrullinemia. Tohoku J Exp Med 137:213–220
34. Ohura T, Kobayashi K, Tazawa Y, Abukawa D, Sakamoto O, Tsuchiya S, Saheki T (2007) Clinical pictures of 75 patients with neonatal intrahepatic cholestasis caused by citrin deficiency (NICCD). J Inherit Metab Dis 30:139–144

35. Saheki T, Kobayashi K, Miura T, Hashimoto S, Ueno Y, Yamasaki T, Araki H, Nara H, Shiozaki Y, Sameshima Y, Suzuki M, Yamauchi Y, Sakazume Y, Akiyama K, Yamamura Y (1986) Serum amino acid pattern of type II citrullinemic patients and effect of oral administration of citrulline. J Clin Biochem Nutr 1:129–142
36. Saheki T, Inoue K, Ono H, Tushima A, Katsura N, Yokogawa M, Yoshidumi Y, Kuhara T, Ohse M, Eto K, Kadowaki T, Sinasac DS, Kobayashi K (2011) Metabolomic analysis reveals hepatic metabolite perturbations in citrin/mitochondrial glycerol-3-phosphate dehydrogenase double-knockout mice, a model of human citrin deficiency. Mol Genet Metab 104:492–500
37. Kamoun P, Richard V, Rabier D, Saudubray JM (2002) Plasma lysine concentration and availability of 2-ketoglutarate in liver mitochondria. J Inherit Metab Dis 25:1–6
38. Mutoh K, Kurokawa K, Kobayashi K, Saheki T (2008) Treatment of a citrin-deficient patient at the early stage of adult-onset type II citrullinaemia with arginine and sodium pyruvate. J Inherit Metab Dis 31:S343–S347
39. Saheki T, Inoue K, Ono H, Katsura N, Yokogawa M, Yoshidumi Y, Furuie S, Kuroda E, Ushikai M, Asakawa A, Inui A, Eto K, Kadowaki T, Sinasac DS, Yamamura K, Kobayashi K (2012) Effects of supplementation on food intake, body weight and hepatic metabolites in the citrin/mitochondrial glycerol-3-phosphate dehydrogenase double-knockout mouse model of human citrin deficiency. Mol Genet Metab 107:322–329
40. Moriyama M, Li MX, Kobayashi K, Sinasac DS, Kannan Y, Iijima M, Horiuchi M, Tsui LC, Tanaka M, Nakamura Y, Saheki T (2006) Pyruvate ameliorates the defect in ureogenesis from ammonia in citrin-deficient mice. J Hepatol 44:930–938
41. Hirai I, Kimura W, Suto K, Fzjimoto H, Watanabe T, Fuse A, Kobayashi K, Iijima M, Saheki T, Nakatsuka T, Sugawara Y, Makuuchi M (2008) Living donor liver transplantation for type II citrullinemia from a heterozygous donor. Hepatogastroenterology 55:2211–2216
42. Saheki T, Inoue K, Tushima A, Mutoh K, Kobayashi K (2010) Citrin deficiency and current treatment concepts. Mol Genet Metab 100:S59–S64
43. Yazaki M, Kinoshita M, Ogawa S, Fujimi S, Matsushima A, Hineno A, Tazawa KI, Fukushima K, Kimura R, Yanagida M, Matsunaga H, Saheki T, Ikeda SI (2013) A 73-year-old patient with adult-onset type II citrullinemia successfully treated by sodium pyruvate and arginine. Clin Neurol Neurosurg 115:1542–1545
44. Lin WX, Zhang ZH, Deng M, Cai XR, Song YZ (2012) Multiple ovarian antral follicles in a preterm infant with neonatal intrahepatic cholestasis caused by citrin deficiency: a clinical, genetic and transcriptional analysis. Gene 505:269–275
45. Hayasaka K, Numakura C, Toyota K, Kimura T (2012) Treatment with lactose (galactose)-restricted and medium-chain triglyceride-supplemented formula for neonatal intrahepatic cholestasis caused by citrin deficiency. JIMD Rep 2:37–44

Chapter 11
Antioxidant Supplementation and Therapies

Varadarajan Sudhahar and Tohru Fukai

Abbreviations

AFR	Ascorbic free radicals
Asc	Ascorbic acid
ATBC	Alpha-Tocopherol-Beta-Carotene Cancer Prevention study
CARET	β-carotene and retinol efficiency trial
CAT	Catalase
CHAOS	Cambridge Heart Antioxidant Study
CHD	Coronary heart disease
CVD	Cardiovascular diseases
DBCT	Double-blinded Clinical Trial
DHA	Docosahexaenoic acid
EPA	Eicosapentaenoic acid
Fl-O˙	Aroxyl radical
Fl-OH	Flavonoid
GISSI	Gruppo Italiano per lo Studio della Sopravvivenza nell'Infarto Miocardico
GPx	Glutathione peroxidase
GRx	Glutathione reductase
GSH	Glutathione

V. Sudhahar
Departments of Medicine (Section of Cardiology) and Pharmacology, Center for Cardiovascular Research, University of Illinois at Chicago, 835 S. Wolcott, M/C868, Chicago, IL 60612, USA

Jesse Brown Veterans Affairs Medical Center, 820 S. Damen Ave., Chicago, IL 60612, USA

T. Fukai, M.D., Ph.D. (✉)
Departments of Medicine (Section of Cardiology) and Pharmacology, Center for Cardiovascular Research, University of Illinois at Chicago, 835 S. Wolcott, M/C868, Chicago, IL 60612, USA
e-mail: tfukai@uic.edu

H. Tsukahara and K. Kaneko (eds.), *Studies on Pediatric Disorders*, Oxidative Stress in Applied Basic Research and Clinical Practice, DOI 10.1007/978-1-4939-0679-6_11,

GSSG	Glutathione disulfide
GST	Glutathione-S-transferase
H2O2	Hydrogen peroxide
HOCl	Hypochlorous acid
HOPE	Heart Outcomes Protection Evaluation
HOPE-TOO	Heart Outcomes Prevention Evaluation-The Ongoing Outcomes
ICARE	Israel Cardiovascular Events Reduction with Vitamin E
$L^{\bullet}$	Carbon-centered lipid radical
LDL	Low-density lipoprotein
LH	Polyunsaturated fatty acid
LOH	Lipid hydroxides
$LOO^{\bullet}$	Lipid peroxyl radical
LOOH	LIPID hydroperoxides
MPO	Myeloperoxidases
NADPH	Nicotinamide adenine dinucleotide phosphate
NHANES	National Health and Nutrition Examination Survey
NO	Nitric oxide
NOS	Nitric oxide synthase
$O2^{\bullet-}$	Superoxide
$^{\bullet}OH$	Hydroxyl radical
$ONOO^{-}$	Peroxynitrite
PHS	Physicians' Health Study
ROS	Reactive oxygen species
ROS/RNS	Reactive oxygen/nitrogen species
Se	Selenium
SELECT	The Selenium and Vitamin E Cancer Prevention Trial
SOD	Superoxide dismutase
SPACE	Secondary prevention with antioxidants of cardiovascular disease in end-stage renal disease
SVCT1	Sodium-dependent vitamin C transporter 1
SVCT2	Sodium-dependent vitamin C transporter 2
Ts	Tocopherols
VEAPS	Vitamin E Atherosclerosis Prevention Study
vit C	Vitamin C
vit E	Vitamin E
VLDL	Very low-density lipoprotein
WHS	Women Health Study
α-T or αTOH	α-Tocopherol

11.1 Introduction

Reactive oxygen species (ROS) such as superoxide and hydrogen peroxide (H_2O_2) play a crucial role in physiological and pathophysiological processes (Table 11.1). Low levels of ROS are indispensable in many biochemical processes, including

Table 11.1 Double-edged sword of ROS in disease promotion and prevention

Low/physiological ROS	High/pathological ROS
Intracellular signaling	Endothelial dysfunction
Defense against microorganism	Promote tissue injury
Cell function	Cause apoptosis/cell death
Recovery of tissue injury	

intracellular signaling, defense against microorganisms, and cell function [1–3]. In contrast, oxidative stress caused by high dose and/or inadequate removal of ROS can lead to modification of macromolecules such as DNA, lipids, and proteins, which has been implicated in the pathogenesis of the development of cardiovascular disease (CVD) (hypertension, diabetes, atherosclerosis, etc), cancer, aging, and neurodegenerative diseases. These ROS are extensively counteracted by antioxidants which are either naturally produced or externally supplied through foods or supplements. This chapter will review and update the role of antioxidant supplements in health and disease.

11.2 Free Radicals in Biology and Health

11.2.1 Generation of Reactive Oxygen Species (Fig. 11.1)

Free radicals are derived either from normal essential metabolic processes in the human body or external sources. Potential enzymatic internal sources of ROS include components of the mitochondrial electron transport chain, xanthine oxidase, the cytochrome p450 monooxygenases, lipoxygenase, uncoupled nitric oxide synthase (NOS), and the nicotinamide adenine dinucleotide phosphate (NADPH) oxidase [4–6] as well as from external sources such as exposure to X-rays, ozone, cigarette smoking, air pollutants, and industrial chemicals [7]. These internal and external factors induced free radical formation such as superoxide ($O_2^{\bullet-}$). $O_2^{\bullet-}$ is formed by the univalent reduction of molecular oxygen and is enzymatically converted to H_2O_2 by superoxide dismutase (SOD) [8]. When $O_2^{\bullet-}$ reacts with nitric oxide (NO), the highly reactive peroxynitrite ($ONOO^-$) is formed. Peroxidases catalyze reactions involving H_2O_2, resulting in the generation of hypochlorous acid (HOCl) and singlet oxygen. Finally, in the presence of reduced transition metals (e.g., ferrous or cuprous ions), H_2O_2 can be converted into the highly reactive hydroxyl radical ($^{\bullet}OH$).

11.2.2 ROS in Disease

High dose and/or inadequate removal of ROS (i.e., oxidative stress) are associated with diseases including diabetes, neurodegenerative diseases, as well as atherosclerosis contributing to CVD and cancer. Pathophysiological events related to ROS in CVD and cancer include oxidation of lipids (e.g., oxidized low-density lipoprotein

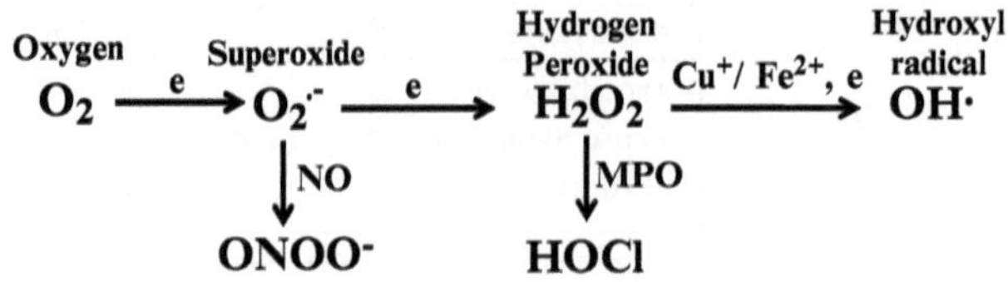

Fig. 11.1 Generation of free radicals and oxidants. Molecular oxygen (O_2) generates ROS by successive one-electron reduction. A single-electron transfer converts molecular oxygen to the superoxide anion ($O_2^{\bullet-}$). Superoxide can then rapidly react with nitric oxide (NO), forming peroxynitrite (ONOO–). Superoxide anion further converts hydrogen peroxide (H_2O_2) and then hydroxyl radicals (•OH) by reduced transition metals (e.g., Fe^{2+}/Cu^{+} ions). Myeloperoxidases (MPO) catalyze reactions involving H_2O_2, resulting in the generation of hypochlorous acid (HOCl). These radicals increase tissue damage

(LDL)) and ROS-mediated DNA damage [9]. Thus, a great deal of effort has been exerted to determine if antioxidant supplementation can be beneficial in preventing or treating these various diseases.

11.3 Antioxidants

Antioxidant is defined as any substance that significantly delays or prevents oxidation of the substrate and can be classified as enzymatic antioxidants and nonenzymatic antioxidants as follows (Fig. 11.2 and Table 11.2).

11.3.1 Enzymatic Antioxidant

Enzymatic antioxidants serve as the body's most potent defense system against free radical reactions. The major antioxidant enzymes involved against free radicals are as follows: superoxide dismutase (SOD), catalase (CAT), glutathione peroxidase (GPx), glutathione reductase (GRx), and thioredoxin peroxidase [10]. SOD exists in several isoforms, differing in the nature of the active metal center and amino acid constituency, as well as their number of subunits, cofactors, and other features [11]. In humans, there are three forms of SOD: cytosolic Cu/Zn-SOD, mitochondrial Mn-SOD, and extracellular SOD (ecSOD). SOD catalyzes the dismutation of superoxide anion radical into H_2O_2 by reduction. CAT is an enzyme present in the cells of plants, animals, and aerobic (oxygen requiring) bacteria [12]. CAT is located in a cell organelle called the peroxisome. The enzyme very efficiently promotes the conversion of H_2O_2 to water and molecular oxygen. There are two forms of the enzyme GPx, one of which is selenium-independent (glutathione-*S*-transferase, GST) while the other is selenium-dependent (GPx) [13]. These two enzymes differ in the

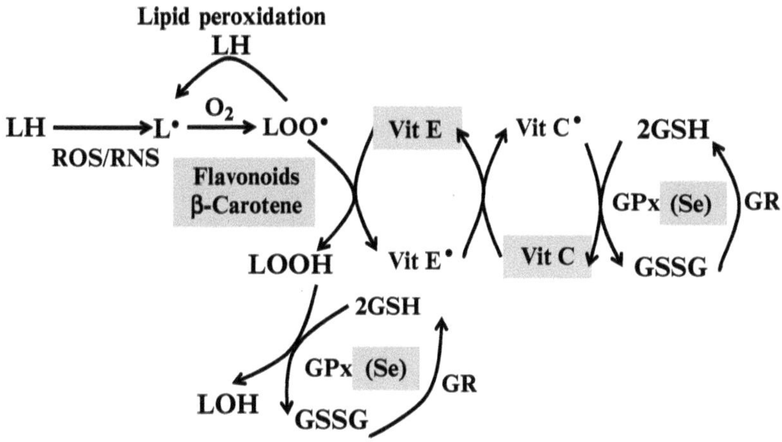

Fig. 11.2 Role of antioxidants on lipid peroxidation in mammalian cells. Reactive oxygen/nitrogen species (ROS/RNS) generated by Fig. 11.1 participate in lipid peroxidation by reacting with polyunsaturated fatty acid (LH) to give rise to a carbon-centered lipid radical (L•). The lipid radical (L•) can further interact with molecular oxygen to give a lipid peroxyl radical (LOO•). If the resulting LOO• is not reduced by antioxidants, the lipid peroxidation process occurs. The LOO• is reduced within the membrane by various antioxidants such as flavonoids, beta-catenin, and vitamin E (vit E), which can be regenerated by vitamin C (vit C). Lipid hydroperoxides (LOOH) are reduced to lipid hydroxides (LOH) and dioxygen by glutathione peroxidase (GPx), which contains selenium (Se) as a catalytic cofactor

Table 11.2 List of enzymatic and nonenzymatic (endogenous or exogenous) antioxidants in mammalian cells

Enzymatic antioxidant	Nonenzymatic antioxidant	
Endogenous enzymes	Endogenous or metabolic factors	Exogenous or nutritional factors
Superoxide dismutase	Glutathione	Vitamin C
Catalase	Lipoic acid	Vitamin E
Glutathione peroxidase	Coenzyme Q	β-Carotenoids
Glutathione reductase	Bilirubin	Flavonoids
Thioredoxin peroxidase	Uric acid	Selenium
	L-Arginine	

number of subunits, the bonding nature of the selenium at the active center and their catalytic mechanisms. GPx also removes H_2O_2 by using it to oxidize reduced glutathione (GSH) into oxidized glutathione (GSSG). GRx, a flavoprotein enzyme, reduces glutathione disulfide (GSSG) to the sulfhydryl form GSH. Besides H_2O_2, GPx also reduces lipid hydroperoxides (LOOH) while oxidizing GSH. In addition, thioredoxin peroxidase (peroxiredoxin) is another antioxidant enzyme which also removes H_2O_2 and phospholipid hydroperoxides [14].

11.3.2 Nonenzymatic Antioxidant

The nonenzymatic antioxidants can be classified as metabolic or endogenous antioxidants and nutrient or exogenous antioxidants (Table 11.2). Metabolic antioxidants are produced in the body through metabolism (e.g., GSH, L-arginine, lipoic acid, etc), whereas nutrient antioxidants are compounds which are provided through food and supplementation and are not synthesized in body (e.g., Vitamins E & C, carotenoids, selenium, flavonoids, etc). Nutritional or exogenous antioxidants act through different mechanisms to help overcome oxidative damage (Fig. 11.2): (1) they directly neutralize free radicals, (2) they reduce the peroxide concentrations and repair oxidized membranes, and (3) they support the stimulation of antioxidant enzymes including GPx, CAT, and SOD. When an antioxidant destroys a free radical, this antioxidant itself becomes oxidized (e.g., vitamin C). Therefore, the antioxidant resources must be constantly restored in the body. The antioxidant process can function in one of two ways: chain breaking or prevention. For the chain-breaking, when a radical releases or steals an electron, a second radical is formed. The last one exerts the same action on another molecule and continues until either the free radical formed is stabilized by a chain-breaking antioxidant (vitamin C, E, carotenoids, etc), or it simply disintegrates into an inoffensive product. The classic example of such a chain reaction is lipid peroxidation (Fig. 11.2). For the preventive function, antioxidant enzymes such as SOD, CAT, and GPx can prevent lipid oxidation by reducing the rate of chain initiation, e.g., either by scavenging initiating free radicals or by stabilizing transition metal radicals such as copper and iron [15, 16].

11.4 Antioxidant Supplements

The structure and antioxidant action of each dietary antioxidants and their role in CVD and cancer are shown below.

11.4.1 Vitamin E

Vitamin E is a fat-soluble vitamin with high antioxidant potency. The tocochromanol vitamin E homologues with the largest diffusion in nature are four tocopherols (Ts) and four tocotrienols: α, β, γ, and δ-T and α, β, γ ,and δ-tocotrienol [17, 18]. Depending on the nature of the isoprenoid chain, Ts (containing a phytyl chain) or tocotrienols (geranylgeranyl chain) can be distinguished (Fig. 11.3a). Tocotrienols differ from Ts by having a farnesyl (isoprenoid) structure compared to a saturated phytyl side chain. Because it is fat-soluble, tocopherol safeguards cell membranes from damage by free radicals. Its antioxidant function mainly resides in the protection against lipid peroxidation (Fig. 11.2). α-T is the most abundant form of tocopherols in blood and tissues and has the highest activity in the classical fertility-restoration assay. α-T is generally considered to be vitamin E. Therefore, many

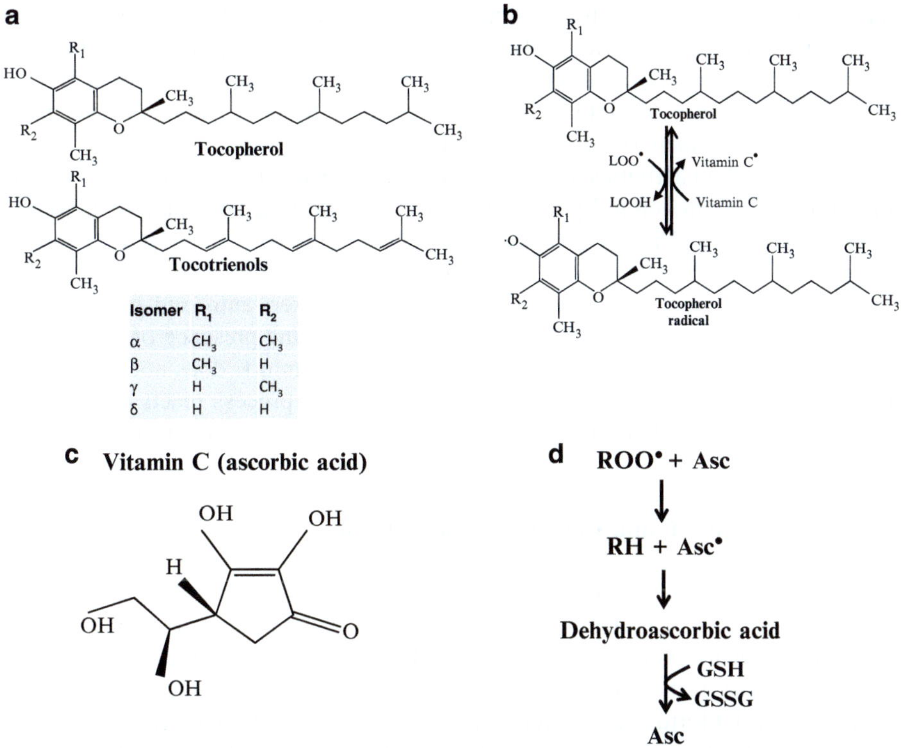

Isomer	R_1	R_2
α	CH_3	CH_3
β	CH_3	H
γ	H	CH_3
δ	H	H

Fig. 11.3 General structure and antioxidant activity of Vitamin E and Vitamin C. (**a**) Chemical structures of various isomers of tocopherol and tocotrienols. The *table* indicates the number and position of ring methyls in α-, β-, γ-, and δ-tocopherols and tocotrienols. (**b**) The tocopherol radical produced upon scavenging of LOO• is reduced back to tocopherol by vitamin C. (**c**) General structure of vitamin C (ascorbic acid). (**d**) Ascorbic acid (Asc, vitamin C) is attacked by a radical species (ROO•) that removes an electron and converts it to the acrobate free radicals (Asc•). Dehydroascorbic acid is formed from the Asc•. The resulting Dehydroascorbic acid can then undergo two-electron reduction back to ascorbate by GSH

studies on vitamin E have been conducted with α-tocopheryl acetate. However, the results of many of the animal studies are inconsistent, and the results of some of the human intervention studies are disappointing [19, 20]. In recent years it has been recognized that δ-T, γ-T, and tocotrienol have beneficial health effects beyond α-T [18, 21–23]. This chapter discusses role of vitamin E in CVD and cancer as well.

11.4.1.1 Dietary Sources and Mechanism of Action

The major dietary sources of Ts are vegetable oils, such as oils from corn, soybean, sesame, cottonseeds, and nuts. In these oils, γ-T is three to five times more abundant than α-T, and δ-T is as abundant as α-T, whereas β-T exists in only minute amounts. Upon ingestions, these Ts are incorporated into the chylomicrons and transported to the liver via the lymphatic system. The transfer of Ts in the liver to VLDL is mediated by a specific α-T transfer protein, which preferentially transfers α-T over γ-T, and δ-T is even less effectively transferred [24]. As a consequence, α-T is efficiently secreted into the circulation and transported to non-hepatic tissues, and is the most abundant form of vitamin E in the blood and tissues. The blood and tissue levels of γ-T are much lower, and those of δ-T are even lower.

The antioxidant activities of T and tocotrienols are due to their abilities to donate their phenolic hydrogen to lipid free radicals and retard the autocatalytic lipid peroxidation process. Vitamin E is a chain-reaction breaking antioxidant since it quenches the intermediate in the chain reaction [17]. It has to be considered that vitamin E, like every redox-active compound, may exert anti- and pro-oxidative effects depending on the reaction partners present. In the presence of other coantioxidants, including ascorbic acid and ubiquinol, vitamin E does not have a pro-oxidant function. The ascorbate radical formed in this process reacts rapidly with the reduced GSH pool or with a specific vitamin C reductase enzyme (Fig. 11.3d).

11.4.1.2 Vitamin E and Cardiovascular Disease

Although initial studies in humans provided encouraging results, clinical trials with vitamin E (α-T) supplementation showed contrasting findings. Two small trials, Cambridge Heart Antioxidant Study (CHAOS) and Alpha-Tocopherol, Beta-Carotene (ATBC), showed the benefit of vitamin E in reducing myocardial infarction [25, 26]. Unfortunately, several larger studies, including the HOPE trial [27], the heart protection study [28], and the GISSI prevenzione [29] (involving many thousands of subjects), have failed to show any benefits of antioxidant vitamins in the prevention of CVD. Surprisingly, some studies have shown that vitamin E is harmful. The incidence and hospitalization for heart failure were increased in subjects treated with vitamin E in the HOPE-TOO trial [30]. A recent large meta-analysis has suggested that high dose (>400 IU/day) of vitamin E might increase mortality and should be avoided [31]. The reasons for the lack of beneficial clinical effects of vitamin E supplementation are unclear and will be discussed later [32]. The effect of vitamin E is summarized in Table 11.3.

Although most of the vitamin E research has focused on α-Ts, current developments in vitamin E research clearly indicate that members of the vitamin E family are not redundant with respect to their biological functions. In particular, tocotrienols may have more potent antioxidant and anticancer effects than Ts [18, 22, 23]. However, the effects in humans must be well established before tocotrienols are used as therapeutic agents in various disease conditions.

11.4.1.3 Vitamin E and Cancer

Many epidemiological studies have suggested that a low vitamin E nutritional status is associated with increased cancer risk, which may be partly through regulating ROS. In cohort studies on lung cancer, two studies out of three found a significant

Table 11.3 Selected controlled clinical trials of antioxidant supplements on CVD

Study	Biological effect
Studies showed beneficial effect	
CHAOS (1996) [25]	Vit E reduced myocardial infarction in patients with coronary atherosclerosis
SPACE (2000) [152]	Vit E reduced composite cardiovascular disease endpoints and myocardial infarction
DBCT (2008) [153]	Vit E reduced the cardiovascular events in diabetic patients
HOPE and ICARE (2010) [154]	Vit E reduced cardiovascular events in diabetic patients
ATBC (2011) [155]	Vit E and β-carotene influenced cancer progression through suppressing VEGF-mediated lymphangiogenesis
ATBC (2004) [156]	Vit E with β-carotene reduced lung cancer risk in male smokers
Studies showed no beneficial effect	
HOPE (2000) [149]	Vit E for 4.5 years had no effect on cardiovascular event
VEAPS (2002) [157]	Vitamin E for 3 years had no effect on atherosclerosis
WHS (2005) [37]	An average of 10.5 years Vit E treatment had no effect on CVD and cancer
PHS II (2009) [38]	Vitamin E did not reduce the risk of prostate cancer or other cancer

CHAOS Cambridge Heart Antioxidant Study, *SPACE* Secondary prevention with antioxidants of cardiovascular disease in end-stage renal disease, *DBCT* Double-blinded Clinical Trial, *HOPE and ICARE* heart outcomes protection evaluation and Israel cardiovascular events reduction with vitamin E, *ATBC* Alpha-Tocopherol-Beta-Carotene Cancer Prevention study, *VEAPS* Vitamin E Atherosclerosis Prevention Study, *WHS* Women Health Study, *PHS* Physicians' Health Study

inverse association between dietary intake of vitamin E and risk of lung cancer [33–35] in current smokers, suggesting a protective effect of vitamin E against insults from cigarette smoking. From 24 studies in breast cancer, 11 studies found a risk reduction; however, 13 studies did not find an association [36].

In contrast to epidemiological studies, several recent large-scale human trials with high doses of α-T have produced disappointing results [37–40]. The ATBC Cancer Prevention Study showed that α-T or β-carotene or both for 5–8 years did not produce a significant effect on the incidence of lung cancer [41]. The Selenium and Vitamin E Cancer Prevention Trial (SELECT) even showed higher prostate cancer incidence in subjects who took α-T supplementation [40]. In addition, previous cancer prevention studies in different animal models with pure α-T have obtained inconsistent results [19]. But the possible reasons for these contradictions are not well known.

Tocotrienols, the other vitamin E isomer with unsaturated side chains, have been shown to display stronger anticancer activities in vitro than γ- and δ-Ts, which exhibit more anticancer activities than α-T [42–45]. This subject has been reviewed recently [46]. Although tocotrienols possess antioxidant activity, the anticancer activity of tocotrienols may be independent from its antioxidant activity because some redox-silent tocotrienol derivatives still exhibit anticarcinogenic properties [47, 48]. The anticancer effect of tocotrienols in animal studies requires further exploration.

11.4.2 Vitamin C

Vitamin C (ascorbic acid) is a potent water-soluble antioxidant in biological fluids, and it is required for multiple biological functions [49, 50]. The general structure of Vitamin C is illustrated in Fig. 11.3c. All known physiological and biochemical actions of vitamin C are due to its action as an electron donor. First, vitamin C functions as antioxidant because, by donating its electrons, it prevents other compounds from being oxidized. However, by the very nature of this reaction, vitamin C itself is oxidized in the process. Its primary antioxidant partners are Vitamin E and the carotenoids, as well as other antioxidant enzymes. Second, vitamin C is a cofactor for several enzymes participating in the posttranslational hydroxylation of collagen, in the biosynthesis of carnitine, in the conversion of the neurotransmitter dopamine to norepinephrine, in peptide amidation, and in tyrosine metabolism. Third, vitamin C is an important regulator of iron uptake by reducing ferric Fe^{3+} to ferrous Fe^{2+} ions. Level of vitamin C is regulated by mechanisms to recycle and accumulate it against a concentration gradient, which involves tissue vitamin C transporters (SVCT1 and SVCT2) [51–53]. Most animals are able to synthesize vitamin C from glucose, but humans, primates, and guinea pigs cannot synthesize vitamin C since they lack the last enzyme involved in the synthesis of vitamin C (gulonolactone oxidase). As a result, they must obtain vitamin C in their diet. Thus, the prolonged deprivation of vitamin C generates defects in the posttranslational modification of collagen that cause scurvy and eventually death. Epidemiological studies show that fruit- and vegetable-rich diet reduces CVD and cancer, but whether these protective effects are directly attributable to vitamin C is not known.

11.4.2.1 Dietary Sources and Mechanism of Action

Vitamin C is mainly found in fruits and vegetables [54]. Rich fruit sources include cantaloupe, grapefruit, honeydew, kiwi, mango, orange, papaya, strawberries, tangelo, tangerine, and watermelon. Citrus fruits and juices containing vitamin C in abundance include grapefruit and orange juices. Data from the third U.S. National Health and Nutrition Examination Survey (NHANES III Part 1 1988–91) suggest that the median vitamin C consumption from diet in adult males and females is 84 and 73 mg daily, respectively.

Vitamin C is a donor antioxidant and reacts with radicals to produce the ascorbate free radicals (Asc$^•$) as shown in Fig. 11.3d. Pairs of Asc$^•$ disproportionate to form one molecule of dehydroascorbic acid and one Asc. Dehydroascorbic acid can be reduced to the useful ascorbate [55]. This property explains why ascorbate may be a preferred antioxidant. In addition to an antioxidant role in biology, ascorbate has also been shown to have a pro-oxidant in vitro [56]. Ascorbate in the presence of iron is used to initiate lipid peroxidation, and ascorbate with copper is a chemical system for generating hydroxyl radicals. Little is known about the extent to which there is a prooxidant role of ascorbate in vivo.

11.4.2.2 Vitamin C and Cardiovascular Disease

It has been proposed that vitamin C may protect against CVD through preventing endothelial dysfunction associated with atherosclerosis, hypercholesterolemia, hypertension, diabetes, and smoking. This process seems to involve increased endothelial NO by vitamin C by protecting it from oxidation and increasing its synthesis [57] in healthy subjects and in patients with CVD [58]. Furthermore, plasma vitamin C and dietary intake were found to be covariates of blood pressure in the elderly, while supplemental vitamin C intake lowered blood pressure [59]. Several epidemiological studies associate low levels of plasma vitamin C with increased death from CVD [60], but clinical trial data are not conclusive as to whether vitamin C supplements are beneficial in well-nourished individuals.

11.4.2.3 Vitamin C and Cancer

Mutations are the initiating events in neoplasms. Because of this, the ability of vitamin C to modulate oxidative DNA damage in vivo is of particular interest because some oxidative DNA lesions are thought to be pre-mutagenic [61]. Epidemiological studies show that a fruit- and vegetable-rich diet may reduce cancers in general [62], in addition to cancers of specific organs, such as stomach cancer [63]. Consistent with this, low serum levels of Vitamin C in high risk population may contribute to the increased risk of gastric metaplasia or chronic gastritis, which are both precancerous lesions [64]. A consistent protective effect of Vitamin C has also been found in lung and colorectal cancer [65]. Vitamin C-rich food may or may not protect against breast cancer [66]. The evidence on whether vitamin C supplements are effective in decreasing cancers is contradictory. Studies also showed that vitamin C had no protective effect against basal cell cancer of the skin [67], non-Hodgkins lymphoma [68], or colorectal cancer [69]. In addition, Bjelakovic et al. [70] found no evidence that antioxidants can prevent gastrointestinal cancers. On the contrary, certain antioxidant combinations (β-carotene with vitamin A or vitamin E) results in increased patient mortality. Vitamin C, when added alone or in combination with other antioxidants, did not seem to have an effect on the incidence of gastrointestinal cancers. Data from intervention studies are contradictory, and overall there is still not enough evidence that vitamin supplementation is beneficial for preventing cancer in humans consuming adequate amounts of vitamin C from their diet. More studies will be needed to determine the exact antioxidant role of vitamin C in cancer.

11.4.3 Beta-Carotene

Carotenoids are a family of pigmented compounds that are synthesized by plants and microorganisms but not animals. In plants, they contribute to the photosynthetic machinery and protect them against photo-damage. Fruits and vegetables constitute the major sources of carotenoid in human diet [71–73]. There are over 600 carotenoids occurring in nature and 50 of which have vitamin A activity. However, only about 40 are present in a typical human diet. Of these 40, about 20 carotenoids have been identified in human blood and tissues. Close to 90 % of the carotenoids in the diet and human body is represented by α-carotene, β-carotene, lycopene, lutein, and cryptoxanthin [74]. The structure of β-carotene is shown in Fig. 11.4a. Various studies have indicated that carotenoids prevent or inhibit certain types of cancer, atherosclerosis, age-related muscular degeneration, and other diseases [75, 76], but conflicting results also exist. The bioavailability, metabolism, and biological functions of carotenoids are only now beginning to be investigated.

11.4.3.1 Dietary Sources and Mechanism of Action

Although carotenoids are present in many common human foods, deeply pigmented fruits, juices, and vegetables constitute the major dietary sources with yellow-orange vegetables and fruits providing most of the β-carotene and α-carotene, orange fruits providing α-cryptoxanthin, dark green vegetables providing lutein, and tomato products providing lycopene [71].

The antioxidant properties of carotenoids have been suggested as being the main mechanism by which they afford their beneficial effects and are associated with their peroxyl radical scavenging properties and their exceptional singlet oxygen quenching abilities [77]. However, it has also been proposed that carotenoids function as a chain-trapping antioxidant only at the physiological O_2 partial pressure (less than 150 Torr) (reaction B in Fig. 11.4b), while carotenoids may act as a pro-oxidant under conditions of high carotenoid and oxygen concentrations (reaction A in Fig. 11.4b) [77, 78]. Other mechanisms for protective biological effect of carotenoids include gap junction communication, cell growth regulation, modulating gene expression and Phase I and II drug metabolizing enzymes, and immune response [75, 79]. Furthermore, some of carotenoid effect (e.g., α- and β-carotene and β-cryptoxanthin) may be derived from their provitamin A activity.

11.4.3.2 Carotenoids and Cardiovascular Disease

Several epidemiologic studies have suggested that a diet rich in a variety of fruits and vegetables, which is good sources of dietary carotenoids, results in lower risk of coronary heart disease (CHD). In addition, studies have shown a link between higher dietary intake and tissue concentrations of carotenoids and lower risk of chronic diseases [72, 73]. Hozawa et al. [80] demonstrated that circulating serum carotenoids such as α-carotene, β-carotene, zeaxanthin/lutein, β-cryptoxanthin, and lycopene were associated in beneficial directions with markers of inflammation, oxidative stress, and endothelial dysfunction. On the contrary, β-carotene has also been reported to act as a pro-oxidant under certain situations [81]. Supplementation of β-carotene increased mortality from CVD in a group of smokers, former smokers,

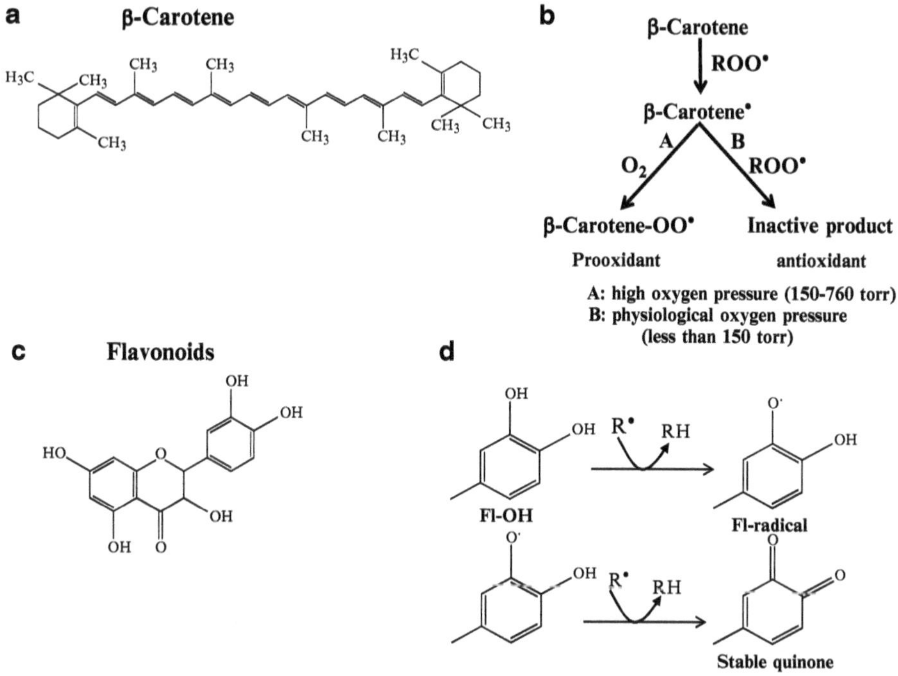

Fig. 11.4 General structure and scavenging reactive oxygen species properties of β-Carotene and Flavonoids. (**a**) Chemical structures of various isomers of β-carotene. (**b**) At the physiological O_2 partial pressure (less than 150 Torr), β-Carotene reacts directly with a peroxyl radical to form a resonance-stabilized carbon-centered radical, thereby contributing to the antioxidant effect of β-carotene on lipid peroxidation induced by peroxyl radicals. Then, the β-carotene radical can be removed from the system in a reaction with another peroxyl radical (reaction B). In contrast, under conditions of high carotenoid and high O_2 partial pressure, the β-carotene radical would combine with oxygen to form a carotenoid-peroxyl radical, which acts as a prooxidant (reaction A). (**c**) Chemical structures of flavonoids. (**d**) Due to their lower redox potentials, flavonoids (Fl-OH) are thermodynamically able to reduce highly oxidizing free radicals (R•) by hydrogen atom donation. The aroxyl radical (Fl-radical) may react with a second radical (R•), acquiring a stable quinone structure

and asbestos-exposed individuals in the β-carotene and retinol efficiency trial (CARET) [82]. Taken together, these observations suggest a possible biphasic response of β-carotene that promotes health when taken at dietary levels, but may have adverse effects when taken in higher amounts. Due to the concerns regarding pro-oxidant property of β-carotene, the focus of research has now shifted to another carotenoid antioxidant lycopene.

A clinical study found an association between higher plasma lycopene concentration (one of the major type of carotenoids) and low risk of CVD in 499 patients with CVD [83]. Kristenson et al., found that the Lithuanian populations showed lower lycopene levels to be associated with increased risk and mortality from CHD than Swedish populations [84]. In addition, lycopene regulates gene function, gap-junction communication, hormone and immune modulation, carcinogen metabolism, and metabolic pathways involving phase II drug-metabolizing enzymes [85, 86]. Overall, carotenoids such as lycopene seem to have benefit for preventing CVD.

11.4.3.3 Carotenoids and Cancer

Carotenoids have beneficial effect in different types of cancer. The results from observational studies is particularly striking for lung cancer, where carotenoid and/or fruit and vegetable intake has been associated with reduced lung cancer risk in eight of eight prospective studies and in 18 of 20 retrospective studies reviewed [87, 88]. Furthermore, several studies reported the beneficial effect of carotenoids in both in vivo and clinical trial. Le Marchand and colleagues [89] found that dietary intake of α-carotene, β-carotene, and lutein was associated with reduced lung cancer risk in both men and women. Jam and colleagues [90] reported that higher intake of saturated fats and lower intake of β-carotene and vitamin C before diagnosis increased the risk of dying of breast cancer. On the contrary, in the ATBC 2 trial, participants receiving β-carotene (alone or in combination with vitamin E) had a significantly higher incidence of lung cancer and total mortality than participants receiving the placebo. Thus, the discrepant results were observed in β-carotene.

As we discussed in previous section, lycopene is emerging as a protective role in cancer in tissue culture studies using human cancer cell lines [91], animal studies [92], and also human clinical trials [93]. They include breast, lung, gastrointestinal, cervical, and ovarian and pancreatic cancers [93]. Overall, carotenoids like lycopene can act as a potent antioxidant and protect cells against oxidative damage and thereby prevent or reduce the risk of several cancers.

11.4.4 Flavonoids

Flavonoids have been proposed to exert beneficial effects in a multitude of disease states, including cancer, CVD, and neurodegenerative disorders. The general structure of Flavonoids is illustrated in Fig. 11.4c. Evidence from cohort studies and randomized trials suggest beneficial effects of food sources of anthocyanidins (berries) and flavan-3-ols (green tea and cocoa) on cardiovascular health. Many of the biological actions of flavonoids have been attributed to their antioxidant properties in vitro through their scavenging several types of radicals and their metal ion chelating abilities. However, there is little evidence of antioxidant effects in vivo [94]. This is probably explained by the generally low concentrations in blood because of low bioavailability and extensive metabolism which reduces their antioxidant activity [94, 95]. The emerging view is that flavonoids are likely to exert beneficial and/or toxic actions on cells not through their potential to act as antioxidants but rather through their effects on signal transduction and different enzyme systems [95].

11.4.4.1 Dietary Sources and Mechanism of Action

Flavonoids are the most common and widely distributed group of plant phenolic compounds, occurring virtually in all plant parts, particularly the photosynthesizing plant cells. They are an integral part of both human and animal diets [96]. Being plant phytochemicals, flavonoids cannot be synthesized by humans and animals [97]. Flavanones are mainly found in citrus fruits and flavones in celery. Catechins are present in large amounts in green and black teas and in red wine, whereas anthocyanins are found in strawberries and other berries. Isoflavones are almost exclusively found in soy foods. Flavonoids are a major coloring component of flowering plants, which are found in all plant foods [98].

Many of the biological actions of flavonoids have been attributed to their antioxidant properties in vitro, since flavonoids are chemically one-electron donors. Due to their lower redox potentials, flavonoids (Fl-OH) are thermodynamically able to reduce highly oxidizing free radicals by hydrogen atom donation [99, 100]. The aroxyl radical (Fl-radical) may react with a second radical, acquiring a stable quinone structure (Fig. 11.4d). The aroxyl radicals could interact with oxygen, generating quinones and superoxide anion, rather than terminating chain reactions. The last reaction may take place in the presence of high levels of transient metal ions and is responsible for the undesired pro-oxidant effect of flavonoids [101]. In addition to antioxidant properties, flavonoids may exert modulatory actions in cells through actions at protein kinase and lipid kinase signaling pathways [95].

11.4.4.2 Flavonoids and Cardiovascular Disease

A possible reason for the inferred protective effects of flavonoids against heart disease is their ability to prevent the oxidation of LDL to an atherogenic form, although antiplatelet aggregation activity and vasodilatory properties are also reported [102]. Flavonoid intake may reduce the risk of death from CHD. Green and black teas are able to protect against NO toxicity, which may be another reason for the beneficial effects observed with flavonoids [103]. In addition, consumption of quercetin may protect against CVD by reducing capillary fragility and inhibiting platelet aggregation [102]. Prospective cohort of US adults demonstrated that flavonoid consumption was associated with lower risk of death from CVD. Most inverse associations appeared with intermediate intakes, suggesting that even relatively small amounts of flavonoid-rich foods may be beneficial [104].

11.4.4.3 Flavonoids and Cancer

A growing number of epidemiological studies suggest that high flavonoid intake may be correlated with a decreased risk of cancer [105]. Dai et al. [106] reported that urinary excretion of total isoflavonoids was substantially lower in breast cancer cases than in controls and strongly suggests a potential role of flavonoids in breast cancer prevention in a population-based control study conducted in China. In addition, studies showed that men with higher quercetin intakes had a lower lung cancer incidence, and men with higher myricetin intakes had a lower prostate cancer risk. These data suggest a protective role of flavonoids against cancer [107]. The research group in Uruguay conducted a case–control study and found that flavonoids displayed a marked reduction by 70 % in the risks of cancer of oral cavity, pharynx, larynx, and esophagus [108]. All these studies provide evidence for a protective role of flavonoids against cancer. There are some contrary reports [109, 110] that may be due to differences in bioavailability of the various flavonoids, and their effects on individual cancer sites cannot be excluded meriting further investigation.

11.4.5 Selenium

As an essential trace element, the importance of selenium (Se) in humans is well established, and its deficiency has caused serious health effects in humans, such as Keshan disease. Se is involved in protection against oxidative damage via Se-dependent GPx and other selenoproteins [111]. Because of its antioxidant properties, it has long been hypothesized that Se may prevent cardiovascular and other chronic diseases. Se supplementation increases enzymatic antioxidant activity and decreases lipid peroxidation [111, 112]. The effect of Se on atherosclerotic CVD, however, is uncertain. Observational studies [113–116] investigating the association of low se concentrations with cardiovascular outcomes as well as randomized trials [117, 118] investigating whether Se supplements prevent CHD have been inconclusive.

11.4.5.1 Dietary Sources and Mechanism of Action

The main food group that provides Se in the diet are bread and cereals, meat, fish, eggs, and milk dairy products [119]. The antioxidant nature of Se is different from that of other antioxidant supplements. Se functions as a component of antioxidant enzymes, for example, GPx and thioredoxin reductases [120], while vitamins E and C are presumed to react nonenzymatically with oxidant molecules.

11.4.5.2 Selenium and Cardiovascular Disease

A previous report demonstrated that the presence of high Se, antioxidant selenoenzymes, and selenoproteins may help to reduce the production of oxidized LDL and, therefore, would reduce the incidence of heart diseases [120]. In animal studies, the presence of selenoprotein P, which is found mainly in plasma, may play a significant role in regulating the plasma cholesterol level by protecting LDL oxidation from ROS [121], thereby preventing atherosclerosis.

In epidemiological studies, the associations of low Se status in humans with increased risk of heart diseases and mortality are still uncertain and controversial [122, 123]. Early supportive evidence from epidemiological studies suggested that a higher mortality of heart diseases was linked to Se-deficient areas [124]. However, subsequent epidemiological studies in Finland, a country of low Se status, gave inconclusive results [125].

One of the organo Se compounds Ebselen has been shown to exhibit a weak GPx-like activity in vivo and could be a promising cardioprotective agent for myocardial ischemia-reperfusion injury. Ebselen has also been shown to have a neuroprotective effect on stroke in a rabbit embolic stroke model [126, 127]. Since, Se has protective effect in animal models, large high-quality randomized controlled trials and observational studies are needed across multiple populations to determine the role of Se in CVD.

11.4.5.3 Selenium and Cancer

There have been numerous animal studies indicating the important role of Se in reducing and preventing the incidence of cancer initiated by a variety of carcinogens, including chemicals and radiation. Evidence from human epidemiological studies has increasingly indicated an inverse relationship between Se status and cancer risk in human populations. Clark and colleagues [128] reported that people who supplemented their diet with selenized yeast had a reduction of nearly 50 % in overall cancer morbidity. Furthermore, other studies showed that Se supplementation was associated with a ~25–60 % reduction in gastrointestinal cancers as well as other types of cancer including esophageal, gastric, small intestine, colorectal, pancreatic, liver, and biliary tract [129]. In a prospective study of case–control design, high toenail Se levels were associated with a reduced risk of advanced prostate cancer [130]. In another study, no association was observed between toenail Se levels and breast cancer in women or prostate cancer in men [131]. These investigators suggested that the effects of smoking and dietary habits of the subjects could have contributed to these inconclusive findings. Even though Se is reported to play a significant role in cancer development, its exact anticancer mechanism of action at molecular levels is not fully understood. However, it has been hypothesized that the most possible mechanistic action of Se is its role in the antioxidant defense systems to reduce oxidative stress and limit DNA damage [120]. Karunasinghe et al. [132] found a significant inverse relationship between the reduction of DNA damage in blood leucocytes and high blood serum Se levels in men who had high prostate-specific antigen and a high prostate cancer risk. The effectiveness of Se in the prevention of DNA damage, however, depends on its chemical forms. In an in vitro study, Battin et al. [133] found that selenocysteine inhibited DNA damage more strongly than the selenomethionine. Other possible anticancer mechanisms of Se include the induction of apoptosis, cell-cycle arrest, and DNA-repair genes; inhibition of protein kinase C activity and cell growth; and effect on estrogen- and androgen-receptor expression [134].

11.4.6 Other Antioxidant Supplements

11.4.6.1 Minerals

Zinc

Zinc is required in over 200 enzymes and so a zinc deficiency is likely to affect a number of different systems. Bettger et al. [135] demonstrated that Zinc has a stabilizing effect on membranes by displacing bound transition metal ions and thereby preventing peroxidation of membrane lipids. Zinc-deficient rats have lower levels of alkaline phosphatase and erythrocytes with raised GST levels and lowered GSH concentrations [136]. Some reports propose zinc to be anti-atherogenic through inhibition of LDL oxidation [137]. Zinc, like copper, is a component of Cu/Zn SOD, an important antioxidant enzyme. Deficiencies of zinc affect the activity of this enzyme.

Manganese

Manganese is a component of several enzymes involved in fatty acid and cholesterol biosynthesis as well as mitochondrial Mn-SOD. Previous study measured MnSOD activities in manganese-sufficient and -deficient rats and showed increased lipid peroxidation which is associated with decreased MnSOD activity [138]. Gene knockdown of MnSOD showed cardiac abnormalities, fat accumulation in the liver and skeletal muscle, and metabolic acidosis [139] and died within 10 days. The lethal nature of MnSOD knockout shows how important this enzyme is in removing mitochondrial superoxide.

Copper

Oxidative stress has been postulated to contribute to the pathology associated with either copper deficiency or excess. Although Cu in excess is toxic, Cu is an essential micronutrient, since Cu functions as a cofactor and is required for structural and catalytic properties of a variety of important enzymes, including cytochrome *c* oxidase, tyrosinase, *p*-hydroxyphenyl pyruvate hydrolase, dopamine beta hydroxylase, lysyl oxidase, Cu/Zn-SOD, and ecSOD [11, 140]. Thus, Cu deficiency may increase cellular susceptibility to oxidative damage likely due to decrease in antioxidant Cu enzymes. Previous study has shown that Cu-deficient Jurkat T-lymphocytes sustain greater oxidative DNA damage than control cells when challenged with H_2O_2 [141]. In addition, supplemental Cu, but not Zn or Fe, prevents H_2O_2-induced DNA damage caused by Cu chelator. Previous studies demonstrated that deficiency of Cu-transporter such as ATP7A or Atox1 in response to angiotensin II increased blood pressure by decreasing ecSOD activity [142, 143] and also impaired cell proliferation [144] and smooth muscle cell migration [145]. Furthermore, erythrocyte Cu/ZnSOD has been reported to be decreased in copper deficiency with a rise in spectrin protein carbonyls [146]. These data suggest that Cu deficiency compromises cellular antioxidant defenses via decreased capability to produce SOD, thereby increasing their susceptibility to oxidative DNA damage.

11.4.6.2 Omega-3 and Omega-6 Fatty Acids

Omega-3 and omega-6 fatty acids are essential long-chain polyunsaturated fatty acids because the human body cannot synthesize them. There are three major dietary types of omega-3 fatty acids: eicosapentaenoic acid (EPA), docosahexaenoic acid (DHA), and alpha-linolenic acid. These Omega-3 fatty acids can be found in fish, krill, algae, walnut, and flaxseed, while Omega-6 fatty acids are found in vegetable oils, nuts, cereals, eggs, and poultry [147]. It is important to maintain an appropriate balance of omega-3s and omega-6s in the diet, as these two substances work together to promote health [147]. Omega-3s reduce inflammation and prevent chronic ailments such as heart disease, stroke, memory loss, depression, arthritis, cataract, and cancer. Omega-6s improve diabetic neuropathy, eczema, psoriasis, osteoporosis, and aid in cancer treatment [147].

11.5 Conclusion and Future Direction

Free radical-induced damage contributes to the etiology of many chronic health problems such as cardiovascular and inflammatory disease and cancer, where antioxidant can play very important role for protective therapy. A therapeutic strategy to increase the antioxidant capacity of cells may be used to reinforce the long-term effective treatment. Synthetic antioxidants are reported to be dangerous to human health [148]. Thus, the search for effective, nontoxic natural compounds with antioxidant activity has been intensified. There is a substantial body of evidence that a diet rich in plant foods (particularly fruit and vegetables) conveys health benefits. However, some clinical trials reported that antioxidant supplements may not prevent disease and are even associated with a poor health outcome.

Several reasons or concerns have been given to explain why the findings of the observational studies have differed from those of the large clinical trials. (1) Most studies have used synthetic forms that may have different biological activity or potency from the natural forms of these vitamins. For example, trials using the natural forms have not suggested different clinical effects from those supplementing with synthetic forms [25, 117, 149]. (2) Antioxidant food supplements may also have pro-oxidant activity under certain circumstances. For example, the tocopheroxyl radical can paradoxically contribute to LDL oxidation under certain circumstances [150]. The tocopheroxyl radical can be recycled to tocopherol by coantioxidants such as vitamin C; however, the efficiency of this process has not been proven in vivo. The use of combined antioxidant interventions, including vitamin C, vitamin E, and beta-carotene (e.g., the heart protection study) has not proven to be effective either. (3) Concerns have been expressed about whether the correct antioxidant supplement has been used. For example, treatment with α-tocopherol, the form often used in clinical trials and the predominant form in vitamin E from natural sources, displaces γ-tocopherol from tissues. This is troubling because γ-tocopherol is especially effective in scavenging the peroxynitrate anion, a molecule that plays an important role in oxidant injury [151]. (4) Tissue concentrations that can be achieved with antioxidants might be far below levels required to counteract a ROS-generating system. (5) Antioxidants scavenge ROS after their production. They are incapable of preventing oxidation of molecules that have a very high affinity for ROS such as NO. (6) Single supplement may also interfere with the uptake, transport, distribution, and metabolism of other antioxidant nutrients. (7) The concern is also to select dose and duration of antioxidants therapy which is appropriately required to achieve the maximum beneficial effects. (8) As noted, ROS have important signaling properties, and the nonselective approach of scavenging all ROS could have deleterious effects. Further research is needed to clarify above facts before this antioxidant supplementation could be authoritatively recommended as an adjuvant therapy.

Acknowledgments This research was supported by NIH R01 HL070187 (to T.F.), Department of Veterans Affairs Merit Review Grant 1I01BX001232 (to T.F.), and 11POST5740006 (to V.S).

References

1. Go YM, Jones DP (2011) Cysteine/cystine redox signaling in cardiovascular disease. Free Radic Biol Med 50:495–509
2. Lassegue B, Griendling KK (2010) NADPH oxidases: functions and pathologies in the vasculature. Arterioscler Thromb Vasc Biol 30:653–661
3. Ushio-Fukai M, Urao N (2009) Novel role of NADPH oxidase in angiogenesis and stem/progenitor cell function. Antioxid Redox Signal 11:2517–2533
4. Orient A, Donko A, Szabo A, Leto TL, Geiszt M (2007) Novel sources of reactive oxygen species in the human body. Nephrol Dial Transplant 22:1281–1288
5. Guzik TJ, Harrison DG (2006) Vascular NADPH oxidases as drug targets for novel antioxidant strategies. Drug Discov Today 11:524–533
6. Brieger K, Schiavone S, Miller FJ Jr, Krause KH (2012) Reactive oxygen species: from health to disease. Swiss Med Wkly 142:w13659
7. Lobo V, Patil A, Phatak A, Chandra N (2010) Free radicals, antioxidants and functional foods: impact on human health. Pharmacogn Rev 4:118–126
8. Fridovich I (1978) The biology of oxygen radicals. Science 201:875–880

9. Valko M, Leibfritz D, Moncol J et al (2007) Free radicals and antioxidants in normal physiological functions and human disease. Int J Biochem Cell Biol 39:44–84
10. Halliwell B (1999) Antioxidant defence mechanisms: from the beginning to the end (of the beginning). Free Radic Res 31:261–272
11. Fukai T, Ushio-Fukai M (2011) Superoxide dismutases: role in redox signaling, vascular function, and diseases. Antioxid Redox Signal 15:1583–1606
12. Chelikani P, Fita I, Loewen PC (2004) Diversity of structures and properties among catalases. Cell Mol Life Sci 61:192–208
13. Brigelius-Flohe R, Maiorino M (1830) Glutathione peroxidases. Biochim Biophys Acta 2013:3289–3303
14. Rhee SG, Woo HA, Kil IS, Bae SH (2012) Peroxiredoxin functions as a peroxidase and a regulator and sensor of local peroxides. J Biol Chem 287:4403–4410
15. Young IS, Woodside JV (2001) Antioxidants in health and disease. J Clin Pathol 54:176–186
16. Valko M, Morris H, Cronin MT (2005) Metals, toxicity and oxidative stress. Curr Med Chem 12:1161–1208
17. Pryor WA (2000) Vitamin E, and heart disease: basic science to clinical intervention trials. Free Radic Biol Med 28:141–164
18. Sen CK, Khanna S, Roy S (2007) Tocotrienols in health and disease: the other half of the natural vitamin E family. Mol Aspects Med 28:692–728
19. Ju J, Picinich SC, Yang Z et al (2010) Cancer-preventive activities of tocopherols and tocotrienols. Carcinogenesis 31:533–542
20. Kris-Etherton PM, Lichtenstein AH, Howard BV, Steinberg D, Witztum JL (2004) Antioxidant vitamin supplements and cardiovascular disease. Circulation 110:637–641
21. Hensley K, Benaksas EJ, Bolli R et al (2004) New perspectives on vitamin E: gamma-tocopherol and carboxyelthylhydroxychroman metabolites in biology and medicine. Free Radic Biol Med 36:1–15
22. Sen CK, Khanna S, Roy S (2006) Tocotrienols: vitamin E beyond tocopherols. Life Sci 78:2088–2098
23. Wong RS, Radhakrishnan AK (2012) Tocotrienol research: past into present. Nutr Rev 70:483–490
24. Qian J, Morley S, Wilson K et al (2005) Intracellular trafficking of vitamin E in hepatocytes: the role of tocopherol transfer protein. J Lipid Res 46:2072–2082
25. Stephens NG, Parsons A, Schofield PM et al (1996) Randomised controlled trial of vitamin E in patients with coronary disease: Cambridge Heart Antioxidant Study (CHAOS). Lancet 347:781–786
26. Virtamo J, Rapola JM, Ripatti S et al (1998) Effect of vitamin E and beta carotene on the incidence of primary nonfatal myocardial infarction and fatal coronary heart disease. Arch Intern Med 158:668–675
27. Jialal I, Devaraj S (2000) Vitamin E supplementation and cardiovascular events in high-risk patients. N Engl J Med 342:1917–1918
28. Heart Protection Study Collaborative Group (2002) MRC/BHF Heart Protection Study of antioxidant vitamin supplementation in 20,536 high-risk individuals: a randomised placebo-controlled trial. Lancet 360:23–33
29. Investigators GP (1999) Dietary supplementation with n-3 polyunsaturated fatty acids and vitamin E after myocardial infarction: results of the GISSI-Prevenzione trial. Gruppo Italiano per lo Studio della Sopravvivenza nell'Infarto miocardico. Lancet 354:447–455
30. Lonn E, Bosch J, Yusuf S et al (2005) Effects of long-term vitamin E supplementation on cardiovascular events and cancer: a randomized controlled trial. JAMA 293:1338–1347
31. Miller ER III, Pastor-Barriuso R, Dalal D et al (2005) Meta-analysis: high-dosage vitamin E supplementation may increase all-cause mortality. Ann Intern Med 142:37–46
32. Widder JD, Harrison DG (2005) Can vitamin E prevent cardiovascular events and cancer? Nat Clin Pract Cardiovasc Med 2:510–511

33. Woodson K, Albanes D, Tangrea JA et al (1999) Association between alcohol and lung cancer in the alpha-tocopherol, beta-carotene cancer prevention study in Finland. Cancer Causes Control 10:219–226
34. Yong LC, Brown CC, Schatzkin A et al (1997) Intake of vitamins E, C, and A and risk of lung cancer. The NHANES I epidemiologic followup study. First National Health and Nutrition Examination Survey. Am J Epidemiol 146:231–243
35. Shibata A, Paganini-Hill A, Ross RK, Henderson BE (1992) Intake of vegetables, fruits, beta-carotene, vitamin C and vitamin supplements and cancer incidence among the elderly: a prospective study. Br J Cancer 66:673–679
36. Smolarek AK, Suh N (2011) Chemopreventive activity of vitamin E in breast cancer: a focus on gamma- and delta-tocopherol. Nutrients 3:962–986
37. Lee IM, Cook NR, Gaziano JM et al (2005) Vitamin E in the primary prevention of cardiovascular disease and cancer: the Women's Health Study: a randomized controlled trial. JAMA 294:56–65
38. Gaziano JM, Glynn RJ, Christen WG et al (2009) Vitamins E and C in the prevention of prostate and total cancer in men: the Physicians' Health Study II randomized controlled trial. JAMA 301:52–62
39. Lippman SM, Klein EA, Goodman PJ et al (2009) Effect of selenium and vitamin E on risk of prostate cancer and other cancers: the Selenium and Vitamin E Cancer Prevention Trial (SELECT). JAMA 301:39–51
40. Klein EA, Thompson IM Jr, Tangen CM et al (2011) Vitamin E and the risk of prostate cancer: the Selenium and Vitamin E Cancer Prevention Trial (SELECT). JAMA 306:1549–1556
41. Albanes D, Heinonen OP, Taylor PR et al (1996) Alpha-tocopherol and beta-carotene supplements and lung cancer incidence in the alpha-tocopherol, beta-carotene cancer prevention study: effects of base-line characteristics and study compliance. J Natl Cancer Inst 88: 1560–1570
42. Nesaretnam K, Guthrie N, Chambers AF, Carroll KK (1995) Effect of tocotrienols on the growth of a human breast cancer cell line in culture. Lipids 30:1139–1143
43. Goh SH, Hew NF, Norhanom AW, Yadav M (1994) Inhibition of tumour promotion by various palm-oil tocotrienols. Int J Cancer 57:529–531
44. Yu W, Simmons-Menchaca M, Gapor A, Sanders BG, Kline K (1999) Induction of apoptosis in human breast cancer cells by tocopherols and tocotrienols. Nutr Cancer 33:26–32
45. Guthrie N, Gapor A, Chambers AF, Carroll KK (1997) Inhibition of proliferation of estrogen receptor-negative MDA-MB-435 and -positive MCF-7 human breast cancer cells by palm oil tocotrienols and tamoxifen, alone and in combination. J Nutr 127:544S–548S
46. Constantinou C, Papas A, Constantinou AI (2008) Vitamin E and cancer: an insight into the anticancer activities of vitamin E isomers and analogs. Int J Cancer 123:739–752
47. Yano Y, Satoh H, Fukumoto K et al (2005) Induction of cytotoxicity in human lung adenocarcinoma cells by 6-O-carboxypropyl-alpha-tocotrienol, a redox-silent derivative of alpha-tocotrienol. Int J Cancer 115:839–846
48. Kashiwagi K, Harada K, Yano Y et al (2008) A redox-silent analogue of tocotrienol inhibits hypoxic adaptation of lung cancer cells. Biochem Biophys Res Commun 365:875–881
49. Duarte TL, Lunec J (2005) Review: when is an antioxidant not an antioxidant? A review of novel actions and reactions of vitamin C. Free Radic Res 39:671–686
50. Padayatty SJ, Katz A, Wang Y et al (2003) Vitamin C as an antioxidant: evaluation of its role in disease prevention. J Am Coll Nutr 22:18–35
51. Tsukaguchi H, Tokui T, Mackenzie B et al (1999) A family of mammalian Na+-dependent L-ascorbic acid transporters. Nature 399:70–75
52. Vera JC, Rivas CI, Fischbarg J, Golde DW (1993) Mammalian facilitative hexose transporters mediate the transport of dehydroascorbic acid. Nature 364:79–82
53. Nualart FJ, Rivas CI, Montecinos VP et al (2003) Recycling of vitamin C by a bystander effect. J Biol Chem 278:10128–10133
54. Haytowitz DB (1995) Information from USDA's Nutrient Data Bank. J Nutr 125: 1952–1955

55. Rose RC, Bode AM (1993) Biology of free radical scavengers: an evaluation of ascorbate. FASEB J 7:1135–1142
56. Podmore ID, Griffiths HR, Herbert KE et al (1998) Vitamin C exhibits pro-oxidant properties. Nature 392:559
57. Heller R, Unbehaun A, Schellenberg B et al (2001) L-ascorbic acid potentiates endothelial nitric oxide synthesis via a chemical stabilization of tetrahydrobiopterin. J Biol Chem 276:40–47
58. Brown AA, Hu FB (2001) Dietary modulation of endothelial function: implications for cardiovascular disease. Am J Clin Nutr 73:673–686
59. Duffy SJ, Gokce N, Holbrook M et al (1999) Treatment of hypertension with ascorbic acid. Lancet 354:2048–2049
60. Khaw KT, Bingham S, Welch A et al (2001) Relation between plasma ascorbic acid and mortality in men and women in EPIC-Norfolk prospective study: a prospective population study. European Prospective Investigation into Cancer and Nutrition. Lancet 357:657–663
61. Wang D, Kreutzer DA, Essigmann JM (1998) Mutagenicity and repair of oxidative DNA damage: insights from studies using defined lesions. Mutat Res 400:99–115
62. Block G (1991) Epidemiologic evidence regarding vitamin C and cancer. Am J Clin Nutr 54:1310S–1314S
63. McCullough ML, Robertson AS, Jacobs EJ et al (2001) A prospective study of diet and stomach cancer mortality in United States men and women. Cancer Epidemiol Biomarkers Prev 10:1201–1205
64. You WC, Zhang L, Gail MH et al (2000) Gastric dysplasia and gastric cancer: Helicobacter pylori, serum vitamin C, and other risk factors. J Natl Cancer Inst 92:1607–1612
65. Knekt P, Jarvinen R, Seppanen R et al (1991) Dietary antioxidants and the risk of lung cancer. Am J Epidemiol 134:471–479
66. Hunter DJ, Manson JE, Colditz GA et al (1993) A prospective study of the intake of vitamins C, E, and A and the risk of breast cancer. N Engl J Med 329:234–240
67. Hunter DJ, Colditz GA, Stampfer MJ et al (1992) Diet and risk of basal cell carcinoma of the skin in a prospective cohort of women. Ann Epidemiol 2:231–239
68. Zhang SM, Hunter DJ, Rosner BA et al (2000) Intakes of fruits, vegetables, and related nutrients and the risk of non-Hodgkin's lymphoma among women. Cancer Epidemiol Biomarkers Prev 9:477–485
69. Jacobs EJ, Connell CJ, Patel AV et al (2001) Vitamin C and vitamin E supplement use and colorectal cancer mortality in a large American Cancer Society cohort. Cancer Epidemiol Biomarkers Prev 10:17–23
70. Bjelakovic G, Nikolova D, Simonetti RG, Gluud C (2004) Antioxidant supplements for prevention of gastrointestinal cancers: a systematic review and meta-analysis. Lancet 364: 1219–1228
71. Mangels AR, Holden JM, Beecher GR, Forman MR, Lanza E (1993) Carotenoid content of fruits and vegetables: an evaluation of analytic data. J Am Diet Assoc 93:284–296
72. Johnson EJ (2002) The role of carotenoids in human health. Nutr Clin Care 5:56–65
73. Agarwal S, Rao AV (2000) Carotenoids and chronic diseases. Drug Metabol Drug Interact 17:189–210
74. Gerster H (1997) The potential role of lycopene for human health. J Am Coll Nutr 16: 109–126
75. Paiva SA, Russell RM (1999) Beta-carotene and other carotenoids as antioxidants. J Am Coll Nutr 18:426–433
76. Astorg P, Gradelet S, Berges R, Suschetet M (1997) Dietary lycopene decreases the initiation of liver preneoplastic foci by diethylnitrosamine in the rat. Nutr Cancer 29:60–68
77. Burton GW (1989) Antioxidant action of carotenoids. J Nutr 119:109–111
78. Burton GW, Ingold KU (1984) Beta-carotene: an unusual type of lipid antioxidant. Science 224:569–573
79. Bertram JS (1999) Carotenoids and gene regulation. Nutr Rev 57:182–191

80. Hozawa A, Jacobs DR Jr, Steffes MW et al (2007) Relationships of circulating carotenoid concentrations with several markers of inflammation, oxidative stress, and endothelial dysfunction: the Coronary Artery Risk Development in Young Adults (CARDIA)/Young Adult Longitudinal Trends in Antioxidants (YALTA) study. Clin Chem 53:447–455
81. Palozza P, Serini S, Torsello A et al (2003) Beta-carotene regulates NF-kappaB DNA-binding activity by a redox mechanism in human leukemia and colon adenocarcinoma cells. J Nutr 133:381–388
82. Omenn GS, Goodman GE, Thornquist MD et al (1996) Effects of a combination of beta carotene and vitamin A on lung cancer and cardiovascular disease. N Engl J Med 334:1150–1155
83. Sesso HD, Buring JE, Norkus EP, Gaziano JM (2005) Plasma lycopene, other carotenoids, and retinol and the risk of cardiovascular disease in men. Am J Clin Nutr 81:990–997
84. Kristenson M, Zieden B, Kucinskiene Z et al (1997) Antioxidant state and mortality from coronary heart disease in Lithuanian and Swedish men: concomitant cross sectional study of men aged 50. BMJ 314:629–633
85. Heber D, Lu QY (2002) Overview of mechanisms of action of lycopene. Exp Biol Med (Maywood) 227:920–923
86. Agarwal S, Rao AV (2000) Tomato lycopene and its role in human health and chronic diseases. CMAJ 163:739–744
87. Ziegler RG, Mayne ST, Swanson CA (1996) Nutrition and lung cancer. Cancer Causes Control 7:157–177
88. Mayne ST, Janerich DT, Greenwald P et al (1994) Dietary beta carotene and lung cancer risk in U.S. nonsmokers. J Natl Cancer Inst 86:33–38
89. Le Marchand L, Hankin JH, Kolonel LN et al (1993) Intake of specific carotenoids and lung cancer risk. Cancer Epidemiol Biomarkers Prev 2:183–187
90. Jain M, Miller AB, To T (1994) Premorbid diet and the prognosis of women with breast cancer. J Natl Cancer Inst 86:1390–1397
91. Kim L, Rao AV, Rao LG (2002) Effect of lycopene on prostate LNCaP cancer cells in culture. J Med Food 5:181–187
92. Lingen C, Ernster L, Lindberg O (1959) The promoting effect of lycopene on the non-specific resistance of animals. Exp Cell Res 16:384–393
93. Giovannucci E (1999) Tomatoes, tomato-based products, lycopene, and cancer: review of the epidemiologic literature. J Natl Cancer Inst 91:317–331
94. Hollman PC, Cassidy A, Comte B et al (2011) The biological relevance of direct antioxidant effects of polyphenols for cardiovascular health in humans is not established. J Nutr 141:989S–1009S
95. Williams RJ, Spencer JP, Rice-Evans C (2004) Flavonoids: antioxidants or signalling molecules? Free Radic Biol Med 36:838–849
96. de Pascual-Teresa S, Moreno DA, Garcia-Viguera C (2010) Flavanols and anthocyanins in cardiovascular health: a review of current evidence. Int J Mol Sci 11:1679–1703
97. Peterson J, Dwyer J (1998) Flavonoids: dietary occurrence and biochemical activity. Nutr Res 18:1995–2018
98. Cook NC, Samman S (1996) Flavonoids—chemistry, metabolism, cardioprotective effects, and dietary sources. J Nutr Biochem 7:66–76
99. Robak J, Gryglewski RJ (1988) Flavonoids are scavengers of superoxide anions. Biochem Pharmacol 37:837–841
100. Chen JW, Zhu ZQ, Hu TX, Zhu DY (2002) Structure-activity relationship of natural flavonoids in hydroxyl radical-scavenging effects. Acta Pharmacol Sin 23:667–672
101. Halliwell B (2008) Are polyphenols antioxidants or pro-oxidants? What do we learn from cell culture and in vivo studies? Arch Biochem Biophys 476:107–112
102. Duthie GG, Duthie SJ, Kyle JA (2000) Plant polyphenols in cancer and heart disease: implications as nutritional antioxidants. Nutr Res Rev 13:79–106
103. Paquay JB, Haenen GR, Stender G et al (2000) Protection against nitric oxide toxicity by tea. J Agric Food Chem 48:5768–5772

104. McCullough ML, Peterson JJ, Patel R et al (2012) Flavonoid intake and cardiovascular disease mortality in a prospective cohort of US adults. Am J Clin Nutr 95:454–464
105. Le Marchand L (2002) Cancer preventive effects of flavonoids—a review. Biomed Pharmacother 56:296–301
106. Dai Q, Franke AA, Jin F et al (2002) Urinary excretion of phytoestrogens and risk of breast cancer among Chinese women in Shanghai. Cancer Epidemiol Biomarkers Prev 11: 815–821
107. Knekt P, Kumpulainen J, Jarvinen R et al (2002) Flavonoid intake and risk of chronic diseases. Am J Clin Nutr 76:560–568
108. De Stefani E, Ronco A, Mendilaharsu M, Deneo-Pellegrini H (1999) Diet and risk of cancer of the upper aerodigestive tract–II. Nutrients. Oral Oncol 35:22–26
109. Garcia R, Gonzalez CA, Agudo A, Riboli E (1999) High intake of specific carotenoids and flavonoids does not reduce the risk of bladder cancer. Nutr Cancer 35:212–214
110. Garcia-Closas R, Agudo A, Gonzalez CA, Riboli E (1998) Intake of specific carotenoids and flavonoids and the risk of lung cancer in women in Barcelona, Spain. Nutr Cancer 32:154–158
111. Rayman MP (2000) The importance of selenium to human health. Lancet 356:233–241
112. Salonen JT, Salonen R, Seppanen K et al (1991) Effects of antioxidant supplementation on platelet function: a randomized pair-matched, placebo-controlled, double-blind trial in men with low antioxidant status. Am J Clin Nutr 53:1222–1229
113. Salonen JT, Alfthan G, Huttunen JK, Pikkarainen J, Puska P (1982) Association between cardiovascular death and myocardial infarction and serum selenium in a matched-pair longitudinal study. Lancet 2:175–179
114. Suadicani P, Hein HO, Gyntelberg F (1992) Serum selenium concentration and risk of ischaemic heart disease in a prospective cohort study of 3000 males. Atherosclerosis 96:33–42
115. Wei WQ, Abnet CC, Qiao YL et al (2004) Prospective study of serum selenium concentrations and esophageal and gastric cardia cancer, heart disease, stroke, and total death. Am J Clin Nutr 79:80–85
116. Akbaraly NT, Arnaud J, Hininger-Favier I et al (2005) Selenium and mortality in the elderly: results from the EVA study. Clin Chem 51:2117–2123
117. Brown BG, Zhao XQ, Chait A et al (2001) Simvastatin and niacin, antioxidant vitamins, or the combination for the prevention of coronary disease. N Engl J Med 345:1583–1592
118. Stranges S, Marshall JR, Trevisan M et al (2006) Effects of selenium supplementation on cardiovascular disease incidence and mortality: secondary analyses in a randomized clinical trial. Am J Epidemiol 163:694–699
119. Kotrebai M, Birringer M, Tyson JF, Block E, Uden PC (2000) Selenium speciation in enriched and natural samples by HPLC-ICP-MS and HPLC-ESI-MS with perfluorinated carboxylic acid ion-pairing agents. Analyst 125:71–78
120. Papp LV, Holmgren A, Khanna KK (2010) Selenium and selenoproteins in health and disease. Antioxid Redox Signal 12:793–795
121. Traulsen H, Steinbrenner H, Buchczyk DP, Klotz LO, Sies H (2004) Selenoprotein P protects low-density lipoprotein against oxidation. Free Radic Res 38:123–128
122. Alissa EM, Bahijri SM, Ferns GA (2003) The controversy surrounding selenium and cardiovascular disease: a review of the evidence. Med Sci Monit 9:RA9–RA18
123. Flores-Mateo G, Navas-Acien A, Pastor-Barriuso R, Guallar E (2006) Selenium and coronary heart disease: a meta-analysis. Am J Clin Nutr 84:762–773
124. Shamberger RJ, Gunch MS, Willis CE, McCormak LJ (1978) Selenium and heart disease. II. Selenium and other trace metal intakes and heart disease in 25 countries. In: Hemphill DD (ed) Trace substances in environmental health - XII. Columbia, Missouri, University of Missouri Press, pp 48–52
125. Huttunen JK (1997) Selenium and cardiovascular diseases–an update. Biomed Environ Sci 10:220–226

126. Lapchak PA, Zivin JA (2003) Ebselen, a seleno-organic antioxidant, is neuroprotective after embolic strokes in rabbits: synergism with low-dose tissue plasminogen activator. Stroke 34:2013–2018
127. Yamaguchi T, Sano K, Takakura K et al (1998) Ebselen in acute ischemic stroke: a placebo-controlled, double-blind clinical trial. Ebselen Study Group. Stroke 29:12–17
128. Clark LC, Combs GF Jr, Turnbull BW et al (1996) Effects of selenium supplementation for cancer prevention in patients with carcinoma of the skin. A randomized controlled trial. Nutritional Prevention of Cancer Study Group. JAMA 276:1957–1963
129. Bjelakovic G, Nikolova D, Simonetti RG, Gluud C (2008) Antioxidant supplements for preventing gastrointestinal cancers. Cochrane Database Syst Rev (3):CD004183
130. Yoshizawa K, Willett WC, Morris SJ et al (1998) Study of prediagnostic selenium level in toenails and the risk of advanced prostate cancer. J Natl Cancer Inst 90:1219–1224
131. Ghadirian P, Maisonneuve P, Perret C et al (2000) A case-control study of toenail selenium and cancer of the breast, colon, and prostate. Cancer Detect Prev 24:305–313
132. Karunasinghe N, Ryan J, Tuckey J et al (2004) DNA stability and serum selenium levels in a high-risk group for prostate cancer. Cancer Epidemiol Biomarkers Prev 13:391–397
133. Battin EE, Perron NR, Brumaghim JL (2006) The central role of metal coordination in selenium antioxidant activity. Inorg Chem 45:499–501
134. Papp LV, Lu J, Holmgren A, Khanna KK (2007) From selenium to selenoproteins: synthesis, identity, and their role in human health. Antioxid Redox Signal 9:775–806
135. Bettger WJ, Reeves PG, Savage JE, O'Dell BL (1980) Interaction of zinc and vitamin E in the chick. Proc Soc Exp Biol Med 163:432–436
136. Kraus A, Roth HP, Kirchgessner M (1997) Influence of vitamin C, vitamin E and beta-carotene on the osmotic fragility and the primary antioxidant system of erythrocytes in zinc-deficient rats. Arch Tierernahr 50:257–269
137. Wilkins GM, Leake DS (1994) The oxidation of low density lipoprotein by cells or iron is inhibited by zinc. FEBS Lett 341:259–262
138. Zidenberg-Cherr S, Keen CL, Lonnerdal B, Hurley LS (1983) Superoxide dismutase activity and lipid peroxidation in the rat: developmental correlations affected by manganese deficiency. J Nutr 113:2498–2504
139. Melov S, Schneider JA, Day BJ et al (1998) A novel neurological phenotype in mice lacking mitochondrial manganese superoxide dismutase. Nat Genet 18:159–163
140. Uauy R, Olivares M, Gonzalez M (1998) Essentiality of copper in humans. Am J Clin Nutr 67:952S–959S
141. Pan Y, Loo G (2000) Effect of copper deficiency on oxidative DNA damage in Jurkat T-lymphocytes. Free Radic Biol Med 28:824–830
142. Ozumi K, Sudhahar V, Kim HW et al (2012) Role of copper transport protein antioxidant 1 in angiotensin II-induced hypertension: a key regulator of extracellular superoxide dismutase. Hypertension 60:476–486
143. Qin Z, Gongora MC, Ozumi K et al (2008) Role of Menkes ATPase in angiotensin II-induced hypertension: a key modulator for extracellular superoxide dismutase function. Hypertension 52:945–951
144. Itoh S, Kim HW, Nakagawa O et al (2008) Novel role of antioxidant-1 (Atox1) as a copper-dependent transcription factor involved in cell proliferation. J Biol Chem 283:9157–9167
145. Ashino T, Sudhahar V, Urao N et al (2010) Unexpected role of the copper transporter ATP7A in PDGF-induced vascular smooth muscle cell migration. Circ Res 107:787–799
146. Sukalski KA, LaBerge TP, Johnson WT (1997) In vivo oxidative modification of erythrocyte membrane proteins in copper deficiency. Free Radic Biol Med 22:835–842
147. Simopoulos AP (2008) The importance of the omega-6/omega-3 fatty acid ratio in cardiovascular disease and other chronic diseases. Exp Biol Med (Maywood) 233:674–688
148. Le Coz CJ, Schneider GA (1998) Contact dermatitis from tertiary-butylhydroquinone in a hair dye, with cross-sensitivity to BHA and BHT. Contact Dermatitis 39:39–40
149. Yusuf S, Dagenais G, Pogue J, Bosch J, Sleight P (2000) Vitamin E supplementation and cardiovascular events in high-risk patients. The Heart Outcomes Prevention Evaluation Study Investigators. N Engl J Med 342:154–160

150. Witting PK, Upston JM, Stocker R (1997) Role of alpha-tocopheroxyl radical in the initiation of lipid peroxidation in human low-density lipoprotein exposed to horse radish peroxidase. Biochemistry 36:1251–1258
151. Christen S, Woodall AA, Shigenaga MK et al (1997) Gamma-tocopherol traps mutagenic electrophiles such as NO(X) and complements alpha-tocopherol: physiological implications. Proc Natl Acad Sci U S A 94:3217–3222
152. Boaz M, Smetana S, Weinstein T et al (2000) Secondary prevention with antioxidants of cardiovascular disease in endstage renal disease (SPACE): randomised placebo-controlled trial. Lancet 356:1213–1218
153. Milman U, Blum S, Shapira C et al (2008) Vitamin E supplementation reduces cardiovascular events in a subgroup of middle-aged individuals with both type 2 diabetes mellitus and the haptoglobin 2-2 genotype: a prospective double-blinded clinical trial. Arterioscler Thromb Vasc Biol 28:341–347
154. Blum S, Vardi M, Brown JB et al (2010) Vitamin E reduces cardiovascular disease in individuals with diabetes mellitus and the haptoglobin 2-2 genotype. Pharmacogenomics 11: 675–684
155. Mondul AM, Rager HC, Kopp W, Virtamo J, Albanes D (2011) Supplementation with alpha-tocopherol or beta-carotene reduces serum concentrations of vascular endothelial growth factor-D, but Not -A or -C, in male smokers. J Nutr 141:2030–2034
156. Wright ME, Mayne ST, Stolzenberg-Solomon RZ et al (2004) Development of a comprehensive dietary antioxidant index and application to lung cancer risk in a cohort of male smokers. Am J Epidemiol 160:68–76
157. Hodis HN, Mack WJ, LaBree L et al (2002) Alpha-tocopherol supplementation in healthy individuals reduces low-density lipoprotein oxidation but not atherosclerosis: the Vitamin E Atherosclerosis Prevention Study (VEAPS). Circulation 106:1453–1459

Chapter 12
Edaravone Therapy: From Bench to Bedside

Toru Yamashita and Koji Abe

Abbreviations

8-oxodG	8-Oxo-7,8-dihydro-2-deoxyguanosine
BBB	Blood–brain barrier
HT	Hemorrhagic transformation
MCAO	Middle cerebral artery occlusion
MMP-9	Matrix metalloproteinase-9
OPB	2-Oxo-3-(phenylhydrazono)-butanoic acid
tPA	Tissue plasminogen activator

12.1 Introduction

The free radical scavenger edaravone (3-methyl-1-phenyl-2-pyrazolin-5-one, MCI-186, Radicut; Mitsubishi Tanabe Pharma Corp., Osaka, Japan) was approved in Japan in April 2001 as a neuroprotective drug for the treatment of acute cerebral infarction. Since then, edaravone has been widely used for many adult stroke patients in Japan.

Free radicals play crucial pathological roles in diseases of various kinds. Therefore, edaravone might be useful in the treatment of diseases other than cerebral infarction. In this chapter, we review the pharmacological characteristics and clinical effects of edaravone while incorporating our recent findings.

T. Yamashita • K. Abe, M.D., Ph.D. (✉)
Department of Neurology, Okayama University Graduate School of Medicine, Dentistry and Pharmaceutical Sciences, 2-5-1 Shikata-cho, Okayama 700-8558, Japan
e-mail: abekabek@cc.okayama-u.ac.jp

H. Tsukahara and K. Kaneko (eds.), *Studies on Pediatric Disorders*, Oxidative Stress in Applied Basic Research and Clinical Practice, DOI 10.1007/978-1-4939-0679-6_12,

12.2 Development of Edaravone

In the 1970s, it was proposed that oxygen-free radical species can damage the lipid membrane of cells and play a key role in pathophysiological aspects of diseases of various kinds, including stroke [1, 2]. In the stroke brain, depletion of ischemic energy engenders increased cytosolic Ca^{2+} through pump failure and cell depolarization, activating phospholipase A2. Phospholipases liberate free fatty acids, particularly arachidonate, from cell membranes. This arachidonate causes a burst of free radicals in the ischemic penumbra, which is a therapeutic target area. Free radicals are increased drastically after reperfusion [3].

Many researchers have sought a free radical scavenger without side-effects (such as narcotizing or suppressing cerebral metabolism) to establish a novel effective drug for stroke patients [4]. In early stages of investigation, edaravone was found to have promising effects by quenching the hydroxyl radical (•OH) and by inhibiting both •OH-dependent and •OH-independent lipid peroxidation (Fig. 12.1) [5]. Edaravone was indicated as being biphasic, water soluble and lipid soluble, with the $c\log P$ value of 1.33 (the $\log P$ value of a compound, the logarithm of its partition coefficient between n-octanol and water $\log(c_{\mathrm{octanol}}/c_{\mathrm{water}})$, is a well-established

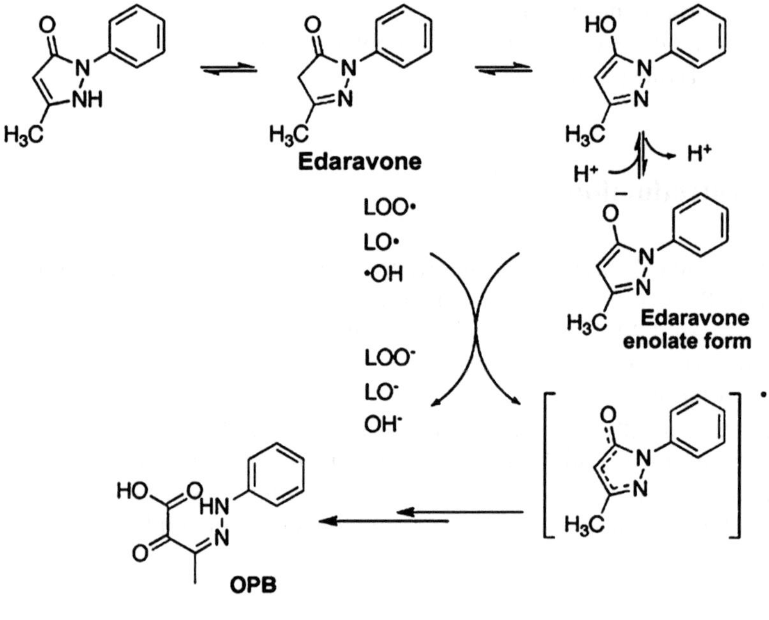

Fig. 12.1 Reaction mechanisms of edaravone with free radicals (revised from Nakagawa, et al. [5]). Edaravone enolate form can interact with both peroxyl (LOO•) and hydroxyl radicals (•OH) with subsequent formation of a stable oxidation product (2-oxo-3-(phenylhydrazono)-butanoic acid; OPB)

measure of the compound's hydrophilicity). Edaravone showed inhibitory effects on both water-soluble and lipid-soluble peroxyl radical-induced peroxidation systems, which differ from the inhibitory effects of vitamin E ($c\log P = 9.96$) and vitamin C ($c\log P = -2.15$) [6]. It was also proposed that approximately half of edaravone exists as the edaravone anion under physiological conditions and that this edaravone anion can donate an electron to free radicals of various kinds and scavenge them [7].

12.3 Effects of Edaravone on Stroke (Animal Model and Clinical Trials)

As described above, activation of the arachidonate cascade is a major cause of edema and tissue injury in cerebral ischemia. Watanabe et al. [8] directly injected arachidonate into the rat cortex and evaluated the anti-edemic effect of edaravone on this animal model. Treatment with edaravone (0.1–3.0 mg/kg i.v.) remarkably inhibited brain swelling observed 24 h after cortical injection of arachidonic acid in rats. To evaluate the effect of edaravone on brain edema in the post-stroke brain, we administered edaravone in the middle cerebral artery occlusion (MCAO) rat model [9]. In this model, the water content, reflecting blood–brain barrier (BBB) disruption, increased markedly after 3 and 6 h of ischemia. A further increase was found after 3 h of reperfusion following 3 h of ischemia. Results showed that edaravone markedly suppressed ischemic and post-ischemic brain swelling (Fig. 12.2). In addition, post-ischemic treatment with edaravone decreased the size of cerebral infarcts considerably and improved the neurological deficits 1 day after MCAO [10].

Another research group reported that edaravone markedly suppressed the accumulation of a product of nucleic acid oxidation, 8-oxo-7,8-dihydro-2-deoxyguanosine (8-oxodG), and sequential inflammatory responses at the peri-infarct lesion in the mouse stroke model [11]. We have also reported that edaravone showed strong neuroprotection after cerebral ischemia, which was confirmed by in vivo and ex vivo optical imaging for the apoptosis marker, annexin V as well as reducing cerebral infarct (Fig. 12.3) [12]. These results demonstrate that edaravone can decrease ischemic brain damage in the animal stroke model by quenching free radical generation and by suppressing arachidonic cascades or inflammatory responses.

A multicenter clinical trial demonstrated that edaravone attenuated the resulting disability in humans 90 days after acute ischemic stroke without severe adverse effects [13]. Consequently, edaravone has been used widely in Japan as an effective neuroprotective drug for acute stroke patients. It is noteworthy that acute renal failure occurs only occasionally in elder patients with renal dysfunction [14]. Renal function should be monitored sequentially in patients undergoing edaravone treatment.

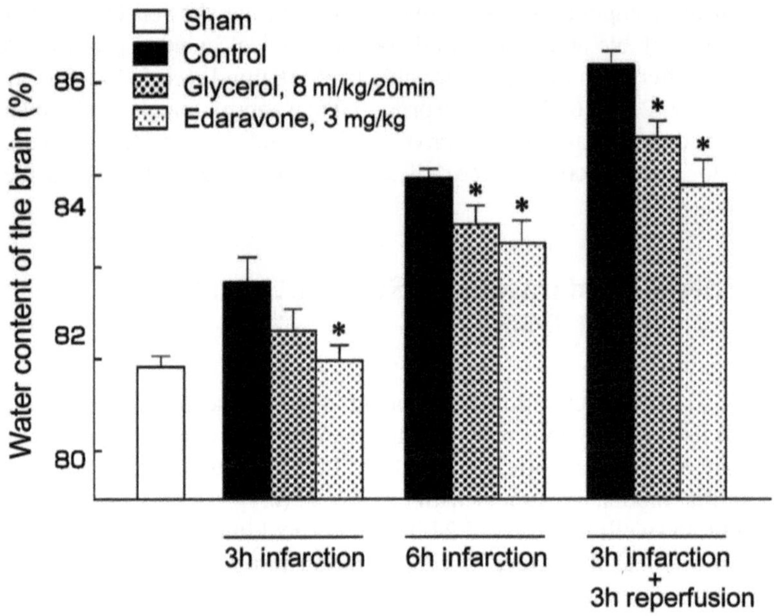

Fig. 12.2 Effects on brain edema in the rat middle cerebral artery occlusion or occlusion–recirculation model (revised from Abe et al. [9]). Rat middle cerebral artery occlusion and recirculation were obtained, respectively, by insertion and removal of a nylon thread. *$p<0.05$ vs. control (Dunn's multiple comparison test). Data represent the mean ± SD

12.4 Edaravone as a Good Partner of Thrombolytic Therapy with Tissue Plasminogen Activator

Free radical scavengers, including edaravone, have been developed for the treatment of acute stroke patients. However, the emergence of thrombolytic therapy with tissue plasminogen activator (tPA) for stroke now highlights the importance of free radical scavenger therapy, augmenting the important role of free radical scavengers in clinical settings.

If tPA restores cerebral blood flow, then ischemic brain damage can be ameliorated [15]. Intravenous injection of tPA has been used clinically around the world as a thrombolytic therapy. However, delayed reperfusion with tPA can cause hemorrhagic transformation (HT) [16], mainly through matrix metalloproteinase-9 (MMP-9) activation. Therefore, the application of tPA is strictly limited in the clinical setting. Reportedly, another free radical scavenger, NXY-059, potentially inhibited symptomatic HT after tPA treatment [17], but this effect was not reproduced [18].

We used a spontaneously hypertensive rat model of MCAO and tested the efficacy of the free radical scavenger, edaravone, in preventing HT. Administration of tPA alone significantly worsened the survival rate compared with those rats treated

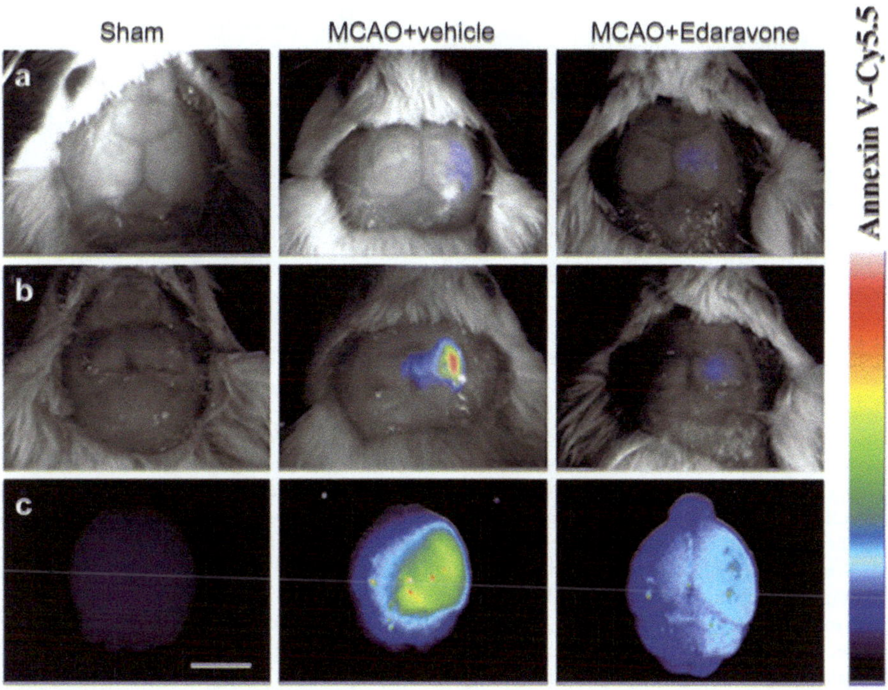

Fig. 12.3 In vivo imaging of Annexin V-Cy5.5 (**a**) with removal of head skin, (**b**) removal of the skull bone, and (**c**) ex vivo imaging of the brain (revised from Liu et al. [12]). This optical imaging method demonstrated that edaravone treatment suppressed apoptosis in post-stroke mice brains at 48 h after middle cerebral artery occlusion

with vehicle, but treatment with edaravone plus tPA increased the survival rate significantly, improved motor function, and decreased HT dramatically [19]. We also demonstrated that treatment with edaravone suppressed MMP-9 expression at and around cerebral microvessels, inhibited the degradation of basement membrane protein, and prevented microvessels from dissociating. These results indicate that edaravone can protect cerebral microvascular integrity because it safeguards the basement membrane from excessive free radicals and MMP-9, leading to a subsequent decrease in HT and improvement in the survival rate and neurological outcome (Fig. 12.4).

NXY-059 is water soluble with $c\log P$ of −2.09. In contrast, edaravone is biphasic, water soluble, and lipid soluble ($c\log P = 1.33$). Apparently, it passes easily through the BBB to enter the brain parenchyma and cerebral fluid [6]. Because the site most vulnerable to free radical damage is on the outer side of the vascular endothelium (such as the basal membrane), this unique chemical property of edaravone might be advantageous for its delivery to the basement membrane. Therefore, combination therapy with edaravone and tPA is a promising therapeutic strategy for acute stroke patients, not only in reducing infarct size but also in minimizing catastrophic HT. One clinical trial has already started in Japan to confirm the effects of this combination therapy in acute stroke patients [20].

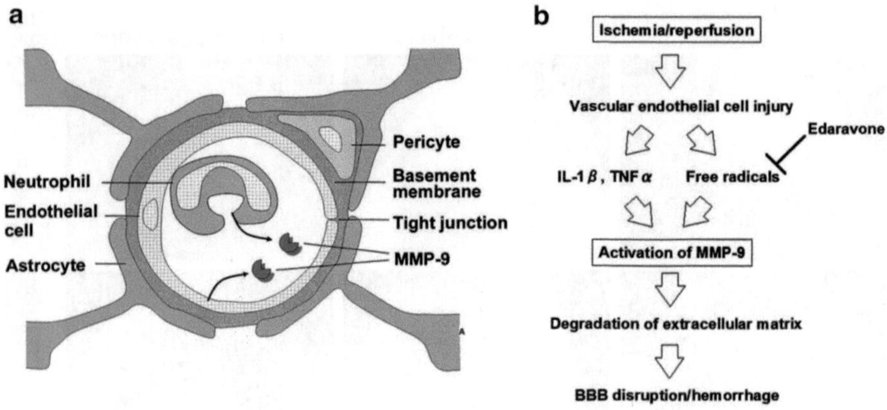

Fig. 12.4 Free radicals and matrix metalloproteinase-9 (MMP-9) play crucial roles in blood–brain barrier (BBB) disruption in the acute phase of stroke. (**a**) Schematic diagram of vascular unit comprising endothelial cells, astrocytes, and pericytes. MMP-9 is derived mainly from brain endothelial cells and infiltrating neutrophils in the acute phase of stroke. MMP-9 can degrade the basement membrane, which links the endothelial cells, and which maintains the vascular unit integrity. (**b**) Possible mechanism of vascular unit disruption after ischemia and reperfusion with tissue plasminogen activator. MMP-9 can disrupt the BBB by degrading the basal membrane/extracellular matrix proteins, thereby leading to BBB leakage and hemorrhaging. In the acute phase of stroke, MMP-9 can be activated by pro-inflammatory factors (e.g., IL-1β and TNF-α) and free radicals. MMP-9 activation can be inhibited by several reagents including the free radical scavenger, edaravone

In addition, plenty of free radicals are generated and play an important role in cerebral hyperperfusion syndrome after carotid endarterectomy, just as seen in reperfusion with tPA therapy. Ogasawara et al. [21] reported that pretreatment with edaravone prevented the occurrence of cerebral hyperperfusion syndrome after carotid endarterectomy in patients with ipsilateral internal carotid artery stenosis.

12.5 Concluding Remarks

This chapter briefly highlighted the pharmacologic characteristics and clinical effects of edaravone together with our recent clinical and experimental findings. Free radicals are regarded as a key regulator of disease progression not only in ischemic disease but also in other various diseases. Therefore, in the near future, edaravone might be incorporated into multidisciplinary treatment for critical diseases of various kinds.

In the pediatric field, edaravone has been used as an antioxidative and anti-inflammatory drug for a small fraction of patients in Japan: brain infarction [22], the brain, and renal infarction [23], and influenza encephalopathy of severe type [24]. A multicenter survey conducted in Japan [25] revealed that 8 (73 %) of 11 pediatric stroke patients who were treated with edaravone recovered to preattack status, whereas only 4 (31 %) of 13 patients who were treated with other drugs exhibited the same level of recovery. Edaravone treatment must be tested further for pediatric patients before any conclusion about its safety or efficacy can be made.

References

1. Demopoulos HB, Flamm ES, Pietronigro DD, Seligman ML (1980) The free radical pathology and the microcirculation in the major central nervous system disorders. Acta Physiol Scand Suppl 492:91–119
2. Siesjo BK (1981) Cell damage in the brain: a speculative synthesis. J Cereb Blood Flow Metab 1:155–185
3. White BC, Sullivan JM, DeGracia DJ, O'Neil BJ, Neumar RW, Grossman LI, Rafols JA, Krause GS (2000) Brain ischemia and reperfusion: molecular mechanisms of neuronal injury. J Neurol Sci 179:1–33
4. Asano T, Sano K (1979) Cerebral protection by pharmacological agents. No Shinkei Geka 7:549–554 [in Japanese]
5. Nakagawa H, Ohyama R, Kimata A, Suzuki T, Miyata N (2006) Hydroxyl radical scavenging by edaravone derivatives: efficient scavenging by 3-methyl-1-(pyridin-2-yl)-5-pyrazolone with an intramolecular base. Bioorg Med Chem Lett 16:5939–5942
6. Yamamoto Y, Kuwahara T, Watanabe K, Watanabe K (1996) Antioxidant activity of 3-methyl-1-phenyl-2-pyrazolin-5-one. Redox Rep 2:333–338
7. Watanabe T, Tanaka M, Watanabe K, Takamatsu Y, Tobe A (2004) Research and development of the free radical scavenger edaravone as a neuroprotectant. Yakugaku Zasshi 124:99–111 [in Japanese]
8. Watanabe T, Egawa M (1994) Effects of an antistroke agent MCl-186 on cerebral arachidonate cascade. J Pharmacol Exp Ther 271:1624–1629
9. Abe K, Yuki S, Kogure K (1988) Strong attenuation of ischemic and postischemic brain edema in rats by a novel free radical scavenger. Stroke 19:480–485
10. Kawai H, Nakai H, Suga M, Yuki S, Watanabe T, Saito KI (1997) Effects of a novel free radical scavenger, MCl-186, on ischemic brain damage in the rat distal middle cerebral artery occlusion model. J Pharmacol Exp Ther 281:921–927
11. Zhang N, Komine-Kobayashi M, Tanaka R, Liu M, Mizuno Y, Urabe T (2005) Edaravone reduces early accumulation of oxidative products and sequential inflammatory responses after transient focal ischemia in mice brain. Stroke 36:2220–2225
12. Liu N, Shang J, Tian F, Nishi H, Abe K (2011) In vivo optical imaging for evaluating the efficacy of edaravone after transient cerebral ischemia in mice. Brain Res 1397:66–75
13. The Edaravone Acute Brain Infarction Study Group (2003) Effect of a novel free radical scavenger, edaravone (MCI-186), on acute brain infarction. Randomized, placebo-controlled, double-blind study at multicenters. Cerebrovasc Dis 15:222–229
14. Hishida A (2007) Clinical analysis of 207 patients who developed renal disorders during or after treatment with edaravone reported during post-marketing surveillance. Clin Exp Nephrol 11:292–296
15. The National Institute of Neurological Disorders and Stroke rt-PA Stroke Study Group (1995) Tissue plasminogen activator for acute ischemic stroke. N Engl J Med 333:1581–1587
16. The NINDS t-PA Stroke Study Group (1997) Intracerebral hemorrhage after intravenous t-PA therapy for ischemic stroke. Stroke 28:2109–2118
17. Lees KR, Zivin JA, Ashwood T, Davalos A, Davis SM, Diener HC, Grotta J, Lyden P, Shuaib A, Hårdemark HG, Wasiewski WW, Stroke-Acute Ischemic NXY Treatment (SAINT I) Trial Investigators (2006) NXY-059 for acute ischemic stroke. N Engl J Med 354:588–600

18. Shuaib A, Lees KR, Lyden P, Grotta J, Davalos A, Davis SM, Diener HC, Ashwood T, Wasiewski WW, Emeribe U, SAINT II Trial Investigators (2007) NXY-059 for the treatment of acute ischemic stroke. N Engl J Med 357:562–571
19. Yamashita T, Kamiya T, Deguchi K, Inaba T, Zhang H, Shang J, Miyazaki K, Ohtsuka A, Katayama Y, Abe K (2009) Dissociation and protection of the neurovascular unit after thrombolysis and reperfusion in ischemic rat brain. J Cereb Blood Flow Metab 29:715–725
20. Tanahashi N (2009) Thrombolysis by intravenous tissue plasminogen activator (t-PA)—current status and future direction. Brain Nerve 61:41–52 [in Japanese]
21. Ogasawara K, Inoue T, Kobayashi M, Endo H, Fukuda T, Ogawa A (2004) Pretreatment with the free radical scavenger edaravone prevents cerebral hyperperfusion after carotid endarterectomy. Neurosurgery 55:1060–1067
22. Baba H, Sugimori H, Nanishi E, Nagata H, Lee S, Kuwashiro T, Hashizume M (2012) Stroke in a child safely treated with intravenous tissue plasminogen activator and edaravone, a free radical scavenger. J Stroke Cerebrovasc Dis 21:903.e5–903.e8
23. Sugimoto K, Iba Y, Fujita S, Sakata N, Okada M, Takemura T (2012) Nephrotic syndrome complicated by renal and cerebral infarctions in a 14-year-old girl. Pediatr Int 54:549–552
24. Okumura A, Nakagawa S, Kawashima H, Morichi S, Muguruma T, Saito O, Fujimoto J, Toida C, Kuga S, Imamura T, Shimizu T, Kondo N, Morishima T (2013) Severe form of encephalopathy associated with 2009 pandemic influenza A (H1N1) in Japan. J Clin Virol 56:25–30
25. Yamamoto H, Hayashi M (2008) Multi-center survey of edaravone treatment in children. No To Hattatsu 40:333–334 [in Japanese]

Chapter 13
The Use of Melatonin against Oxidative Stress in Pediatric Disorders

You-Lin Tain and Li-Tung Huang

Abbreviations

8-OHdG	8-Hydroxydeoxyguanosine
ADMA	Asymmetric dimethylarginine
AFMK	N^1-acetyl-N^2-formyl-5-methoxykynuramine
AMK	N^1-acetyl-5-methoxykynuramine
BDL	Bile duct ligation
CLD	Chronic lung disease
DDAH	Dimethylarginine dimethylaminohydrolase
DNMT	DNA methyltransferases
HI	Hypoxia-ischemia
ICSD	International Classification of Sleep Disorders
IUGR	Intrauterine growth restriction
NOS	Nitric oxide synthase
PKC	Protein kinase C
PVL	Periventricular leukomalacia
RDS	Respiratory distress syndrome
RNS	Reactive nitrogen species
ROS	Reactive oxygen species
SHR	Spontaneously hypertensive rat

Y.-L. Tain (✉) • L.-T. Huang
Department of Pediatrics, Kaohsiung Chang Gung Memorial Hospital and Chang Gung University College of Medicine, Kaohsiung, Taiwan
e-mail: tainyl@hotmail.com

H. Tsukahara and K. Kaneko (eds.), *Studies on Pediatric Disorders*, Oxidative Stress in Applied Basic Research and Clinical Practice, DOI 10.1007/978-1-4939-0679-6_13,

13.1 Introduction

Melatonin was identified 50 years ago [1] and in the last 20 years, numerous clinical trials were proposed to study its therapeutic effects in different fields of medicine [2–4].

The objective of this chapter is to provide an overview of potential clinical applications of melatonin in pediatric practice, with an emphasis on the following areas: the biochemistry and physiology of melatonin; the clinical impact of oxidative stress in pediatric population; the use of melatonin as an antioxidant against oxidative stress; current use of melatonin in pediatrics; and new perspectives in melatonin uses in pediatric medicine.

13.2 Biochemistry and Physiology of Melatonin

13.2.1 Structure, Synthesis, and Metabolism

Melatonin (*N*-acetyl-5-methoxytryptamine) is an endogenously produced indoleamine containing two functional groups, which are for the receptor binding and capacity to enter cells. Tryptophan and serotonin are precursors of melatonin. At least four enzymes are involved in melatonin's synthesis (Fig. 13.1). Among them, serotonin *N*-acetyltransferase is considered the rate-limiting enzyme in the regulation of melatonin biosynthesis [5]. The half-life of melatonin in serum varies in the range 30–60 min [6]. Its lipophilic nature makes melatonin easily cross all membranes. Once released into the blood, 70 % of melatonin is bound to albumin and another 30 % diffuses to the surrounding tissues [7]. Circulating melatonin is mainly synthesized in the pineal gland. Melatonin can be locally found in various tissues and organs, including lymphocytes, bone marrow, the thymus, the gastrointestinal tract, skin, and retina. On the other hand, melatonin is metabolized primarily in the liver, and secondarily in the kidney [7]. The catabolism of melatonin is exclusively done by the hepatic P450 monooxygenase, followed by conjugation of the resulting 6-sulfatoxy-melatonin to give the main urinary metabolite 6-sulftory-melatonin. In contrast, N^1-acetyl-N^2-formyl-5-methoxykynuramine (AFMK) and N^1-acetyl-5-methoxykynuramine (AMK) as major brain metabolites of melatonin [8]. AFMK and AMK form metabolites by interaction with reactive oxygen and nitrogen species (Fig. 13.1).

13.2.2 Physiological Function of Melatonin

In mammals, the melatonin rhythm is generated by an endogenous circadian clock in the suprachiasmatic nucleus of the hypothalamus. The most important regulating metabolism of the mammalian pineal gland is the light–dark cycle.

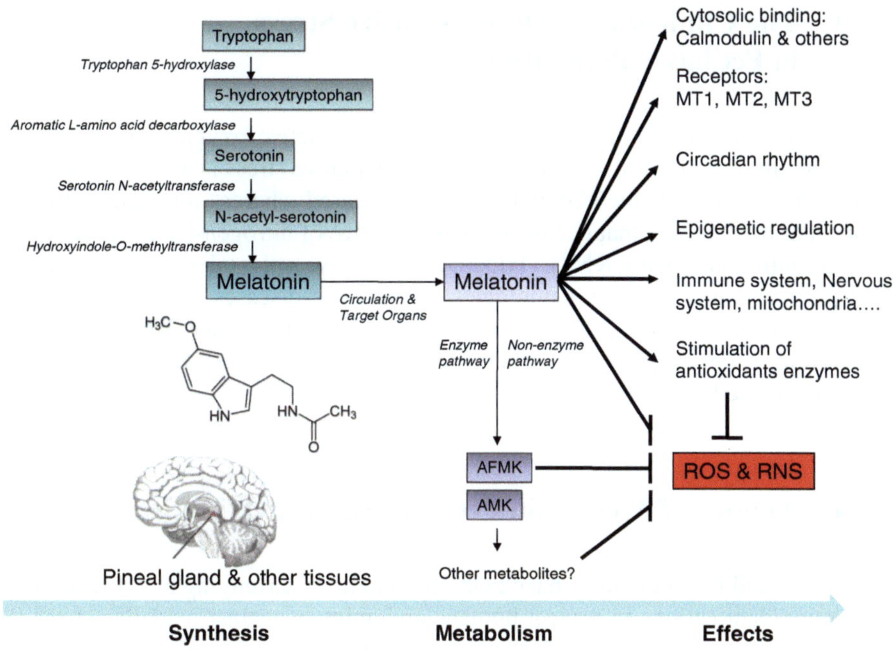

Fig. 13.1 Synthesis, metabolism, and effects of melatonin

Three subtypes of mammalian melatonin receptors have been cloned. Two of these, MT1 and MT2, which belong to the G-protein-coupled receptor superfamily. The melatonin MT1 and MT2 membrane-bound receptors show 60 % homology at the amino acid level. MT3, the third receptor, is an enzyme identified as quinine reductase 2. Next, melatonin has several important physiological functions, including stimulatory action in the immune system and mitochondrial biogenesis, induction of sleep initiation, resetting of circadian clock, regulation of intestinal motility and pancreas function, and prevention of tumor progression [5].

Moreover, melatonin is important in pregnancy and parturition [9, 10]. Maternal plasma melatonin levels are elevated during pregnancy, reaching a maximum at term and then returning to basal levels soon after delivery. Melatonin can readily cross the placenta and the fetal blood–brain barrier [11]. After birth, the circadian rhythm of melatonin biosynthesis develops between the sixth and the eighth week of life [12]. Detectable circadian variations in melatonin level are observed from the third to sixth month of life, which coincides chronologically with the development of normal sleep rhythm. The maximal rise of the amplitude of this nocturnal melatonin secretion occurs between the fourth and seventh year of life. At puberty, there is a drop in melatonin concentrations, and thereafter plasma concentrations of melatonin diminish gradually. In adulthood, melatonin production in the pineal gland declines progressively with age.

13.3 The Clinical Impact of Oxidative Stress in Pediatric Population

Oxygen-derived metabolites, collectively termed reactive oxygen species (ROS), are normally produced in aerobic organisms. Oxidative stress is a result of damage to essential molecules by ROS that are oxygen-based whereas nitrosative stress is molecular dysfunction that occurs as a consequence of nitrogen-based species (i.e., reactive nitrogen species, RNS). Oxidative stress plays a pivotal role in the initiation and progression of various pathologies in pediatric diseases. To understand the effect of melatonin on oxidative stress is the focus of this chapter. Thus, readers are referred to recent chapters for further information on the impact of oxidative stress in pediatric medicine [13–16].

13.4 Actions of Melatonin Protects Against Oxidative Stress

Melatonin and its metabolites have both direct radical scavenging effects and indirect antioxidant actions to protect humans and animals against oxidative stress [8, 17]. First, there is growing evidence that melatonin is a highly effective scavenger of both oxygen and nitrogen-based free radicals. Melatonin and its metabolites are capable of scavenging both ROS and RNS including the hydroxyl radical ($^{\bullet}OH$) [18], hydrogen peroxide (H_2O_2) [19], singlet oxygen ($^1O_2$) [20], hypochlorous acid (HClO) [21], peroxynitrite anion ($ONOO^-$) [22], superoxide anion radical, nitric oxide [23], and others [24].

In addition, melatonin stimulates gene expression of antioxidative enzymes [25]. The activity and expression of antioxidant enzymes, such as superoxide dismutase, glutathione, catalase, glutathione peroxidase, and glutathione reductase have been shown to be increased by melatonin, supporting its indirect antioxidant action. Another advantage of melatonin over classical antioxidants is its lack of prooxidative actions. All classical antioxidants are potential electron donors and they exhibit both reduced and oxidized forms. The oxidized form is transformed from the reduced state through the redox recycling. Therefore, classical antioxidants are prooxidants. In contrast, melatonin sacrifices itself and does not participate in redox recycling after scavenging free radicals. Most importantly, melatonin's metabolites (e.g., AFMK) still have the ability to scavenge free radicals (Fig. 13.1).

Furthermore, melatonin is an effective metal chelating agent. A 60 mg/mL concentration of melatonin exhibited a 95 % chelating effect on ferrous ions [26]. Also, melatonin can bind effectively to both copper (II) and copper (I) ions [27]. Therefore, the ability of melatonin to serve as a metal chelator provides further evidence supporting its protective role against oxidative stress.

Asymmetric dimethylarginine (ADMA) is an endogenous inhibitor of nitric oxide synthase (NOS). ADMA can reduce the synthesis of NO but induces superoxide production by uncoupling NOS. Thus, ADMA concentrations tightly regulate local status of oxidative stress [28]. Melatonin can prevent the development of hypertension in young spontaneously hypertensive rat (SHR) by reduction of plasma ADMA and attenuation of oxidative stress. Our data demonstrated that melatonin restored the decreased renal dimethylarginine dimethylaminohydrolase (DDAH, ADMA-metabolizing enzyme) activity to reduce ADMA and reduced the degree of oxidative damaged DNA product, 8-hydroxydeoxyguanosine (8-OHdG) immunostaining in SHR [29]. We also found that melatonin lessens bile duct ligation (BDL)-induced mortality and kidney injury in young rats through suppressing increased ADMA and oxidative stress [30, 31]. In addition to ADMA, we found melatonin regulates the expression and activity of NADPH oxidase, a major oxidative enzyme. Our previous study showed melatonin lowered the prefrontal cortex NADPH-dependent superoxide production to improve brain dysfunction in young BDL rats [32]. Another study in our hands moreover demonstrated that melatonin inhibits protein kinase C (PKC)–β to reduce NADPH-dependent superoxide production, which protects the BDL kidney against damage [33]. Additionally, melatonin can prevent oxidative stress-induced mitochondrial dysfunction by reducing the leakage of electrons from the electron transport chain [34].

Conclusively, the protective effects of melatonin against oxidative stress seem to be wide, although more investigation is needed to know better the entire aspect of melatonin.

13.5 Current Use of Melatonin in Pediatrics

13.5.1 Safety Profiles and Possible Side-Effects of Melatonin

There is general agreement that melatonin therapy has a remarkably benign safety profile in experimental and human studies. In pregnant rats, high doses of melatonin (200 mg/kg/day) do not adversely influence the development of rat pups [35]. Very few significant adverse events have been reported in children under melatonin treatment. Human trials have shown melatonin toxicity to be remarkably low with no serious negative side effects even at high doses [36]. Safety data are also available from use in children with various neurologically disabling disorders [37] and in neonates with sepsis [38]. The commonly used dose of melatonin in pediatric disorder is summarized in Table 13.1.

13.5.2 Use of Melatonin in the Prenatal Period

Preeclampsia and maternal malnutrition are two common entities of compromised pregnancy prone to oxidative stress-induced organ damages.

Table 13.1 Effect of melatonin in the treatment of different pediatric disorders

Condition	Dose and duration	Effects	References
Conditions related to prematurity			
Respiratory distress syndrome (RDS)	10 mg/kg IVD 2 h for 10 times (q2h–q12h)	Melatonin prevented progression to CLD and decreased mortality	[47]
Bronchopulmonary dysplasia (BPD)	10 mg/kg IVD 2 h for 10 times (q2h–q12h)	Melatonin reduced the proinflammatory cytokines and improved the clinical outcome	[44]
Neonatal events			
Melatonin in surgical neonate	10 mg/kg IVD 2 h for 10 times	Melatonin exerted antioxidative activity and improved clinical outcome	[57]
Neonatal sepsis	10 mg PO q1h for 2 doses	Melatonin improved the clinical outcome	[38]
Children			
Sleep disorders	2–20 mg/day PO	>80 % clinical response	[59]
Seizure disorders	5–10 mg/day PO	Not support for improvement of seizure control	[64]

In preeclampsia, lipid peroxidation is increased whereas total antioxidant activity is decreased in maternal blood and placental tissue. Endogenous melatonin level is significantly decreased in severe preeclampsia [39]. Although most studies for supplement of melatonin in pregnancy are from experimental studies, it has been proposed that melatonin supplementation serves as a way to prevent preeclampsia in humans [40].

Maternal undernutrition can impair development via intrauterine growth restriction (IUGR), permanent structural changes, and the resetting of physiological function, namely developmental programming, which are associated with increased oxidative stress [41]. In humans, fetal undernutrition is associated with significant oxidative stress in small for gestational age neonates born at term to malnourished mothers [42]. Although maternal melatonin supplementation has beneficial effects on the offspring in malnourished pregnancy in animals [39], there is lack of direct evidence to support its use in clinical practice yet.

13.5.3 Use of Melatonin in the Neonatal Period

Newborns and those preterm labors have less protection against oxidative stress and are susceptible to oxidative damage. Melatonin has been shown to reduce oxidative stress in neonates with sepsis, asphyxia, respiratory distress, and surgical stress [43]. These clinical trials have been studied in the Neonatal Intensive Care Unit of the University of Mesina in Italy [38, 44–47]. Although hyperoxia is crucial for promoting survival of infants with respiratory distress syndrome (RDS), it causes excessive production of ROS and depletion of antioxidants in the respiratory system. With advances in neonatal care, an increased number of infants with chronic lung disease (CLD) occur. Even though a recent report shows melatonin administration reduces oxidative stress and improves dyspnea in CLD in adults [48], further studies are necessary to determine the potential role for melatonin in the long-term benefit on neonates with CLD.

Next, the periventricular white matter, due to immaturity of blood vessels and a decrease in the vessel density, is highly susceptible to oxidative damage during hypoxic–ischemic episodes. Periventricular leukomalacia (PVL) contributes significantly to neonatal mortality and long-term neurodevelopmental deficits in premature infants and is one of the main causes of cerebral palsy. PVL is associated with increased lipid peroxidation and free radical products [49]. Recent experimental studies demonstrated the beneficial effects of melatonin on PVL [50–52]. Nevertheless, it remains not to be confirmed in clinical trials yet.

Another example of oxidative stress is hypoxia-ischemia (HI). Neonatal HI is the major cause of neonatal brain injury, resulting in cerebral palsy, learning disabilities, visual field deficits, and epilepsy. In rats, melatonin has long-lasting beneficial effects on HI-induced neurologic damage and the oxidative stress damage is prevented by melatonin treatment [53, 54]. A small randomized trial of 20 neonates with birth asphyxia has demonstrated that melatonin significantly reduces circulating levels of malondialdehyde and nitrate/nitrite compared to control [55]. Although this is a small study with only limited outcome measures, it does provide proof-of-principle that melatonin can effectively reduce asphyxia-related oxidative stress in newborns without apparent harm to the baby. Therefore, melatonin could be considered a potential therapy for the reduction of oxidative stress injury and prevention of the ongoing neurological complication in HI injury in neonates.

Next, sepsis represents a serious problem in newborns. There is no doubt that ROS/RNS-related oxidative stress involved in neonatal sepsis and its complications. In a number of animal models of septic shock, as well as in patients with septic disease, melatonin reportedly exerts beneficial effects to arrest cellular damage and multiorgan failure [56]. Several clinical reports have demonstrated the positive effects of melatonin in neonates with sepsis [38, 55]. The use of melatonin to modify the operative stress response has only been tested in one trial [57]; ten neonates received lung or abdomen operation and showed a significant clinical improvement after melatonin therapy, which were associated with lower levels of interleukin-6 and interleukin-8.

13.5.4 Use of Melatonin in Children and Adolescents

The impact of oxidative stress on sleep disorder is well known [58]. However, the reason of use of melatonin to treat various sleep disorders is not only its antioxidant property but also its effects on circadian clock. Sleep disturbances are a common problem in children and adolescents. Children with neurodevelopmental disabilities are at increased risk of having a sleep disorder. Sanchez-Barcelo et al. reviewed that melatonin is considered to treat children with different categories of sleep disorders as defined by the International Classification of Sleep Disorders (ICSD), including dyssomnias, parasomnias, and sleep disorders associated with mental, neurological, or other mental disorders [3]. The dose of melatonin used for children with sleep disorders ranges from 2 to 20 mg/day [59]. A recent meta-analysis report from nine randomized placebo-controlled trials published between 1990 and 2008 concluded that melatonin, at a dose ranging from 0.5 to 9 mg/day, increases total sleep time, and reduces the numbers of awakes per night in children with intellectual disabilities [60].

Next, melatonin has been shown to have antiepileptic activity in experimental animal studies as well as in cases of childhood epilepsy [61]. Epilepsy is one of the most common chronic neurologic disorders in children. Oxidative stress is considered as a crucial mechanism involved in epileptogenesis [62]. Experimental studies suggest that oxidative stress is a contributing factor to the onset and evolution of epilepsy. A small randomized trial of 31 epileptic children has demonstrated that melatonin add-on therapy exerts neuroprotection due to its antioxidant and free radical scavenging properties within the central nervous system [63]. Nevertheless, a recent meta-analysis study failed to make a conclusion because included studies were of poor methodologic quality, and did not systematically evaluate seizure frequency and adverse events [64].

Melatonin-induced sleep is a good alternative to sedation, especially in children younger than 3 years [65]. Three clinical trials involving children reported that melatonin was as effective as midazolam in sedation [66–68]. Importantly, melatonin was associated with a rapid recovery, a reduced incidence of postoperative delirium, and there is no risk of respiratory compromise compared with midazolam.

13.6 New Perspectives in Melatonin Uses in Pediatric Medicine

In adulthood, melatonin has been combined with other drugs to treat different diseases, including dementia senile, Parkinson disease, Alzheimer disease, hepatoma, gastric ulcer, Chagas disease, and tuberculosis [5]. In addition, several antihypertensive mechanisms of melatonin have been discovered, including activation of MT2 receptor to increase cytosolic Ca^{2+} leading to vasodilatation [69], modulation of central sympathetic tone to improve baroreflex responses and decrease sympathetic output [70], and reduction of oxidative stress in the renal and cardiovascular systems [29, 71]. Thus, melatonin is proposed as a potential antihypertensive treatment [72]. These observations indicate the potential use of melatonin for prevention of tumor progression, regulation of immune system, restoration of intestinal motility, and antihypertension in pediatric medicine.

Moreover, melatonin may be applied to a new research area, namely epigenetic programming [73]. Growing evidence suggests that oxidative stress is involved in epigenetic fetal programming. DNA methyltransferases (DNMTs) are a family of enzymes that methylate DNA and they play a crucial role on epigenetic regulation. Interestingly, melatonin has a similar structure and hypothetically could inhibit DNMT either by masking target sequences or by blocking the active site of the enzyme [73]. Hence, a novel research area for mechanisms of epigenetic regulation on developmental programming by melatonin may have emerged.

13.7 Certain Questions and Problems

The major block to future clinical application of melatonin on pediatric practice is the numbers of randomized controlled trials are still small. Therefore, large, prospective, multicenter collaborations are required to conduct meaningful clinical studies in children to decide appropriate dosage, formulations, and length of therapy for various pediatric disorders. In addition, problems in regard to the high interindividual variability of diurnal melatonin concentrations that must be addressed before melatonin can be used in pediatric medicine. Although previous studies with small case numbers suggested that melatonin concentrations in saliva and the urinary levels of melatonin metabolites can be used to evaluate melatonin diurnal patterns in infants and children [74, 75], further investigations are needed to validate ideal measures for the detection of diurnal melatonin patterns in children and adolescents.

13.8 Conclusions

Melatonin is a potent-free radical scavenger, a metal chelator, and a broad-spectrum antioxidant. Basic and clinical research performed during the last 20 years has implicated melatonin as a novel treatment against oxidative stress in adult populations. Additional studies exploring its role as a target for pediatric diseases are warranted.

Acknowledgments This work was supported by grant 101-2314-B-182A-021-MY3 (to Tain, Y.L.) and grant 99-2314-B-182A-003-MY3 (to Huang, L.T.) from National Science Council, Taiwan, and grant CMRPG8B0171 (to Tain, Y.L.) from Kaohsiung Chang Gung Memorial Hospital, Taiwan.

References

1. Lerner AB, Case JD, Takahashi Y (1960) Isolation of melatonin and 5-methoxyindole-3-acetic acid from bovine pineal glands. J Biol Chem 235:1992–1997
2. Sanchez-Barcelo EJ, Mediavilla MD, Tan DX, Reiter RJ (2010) Clinical uses of melatonin: evaluation of human trials. Curr Med Chem 17:2070–2095
3. Sanchez-Barcelo EJ, Mediavilla MD, Reiter RJ (2011) Clinical uses of melatonin in pediatrics. Int J Pediatr 2011:892624

4. Chen YC, Tain YL, Sheen JM, Huang LT (2012) Melatonin utility in neonates and children. J Formos Med Assoc 111:57–66
5. Carpentieri A, Diaz de Barboza G, Areco V, Peralta Lopez M, Tolosa de Talamoni N (2012) New perspectives in melatonin uses. Pharmacol Res 65:437–444
6. Lane EA, Moss HB (1985) Pharmacokinetics of melatonin in man: first pass hepatic metabolism. J Clin Endocrinol Metab 61:1214–1216
7. Hardeland R, Pandi-Perumal SR, Cardinali DP (2006) Melatonin. Int J Biochem Cell Biol 38:313–316
8. Hardeland R, Tan DX, Reiter RJ (2009) Kynuramines, metabolites of melatonin and other indoles: the resurrection of an almost forgotten class of biogenic amines. J Pineal Res 47:109–126
9. Tamura H, Nakamura Y, Terron MP, Flores LJ, Manchester LC et al (2008) Melatonin and pregnancy in the human. Reprod Toxicol 25:291–303
10. Tamura H, Nakamura Y, Korkmaz A, Manchester LC, Tan DX et al (2009) Melatonin and the ovary: physiological and pathophysiological implications. Fertil Steril 92:328–343
11. Sadowsky DW, Yellon S, Mitchell MD, Nathanielsz PW (1991) Lack of effect of melatonin on myometrial electromyographic activity in the pregnant sheep at 138–142 days gestation (term = 147 days gestation). Endocrinology 128:1812–1818
12. Seron-Ferre M, Torres-Farfan C, Forcelledo ML, Valenzuela GJ (2001) The development of circadian rhythms in the fetus and neonate. Semin Perinatol 25:363–370
13. Tsukahara H (2007) Biomarkers for oxidative stress: clinical application in pediatric medicine. Curr Med Chem 14:339–351
14. Tsukahara H, Jiang MZ, Ohta N, Sato S, Tamura S et al (2004) Oxidative stress in neonates: evaluation using specific biomarkers. Life Sci 75:933–938
15. Dennery PA (2010) Oxidative stress in development: nature or nurture? Free Radic Biol Med 49:1147–1151
16. Shoji H, Koletzko B (2007) Oxidative stress and antioxidant protection in the perinatal period. Curr Opin Clin Nutr Metab Care 10:324–328
17. Reiter RJ, Korkmaz A, Paredes SD, Manchester LC, Tan DX (2008) Melatonin reduces oxidative/nitrosative stress due to drugs, toxins, metals, and herbicides. Neuro Endocrinol Lett 29:609–613
18. Tan DX, Poeggeler B, Reiter RJ, Chen LD, Chen S et al (1993) The pineal hormone melatonin inhibits DNA-adduct formation induced by the chemical carcinogen safrole in vivo. Cancer Lett 70:65–71
19. Tan DX, Manchester LC, Reiter RJ, Plummer BF, Limson J et al (2000) Melatonin directly scavenges hydrogen peroxide: a potentially new metabolic pathway of melatonin biotransformation. Free Radic Biol Med 29:1177–1185
20. Matuszak Z, Bilska MA, Reszka KJ, Chignell CF, Bilski P (2003) Interaction of singlet molecular oxygen with melatonin and related indoles. Photochem Photobiol 78:449–455
21. Zavodnik IB, Lapshina EA, Zavodnik LB, Labieniec M, Bryszewska M et al (2004) Hypochlorous acid-induced oxidative stress in Chinese hamster B14 cells: viability, DNA and protein damage and the protective action of melatonin. Mutat Res 559:39–48
22. Reiter RJ, Tan DX, Manchester LC, Qi W (2001) Biochemical reactivity of melatonin with reactive oxygen and nitrogen species: a review of the evidence. Cell Biochem Biophys 34:237–256
23. Aydogan S, Yerer MB, Goktas A (2006) Melatonin and nitric oxide. J Endocrinol Invest 29:281–287
24. Reiter RJ, Tan DX, Burkhardt S (2002) Reactive oxygen and nitrogen species and cellular and organismal decline: amelioration with melatonin. Mech Ageing Dev 123:1007–1019
25. Rodriguez C, Mayo JC, Sainz RM, Antolin I, Herrera F et al (2004) Regulation of antioxidant enzymes: a significant role for melatonin. J Pineal Res 36:1–9
26. Gulcin I, Buyukokuroglu ME, Kufrevioglu OI (2003) Metal chelating and hydrogen peroxide scavenging effects of melatonin. J Pineal Res 34:278–281

27. Parmar P, Limson J, Nyokong T, Daya S (2002) Melatonin protects against copper-mediated free radical damage. J Pineal Res 32:237–242
28. Tain YL, Huang LT (2011) Asymmetric dimethylarginine: clinical applications in pediatric medicine. J Formos Med Assoc 110:70–77
29. Tain YL, Huang LT, Lin IC, Lau YT, Lin CY (2010) Melatonin prevents hypertension and increased asymmetric dimethylarginine in young spontaneous hypertensive rats. J Pineal Res 49:390–398
30. Tain YL, Hsieh CS, Chen CC, Sheen JM, Lee CT et al (2010) Melatonin prevents increased asymmetric dimethylarginine in young rats with bile duct ligation. J Pineal Res 48:212–221
31. Tain YL, Kao YH, Hsieh CS, Chen CC, Sheen JM et al (2010) Melatonin blocks oxidative stress-induced increased asymmetric dimethylarginine. Free Radic Biol Med 49:1088–1098
32. Chen YC, Sheen JM, Tain YL, Chen CC, Tiao MM et al (2012) Alterations in NADPH oxidase expression and blood–brain barrier in bile duct ligation-treated young rats: effects of melatonin. Neurochem Int 60:751–758
33. Tain YL, Chen CC, Lee CT, Kao YH, Sheen JM et al (2013) Melatonin regulates L-arginine transport and NADPH oxidase in young rats with bile duct ligation: role of protein kinase C. Pediatr Res 73:395–401
34. Srinivasan V, Spence DW, Pandi-Perumal SR, Brown GM, Cardinali DP (2011) Melatonin in mitochondrial dysfunction and related disorders. Int J Alzheimers Dis 2011:326320
35. Jahnke G, Marr M, Myers C, Wilson R, Travlos G et al (1999) Maternal and developmental toxicity evaluation of melatonin administered orally to pregnant Sprague–Dawley rats. Toxicol Sci 50:271–279
36. Seabra ML, Bignotto M, Pinto LR Jr, Tufik S (2000) Randomized, double-blind clinical trial, controlled with placebo, of the toxicology of chronic melatonin treatment. J Pineal Res 29:193–200
37. Gordon N (2000) The therapeutics of melatonin: a paediatric perspective. Brain Dev 22: 213–217
38. Gitto E, Karbownik M, Reiter RJ, Tan DX, Cuzzocrea S et al (2001) Effects of melatonin treatment in septic newborns. Pediatr Res 50:756–760
39. Richter HG, Hansell JA, Raut S, Giussani DA (2009) Melatonin improves placental efficiency and birth weight and increases the placental expression of antioxidant enzymes in undernourished pregnancy. J Pineal Res 46:357–364
40. Briceno-Perez C, Briceno-Sanabria L, Vigil-De Gracia P (2009) Prediction and prevention of preeclampsia. Hypertens Pregnancy 28:138–155
41. Wu G, Bazer FW, Cudd TA, Meininger CJ, Spencer TE (2004) Maternal nutrition and fetal development. J Nutr 134:2169–2172
42. Gupta P, Narang M, Banerjee BD, Basu S (2004) Oxidative stress in term small for gestational age neonates born to undernourished mothers: a case control study. BMC Pediatr 4:14
43. Gitto E, Pellegrino S, Gitto P, Barberi I, Reiter RJ (2009) Oxidative stress of the newborn in the pre- and postnatal period and the clinical utility of melatonin. J Pineal Res 46:128–139
44. Gitto E, Reiter RJ, Sabatino G, Buonocore G, Romeo C et al (2005) Correlation among cytokines, bronchopulmonary dysplasia and modality of ventilation in preterm newborns: improvement with melatonin treatment. J Pineal Res 39:287–293
45. Gitto E, Reiter RJ, Karbownik M, Tan DX, Gitto P et al (2002) Causes of oxidative stress in the pre- and perinatal period. Biol Neonate 81:146–157
46. Gitto E, Reiter RJ, Cordaro SP, La Rosa M, Chiurazzi P et al (2004) Oxidative and inflammatory parameters in respiratory distress syndrome of preterm newborns: beneficial effects of melatonin. Am J Perinatol 21:209–216
47. Gitto E, Reiter RJ, Amodio A, Romeo C, Cuzzocrea E et al (2004) Early indicators of chronic lung disease in preterm infants with respiratory distress syndrome and their inhibition by melatonin. J Pineal Res 36:250–255
48. de Matos Cavalcante AG, de Bruin PF, de Bruin VM, Nunes DM, Pereira ED et al (2012) Melatonin reduces lung oxidative stress in patients with chronic obstructive pulmonary disease: a randomized, double-blind, placebo-controlled study. J Pineal Res 53:238–244

49. Haynes RL, Folkerth RD, Keefe RJ, Sung I, Swzeda LI et al (2003) Nitrosative and oxidative injury to premyelinating oligodendrocytes in periventricular leukomalacia. J Neuropathol Exp Neurol 62:441–450
50. Welin AK, Svedin P, Lapatto R, Sultan B, Hagberg H et al (2007) Melatonin reduces inflammation and cell death in white matter in the mid-gestation fetal sheep following umbilical cord occlusion. Pediatr Res 61:153–158
51. Kaur C, Sivakumar V, Ling EA (2010) Melatonin protects periventricular white matter from damage due to hypoxia. J Pineal Res 48:185–193
52. Olivier P, Fontaine RH, Loron G, Van Steenwinckel J, Biran V et al (2009) Melatonin promotes oligodendroglial maturation of injured white matter in neonatal rats. PLoS One 4:e7128
53. Carloni S, Perrone S, Buonocore G, Longini M, Proietti F et al (2008) Melatonin protects from the long-term consequences of a neonatal hypoxic-ischemic brain injury in rats. J Pineal Res 44:157–164
54. Balduini W, Carloni S, Perrone S, Bertrando S, Tataranno ML et al (2012) The use of melatonin in hypoxic-ischemic brain damage: an experimental study. J Matern Fetal Neonatal Med 25(Suppl 1):119–124
55. Fulia F, Gitto E, Cuzzocrea S, Reiter RJ, Dugo L et al (2001) Increased levels of malondialdehyde and nitrite/nitrate in the blood of asphyxiated newborns: reduction by melatonin. J Pineal Res 31:343–349
56. Srinivasan V, Pandi-Perumal SR, Spence DW, Kato H, Cardinali DP (2010) Melatonin in septic shock: some recent concepts. J Crit Care 25(656):e651–e656
57. Gitto E, Romeo C, Reiter RJ, Impellizzeri P, Pesce S et al (2004) Melatonin reduces oxidative stress in surgical neonates. J Pediatr Surg 39:184–189, discussion 184–189
58. Pack AI, Pien GW (2011) Update on sleep and its disorders. Annu Rev Med 62:447–460
59. Jan JE, Freeman RD, Fast DK (1999) Melatonin treatment of sleep-wake cycle disorders in children and adolescents. Dev Med Child Neurol 41:491–500
60. Braam W, Smits MG, Didden R, Korzilius H, Van Geijlswijk IM et al (2009) Exogenous melatonin for sleep problems in individuals with intellectual disability: a meta-analysis. Dev Med Child Neurol 51:340–349
61. Banach M, Gurdziel E, Jedrych M, Borowicz KK (2011) Melatonin in experimental seizures and epilepsy. Pharmacol Rep 63:1–11
62. Shin EJ, Jeong JH, Chung YH, Kim WK, Ko KH et al (2011) Role of oxidative stress in epileptic seizures. Neurochem Int 59:122–137
63. Gupta M, Gupta YK, Agarwal S, Aneja S, Kohli K (2004) A randomized, double-blind, placebo controlled trial of melatonin add-on therapy in epileptic children on valproate monotherapy: effect on glutathione peroxidase and glutathione reductase enzymes. Br J Clin Pharmacol 58:542–547
64. Brigo F, Del Felice A (2012) Melatonin as add-on treatment for epilepsy. Cochrane Database Syst Rev (6):CD006967
65. Schmidt CM, Knief A, Deuster D, Matulat P, am Zehnhoff-Dinnesen AG (2007) Melatonin is a useful alternative to sedation in children undergoing brainstem audiometry with an age dependent success rate—a field report of 250 investigations. Neuropediatrics 38:2–4
66. Schmidt CM, Bohlender JE, Deuster D, Knief A, Matulat P et al (2004) The use of melatonin as an alternative to sedation in children undergoing brainstem audiometry. Laryngorhinootologie 83:523–528
67. Johnson K, Page A, Williams H, Wassemer E, Whitehouse W (2002) The use of melatonin as an alternative to sedation in uncooperative children undergoing an MRI examination. Clin Radiol 57:502–506
68. Isik B, Baygin O, Bodur H (2008) Premedication with melatonin vs midazolam in anxious children. Paediatr Anaesth 18:635–641
69. Pogan L, Bissonnette P, Parent L, Sauve R (2002) The effects of melatonin on Ca(2+) homeostasis in endothelial cells. J Pineal Res 33:37–47
70. Williams LM, Hannah LT, Hastings MH, Maywood ES (1995) Melatonin receptors in the rat brain and pituitary. J Pineal Res 19:173–177

71. Ersahin M, Sehirli O, Toklu HZ, Suleymanoglu S, Emekli-Alturfan E et al (2009) Melatonin improves cardiovascular function and ameliorates renal, cardiac and cerebral damage in rats with renovascular hypertension. J Pineal Res 47:97–106
72. Simko F, Paulis L (2007) Melatonin as a potential antihypertensive treatment. J Pineal Res 42:319–322
73. Korkmaz A, Reiter RJ (2008) Epigenetic regulation: a new research area for melatonin? J Pineal Res 44:41–44
74. Ardura J, Gutierrez R, Andres J, Agapito T (2003) Emergence and evolution of the circadian rhythm of melatonin in children. Horm Res 59:66–72
75. Praninskiene R, Dumalakiene I, Kemezys R, Mauricas M, Jucaite A (2012) Diurnal melatonin patterns in children: ready to apply in clinical practice? Pediatr Neurol 46:70–76

Chapter 14
Thioredoxin Therapy: Challenges in Translational Research

Masato Yashiro, Hirokazu Tsukahara, and Tsuneo Morishima

Abbreviations

AMP	Adenosine monophosphate
ASDs	Autism spectrum disorders
ELISA	Enzyme-linked immunosorbent assay
KD	Kawasaki disease
NADPH	Nicotinamide adenine dinucleotide phosphate
8-OHdG	8-Hydroxy-2′-deoxyguanosine
PBS	Phosphate-buffered saline
rHSA	Recombinant human serum albumin
rhTRX-1	Recombinant human thioredoxin-1
ROS	Reactive oxygen species
RP	Redox potential
TH	Total hydroperoxides
TNF	Tumor necrosis factor
TRX	Thioredoxin

14.1 Introduction

Cellular reduction–oxidation (redox) status is an important regulator of various cellular responses, including adaptation to cellular stress [1, 2]. To date, the glutathione and thioredoxin (TRX) pathways have been identified as the major cellular

M. Yashiro, M.D., Ph.D. • H. Tsukahara, M.D., Ph.D. (✉) • T. Morishima, M.D., Ph.D.
Department of Pediatrics, Okayama University Graduate School of Medicine,
Dentistry and Pharmaceutical Sciences, Okayama 700-8558, Japan
e-mail: hirokazu010661@yahoo.co.jp

H. Tsukahara and K. Kaneko (eds.), *Studies on Pediatric Disorders*, Oxidative Stress in Applied Basic Research and Clinical Practice, DOI 10.1007/978-1-4939-0679-6_14,

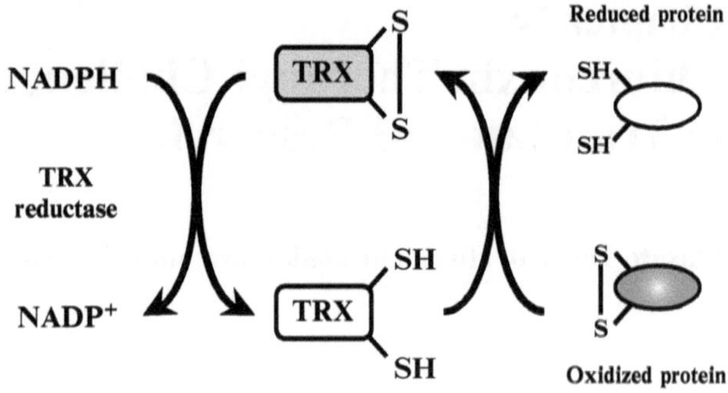

Fig. 14.1 Redox cycling of thioredoxin and thioredoxin reductase. Thioredoxin (TRX) is oxidized as it transfers reducing equivalents to disulfide groups in target proteins. TRX is then reduced back to the dithiol form by NADPH-dependent flavoprotein, TRX reductase

redox systems, shown to modulate cellular functions such as gene expression, intracellular signaling, and various stress responses.

TRX was originally identified in 1964 in *Escherichia coli* as an electron donor to ribonucleotide reductase, a necessary enzyme for DNA synthesis [3, 4]. The TRX system comprises nicotinamide adenine dinucleotide phosphate (NADPH), TRX reductase, and TRX. Oxidized TRX is reduced reversibly by NADPH and TRX reductase (Fig. 14.1). Mammalian TRX reductase, which contains selenocystine in the conserved C-terminal sequence, has broad substrate specificity.

Human TRX was cloned as an adult T cell leukemia-derived factor produced by human T cell leukemia virus type-1-transformed T cell line ALT-2 cells [3]. Human TRX, which means cytosolic TRX-1 here (12 kDa molecular weight), consists of 105 amino acids, although the first N-terminal methionine is mostly removed after its translocation by N-terminal methionine excision processes and TRX-1 in the human body consists largely of 104 amino acids starting from the second N-terminal valine [4].

This chapter presents a review of the physiological and pathophysiological roles of TRX-1 in animals, with explanations of the TRX system's involvement in the oxidative environment, its modulation of human diseases via redox regulation, and TRX-based strategies for oxidative stress and inflammation in humans. In addition, this report presents new and promising data demonstrating the efficacy of TRX-1 treatment for severe influenza pneumonia.

14.2 Redox Activity and Biological Properties of Thioredoxin-1

TRX-1 gene promoter contains the oxidative stress responsive element, specificity protein 1, antioxidant responsive element, and cyclic adenosine monophosphate (AMP) responsive elements, and TRX-1 is induced by various oxidative conditions [3, 4].

The biological properties of TRX-1 rely to a great degree on redox activity, which is the ability to transfer "reducing equivalents" to disulfide groups in target

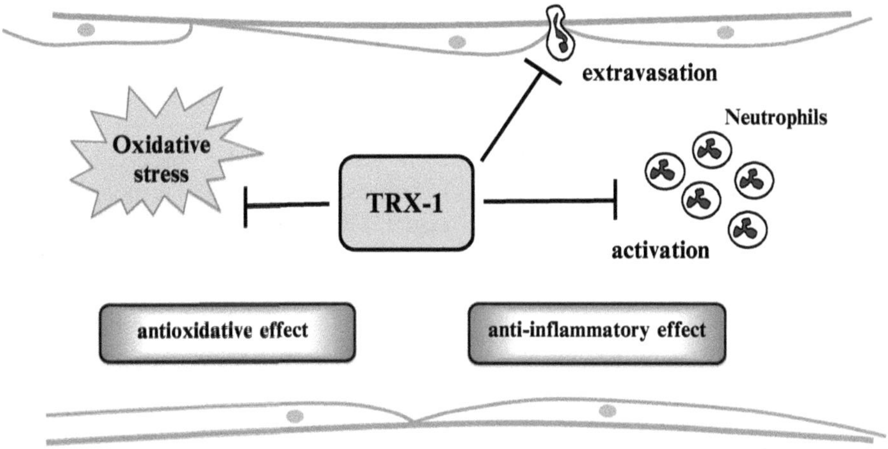

Fig. 14.2 Antioxidative and anti-inflammatory effects of circulating thioredoxin-1. Elevated levels of thioredoxin (TRX)-1 in circulation maintain the redox environment in the tissue and suppress neutrophil infiltration in the inflammatory site

proteins (Fig. 14.1). The key to the redox activity of TRX-1 is the presence of two cysteine residues (^{32}Cys and ^{35}Cys) separated by the two amino acids (Gly-Pro) in its active site. These cysteines exist as a dithiol [$(\text{-SH})_2$] in the reduced form and a disulfide (-S-S-) in the oxidized form [4, 5].

Key biological activities of TRX-1 that are applicable to human disease can be classified as antioxidant, anti-apoptotic, growth-promoting, and inflammation modulating [4, 5]. Several of these activities result from redox regulation by TRX-1 of signal transduction pathways and gene expression. Of particular importance, TRX-1 suppresses inflammation by regulating neutrophil activation and extravasation and exerts anti-inflammatory effects [6] (Fig. 14.2).

Human-TRX-1-overexpressing transgenic mice survive longer and are more resistant to various oxidative conditions than control mice are [7, 8]. Overexpression of human TRX-1 in mice induces resistance to harmful conditions including ischemic brain damage, adriamycin-induced cardiotoxicity, ischemia-reperfusion renal injury, Helicobacter gastritis, cerulein-induced pancreatitis, dextran sulfate sodium-induced colitis, and indomethacin-induced gastric mucosal injury [9–15]. Furthermore, human TRX-1 transgenic mice are more resistant than control mice to proinflammatory cytokine-, bleomycin-, diesel exhaust particle-, or cigarette smoke-induced lung injury [16–18] and to influenza virus-induced pneumonia [19]. These results are explained by the antioxidative, anti-inflammatory, and anti-apoptotic effects of twofold to tenfold overexpression of human TRX-1. It is worth mentioning that complete deficiency of TRX-1 results in early embryonic lethality in mice [20]. All these results demonstrate that endogenous TRX-1 has potent protective effects on oxidative stress-associated or inflammation-associated organ injuries in animals.

14.3 Thioredoxin-1 in Human Disease

As expected from the initial finding that TRX-1 was purified and cloned as a cytokine-like factor from the cellular supernatants, TRX-1 has several important biological functions (as described above) in the extracellular space (Fig. 14.2). Although it has no signal peptide, TRX-1 is released from cells in response to oxidative stress [4, 5]. Transfected flag-tagged TRX-1 is released from cells in response to a sublethal amount of hydrogen peroxide (H_2O_2), which is inhibited by an excess amount of recombinant human thioredoxin-1 (rhTRX-1) protein [21]. These results suggest that the oxidative stress-induced TRX-1 release might be regulated by negative feedback loops using reactive oxygen species (ROS)-mediated signal transduction.

TRX-1 is usually measured using a sensitive sandwich enzyme-linked immunosorbent assay (ELISA) system produced by Redox Bioscience Inc. (Kyoto, Japan) [4, 5]. This system uses two specific mouse monoclonal antibodies to non-overlapping epitopes of human TRX-1 (ADF 11 and ADF 21). These antibodies do not cross-react to mitochondrial TRX: TRX-2. In the clinical field, it is recognized that TRX-1 levels in human serum/plasma are reliable markers for oxidative stress. Serum/plasma concentrations of TRX-1 are 10–30 ng/mL in healthy adults, rising to 40–140 ng/mL in patients presenting with diseases characterized by enhancement of oxidative stress [5]. In our laboratory, serum TRX-1 concentrations (mean $\pm$ SD) in 13 healthy Japanese adults (female/male: 7/6) aged 32 ± 7 years (range: 18–43 years) determined using this methodology are 20 ± 17 ng/mL (range: 5–53 ng/mL). In the kidney, TRX-1 is excreted through the glomerulus and is reabsorbed primarily by the proximal tubules. Therefore, TRX-1 concentrations in the urine of healthy adults are quite low; usually they are undetectable.

Extracellular concentrations of TRX-1 have been measured in various conditions characterized by oxidative stress and inflammation, including sepsis, viral infection, severe burns, acute lung injury, asthma exacerbation, coronary angina, heart failure, hepatitis, pancreatitis, inflammatory bowel disease, diabetes mellitus, renal failure, rheumatoid arthritis, obstructive sleep apnea, schizophrenia, and open-heart surgery (Table 14.1) [14, 22–39]. These studies documented that the TRX-1 concentrations are elevated in patients with these diseases and that they are correlated with the activity of such diseases. High levels of TRX-1 are likely to be effective in counteracting oxidative stress and augmenting host defenses against the disease conditions explained above.

As described above, blood TRX-1 concentrations are elevated in many oxidative and inflammatory disorders, but TRX-1 concentrations were lower in patients with neutropenia/sepsis than in patients with systemic inflammatory response syndrome/sepsis [39]. Although the idea is speculative, the low TRX-1 condition might contribute to the devastating nature of sepsis in those patients. In this context, it is noteworthy that the cecal ligation and puncture septic murine model revealed that neutralization of endogenous TRX-1 by anti-TRX-1 antiserum impaired survival but that treatment with rhTRX-1 enhanced survival [37]. Furthermore, notably, the anti-TRX-1 treatment exacerbated dextran sulfate sodium-induced colitis in mice [14].

Table 14.1 Pathological conditions associated with increased blood concentrations of thioredoxin-1

Inflammatory bowel disease (ulcerative colitis, Crohn's disease) [14]
HIV/AIDS [22]
Cardiac surgery with cardiopulmonary bypass [23]
Severe burn injury [24]
Hepatitis C [25, 31]
Heart failure (acute coronary syndrome, dilated cardiomyopathy, ischemic heart disease) [26, 33]
Type 2 diabetes mellitus [27]
Rheumatoid arthritis [28]
Asthma exacerbation [29]
Nonalcoholic steatohepatitis [30]
Renal failure [31]
Coronary spastic angina [32]
Acute pancreatitis [34]
Acute lung injury (acute respiratory distress syndrome) [35]
Obstructive sleep apnea [36]
Sepsis [37, 39]
Schizophrenia [38]
Kawasaki disease (unpublished results)
Autism spectrum disorders [47]
Severe psychosomatic disability [48]
Maternal preeclampsia [53]

The numbers in the parentheses indicate the references that are cited in this chapter

14.4 Thioredoxin-1 in Pediatric Field

Only sparse results are available in the literature for determination of blood TRX-1 levels in clinical pediatrics. Kawasaki disease (KD), a systemic vasculitis, is the leading cause of acquired heart disease in children. In the acute stage, coronary aneurysm formation might occur, which is associated with myocardial infarction and death. High-dose intravenous immunoglobulin together with oral aspirin is effective for resolving inflammation that occurs with KD. Results of recent studies indicate that oxidative stress is a key event in the progression of coronary artery lesions to arteriosclerosis or atherosclerosis [40].

Our interesting results of serum TRX-1 determination in ten Japanese children with KD can be explained briefly here (unpublished results). The initial concentrations (range (median); mean ± SD) of serum TRX-1 were high (15–165 (48) ng/mL; 64 ± 47 ng/mL), but they decreased to half values (13–66 (31) ng/mL; 32 ± 15 ng/mL) after intravenous immunoglobulin treatment ($p < 0.05$ using Wilcoxon test). Other oxidative stress markers (e.g., nitrite/nitrate and 8-isoprostane in urine, total hydroperoxides (TH) in blood) reportedly showed a similar trend [41–43]. These findings imply that intensive antioxidative interventions (e.g., intravenous immunoglobulin plus corticosteroids) at an early stage can show beneficial effects for clinical outcomes in children with severe KD. In addition, fluvastatin (3-hydroxy-3-methylglutaryl coenzyme A reductase inhibitor) for 12 months might be useful for reducing post-inflammatory vascular remodeling in the later stage of KD [44].

Autism spectrum disorders (ASDs) are prevalent neurodevelopmental disorders that affect not fewer than 1 in 150 children born [45]. These disorders are associated with increased oxidative stress, at least in some cases. Results showed that urinary concentrations of acrolein-lysine (a marker of lipid peroxidation) in Japanese children with ASDs were higher than those of the control subjects [46]. Recently, several parameters related to sulfur-dependent detoxification pathways were examined in plasma of autistic children in Saudi Arabia [47]. Reduced glutathione (GSH), total glutathione (GSH + GSSG), glutathione status (GSH/GSSG), and glutathione-s-transferase activity were found, although TRX-1, TRX reductase, peroxiredoxins (I and III), and glutathione reductase activity were higher in autistic children than in control subjects. Additionally, our report described that severely psychosomatic disabled children under intubation feeding showed high serum concentrations of TRX-1 and that the TRX-1 concentrations were linked with lower selenium status [48]. Our findings suggest the presence of systemic oxidant loads in severely disabled children.

However, plasma TRX-1 concentrations were persistently low in children with meningococcal septic shock [49]. In these patients, plasma TRX-1 was higher in non-survivors than in survivors. It is possible that some patients had a genetic predisposition that caused low levels of TRX-1. Further studies are warranted to determine the nature and significance of the low TRX-1 levels in septic shock.

14.5 Thioredoxin-1 in Fetal and Neonatal Field

Increased generation of ROS during growth of the fetal–placental unit is a prominent feature of pregnancy [50]. Kuroda and coworkers [51] reported that serum TRX-1 concentrations were elevated during pregnancy. The TRX-1 concentrations were, respectively, 57 ± 26, 66 ± 25, and 66 ± 23 ng/mL in the first, second, and third trimesters of pregnancy. We recently determined serum concentrations of TRX-1, TH, and redox potential (RP) (Free Radical Analytical System; Diacron International, Grosseto, Italy) [43] in 60 pregnant women at the early third trimester: gestational age of 27–29 weeks [52]. The TRX-1 concentration was 90 ± 42 ng/mL. TH was 471 ± 105 U.CARR (1 U.CARR = 0.08 mg/dL H_2O_2) and RP was $2{,}142 \pm 273$ μmol/L. TRX-1 and TH were higher and RP was lower than in the blood of healthy adults. TH and RP were mutually correlated significantly and negatively. TRX-1 correlated significantly and negatively and RP correlated significantly and positively with maternal body weight. Notably, TH correlated significantly and negatively and RP significantly and positively with neonatal birth weight. These results suggest that high concentrations of TRX-1 are linked to high oxidative stress status in pregnant women and that neonatal birth weight is affected by maternal oxidative conditions during later pregnancy.

Table 14.2 Concentrations of thioredoxin-1 in body fluid samples

Serum from healthy adults: 20±17 ng/mL
Umbilical cord blood: 127±81 ng/mL
Early breast milk: 268±149 ng/mL
Serum from lactating women: 35±19 ng/mL
Amniotic fluid: 346±221 ng/mL (unpublished results)

Data are expressed as mean±SD. The results are derived primarily from our earlier report [53]. Note the remarkably high concentrations of TRX-1 in early breast milk and amniotic fluid

Our earlier report described that the TRX-1 concentrations in umbilical cord blood (127±81 ng/mL) and those in early breast milk (268±149 ng/mL) were remarkably elevated [53]. Those of healthy lactating women during postpartum days 3–7 were 35±19 ng/mL (Table 14.2). It appears that the systemic release of TRX-1 is enhanced in neonates at birth and that early breast milk is a rich source of this protein. Our results of TRX-1 in amniotic fluid samples from 14 pregnant women at delivery (gestational age of 35–40 weeks) are also intriguing. The concentrations of TRX-1 (346±221 ng/mL) were several times higher than the maternal levels (unpublished results) (Table 14.2).

Collectively, the findings from the clinical studies reported above indicate that TRX-1 provides a unique protective mechanism that allows maintenance of the redox balance in female reproductive processes and outcomes.

14.6 Pharmaceutical Effects of Thioredoxin-1 in Animal Models

Based on the host-defensive effects of TRX-1, administration of exogenous rhTRX-1 has been tested in widely various animal models. The results showed its potent anti-oxidative and anti-inflammatory effects [13, 14, 16, 18, 54–60] (Table 14.3). When rhTRX-1 was injected intravenously, the half life of rhTRX-1 in plasma was estimated as roughly 1 h in mice, 2 h in rats, and 8 h in monkeys [5]. When a large amount of rhTRX-1 (10 mg/kg body weight) was injected into animals, rhTRX-1 protein was excreted into the urine as an immunologically intact form, suggesting that this protein is not metabolized substantially. Judging from the measurement of tissue deposition of rhTRX-1 using ELISA (Redox Bioscience Inc.), the tissue level in the kidney was the highest (as expected). The second highest tissue was the lung, indicating that lung disease might be a good target for this protein [5].

Actually, administration of rhTRX-1 is effective in animal models, especially for lung injury, including proinflammatory cytokine-, bleomycin-, or cigarette smoke-induced inflammatory injury, ovalbumin-induced airway hyperresponsiveness and inflammation, and lipopolysaccharide-induced bronchoalveolar neutrophil infiltration [16, 18, 57, 58]. All these results demonstrate that TRX-1 has potent protective effects on oxidative stress-associated or inflammation-associated lung disorders in animals.

Table 14.3 Beneficial effects of recombinant human thioredoxin-1 in experimental disease models

Cerulein/lipopolysaccharide-induced pancreatitis in mice [13]
Dextran sulfate sodium induced-colitis in mice [14]
Proinflammatory cytokine- or bleomycin-induced lung injury in mice [16]
Cigarette smoke induced-lung injury in mice [18]
Retinal ischemia-reperfusion injury in rats [54]
Cerebral ischemia/reperfusion injury in mice [55]
Myosin-induced autoimmune myocarditis in mice [56]
Ovalbumin-induced airway hyperresponsiveness and inflammation in mice [57]
Lipopolysaccharide-induced lung injury in rats [58]
Ethanol-induced liver injury [59]
Type 1 diabetes mellitus in nonobese diabetic mice [60]

The numbers in the parentheses indicate the references that are cited in this chapter

14.7 Animal Model of Influenza Pneumonia and Redox Modulation Strategies

Influenza virus infections are responsible for numerous pneumonia cases every year [61–63]. Under treatment with antiviral drugs, influenza virus infection occasionally causes severe pneumonia, necessitating intensive care and mechanical ventilation in clinical settings. In severe cases, they can cause death. The discovery of a novel anti-influenza therapeutic approach can increase the effectiveness of traditional virus-based strategies.

The pathogenesis of influenza pneumonia involves not only apoptotic cell death mediated through viral replication in the infected cells, but also the injury of not-infected cells by ROS derived from infiltrating neutrophils and macrophages and respiratory tract epithelium [64]. The extremely important role of ROS as mediators of influenza virus-induced lung injury is supported by previous studies. In murine models of influenza pneumonia, excessive generation of ROS contributed to lung injury in infected animals; treatment with superoxide dismutase, catalase (antioxidative enzymes), *N*-monomethyl-L-arginine (nitric oxide synthase inhibitor) or allopurinol (xanthine oxidase inhibitor), and overexpression of extracellular superoxide dismutase or heme oxygenase-1 suppressed lung injury and inflammation and improved the survival rate [65–70]. Virus-infected selenium-deficient mice developed more severe influenza pneumonia than did selenium-adequate mice, implying the importance of selenium-dependent glutathione peroxidase and TRX reductase for protection against influenza virus-induced inflammatory processes [71].

Of greater importance is the fact that TRX-1 overexpression is also effective in augmenting the host defense against influenza pneumonia, thereby reducing mortality [19]. Overexpression of TRX-1 might modulate the ROS generation induced by influenza virus infection and regulate the redox-dependent signal transductions in the host defense responses against influenza pneumonia.

14.8 Thioredoxin-1 Administration for the Murine Model of Influenza Pneumonia

The findings presented above have prompted us to evaluate the protective effects of rhTRX-1 administration on acute lung injury in mice with influenza pneumonia. The results are promising and are illustrated in the following [72].

14.8.1 Materials and Methods

Animals and protocol: rhTRX-1 was supplied by Redox Bio Science Inc. The C57BL/6 mice were divided into a vehicle-treated group (control group) and an rhTRX-1-treated group (treatment group). The doses of influenza virus (H1N1) were 300 plaque-forming units for survival rate analysis and 1,000 plaque-forming units for pathological and lung lavage analyses. The day of virus inoculation was defined as Day 0.

Survival rate analysis: Vehicle (phosphate-buffered saline (PBS)) or rhTRX-1 (40 μg in PBS) was administered intraperitoneally every second day from Day-1 to Day 13 (ten mice per group). Survival was observed until Day 14. We had confirmed that a bolus intraperitoneal injection of 40 μg of rhTRX-1 led to systemic delivery and lung deposition in mice [16].

Lung pathological and lavage analyses: Vehicle (PBS) or rhTRX-1 (40 μg in PBS) was administered intraperitoneally on Days -1, 1, and 3. Lung pathological and lavage analyses were performed on Days 1 (24 h), 3 (72 h), and 5 (120 h) after H1N1 inoculation (five mice per group). The lung tissues were analyzed for the influenza virus copies using the PCR technique. The lung histological examination followed the procedure described by Xu et al. [73]. Four easily identifiable pathological processes were graded semiquantitatively on a scale of 0–4 (0 = normal, 1 = slight, 2 = moderate, 3 = severe, 4 = very severe): alveolar and interstitial edema, hemorrhage, margination and infiltration of inflammatory cells, and formation of bronchiolitis. For each mouse, the lung injury score was calculated by adding the individual grades for each category. Concentrations of tumor necrosis factor (TNF)-α (a proinflammatory cytokine participating in important processes involved in the inflammatory response) and CXCL1 (a CXC chemotactic cytokine playing a pivotal role in the activation and extravasation of neutrophils) were measured using ELISA [74]. Immunohistochemical detection of granulocyte-differentiation antigen (for neutrophils) [75] and 8-hydroxy-2′-deoxyguanosine (8-OHdG; for cellular DNA oxidation), also known as 8-oxo-7,8-dihydro-2-deoxyguanosine [8, 12, 18, 56], was also conducted.

Blood analyses: Blood serum was obtained from vehicle-treated and TRX-treated mice on Days 1, 3, and 5 after H1N1 inoculation (five mice per group). The serum concentration of hydroperoxides was measured using the Free Radical Analytical System [43, 52].

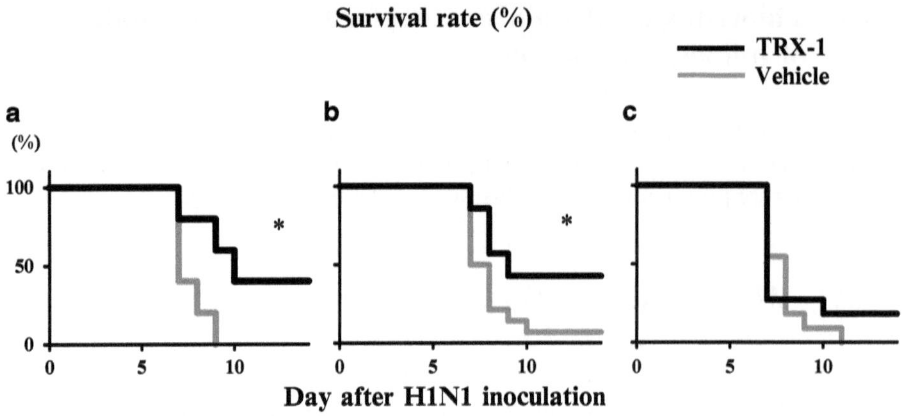

Fig. 14.3 Effects of recombinant human thioredoxin-1 treatment on survival rate in H1N1-inoculated mice. (**a**) The treatment is started the day before inoculation (day-1) (ten mice per group). (**b**) The treatment is started 30 min after inoculation (14 mice per group). (**c**) The treatment is started 4 h after inoculation (11 mice per group). *$p<0.05$ vs. vehicle

Thioredoxin treatment after influenza virus inoculation (*therapeutic protocols*): The intraperitoneal administration of vehicle (PBS) or rhTRX-1 (40 μg in PBS) was started 30 min (14 mice per group) or 4 h (11 mice per group) after H1N1 inoculation and was repeated every second day until Day 12. Survival was observed until Day 14.

Statistics: All data were expressed as mean ± SEM and compared using unpaired *t*-test, or one-way ANOVA or two-way ANOVA followed by Bonferroni's post-test where appropriate. Survival curves were analyzed using the Kaplan–Meier log-rank test. Differences for which $p<0.05$ were considered significant.

14.8.2 Results

Effects of rhTRX-1 on the survival rate and viral load in the lung after H1N1 inoculation: Survival curves using the Kaplan–Meier log-rank test showed a significant difference between the vehicle-treated group and the rhTRX-1-treated group (Fig. 14.3a). The rhTRX-1 treatment significantly improved the survival rate of H1N1-inoculated mice. The viral load in rhTRX-1-treated mice was comparable to that in control mice at each time point. These results indicate that rhTRX-1 treatment improves the survival rate of H1N1-inoculated mice significantly, although the treatment does not affect propagation of the influenza virus in the lungs of these animals.

Effects of rhTRX-1 on lung histology after H1N1 inoculation: H1N1-inoculated control mice presented diffuse edema and inflammatory cellular infiltration in

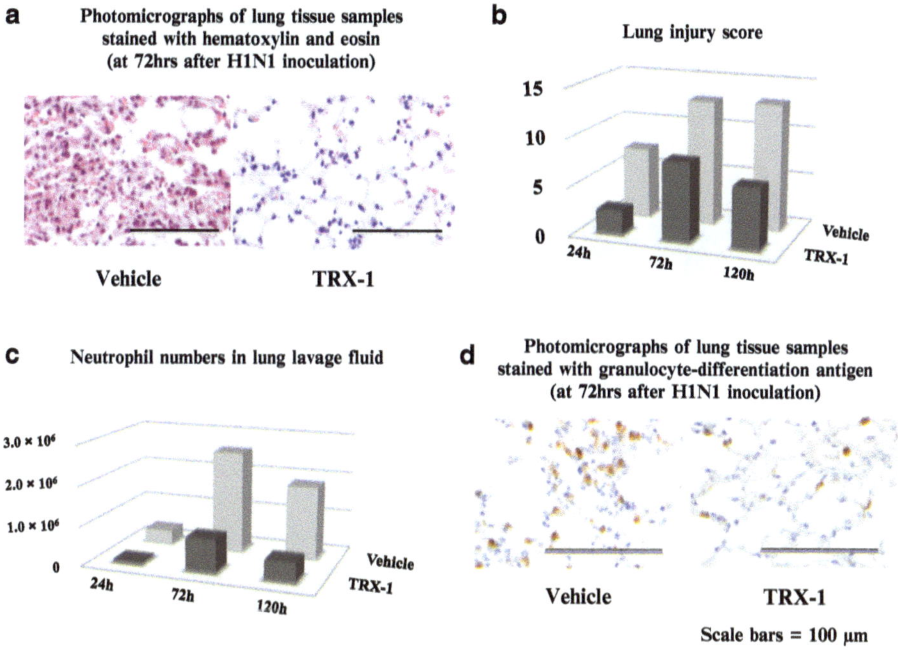

Fig. 14.4 Effects of recombinant human thioredoxin-1 treatment on lung histology and neutrophil infiltration after H1N1 inoculation. (**a**) Photomicrographs of lung tissue samples stained with hematoxylin and eosin at 72 h after H1N1 inoculation. *Left*: vehicle; *Right*: thioredoxin (TRX)-1. (**b**) Lung injury scores. Data represent the mean of five experiments. $p<0.01$ at 24, 72, 120 h after inoculation. (**c**) Neutrophil numbers in lung lavage fluid. Data represent the mean of five experiments. $p<0.05$ at 72, 120 h after inoculation. (**d**) Photomicrographs of lung tissue samples stained with granulocyte-differentiation antigen at 72 h after H1N1 inoculation. *Left*: vehicle; *Right*: TRX-1

alveoli and interstitium of the lung, hemorrhage, and thickened airways at 72 h after inoculation. The rhTRX-1 treatment attenuated the histological changes in the lung (Fig. 14.4a). The lung injury score increased significantly from 24 to 72 h and 120 h in the control group ($p<0.01$ at 24 h vs. 72, 120 h). The lung injury score remained significantly lower in rhTRX-1-treated mice than in control mice (Fig. 14.4b). Results obtained from these analyses indicate that rhTRX-1 treatment significantly attenuates the degree of acute lung injury in H1N1-inoculated mice.

Effects of rhTRX-1 on neutrophil infiltration in the lung after H1N1 inoculation: The neutrophil number in lung lavage fluid increased significantly from 24 to 72 h after virus inoculation in the control group ($p<0.01$). The neutrophil number remained significantly lower in rhTRX-1-treated mice compared with control mice at 72 and 120 h after inoculation (Fig. 14.4c). Histologically, influenza virus inoculation increased neutrophil infiltration in the lung at 72 h. The rhTRX-1 treatment almost reversed this effect (Fig. 14.4d). These results indicate that rhTRX-1 treatment significantly attenuates the neutrophil infiltration in the lung of H1N1-inoculated mice.

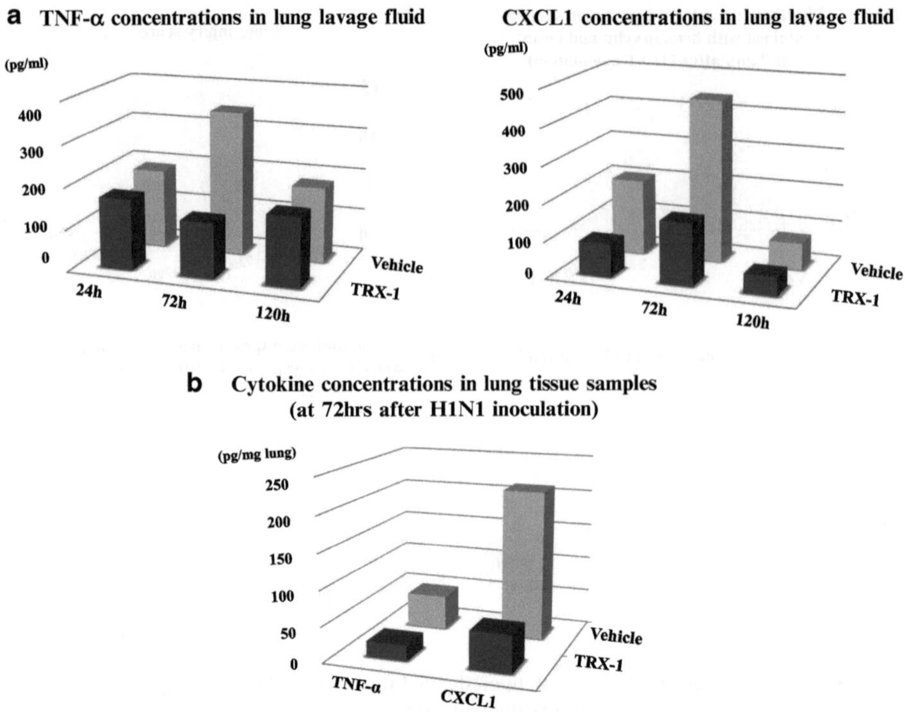

Fig. 14.5 Effects of recombinant human thioredoxin-1 treatment on cytokine production in the lung after H1N1 inoculation. (**a**) TNF-α and CXCL1 concentrations in lung lavage fluid. Data represent the mean of five experiments. $p<0.01$ at 72 h after inoculation. (**b**) Cytokine concentrations in lung tissue samples. Data represent the mean of five experiments. $p<0.01$ at 72 h after inoculation

Effects of rhTRX-1 on cytokine production in the lung after H1N1 inoculation: The TNF-α and CXCL1 concentrations in lung lavage fluid were significantly lower in rhTRX-1-treated mice than in control mice at 72 h after inoculation (Fig. 14.5a). Lung tissue analyses revealed that both cytokine concentrations were lowered significantly by rhTRX-1 treatment at 72 h after inoculation (Fig. 14.5b). These results indicate that rhTRX-1 treatment significantly attenuates the inflammatory cytokine production in the lung of H1N1-inoculated mice.

Effects of rhTRX-1 on oxidative stress markers after H1N1 inoculation: Histologically, H1N1 inoculation increased 8-OHdG formation in the lung at 72 h. The 8-OHdG formation was observed primarily in infiltrating cells and occasionally in lung epithelial cells. The rhTRX-1 treatment almost reversed this effect (Fig. 14.6a). The serum concentration of TH was significantly lower in rhTRX-1-treated mice than in control mice at 72 h after inoculation (Fig. 14.6b). These results indicate that rhTRX-1 treatment significantly attenuates the oxidative stress enhancement that is observed in H1N1-inoculated mice.

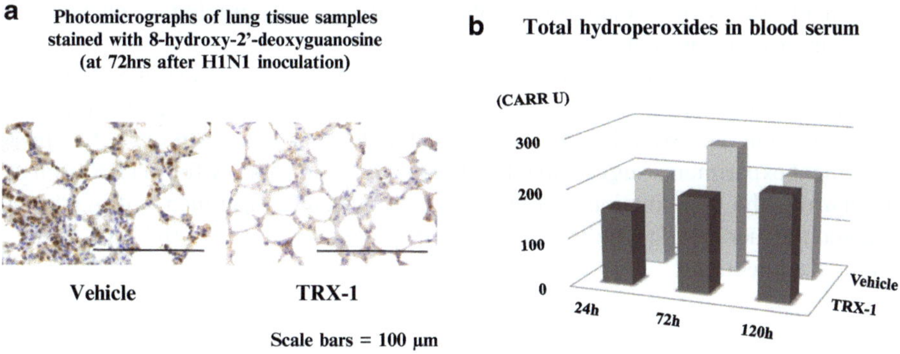

Fig. 14.6 Effects of recombinant human thioredoxin-1 treatment on oxidative stress markers after H1N1 inoculation. (**a**) Photomicrographs of lung tissue samples stained with 8-hydroxy-2′-deoxyguanosine at 72 h after H1N1 inoculation. *Left*: vehicle; *Right*: thioredoxin (TRX)-1. (**b**) Hydroperoxides in blood serum. Data represent the mean of five experiments. $p < 0.05$ at 72 h after inoculation

Treatment with rhTRX-1 after H1N1 inoculation (therapeutic protocols): The rhTRX-1 treatment started 30 min after virus inoculation significantly improved the survival rate of H1N1-inoculated mice ($p < 0.05$ vs. control) (Fig. 14.3b). However, the rhTRX-1 treatment started 4 h after inoculation had no effect on the survival rate (Fig. 14.3c).

14.8.3 Interpretation of the Results

In our murine model, influenza A virus (H1N1)-induced pneumonia is characterized histologically by intense infiltration of inflammatory cells (mainly neutrophils). Activated neutrophils, macrophages, and respiratory tract epithelial cells release excessive amounts of bioactive substances, including cytokines/chemokines, ROS, and tissue degradative enzymes and thereby induce acute lung inflammatory disease [64]. Presumably, oxidative injury by the inflamed cells is targeted to the vascular endothelium, leading to lung edema and hemorrhage.

Preventive rhTRX-1 treatment improved the survival rate of H1N1-inoculated mice significantly and also significantly attenuated the histological changes and neutrophil infiltration in the lung of H1N1-inoculated mice. The treatment also significantly attenuated the production of TNF-α and CXCL1 in the lung and oxidative stress enhancement, which were observed in H1N1-inoculated mice.

We also examined whether treatment with rhTRX-1 after H1N1 inoculation was efficacious for protection. Administration of rhTRX-1, started 30 min (but not 4 h) after H1N1 inoculation, also exerted a significant survival-promoting effect. Therefore, it is conceivable that rhTRX-1 provided 30 min after virus inoculation ameliorated H1N1-induced lung inflammatory injury in mice. Our finding that treatment after virus inoculation is also effective is encouraging when practical applications are considered.

Based on these considerations, it is plausible that exogenous administration of rhTRX-1 ameliorates the lethal effects of influenza A virus-induced pneumonia in mice through antioxidative and anti-inflammatory actions. Although the post-treatment experiments suggest that the therapeutic time window for rhTRX-1 might be narrow, the combination of rhTRX-1 with current anti-influenza drugs might expand the therapeutic window as well as the protective effects of rhTRX-1 against influenza infection.

14.9 Augmentation of Thioredoxin-1 System

Augmentation of TRX-1 might be clinically beneficial for patients. Yodoi and coworkers reported that endogenous TRX-1 expression was induced by natural substances including estrogen, prostaglandins, and cyclic AMP, geranylgeranylacetone (an anti-ulcer drug derived from a natural plant constituent), and temocapril (a non-sulfhydryl-containing angiotensin-converting enzyme inhibitor) [4, 76, 77]. These agents might exert their physiological functions including cytoprotective actions partly through the induction of TRX-1, but without the massive oxidative stress that can induce TRX-1 and other stress proteins substantially.

The short biological half life of TRX-1 limits its potential clinical use. To elongate the half life of TRX-1 in blood and its antioxidative and anti-inflammatory effects, some valid modifications might be advantageous. Conjugation with polyethylene glycol, a common technique used for small proteins including interferon-γ and granulocyte colony-stimulating factor, might be a feasible approach [5]. Because the cysteine residues play crucial roles for the activity of TRX-1, the disulfide bridge-based PEGylation is not recommended. Site-specific modification other than the disulfide bridge might work to maintain and elongate its activity.

Recombinant human serum albumin (rHSA) has been produced using *Pichia pastoris*, a methylotrophic yeast. The safety profile of rHSA has been established, and rHSA also has redox activity. The molecule has a half life in the human body of more than 2 weeks because of its large molecular size (66.5 kDa molecular weight) and negative electric potential, which inhibit the glomerular filtration and vascular permeability or decrease lymphatic clearance. The fusion of TRX-1 to rHSA might extend its half life and allow a bolus intravenous administration with less dosing frequency (e.g., once-weekly dosing) [78]. Further examination of the molecular mechanisms for albumin fusion of TRX-1 will be necessary.

As described in Sect. 14.5, concentrations of TRX-1 (mean: 268 ng/mL) and nitric oxide (as measured using nitrite/nitrate; mean: 479 μmol/L) in early human milk were found to be about ten times higher than those (mean: 20 ng/mL, 40 μmol/L, respectively) in the blood of healthy adults [53, 79]. Ma and coworkers [80] reported also that the cytoprotective action of TRX-1 was augmented after S-nitrosation.

Very recently, a non-genetically modified method to prepare yeast (*Saccharomyces cerevisiae*) TRX-enriched extracts has become applicable on a large-scale [81]. Yodoi and coworkers examined whether yeast TRX can mitigate the allergenicity of ovomucoid by conducting immediate allergy tests on guinea pigs. Treatment with TRX reduced the anaphylactic symptoms induced by ovomucoid in these tests [82]. They also reported that prophylactic oral administration of yeast TRX reduced indomethacin-induced gastric injury in mice [83].

The protective effects of such "TRX-1 inducers," "TRX-1 augmenters," or "TRX-1 donors" against oxidative and inflammatory disorders (such as severe influenza pneumonia), warrant further study for future clinical applications.

14.10 Summary and Conclusions

Studies oriented to oxidative stress and antioxidative defenses are increasing explosively. In vitro and in vivo experimental findings indicate that TRX-1 is a cytoprotective protein with antioxidative and anti-inflammatory properties, which play an important role in the homeostasis of the human body. The TRX system protects cells and tissues through various direct or indirect mechanisms. Although the physiological roles of TRX-1 remain to be clarified, results of recent studies suggest a likely role of this protein for redox signaling and development in infants and children.

"Antioxidant therapy" has been tested widely as a treatment for many kinds of acute and chronic disorders, but it has had limited success in clinical practice. The clinical outcome is likely to be improved by combination therapy and better patient selection. The findings documented in this chapter nominate TRX-1 as a preventive or therapeutic measure. The pharmaceutical modulation of the TRX system might represent an effective and cooperative strategy for the management of several human disorders.

Gaps remain between the basic findings and their clinical application because no large-scale clinical study has provided evidence to prove that treatment with TRX-1 is beneficial to specific patients. However, the rationale for treatment with TRX-1 is solid, and further research to elucidate the benefits of the strategy (including the identification of protective agents that target the TRX system) must be undertaken in experimental and clinical settings.

Acknowledgments The authors thank Dr. Mitsufumi Mayumi (Department of Pediatrics, Faculty of Medical Sciences, University of Fukui, Fukui, Japan), Dr. Junji Yodoi (Department of Biological Responses, Institute for Virus Research, Kyoto University, Kyoto, Japan; Center for Cell Signaling Research and Department of Bioinspired Science, Ewha Womans University, Seoul, Korea), and Dr. Hajime Nakamura (Department of Preventive Medicine, The Tazuke Kofukai Medical Research Institute Kitano Hospital, Osaka, Japan) for scientific encouragement and helpful discussions.

References

1. Toyokuni S, Akatsuka S (2007) Pathological investigation of oxidative stress in the post-genomic era. Pathol Int 57:461–473
2. Tsukahara H (2007) Biomarkers for oxidative stress: clinical application in pediatric medicine. Curr Med Chem 14:339–351
3. Nakamura H, Nakamura K, Yodoi J (1997) Redox regulation of cellular activation. Annu Rev Immunol 15:351–369
4. Burke-Gaffney A, Callister ME, Nakamura H (2005) Thioredoxin: friend or foe in human disease? Trends Pharmacol Sci 26:398–404
5. Nakamura H, Hoshino Y, Okuyama H, Matsuo Y, Yodoi J (2009) Thioredoxin 1 delivery as new therapeutics. Adv Drug Deliv Rev 61:303–309
6. Nakamura H, Herzenberg LA, Bai J, Araya S, Kondo N, Nishinaka Y, Herzenberg LA, Yodoi J (2001) Circulating thioredoxin suppresses lipopolysaccharide-induced neutrophil chemotaxis. Proc Natl Acad Sci U S A 98:15143–15148
7. Mitsui A, Hamuro J, Nakamura H, Kondo N, Hirabayashi Y, Ishizaki-Koizumi S, Hirakawa T, Inoue T, Yodoi J (2002) Overexpression of human thioredoxin in transgenic mice controls oxidative stress and life span. Antioxid Redox Signal 4:693–696
8. Kobayashi-Miura M, Nakamura H, Yodoi J, Shiota K (2002) Thioredoxin, an anti-oxidant protein, protects mouse embryos from oxidative stress-induced developmental anomalies. Free Radic Res 36:949–956
9. Takagi Y, Mitsui A, Nishiyama A, Nozaki K, Sono H, Gon Y, Hashimoto N, Yodoi J (1999) Overexpression of thioredoxin in transgenic mice attenuates focal ischemic brain damage. Proc Natl Acad Sci U S A 96:4131–4136
10. Shioji K, Kishimoto C, Nakamura H, Masutani H, Yuan Z, Oka S, Yodoi J (2002) Overexpression of thioredoxin-1 in transgenic mice attenuates adriamycin-induced cardiotoxicity. Circulation 106:1403–1409
11. Kasuno K, Nakamura H, Ono T, Muso E, Yodoi J (2003) Protective roles of thioredoxin, a redox-regulating protein, in renal ischemia/reperfusion injury. Kidney Int 64:1273–1282
12. Kawasaki K, Nishio A, Nakamura H, Uchida K, Fukui T, Ohana M, Yoshizawa H, Ohashi S, Tamaki H, Matsuura M, Asada M, Nishi T, Nakase H, Toyokuni S, Liu W, Yodoi J, Okazaki K, Chiba T (2005) *Helicobacter felis*-induced gastritis was suppressed in mice overexpressing thioredoxin-1. Lab Invest 85:1104–1117
13. Ohashi S, Nishio A, Nakamura H, Kido M, Ueno S, Uza N, Inoue S, Kitamura H, Kiriya K, Asada M, Tamaki H, Matsuura M, Kawasaki K, Fukui T, Watanabe N, Nakase H, Yodoi J, Okazaki K, Chiba T (2006) Protective roles of redox-active protein thioredoxin-1 for severe acute pancreatitis. Am J Physiol Gastrointest Liver Physiol 290:G772–G781
14. Tamaki H, Nakamura H, Nishio A, Nakase H, Ueno S, Uza N, Kido M, Inoue S, Mikami S, Asada M, Kiriya K, Kitamura H, Ohashi S, Fukui T, Kawasaki K, Matsuura M, Ishii Y, Okazaki K, Yodoi J, Chiba T (2006) Human thioredoxin-1 ameliorates experimental murine colitis in association with suppressed macrophage inhibitory factor production. Gastroenterology 131:1110–1121
15. Tan A, Nakamura H, Kondo N, Tanito M, Kwon YW, Kaimul Ahsan M, Matsui H, Narita M, Yodoi J (2007) Thioredoxin-1 attenuates indomethacin-induced gastric mucosal injury in mice. Free Radic Res 41:861–869
16. Hoshino T, Nakamura H, Okamoto M, Kato S, Araya S, Nomiyama K, Oizumi K, Young HA, Aizawa H, Yodoi J (2003) Redox-active protein thioredoxin prevents proinflammatory cytokine- or bleomycin-induced lung injury. Am J Respir Crit Care Med 168:1075–1083
17. Kaimul Ahsan M, Nakamura H, Tanito M, Yamada K, Utsumi H, Yodoi J (2005) Thioredoxin-1 suppresses lung injury and apoptosis induced by diesel exhaust particles (DEP) by scavenging reactive oxygen species and by inhibiting DEP-induced downregulation of Akt. Free Radic Biol Med 39:1549–1559

18. Sato A, Hoshino Y, Hara T, Muro S, Nakamura H, Mishima M, Yodoi J (2008) Thioredoxin-1 ameliorates cigarette smoke-induced lung inflammation and emphysema in mice. J Pharmacol Exp Ther 325:380–388
19. Nakamura H, Tamura S, Watanabe I, Iwasaki T, Yodoi J (2002) Enhanced resistancy of thioredoxin-transgenic mice against influenza virus-induced pneumonia. Immunol Lett 82: 165–170
20. Matsui M, Oshima M, Oshima H, Takaku K, Maruyama T, Yodoi J, Taketo MM (1996) Early embryonic lethality caused by targeted disruption of the mouse thioredoxin gene. Dev Biol 178:179–185
21. Kondo N, Ishii Y, Kwon YW, Tanito M, Horita H, Nishinaka Y, Nakamura H, Yodoi J (2004) Redox-sensing release of human thioredoxin from T lymphocytes with negative feedback loops. J Immunol 172:442–448
22. Nakamura H, De Rosa S, Roederer M, Anderson MT, Dubs JG, Yodoi J, Holmgren A, Herzenberg LA, Herzenberg LA (1996) Elevation of plasma thioredoxin levels in HIV-infected individuals. Int Immunol 8:603–611
23. Nakamura H, Vaage J, Valen G, Padilla CA, Björnstedt M, Holmgren A (1998) Measurements of plasma glutaredoxin and thioredoxin in healthy volunteers and during open-heart surgery. Free Radic Biol Med 24:1176–1186
24. Abdiu A, Nakamura H, Sahaf B, Yodoi J, Holmgren A, Rosén A (2000) Thioredoxin blood level increases after severe burn injury. Antioxid Redox Signal 2:707–716
25. Sumida Y, Nakashima T, Yoh T, Nakajima Y, Ishikawa H, Mitsuyoshi H, Sakamoto Y, Okanoue T, Kashima K, Nakamura H, Yodoi J (2000) Serum thioredoxin levels as an indicator of oxidative stress in patients with hepatitis C virus infection. J Hepatol 33:616–622
26. Kishimoto C, Shioji K, Nakamura H, Nakayama Y, Yodoi J, Sasayama S (2001) Serum thioredoxin (TRX) levels in patients with heart failure. Jpn Circ J 65:491–494
27. Kakisaka Y, Nakashima T, Sumida Y, Yoh T, Nakamura H, Yodoi J, Senmaru H (2002) Elevation of serum thioredoxin levels in patients with type 2 diabetes. Horm Metab Res 34:160–164
28. Jikimoto T, Nishikubo Y, Koshiba M, Kanagawa S, Morinobu S, Morinobu A, Saura R, Mizuno K, Kondo S, Toyokuni S, Nakamura H, Yodoi J, Kumagai S (2002) Thioredoxin as a biomarker for oxidative stress in patients with rheumatoid arthritis. Mol Immunol 38: 765–772
29. Yamada Y, Nakamura H, Adachi T, Sannohe S, Oyamada H, Kayaba H, Yodoi J, Chihara J (2003) Elevated serum levels of thioredoxin in patients with acute exacerbation of asthma. Immunol Lett 86:199–205
30. Sumida Y, Nakashima T, Yoh T, Furutani M, Hirohama A, Kakisaka Y, Nakajima Y, Ishikawa H, Mitsuyoshi H, Okanoue T, Kashima K, Nakamura H, Yodoi J (2003) Serum thioredoxin levels as a predictor of steatohepatitis in patients with nonalcoholic fatty liver disease. J Hepatol 38:32–38
31. Kato A, Odamaki M, Nakamura H, Yodoi J, Hishida A (2003) Elevation of blood thioredoxin in hemodialysis patients with hepatitis C virus infection. Kidney Int 63:2262–2268
32. Miyamoto S, Kawano H, Sakamoto T, Soejima H, Kajiwara I, Hokamaki J, Hirai N, Sugiyama S, Yoshimura M, Yasue H, Nakamura H, Yodoi J, Ogawa H (2004) Increased plasma levels of thioredoxin in patients with coronary spastic angina. Antioxid Redox Signal 6:75–80
33. Jekell A, Hossain A, Alehagen U, Dahlström U, Rosén A (2004) Elevated circulating levels of thioredoxin and stress in chronic heart failure. Eur J Heart Fail 6:883–890
34. Ohashi S, Nishio A, Nakamura H, Kido M, Kiriya K, Asada M, Tamaki H, Fukui T, Kawasaki K, Watanabe N, Yodoi J, Okazaki K, Chiba T (2006) Clinical significance of serum thioredoxin 1 levels in patients with acute pancreatitis. Pancreas 32:264–270
35. Callister ME, Burke-Gaffney A, Quinlan GJ, Nicholson AG, Florio R, Nakamura H, Yodoi J, Evans TW (2006) Extracellular thioredoxin levels are increased in patients with acute lung injury. Thorax 61:521–527
36. Takahashi K, Chin K, Nakamura H, Morita S, Sumi K, Oga T, Matsumoto H, Niimi A, Fukuhara S, Yodoi J, Mishima M (2008) Plasma thioredoxin, a novel oxidative stress marker,

in patients with obstructive sleep apnea before and after nasal continuous positive airway pressure. Antioxid Redox Signal 10:715–726

37. Hofer S, Rosenhagen C, Nakamura H, Yodoi J, Bopp C, Zimmermann JB, Goebel M, Schemmer P, Hoffmann K, Schulze-Osthoff K, Breitkreutz R, Weigand MA (2009) Thioredoxin in human and experimental sepsis. Crit Care Med 37:2155–2159
38. Zhang XY, da Chen C, Xiu MH, Wang F, Qi LY, Sun HQ, Chen S, He SC, Wu GY, Haile CN, Kosten TA, Lu L, Kosten TR (2009) The novel oxidative stress marker thioredoxin is increased in first-episode schizophrenic patients. Schizophr Res 113:151–157
39. Leaver SK, MacCallum NS, Pingle V, Hacking MB, Quinlan GJ, Evans TW, Burke-Gaffney A (2010) Increased plasma thioredoxin levels in patients with sepsis: positive association with macrophage migration inhibitory factor. Intensive Care Med 36:336–341
40. Straface E, Marchesi A, Gambardella L, Metere A, Tarissi de Jacobis I, Viora M, Giordani L, Villani A, Del Principe D, Malorni W, Pietraforte D (2012) Does oxidative stress play a critical role in cardiovascular complications of Kawasaki disease? Antioxid Redox Signal 17: 1441–1446
41. Tsukahara H, Kikuchi K, Matsuda M, Saito M, Hata I, Tsuchida S, Sudo M (1997) Endogenous nitric oxide production in Kawasaki disease. Scand J Clin Lab Invest 57:43–47
42. Takatsuki S, Ito Y, Takeuchi D, Hoshida H, Nakayama T, Matsuura H, Saji T (2009) IVIG reduced vascular oxidative stress in patients with Kawasaki disease. Circ J 73:1315–1318
43. Yahata T, Suzuki C, Hamaoka A, Fujii M, Hamaoka K (2011) Dynamics of reactive oxygen metabolites and biological antioxidant potential in the acute stage of Kawasaki disease. Circ J 75:2453–2459
44. Hamaoka A, Hamaoka K, Yahata T, Fujii M, Ozawa S, Toiyama K, Nishida M, Itoi T (2010) Effects of HMG-CoA reductase inhibitors on continuous post-inflammatory vascular remodeling late after Kawasaki disease. J Cardiol 56:245–253
45. Frustaci A, Neri M, Cesario A, Adams JB, Domenici E, Dalla Bernardina B, Bonassi S (2012) Oxidative stress-related biomarkers in autism: systematic review and meta-analyses. Free Radic Biol Med 52:2128–2141
46. Kawatani M, Tsukahara H, Mayumi M (2011) Evaluation of oxidative stress status in children with pervasive developmental disorder and attention deficit hyperactivity disorder using urinary-specific biomarkers. Redox Rep 16:45–46
47. Al-Yafee YA, Al-Ayadhi LY, Haq SH, El-Ansary AK (2011) Novel metabolic biomarkers related to sulfur-dependent detoxification pathways in autistic patients of Saudi Arabia. BMC Neurol 11:139
48. Sato S, Kajiwara M, Shiba H, Takeuchi M, Tsukahara H, Deguchi Y, Mayumi M (2006) Serum selenium and thioredoxin levels in severely psychosomatic disabled children. Jpn J Pediatr 59:1147–1151 (Japanese)
49. Callister ME, Burke-Gaffney A, Quinlan GJ, Betts H, Nadel S, Evans TW (2007) Persistently low plasma thioredoxin is associated with meningococcal septic shock in children. Intensive Care Med 33:364–367
50. Burton GJ, Jauniaux E (2011) Oxidative stress. Best Pract Res Clin Obstet Gynaecol 25: 287–299
51. Kuroda S, Watanabe M, Santo T, Shimizuishi Y, Takano T, Hidaka Y, Kimura T, Iwatani Y (2010) Postpartum increase of serum thioredoxin concentrations and the relation to CD8 lymphocytes. Ann Clin Biochem 47:62–66
52. Nakatsukasa Y, Tsukahara H, Tabuchi K, Tabuchi M, Magami T, Yamada M, Fujii Y, Yashiro M, Tsuge M, Morishima T (2013) Thioredoxin-1 and oxidative stress status in pregnant women at early third trimester of pregnancy: relation to maternal and neonatal characteristics. J Clin Biochem Nutr 52:27–31
53. Todoroki Y, Tsukahara H, Ohshima Y, Shukunami K, Nishijima K, Kotsuji F, Hata A, Kasuga K, Sekine K, Nakamura H, Yodoi J, Mayumi M (2005) Concentrations of thioredoxin, a redox-regulating protein, in umbilical cord blood and breast milk. Free Radic Res 39:291–297
54. Shibuki H, Katai N, Kuroiwa S, Kurokawa T, Yodoi J, Yoshimura N (1998) Protective effect of adult T-cell leukemia-derived factor on retinal ischemia-reperfusion injury in the rat. Invest Ophthalmol Vis Sci 39:1470–1477

55. Hattori I, Takagi Y, Nakamura H, Nozaki K, Bai J, Kondo N, Sugino T, Nishimura M, Hashimoto N, Yodoi J (2004) Intravenous administration of thioredoxin decreases brain damage following transient focal cerebral ischemia in mice. Antioxid Redox Signal 6:81–87
56. Liu W, Nakamura H, Shioji K, Tanito M, Oka S, Kaimul Ahsan M, Son A, Ishii Y, Kishimoto C, Yodoi J (2004) Thioredoxin-1 ameliorates myosin-induced autoimmune myocarditis by suppressing chemokine expressions and leukocyte chemotaxis in mice. Circulation 110: 1276–1283
57. Ichiki H, Hoshino T, Kinoshita T, Imaoka H, Kato S, Inoue H, Nakamura H, Yodoi J, Young HA, Aizawa H (2005) Thioredoxin suppresses airway hyperresponsiveness and airway inflammation in asthma. Biochem Biophys Res Commun 334:1141–1148
58. Ueda S, Nakamura T, Yamada A, Teratani A, Matsui N, Furukawa S, Hoshino Y, Narita M, Yodoi J, Nakamura H (2006) Recombinant human thioredoxin suppresses lipopolysaccharide-induced bronchoalveolar neutrophil infiltration in rat. Life Sci 79:1170–1177
59. Cohen JI, Roychowdhury S, DiBello PM, Jacobsen DW, Nagy LE (2009) Exogenous thioredoxin prevents ethanol-induced oxidative damage and apoptosis in mouse liver. Hepatology 49:1709–1717
60. Chernatynskaya AV, Looney B, Hu H, Zhu X, Xia CQ (2011) Administration of recombinant human thioredoxin-1 significantly delays and prevents autoimmune diabetes in nonobese diabetic mice through modulation of autoimmunity. Diabetes Metab Res Rev 27:809–812
61. Subbarao K (2008) Influenza viruses. In: Long SS, Pickering LK, Prober CG (eds) Principles and practice of pediatric infectious diseases, 3rd edn. Churchill Livingstone-Elsevier, Philadelphia, pp 1130–1138
62. Writing Committee of the Second World Health Organization Consultation on Clinical Aspects of Human Infection with Avian Influenza A (H5N1) Virus, Abdel-Ghafar AN, Chotpitayasunondh T, Gao Z, Hayden FG, Nguyen DH, de Jong MD, Naghdaliyev A, Peiris JS, Shindo N, Soeroso S, Uyeki TM (2008) Update on avian influenza A (H5N1) virus infection in humans. N Engl J Med 358:261–273
63. Perez-Padilla R, de la Rosa-Zamboni D, Ponce de Leon S, Hernandez M, Quiñones-Falconi F, Bautista E, Ramirez-Venegas A, Rojas-Serrano J, Ormsby CE, Corrales A, Higuera A, Mondragon E, Cordova-Villalobos JA, INER Working Group on Influenza (2009) Pneumonia and respiratory failure from swine-origin influenza A (H1N1) in Mexico. N Engl J Med 361:680–689
64. Uchide N, Toyoda H (2010) Antioxidant therapy as a potential approach to severe influenza-associated complications. Molecules 16:2032–2052
65. Akaike T, Ando M, Oda T, Doi T, Ijiri S, Araki S, Maeda H (1990) Dependence on O_2-generation by xanthine oxidase of pathogenesis of influenza virus infection in mice. J Clin Invest 85:739–745
66. Choi AM, Knobil K, Otterbein SL, Eastman DA, Jacoby DB (1996) Oxidant stress responses in influenza virus pneumonia: gene expression and transcription factor activation. Am J Physiol 271:L383–L391
67. Akaike T, Noguchi Y, Ijiri S, Setoguchi K, Suga M, Zheng YM, Dietzschold B, Maeda H (1996) Pathogenesis of influenza virus-induced pneumonia: involvement of both nitric oxide and oxygen radicals. Proc Natl Acad Sci U S A 93:2448–2453
68. Suliman HB, Ryan LK, Bishop L, Folz RJ (2001) Prevention of influenza-induced lung injury in mice overexpressing extracellular superoxide dismutase. Am J Physiol Lung Cell Mol Physiol 280:L69–L78
69. Hashiba T, Suzuki M, Nagashima Y, Suzuki S, Inoue S, Tsuburai T, Matsuse T, Ishigatubo Y (2001) Adenovirus-mediated transfer of heme oxygenase-1 cDNA attenuates severe lung injury induced by the influenza virus in mice. Gene Ther 8:1499–1507
70. Shi XL, Shi ZH, Huang H, Zhu HG, Zhou P, Ju D (2010) Therapeutic effect of recombinant human catalase on H1N1 influenza-induced pneumonia in mice. Inflammation 33:166–172
71. Beck MA, Nelson HK, Shi Q, Van Dael P, Schiffrin EJ, Blum S, Barclay D, Levander OA (2001) Selenium deficiency increases the pathology of an influenza virus infection. FASEB J 15:1481–1483

72. Yashiro M, Tsukahara H, Matsukawa A, Yamada M, Fujii Y, Nagaoka Y, Tsuge M, Yamashita N, Ito T, Yamada M, Masutani H, Yodoi J, Morishima T (2013) Redox-active protein thioredoxin-1 administration ameliorates influenza A virus (H1N1)-induced acute lung injury in mice. Crit Care Med 41:166–176
73. Xu T, Qiao J, Zhao L, He G, Li K, Wang J, Tian Y, Wang H (2009) Effect of dexamethasone on acute respiratory distress syndrome induced by the H5N1 virus in mice. Eur Respir J 33:852–860
74. Matsukawa A, Hogaboam CM, Lukacs NW, Lincoln PM, Evanoff HL, Strieter RM, Kunkel SL (2000) Expression and contribution of endogenous IL-13 in an experimental model of sepsis. J Immunol 164:2738–2744
75. Fleming TJ, Fleming ML, Malek TR (1993) Selective expression of Ly-6G on myeloid lineage cells in mouse bone marrow: RB6-8C5 mAb to granulocyte-differentiation antigen (Gr-1) detects members of the Ly-6 family. J Immunol 151:2399–2408
76. Dekigai H, Nakamura H, Bai J, Tanito M, Masutani H, Hirota K, Matsui H, Murakami M, Yodoi J (2001) Geranylgeranylacetone promotes induction and secretion of thioredoxin in gastric mucosal cells and peripheral blood lymphocytes. Free Radic Res 35:23–30
77. Hirota K, Nakamura H, Masutani H, Yodoi J (2002) Thioredoxin superfamily and thioredoxin-inducing agents. Ann N Y Acad Sci 957:189–199
78. Ikuta S, Chuang VT, Ishima Y, Nakajou K, Furukawa M, Watanabe H, Maruyama T, Otagiri M (2010) Albumin fusion of thioredoxin—the production and evaluation of its biological activity for potential therapeutic applications. J Control Release 147:17–23
79. Ohta N, Tsukahara H, Ohshima Y, Nishii M, Ogawa Y, Sekine K, Kasuga K, Mayumi M (2004) Nitric oxide metabolites and adrenomedullin in human breast milk. Early Hum Dev 78:61–65
80. Tao L, Gao E, Bryan NS, Qu Y, Liu HR, Hu A, Christopher TA, Lopez BL, Yodoi J, Koch WJ, Feelisch M, Ma XL (2004) Cardioprotective effects of thioredoxin in myocardial ischemia and reperfusion: role of S-nitrosation. Proc Natl Acad Sci U S A 101:11471–11476
81. Inoue Y, Nomura W, Takeuchi Y, Ohdate T, Tamasu S, Kitaoka A, Kiyokawa Y, Masutani H, Murata K, Wakai Y, Izawa S, Yodoi J (2007) Efficient extraction of thioredoxin from *Saccharomyces cerevisiae* by ethanol. Appl Environ Microbiol 73:1672–1675
82. Taketani Y, Kinugasa K, Furukawa S, Nakamura H, Otsuki R, Yasuda H, Fujita T, Kanzaki K, Masutani H, Yodoi J (2011) Yeast thioredoxin-enriched extracts for mitigating the allergenicity of foods. Biosci Biotechnol Biochem 75:1872–1879
83. Nakajima A, Fukui T, Takahashi Y, Kishimoto M, Yamashina M, Nakayama S, Sakaguchi Y, Yoshida K, Uchida K, Nishio A, Yodoi J, Okazaki K (2012) Attenuation of indomethacin-induced gastric mucosal injury by prophylactic administration of sake yeast-derived thioredoxin. J Gastroenterol 47:978–987

Part III
Clinical Topics

Chapter 15
Allergic and Immunological Disorders

Akihiro Yachie

Abbreviations

AIRE	Autoimmune regulator gene
CAPS	Cryopyrin-associated periodic syndrome
CINCA	Chronic infantile neurological cutaneous and articular syndrome
CO	Carbon monoxide
FDP	Fibrinogen degradation product
Hb	Hemoglobin
HO	Heme oxygenase
Hp	Haptoglobin
IFN	Interferon
LCL	Lymphoblastoid cell line
LPS	Lipopolysaccharide
MAPK	Mitogen-activated protein kinase
MAS	Macrophage activation syndrome
MetHb	Methemoglobin
NSAID	Non-steroidal anti-inflammatory drug
PAMP	Pathogen-associated molecular pattern
PIC	Plasmin-α_2 plasmin inhibitor complex
RBC	Red blood cells
sJIA	Systemic juvenile idiopathic arthritis
TAT	Thrombin–antithrombin complex
TF	Tissue factor
TLRs	Toll-like receptors

A. Yachie, M.D., Ph.D. (✉)
Department of Pediatrics, School of Medicine, Graduate School of Medical Sciences, Kanazawa University, Kanazawa 920-8641, Japan
e-mail: yachie@staff.kanazawa-u.ac.jp

H. Tsukahara and K. Kaneko (eds.), *Studies on Pediatric Disorders*, Oxidative Stress in Applied Basic Research and Clinical Practice, DOI 10.1007/978-1-4939-0679-6_15,

15.1 Introduction

Immune response is an indispensible defense mechanism to fight against pathogens and tumor cells, thereby preventing life-threatening condition and at the same time maintaining the homeostasis of our body's inner milieu. Recent advances in immunology disclosed the critical roles of antigen-independent innate immunity in addition to antigen-specific acquired immunity for optimal immune responses. These two seemingly distinct immune systems have been found to be intimately related to each other. In particular, understanding of the mechanism to recognize pathogen-associated molecular patterns or danger-associated molecular patterns via toll-like receptors provided a new concept of innate immunity [1–4]. Furthermore, elucidation of regulatory mechanisms, including those for thymic epithelial cells expressing the autoimmune regulator gene (AIRE) and regulatory T cells, has shed new light on the understanding of immune tolerance [5–8]. These mechanisms control the quality and quantity of the acquired immune responses to achieve the best defense while avoiding damage to the self. In addition to these two arms of the host defense, the third component of our defenses is the mechanism to protect the battlefield itself, i.e., the mechanism to minimize inevitable cellular and tissue injury, and organ dysfunctions associated with intense immune and inflammatory reactions (Fig. 15.1). This anti-oxidative system supports and strengthens the stability of the host [9–11].

This chapter will focus on the impact of oxidative stresses, together with the importance of anti-oxidative functions in host immune responses, with particular emphasis on HO and the carbon monoxide (CO) system [12].

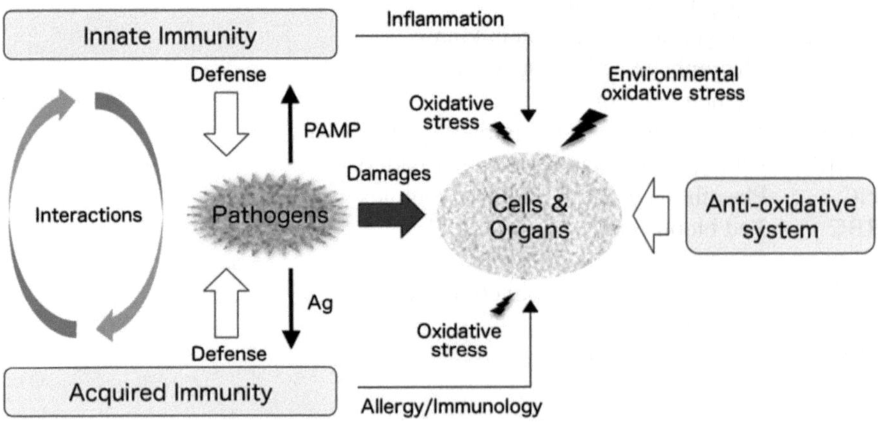

Fig. 15.1 Innate immunity, acquired immunity, and anti-oxidative system. Our defense system is composed of the well-known, innate, and acquired immunity. The third component of our defense is the anti-oxidative system which plays a significant role in protecting cells and organs from accumulating oxidative stress during immune reactions. *PAMP* pathogen-associated molecular pattern

15.2 Oxidative Stress and Heme Oxygenase

15.2.1 Immunity, Inflammation, and Oxidative Stress

Coinciding with the development of immunology, the molecular backgrounds of various primary immunodeficiency diseases have been elucidated at an unexpected rate. Thus, our understanding of the pathogenesis of inflammatory diseases has also advanced significantly. In particular, we now understand that molecular defects in immune tolerance [5, 13], apoptosis [14, 15], and immune regulatory functions [6–8, 16] at various levels may lead to "abnormal immune response to self antigens" in autoimmune diseases.

Even with the orchestrated, well-tuned immune responses, damage to self is unavoidable to some extent. Many inflammatory and immune-related diseases are thought to be the result of exaggerated immune responses or uncontrollable inflammatory reactions, leading to significant tissue injury and organ damage. This seemingly unavoidable impact on ourselves comprises a significant portion of the oxidative stress that is encountered in our daily lives. Additional well-recognized exogenous oxidative stresses include air pollution, UV, pathogens, allergens, food chemicals, and exposure to excess cold or heat [17].

In combating inflammatory diseases, it is not always suggestive to suppress immune reactions and inhibit inflammatory reactions. Although these reactions damage cells and tissues and induce organ dysfunctions, they are also the key mechanisms in fighting against infections and malignant diseases. Indiscriminate use of steroids and immunosuppressive agents may thus lead to paradoxical states where "the inflammation is gone but the disease gets worse."

In this chapter, I propose a novel approach to treat inflammatory diseases that could minimize tissue damage by inducing a protective mechanism to the cells and organs through judicious pharmacological intervention. This approach was developed through the experience of a still rare genetic disease, HO-1 deficiency [18]. In the following sections, the details of the clinical features of a patient with HO-1 deficiency are described. Through the analysis of this patient, I try to shed light on how the deficient anti-oxidative mechanism leads to an exaggerated inflammatory reaction and enhanced tissue damage.

15.2.2 Heme Metabolism and HO/CO System

Heme is a major component of hemoglobin (Hb), the product of erythrocyte destruction. Heme is constantly produced in vivo and it is extremely toxic to the cells [19, 20]. Therefore, constitutive mechanisms exist to cancel the toxic effect of heme. Serum haptoglobin (Hp) binds free Hb efficiently [21, 22] and the Hb:Hp complex is promptly taken up by phagocytes and hepatocytes which express the receptor for the complex, now known as CD163 [23–25]. HO-1 constitutes one of the three

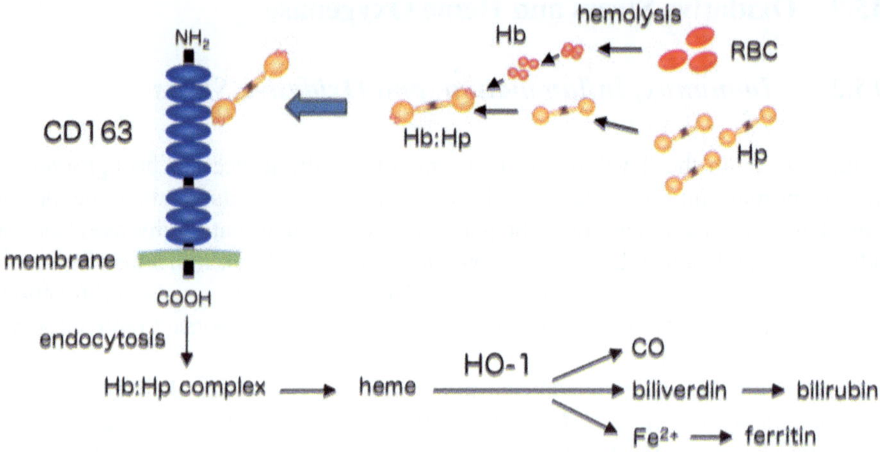

Fig. 15.2 Heme, CD163 and heme metabolism. Free hemoglobin (Hb) binds haptoglobin (Hp) and forms the Hb:Hp complex, which is rapidly taken up by the receptor, CD163. Heme derived from Hb is degraded into carbon monoxide (CO), Fe^{2+} and biliverdin by the action of heme oxygenase (HO). Fe^{2+} induces ferritin production and biliverdin is reduced to bilirubin by biliverdin reductase. CO, ferritin, and bilirubin all exert anti-oxidative roles. *RBC* red blood cells

isozymes of HO which catalyze the degradation of heme into biliverdin, CO, and free iron (Fig. 15.2) [26, 27].

Recent experimental data indicate that one of the heme degradation products, CO, acts on cellular metabolism to protect cells from oxidative stress and regulates production of inflammatory molecules [28–30]. CO directly controls the inflammatory state of a given tissue and at the same time, regulates the level of microcirculation within the target organs acting as a gaseous vasodilator [31, 32]. Among the three isozymes, HO-1 is the only protein that is rapidly induced upon stimulation with various oxidative stresses [31, 33]. Therefore, any defect in its function may lead to uncontrollable inflammation in response to certain exogenous insults, such as infection and hemolysis.

15.3 HO-1 Deficiency

15.3.1 HO-1 Deficiency and Inflammation

We experienced the first case of human HO-1 deficiency and have reported the details of the unique clinical features and laboratory findings of the patient [18, 34]. The patient initially exhibited recurrent fever, erythematous rash, and joint pain at 2 years of age. The triad suggested that the patient suffered from a childhood chronic inflammatory illness, such as systemic juvenile idiopathic arthritis (sJIA) [35], or

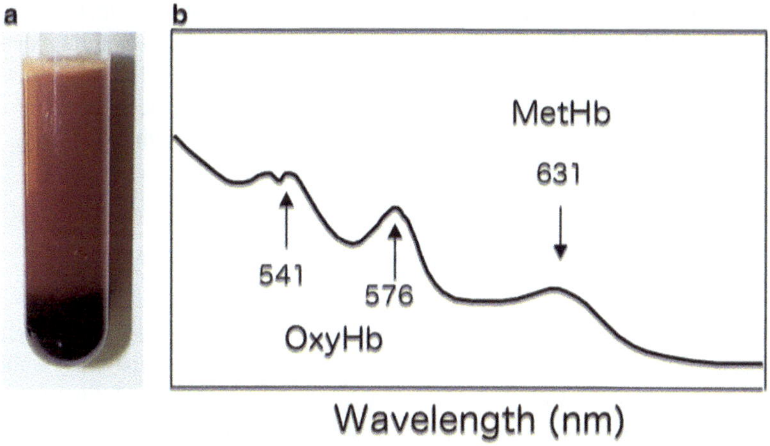

Fig. 15.3 Accumulation of MetHb in HO-1 deficiency. Plasma from the HO-1-deficient patient is characterized with a turbid, *dark brown* appearance, reflecting the massive accumulation of oxyhemoglobin (OxyHb) and MetHb (**a**). Analysis of the patient's serum revealed the 631-nm MetHb peak, in addition to peaks at 541 and 576 nm for OxyHb (**b**)

chronic infantile neurological cutaneous and articular (CINCA) syndrome [36], which is now collectively classified with familial cold urticaria and Muckle-Wells syndrome as cryopyrin-associated periodic syndrome (CAPS) [37]. Both sJIA and CAPS proved unlikely during the course of the illness. Although hepatomegaly was marked and progressive, the spleen was absent. Unlike most of patients with asplenia, this patient did not have any form of congenital heart disease.

There were marked and sustained increases in platelets and leukocytes. Peripheral blood smear showed numerous fragmented erythrocytes and erythroblasts. Both gross appearance of the serum and absorption spectrum analysis indicated that heme in the form of Hb and methemoglobin (MetHb) was markedly increased in the patient's serum (Fig. 15.3a). In contrast to normal fresh hemolysate, which shows two distinct oxyhemoglobin peaks at 541 and 576 nm, the patient's serum showed a third unique peak at 631 nm, which corresponded to MetHb (Fig. 15.3b). These results indicated that either massive hemolysis is constantly taking place in vivo or Hb accumulated in the serum due to a defect in Hb catabolism. Hp concentration was extremely elevated (800–1,200 mg/dL; normally 19–170 mg/dL) and a large amount of the Hb:Hp complex was detected in the patient's urine sample. Repeated measurement of serum bilirubin concentration was always low. Low bilirubin level in the presence of significant intravascular hemolysis suggested an abnormality in the heme degradation pathway.

Immediately before we reported the first human case of HO-1 deficiency, Poss et al. reported monumental reports describing the characteristics of HO-1 deficiency in mice [38, 39]. Their findings share many similarities with our patient, including severe anemia due to defective iron reuse, tissue iron deposits, enhanced cellular injury to oxidative stress, and accelerated inflammation.

Immunohistochemical analysis of the liver biopsy specimen showed that Kupffer cells did not produce HO-1 in the patient's liver. Exposure of monocytes to cadmium or sodium arsenite did not induce HO-1 protein. HO-1 gene analysis revealed that the patient had compound heterozygotes of HO-1 gene mutations. The maternal allele lacked the second exon and the paternal allele showed two base pair deletions within the third exon.

15.3.2 Pathology of HO-1 Deficiency

Pathological examination of the first case of HO-1 deficiency revealed characteristic tissue injury. Notably, cellular injury was not observed ubiquitously, but rather was confined to select organs and cell types, including kidney, liver, circulating monocytes, and vascular endothelial cells. In the kidney, mild mesangial proliferation and thickening of the capillary loop were observed within the glomeruli. Electron microscopy revealed marked swelling of the endothelial cells and their detachment throughout the glomerular capillary. In addition to the glomerular damage, tubulointerstitial injury with tubular atrophy was significant. The liver was massively enlarged and there was a significant amyloid accumulation resulting in marked atrophy of hepatocytes. Scattered foci of iron deposits were observed in both the kidney and liver. Cytoplasm of the circulating monocytes was vacuolated, and monocyte surface antigens were significantly different from normal profiles, as described later in this chapter.

There seem to be several reasons for this selective organ damage in HO-1 deficiency. First of all, these susceptible cells were the targets of constant exposure to oxidative stress, both for anatomical and functional reasons. Vascular endothelial cells are the target of shear stress and are exposed to multiple oxidative stresses, including hemolysis, pH changes, and hypoxemia. In the HO-1-deficient patient, secondary accumulation of heme proteins, cholesterol, and fragmented erythrocytes further aggravated the oxidative stresses. Renal tubular cells are constantly exposed to hematuria, proteinuria, and various other excretory substances in urine. Resident macrophages and circulating monocytes are frequently turned on to deal with scavenger functions. In addition, they function as one of the central players of innate and acquired immunity.

Second, while these cells may be particularly sensitive to oxidative injury, they serve as a high quality sensor of oxidative stress for the organs. Oxidative insults to these susceptible cells rapidly induce anti-oxidative molecules. As a result, the organs are saved from critical damage. Lastly, because these cells have to serve as the oxidative stress sensors, they have the neutralizing capacity to combat these insults. HO-1 may participate as a central molecule in these mechanisms.

In normal rat and human kidneys, renal tubular epithelial cells express HO-1 selectively. Distal tubules express significantly more HO-1 than the proximal tubules in a normal kidney. However, HO-1 expression in proximal tubules is enhanced in patients with hematuria or proteinuria [40, 41]. In vitro renal proximal tubular epithelial cells also produce significant levels of HO-1 upon various oxidative stimuli, indicating that these cells are inherently capable of responding to exogenous

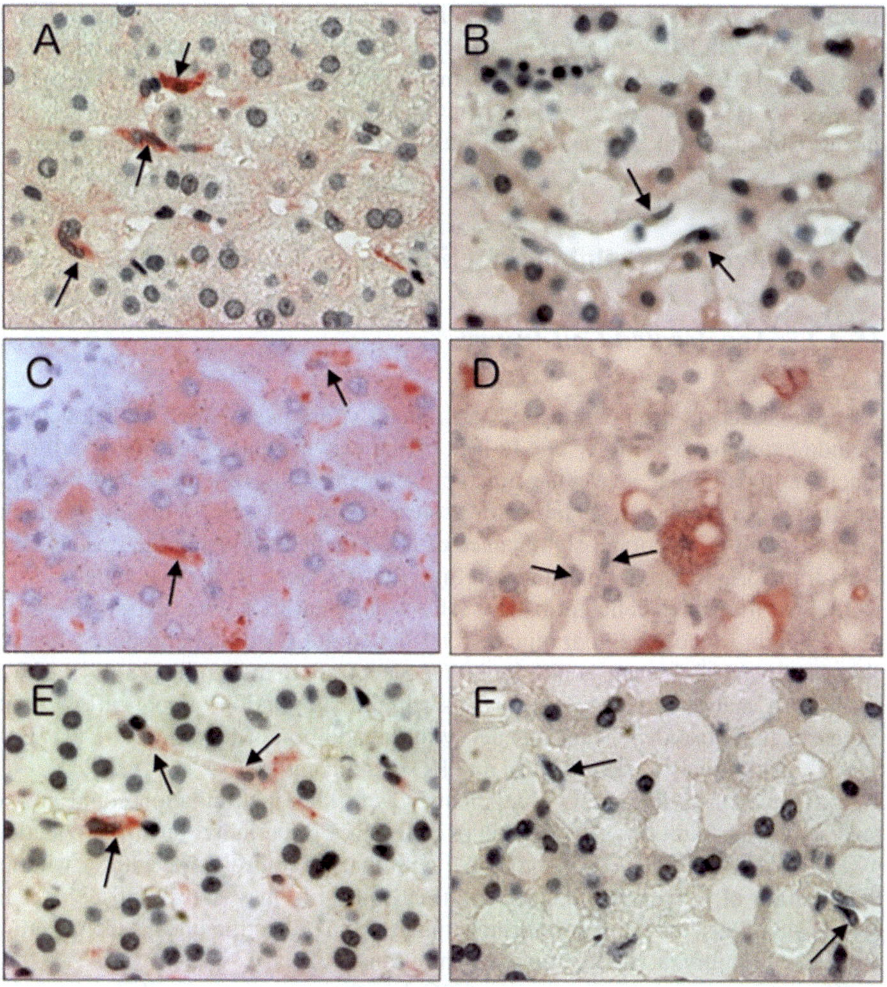

Fig. 15.4 Hepatic expression of HO-1, haptoglobin and CD163 in HO-1 deficiency. Hepatic Kupffer cells normally express significant levels of HO-1 (**a**). In HO-1 deficiency, the hepatic tissue is characterized by massive accumulation of amyloid within the sinusoidal spaces. HO-1 expression is almost completely absent (**b**). Accordingly, haptoglobin (Hp) can be detected in hepatocytes (**c**). Kupffer cells express high levels of Hp (*arrows*). In contrast, Hp is sparsely detected in HO-1-deficient hepatocytes (**d**) and is absent in the Kupffer cells (*arrows*). CD163 is detected on normal Kupffer cells (**e**) but not on HO-1-deficient cells (**f**)

noxious insults [42]. In accordance with these findings, our HO-1-deficient patient showed prominent degeneration of the renal tubular epithelium, with progressive narrowing of the tubular canals [43].

In the patient liver, Kupffer cells were present, but expressed no HO-1 at all, compared to a normal liver which expresses significant levels of HO-1 (Fig. 15.4a, b). In addition, Hp staining was significantly reduced in the patient liver and almost absent in Kupffer cells; Hp was evenly detected in the normal liver, representing the regular uptake of the Hb:Hp complex (Fig. 15.4c, d). Moreover, the receptor for the Hb:Hp complex, CD163, was expressed by normal Kupffer cells but not in the patient liver (Fig. 15.4e, f). Taken together, the patient had progressive accumulation of amyloid and atrophy of hepatocytes, associated with loss of the normal liver architecture.

15.4 Defects in Cellular Functions in HO-1 Deficiency

15.4.1 Accelerated Cell Injury

As summarized in the previous sections, direct consequences of HO-1 deficiency include extensive cell injury due to lack of the enzyme and induction of cell dysfunction, in particular, scavenger functions of macrophages. As we cannot extrapolate that these are the universal features observed in every HO-1-deficient patient, we analyzed the functional significance of HO-1 deficiency using the Epstein Barr virus-transformed lymphoblastoid cell line (LCL) derived from the patient.

LCL of the HO-1-deficient patient (HO-1-deficient LCL) was extremely sensitive to hemin-induced cellular injury. HO-1-deficient LCL did not produce a detectable level of HO-1 even at the highest concentration of hemin [18, 44]; HO-1 production can be detected at low levels without stimulation in control LCL, and increases in a dose-dependent manner with the addition of hemin. HO-2 was produced constitutively in both control and HO-1-deficient LCLs.

Extreme sensitivity of HO-1-deficient LCL to hemin stimulation could not be reversed by the addition of apoferritin or bilirubin. Furthermore, ferritin production by HO-1-deficient LCL was comparable with that by control LCL with or without the addition of hemin. Although ferritin and bilirubin may act as anti-oxidant in certain situations, they do not contribute much to the protection of cells from hemin-induced cell injury, at high hemin concentrations.

We next examined the direct role of HO-1 in the protection of cells from hemin-induced cell injury by transfecting the HO-1 gene into HO-1-deficient LCL using a retrovirus vector. When LCL transfected with pGCsamENHO-1 was cultured with different concentrations of hemin, significant inhibition of cellular injury was observed; LCL transfected with control vector was injured dose-dependently, as found for the patient's original LCL. These results support the notion that degradation of heme by HO-1 is directly responsible for the reversal of hemin-induced cellular injury.

15.4.2 Phagocyte Dysfunction

In addition to cellular injury to the resident macrophages, the patient exhibited peculiar findings indicating a disturbance of macrophage scavenger function. Asplenia was certainly a significant contribution to the reduced scavenging function in the patient, resulting in the overload by other reticuloendothelial systems, including circulating macrophages and the hepatic Kupffer cells. The patient's peripheral blood monocytes exhibited some morphological characteristics, including prominent vacuolation and basophilic cytoplasm. These changes may reflect a persistent, systemic inflammatory reaction. In addition to morphological changes, the surface antigen expression by these monocytes was abnormal, i.e., expressions of HLA-DR and CD36 were significantly reduced as compared with normal monocytes [45]. CD14 expression was comparable to that in the control. Unfortunately, CD163 expression could not be examined at that time. The reduction in the antigen expression was constantly observed after repeated examinations, indicating that these changes reflect abnormal monocyte functions in the patient.

To determine whether the exposure to a high concentration of heme induces downregulation of monocyte antigens, isolated monocytes from peripheral blood were cultured for several hours in the presence of variable concentrations of hemin. Expressions of CD36, HLA-DR, CD16, and CD11b were rapidly downregulated after culturing, whereas CD45 expression showed no change [44]; these surface antigens are all thought to be involved in the receptor-mediated phagocytosis by monocytes. Consistent with the changes in the surface molecules, monocyte phagocytic functions were also impaired dramatically. Although phagocytosis of fluorescence-labeled latex beads did not change significantly, that of opsonized erythrocytes was almost completely abolished. These results indicate that hemin exposure and the subsequent reduction of surface molecules are directly related to the abolishment of receptor-mediated phagocytosis of opsonized erythrocytes by monocytes. Various hematological and biochemical abnormalities seen in the HO-1-deficient patient, including increased Hp concentration, abundance of fragmented erythrocytes and thrombocytosis, and hyperlipidemia, may all be explained by the reduced scavenging functions of phagocytes, which is secondary to heme exposure. Macrophages are known to play pivotal roles in the pathogenesis of atherosclerotic lesions in the presence of major risk factors, including hypercholesterolemia and endothelial injury. Dysfunction of macrophages may also explain why the patient exhibited paradoxically little atherosclerotic changes in the aorta in the presence of endothelial injury and sustained hypercholesterolemia.

A recent report on HO-1−/− mice by Kovtunovych et al. clearly showed that macrophages in these mice died upon exposure to heme [46]. In contrast to splenomegaly seen in HO-1−/− reported by Poss et al., the spleen in these mice underwent progressive atrophy as the mice became aged. Reduction of the Hb:Hp scavenger receptor by macrophages was also noted, further indicating the dysfunction of macrophages as HO-1 producer.

15.4.3 Endothelial Cell Dysfunction

Another distinct feature of the first human HO-1 deficiency case was the defective endothelial function, as represented by extremely abnormal parameters of coagulation/fibrinolysis. Unlike in cases of other hematological illnesses associated with

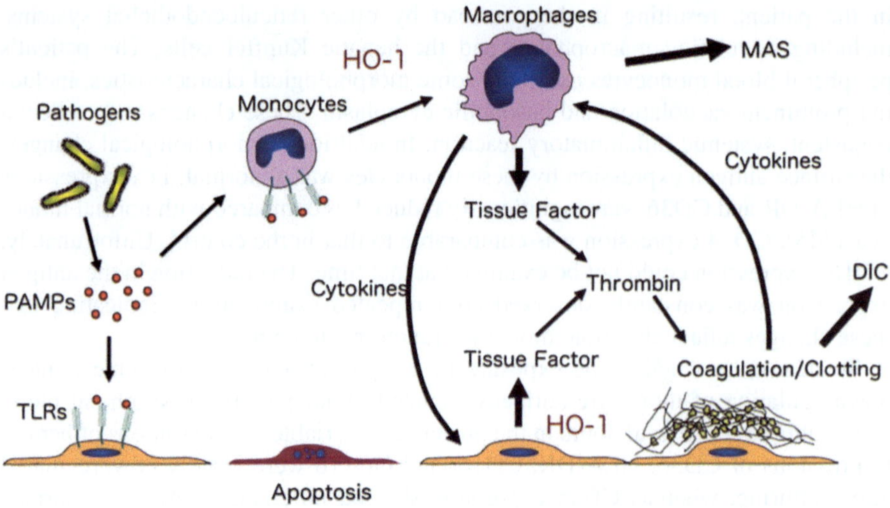

Fig. 15.5 Macrophage activation and endothelial dysfunction in HO-1 deficiency. Pathogens trigger macrophage activation and endothelial cell injury through toll-like receptors (TLRs). Activated macrophages produce various inflammatory cytokines and the endothelial cells produce tissue factor in response to macrophage-derived inflammatory cytokines. In the absence of HO-1, these seemingly protective processes become excessive, leading to cytokinemia and apoptosis and dysfunction of the endothelial cells. Macrophage activation syndrome and disseminated intravascular coagulation may result. *MAS* macrophage activation syndrome, *PAMPs* pathogen-associated molecular patterns

disseminated intravascular coagulation, our patient exhibited extraordinarily elevated values for thrombin–antithrombin complex (TAT), fibrinogen degradation product (FDP), and plasmin-α_2 plasmin inhibitor complex (PIC). Paradoxically, the platelet numbers constantly increased. The data indicated that HO-1 or HO-1 products, such as CO, may be associated with the regulation of the coagulation/fibrinolytic system.

We recently demonstrated in in vitro cultures that a CO-releasing molecule suppressed TNF-α-induced upregulation of tissue factor (TF) and plasminogen activator inhibitor type-1 by human umbilical vein endothelial cells. It also suppressed mitogen-activated protein kinases (MAPKs) and NK-κB signaling pathway activation by TNF-α. Lipopolysaccharide (LPS)-induced TNF-α production by circulating mononuclear cells was also significantly inhibited by the CO-releasing molecules [47]. These results may explain the characteristic findings seen in the HO-1-deficient patient. At the same time, the data support the view that CO-releasing molecules may constitute a novel anti-coagulative and anti-inflammatory therapy [48, 49].

A summary of the macrophage activation and endothelial cell dysfunction is shown in Fig. 15.5. Lack of HO-1 resulted in unregulated activation of macrophages with excess inflammatory cytokine release. At the same time, HO-1 deficiency resulted in overproduction of tissue factor by endothelial cells, leading to the abnormal activation of the coagulation/fibrinolysis system. The figure illustrates the role of HO-1 as an inhibitor of cytokine overproduction and endothelial cell dysfunction associated with catastrophic tissue injury seen in systemic inflammatory response syndrome.

15.5 What We Learned from HO-1 Deficiency

15.5.1 Common Denominators of Human HO-1 Deficiency

Since our report of the first HO-1-deficient patient, there have been five additional patients with HO-1 deficiency identified [50, 51 and personal communications]. All five cases were from India and had an identical homozygous nonsense mutation, indicating the presence of a founder effect in this country. We do not know the exact reason why there have been very few patients reported so far. It is possible that most patients with HO-1 deficiency have significant organ dysfunctions shortly after birth and die undiagnosed. However, most of the patients from India maintained a relatively healthy life until the onset of the inflammatory illness. Another possibility is that HO-1 deficiency is fatal in utero and most HO-1-deficient fetuses die before birth. The mouse model of HO-1 deficiency suggests the latter possibility.

Although the age of onset among the six HO-1 deficiency cases varied, ranging from infancy to 15 years of age, laboratory data and clinical profiles were surprisingly uniform. Fever, hemolytic anemia, and hematuria/proteinuria were consistent findings. In all cases, bilirubin remained within normal range, while serum ferritin and LDH values were invariably high. Notably, the absence or hypoplasia of the spleen seems to be the hallmark of the illness, although its significance in patients with HO-1 deficiency has not been determined.

One hypothesis is that the presence of the spleen may aggravate the destruction of circulating blood cells thereby leading to fatal hemolysis and thrombocytopenia. In this scenario, splenic dysfunction helps the survival of damaged, but still functioning blood cells and prevents the fatal outcome.

The second possibility is that HO-1 deficiency and subsequent vascular endothelial injury results in early (most likely in utero) vascular damage involving the splenic artery. Due to accumulating injured blood cells, the patients' spleens are overworked. The continuing splenic overwork may lead to the enhanced oxidative damage to the splenic arteries which are already vulnerable to oxidative stress. It is intriguing in this respect that hypomorphic mutation of HO-1 in mice resulted in the progressive atrophy of the spleen due to fibrotic changes in the splenic artery. The report by Kovtunovych et al. supports the latter view [47]. These assumptions will remain as such until cases of HO-1 deficiency with intact splenic function are discovered.

15.5.2 *Alternative Macrophage Activation*

Monocytes/macrophages are composed of at least two functionally distinct subsets, M1 and M2 [52–54]. The different subsets of the monocyte/macrophage lineage differentiate in response to environmental stimuli. M1 macrophages are the "classical" macrophages, and they comprise the pro-inflammatory subset, whereas M2 macrophages are "alternatively" activated macrophages. They resolve inflammatory responses, perform scavenger functions, and promote tissue remodeling and repair. Interferon (IFN)-γ is the key cytokine driving the M1 pathway, whereas IL-4, IL-10, and steroids promote monocyte differentiation into M2 macrophages [55, 56].

We reported previously that circulating monocytes produce significant levels of HO-1 during Kawasaki disease and infectious diseases, suggesting their role in anti-inflammation during inflammatory illnesses [57]. Furthermore, we investigated the profiles of cytokine mRNA expression in two subsets of circulating monocytes [58]. In this study, freshly isolated $CD16^{high}/CCR2^{negative}$ monocytes expressed significant levels of HO-1 mRNA in vivo. They produced little IL-10 upon stimulation with LPS. In contrast, the major subset of $CD16^{low}/CCR2^{positive}$ monocytes did not express HO-1 mRNA in vivo, whereas they responded significantly to LPS and produced IL-10. The fractions of $CD16^{high}/CCR2^{negative}$ monocytes increased during various acute inflammatory diseases, such as Kawasaki disease and influenza virus infection, suggesting the anti-inflammatory roles played by monocytes through HO-1 production.

In macrophages and dendritic cells, CO reduces pro-inflammatory and increases anti-inflammatory cytokine secretion in response to LPS [59, 60]. HO-1-mediated anti-inflammatory effects may therefore be closely linked to anti-inflammatory mechanisms, such as the suppression of the immune and inflammatory responses in macrophages via diminished antigen-presenting capacity and cytokine synthesis [61, 62]. This is consistent with our finding in the HO-1-deficient patient, in whom the lack of HO-1 resulted in a marked rise in circulating heme and subsequent oxidative vascular and tissue injury, anemia, and chronic inflammation.

HO-1 is induced by CD163-mediated Hb:Hp complex uptake [63–65]. Schaer et al. reported that macrophages express upregulated levels of CD163 in sepsis-induced hemophagocytic syndrome [66]. These macrophages expressed significant levels of HO-1, suggesting their role as a negative regulator of inflammation. We demonstrated that serum HO-1 concentration correlates closely with serum sCD163 concentrations, and these are extremely high in the macrophage activation syndrome associated with sJIA [67]. These findings indicate that serum HO-1 might be derived from $CD163^{+}$ alternatively activated macrophages, and in particular, hemophagocytic macrophages may be a major source of HO-1 in sJIA. Increased levels of serum HO-1, as well as CD163, suggest that alternative activation of macrophages is switched on in sJIA.

15.6 Paradigm Shift of Anti-inflammatory Therapy

For many years, various anti-inflammatory agents, including steroids, non-steroidal anti-inflammatory drugs (NSAIDs), and immunosuppressants have been the mainstream drugs for controlling excessive inflammation and tissue/organ injury. Although these agents are effective in some cases, conventional anti-inflammatory/immunosuppressive therapy has resulted in therapeutic failure for two main reasons. One important and critical reason is that inflammation and immune reactions are the critical components of the host response to pathogens and the insults, even though they can cause extensive tissue injury and organ damage, together with much discomfort to the suffering patient. Non-judicious use of the conventional drugs may result in the host failure to expel the pathogens and succumb to overwhelming infection or uncontrollable cancer growth. Another reason is the fact that when the tissue and organs have already had significant damage and injury after the pathogen attacks and subsequent host immune responses, the addition of anti-inflammatory or immunosuppressive agents may further aggravate the host status. Based on the understanding of the built-in mechanism to protect cells and organs from oxidative injury, I propose the third alternative of anti-inflammatory therapy for the promotion of anti-oxidative functions of each cell.

The prime target of such therapy is HO-1. Pharmacological induction of cellular HO-1 production, use of CO-releasing molecules, or enhancement of Hb:Hp receptor expression by steroids can be examples of these approaches. Further studies will reveal a novel therapy for chronic recalcitrant inflammatory illnesses. We need to develop the optimum therapy for the novel, and still rare, systemic inflammatory illness, HO-1 deficiency.

Acknowledgements The author would like to express his sincere thanks to all the collaborators for the past works. Special thanks are given to the H. Matsukawa and S. Kouraba for their excellent technical assistance.

References

1. Kawai T, Akira S (2011) The role of pattern-recognition receptors in innate immunity: update on Toll-like receptors. Nat Immunol 11:373–384
2. Gill R, Tsung A, Billiar T (2010) Linking oxidative stress and to inflammation: Toll-like receptors. Free Radic Biol Med 48:1121–1132
3. Auerbach A, Hernandez ML (2012) The effect of environmental oxidative stress on airway inflammation. Curr Opin Allergy Clin Immunol 12:133–139
4. Rubartelli A, Lotze MT (2007) Inside, outside, upside down: damage-associated molecular pattern molecules (DAMPs) and redox. Trends Immunol 28:429–436
5. Mathis D, Benoist C (2009) Aire. Annu Rev Immunol 27:287–312
6. Ziegler SF (2006) FOXP3: of mice and men. Annu Rev Immunol 24:209–226
7. Jiang H, Chess L (2006) Regulation of immune responses by T cells. N Engl J Med 354: 1166–1176

8. Bacchetta R, Gambineri E, Roncarolo MG (2007) Role of regulatory T cells and FOXP3 in human diseases. J Allergy Clin Immunol 120:227–235
9. Davies KJA (2000) Oxidative stress, antioxidant defenses, and damage removal, repair, and replacement systems. IUBMB Life 50:279–289
10. Hybertson BM, Gao B, Bose SK, McCord JM (2011) Oxidative stress in health and disease: the therapeutic potential of Nrf2 activation. Mol Aspects Med 32:234–246
11. Rushworth SA, Ogborne RM, Charalambos CA, O'Connell MA (2006) Role of protein kinase C delta in curcumin-induced antioxidant response element-mediated gene expression in human monocytes. Biochem Biophys Res Commun 341:1007–1016
12. Paine A, Eiz-Vesper B, Blasczyk R, Immenschuh S (2010) Signaling to heme oxygenase-1 and its anti-inflammatory therapeutic potential. Biochem Pharmacol 80:1895–1903
13. Michels AW, Gottlieb PA (2010) Autoimmune polyglandular syndromes. Nat Rev 6:270–277
14. Teachey DT, Seif AE, Grupp SA (2009) Advances in the management and understanding of autoimmune lymphoproliferative syndrome (ALPS). Br J Haematol 148:205–216
15. Lenardo MJ, Oliveira JB, Zheng L, Rao VK (2010) ALPS—ten lessons from an international workshop on a genetic disease of apoptosis. Immunity 32:291–295
16. Sakaguchi S, Ono M, Setoguchi R, Yagi H, Hori S, Fehervari Z, Shimizu J, Takahashi T, Nomura T (2006) Foxp3^{+} CD25^{+} CD4^{+} natural regulatory T cells in dominant self-tolerance and autoimmune disease. Immunol Rev 212:8–27
17. Birben E, Sahiner UM, Sackesen C, Erzurum S, Kalayci O (2012) Oxidative stress and antioxidant defense. World Allergy Organ J 5:9–19
18. Yachie A, Niida Y, Wada T, Igarashi N, Kaneda H, Toma T, Ohta K, Kasahara Y, Koizumi S (1999) Oxidative stress causes enhanced endothelial cell injury in human heme oxygenase-1 deficiency. J Clin Invest 103:129–135
19. Jeney V, Balla J, Yachie A, Varga Z, Vercellotti GM, Eaton JW, Balla G (2002) Pro-oxidant and cytotoxic effects of circulating heme. Blood 100:879–887
20. Balla J, Vercellotti GM, Jeney V, Yachie A, Varga Z, Jacob HS, Eaton JW, Balla G (2007) Heme, heme oxygenase, and ferritin: how the vascular endothelium survives (and dies) in an iron-rich environment. Antioxid Redox Signal 9:2119–2137
21. Nagel RL, Gibson QH (1971) The binding of hemoglobin to haptoglobin and its relation to subunit dissociation to hemoglobin. J Biol Chem 246:69–73
22. McCormick DJ, Atassi MZ (1990) Hemoglobin binding with haptoglobin: delineation of the haptoglobin binding site on the alpha-chain of human hemoglobin. J Protein Chem 9: 735–742
23. Kristiansen M, Graversen JH, Jacobsen C, Sonne O, Hoffman HJ, Law SK, Moestrup SK (2001) Identification of the haemoglobin scavenger receptor. Nature 409:198–201
24. Madsen M, Graversen JH, Moestrup SK (2001) Haptoglobin and CD163: captor and receptor gating hemoglobin to macrophage lysosomes. Redox Rep 6:386–388
25. Madsen M, Møller HJ, Nielsen MJ, Jacobsen C, Graversen JH, van den Berg T, Moestrup SK (2004) Molecular characterization of the haptoglobin-hemoglobin receptor CD163. J Biol Chem 279:51561–51567
26. Maines MD (1988) Heme oxygenase: function, multiplicity, regulatory mechanisms, and clinical application. FASEB J 2:2557–2568
27. Nath KA (2006) Heme oxygenase-1: a provenance for cytoprotective pathways in the kidney and other tissues. Kidney Int 70:432–443
28. Ryter SW, Alam J, Choi AM (2006) Heme oxygenase/carbon monoxide: from basic science to therapeutic applications. Physiol Rev 86:583–650
29. Nakahira K, Kim HP, Geng XH, Nakao A, Wang X, Murase N, Drain PF, Wang X, Sasidhar M, Nabel EG, Takahashi T, Lukacs NW, Ryter SW, Morita K, Choi AM (2006) Carbon monoxide differentially inhibits TLR signaling by regulating ROS-induced trafficking of TLRs to lipid rafts. J Exp Med 203:2377–2389
30. Amano MT, Camara NO (2013) The immunomodulatory role of carbon monoxide during transplantation. Med Gas Res 3:1

31. Maines MD (1997) The heme oxygenase system: a regulator of second messenger gases. Annu Rev Pharmacol Toxicol 37:517–554
32. Leffler CW, Parfenova H, Jaggar JH (2011) Carbon monoxide as an endogenous vascular modulator. Am J Physiol Heart Circ Physiol 301:H1–H11
33. Choi AM, Alam J (1996) Heme oxygenase-1: function, regulation, and implication of a novel stress-inducible protein in oxidant-induced lung injury. Am J Respir Cell Mol Biol 15:9–19
34. Kawashima A, Oda Y, Yachie A, Koizumi S, Nakanishi I (2002) Heme oxygenase-1 deficiency: the first autopsy case. Hum Pathol 33:125–130
35. Martini A (2012) Systemic juvenile idiopathic arthritis. Autoimmun Rev 12:56–59
36. Feldmann J, Prieur AM, Quartier P, Berquin P, Certain S, Cortis E, Teillac-Hamel D, Fischer A, de Saint BG (2002) Chronic infantile neurological cutaneous and articular syndrome is caused by mutations in CIAS1, a gene highly expressed in polymorphonuclear cells and chondrocytes. Am J Hum Genet 71:198–203
37. Chang C (2013) The pathogenesis of neonatal autoimmune and autoinflammatory diseases: a comprehensive review. J Autoimmun 41:100–110
38. Poss KD, Tonegawa S (1997) Heme oxygenase-1 is required for mammalian iron reutilization. Proc Natl Acad Sci U S A 94:10919–10924
39. Poss KD, Tonegawa S (1997) Reduced stress defense in heme oxygenase-1 deficient cells. Proc Natl Acad Sci U S A 94:10925–10930
40. Morimoto K, Ohta K, Yachie A, Yang Y, Shimizu M, Goto C, Toma T, Kasahara Y, Yokoyama H, Miyata T, Seki H, Koizumi S (2001) Cytoprotective role of heme oxygenase (HO)-1 in human kidney with various renal diseases. Kidney Int 60:1858–1866
41. Shimizu M, Ohta K, Yang Y, Nakai A, Toma T, Saikawa Y, Kasahara Y, Yachie A, Yokoyama H, Seki H, Koizumi S (2005) Glomerular proteinuria induces heme oxygenase-1 gene expression within renal epithelial cells. Pediatr Res 58:666–671
42. Yang Y, Ohta K, Shimizu M, Morimoto K, Goto C, Nakai A, Toma T, Kasahara Y, Yachie A, Seki H, Koizumi S (2003) Selective protection of renal tubular epithelial cells by heme oxygenase (HO)-1 during stress-induced injury. Kidney Int 64:1302–1309
43. Ohta K, Yachie A, Fujimoto K, Kaneda H, Wada T, Toma T, Seno A, Kasahara Y, Yokoyama H, Seki H, Koizumi S (2000) Tubular injury as a cardinal pathologic feature in human heme oxygenase-1 deficiency. Am J Kidney Dis 35:863–870
44. Yachie A, Kawashima A, Ohta K, Saikawa Y, Koizumi S (2002) Human HO-1 deficiency and cardiovascular dysfunction. In: Wang R (ed) Carbon monoxide and cardiovascular functions. CRC Press, Boca Raton
45. Yachie A, Toma T, Shimura S, Yue L, Morimoto K, Maruhashi K, Niida Y, Ohta K, Kasahara Y, Saikawa Y, Koizumi S (2002) Human HO-1 deficiency and the oxidative injury of vascular endothelial cells. In: Abraham NG (ed) Heme oxygenase in biology and medicine. Kluwer Academic/Plenum, New York
46. Kovtunovych G, Eckhaus MA, Ghosh MC, Ollivierre-Wilson H, Rouault TA (2010) Dysfunction of the heme recycling system in heme oxygenase 1-deficient mice: effects on macrophage viability and tissue iron distribution. Blood 116:6054–6062
47. Maruyama K, Morishita E, Yuno T, Sekiya A, Asakura H, Ohtake S, Yachie A (2012) Carbon monoxide (CO)-releasing molecule-derived CO regulates tissue factor and plasminogen activator inhibitor type 1 in human endothelial cells. Throm Res 130:e188–e193
48. Motterlini R, Clark JE, Foresti R, Sarathchandra P, Mann BE, Green CJ (2002) Carbon monoxide-releasing molecules: characterization of biochemical and vascular activities. Circ Res 90:E17–E24
49. Motterlini R, Otterbein LE (2010) The therapeutic potential of carbon monoxide. Nat Rev Drug Discov 9:728–743
50. Radhakrishnan N, Yadav SP, Sachdeva A, Pruthi PK, Sawhney S, Piplani T, Wada T, Yachie A (2011) Human heme oxygenase-1 deficiency presenting with hemolysis, nephritis, and asplenia. J Pediatr Hematol Oncol 33:74–78
51. Radhakrishnan N, Yadav SP, Sachdeva A, Wada T, Yachie A (2011) An interesting tetrad of asplenia, inflammation, hemolysis, and nephritis. Pediatr Hematol Oncol 28:723–726

52. Geissmann F, Jung S, Littman DR (2003) Blood monocytes consist of two principal subsets with distinct migratory properties. Immunity 19:71–82
53. Tacke F, Randolph GJ (2006) Migratory fate and differentiation of blood monocyte subsets. Immunobiology 211:609–618
54. Sica A, Mantovani A (2012) Macrophage plasticity and polarization: in vivo veritas. J Clin Invest 122:787–795
55. Martinez FO, Sica A, Mantovani A, Locati M (2008) Macrophage activation and polarization. Front Biosci 13:453–461
56. Auffray C, Sieweke MH, Geismann F (2009) Blood monocytes: development, heterogeneity and relationship with dendritic cells. Annu Rev Immunol 27:669–692
57. Yachie A, Toma T, Mizuno K, Okamoto H, Shimura S, Ohta K, Kasahara Y, Koizumi S (2003) Heme oxygenase-1 production by peripheral blood monocytes during acute inflammatory illnesses of children. Exp Biol Med 228:550–556
58. Mizuno K, Toma T, Tsukiji H, Okamoto H, Yamazaki H, Ohta K, Ohta K, Kasahara Y, Koizumi S, Yachie A (2005) Selective expansion of $CD16^{high}$ $CCR2^{-}$ subpopulation of circulating monocytes with preferential production of haem oxygenase (HO)-1 in response to acute inflammation. Clin Exp Immunol 142:461–470
59. Lee T-S, Chau L-Y (2002) Heme oxygenase-1 mediates the anti-inflammatory effect of interleukin-10 in mice. Nat Med 8:240–246
60. Chauveau C, Rémy S, Royer PJ, Hill M, Tanguy-Royer S, Hubert FX, Tesson L, Brion R, Beriou G, Gregoire M, Josien R, Cuturi MC, Anegon I (2005) Heme oxygenase-1 expression inhibits dendritic cell maturation and proinflammatory function but conserves IL-10 expression. Blood 106:1694–1702
61. Listopad J, Asadullah K, Sievers C, Ritter T, Meisel C, Sabat R, Döcke WD (2007) Heme oxygenase-1 inhibits T cell-dependent skin inflammation and differentiation and function of antigen-presenting cells. Exp Dermatol 16:661–670
62. Kotsch K, Martins PN, Klemz R, Janssen U, Gerstmayer B, Dernier A, Reutzel-Selke A, Kuckelkorn U, Tullius SG, Volk HD (2007) Heme oxygenase-1 ameliorates ischemia/reperfusion injury by targeting dendritic cell maturation and migration. Antioxid Redox Signal 9: 2049–2063
63. Abraham NG, Drummond G (2006) CD163-mediated hemoglobin-heme uptake activates macrophage HO-1, providing an anti-inflammatory function. Circ Res 99:911–914
64. Schaer CA, Schoedon G, Imhof A, Kurrer MO, Schaer DJ (2006) Constitutive endocytosis of CD163 mediates hemoglobin-heme uptake and determines the noninflammatory and protective transcriptional response of macrophages to hemoglobin. Circ Res 99:943–950
65. Yamazaki H, Ohta K, Tsukiji H, Toma T, Hashida Y, Ishizaki A, Saito T, Arai S, Koizumi S, Yachie A (2007) Corticosteroid enhances heme oxygenase-1 production by circulating monocytes by up-regulating hemoglobin scavenger receptor and amplifying the receptor-mediated uptake of hemoglobin-haptoglobin complex. Biochem Biophys Res Commun 358:506–512
66. Schaer DJ, Schaer CA, Schoedon G, Imhof A, Kurrer MO (2006) Hemophagocytic macrophages constitute a major compartment of heme oxygenase expression in sepsis. Eur J Haematol 77:432–436
67. Shimizu M, Yachie A (2012) Compensated inflammation in systemic juvenile idiopathic arthritis: role of alternatively activated macrophages. Cytokine 60:226–232

Chapter 16
Oxidative Stress in Kawasaki Disease

Tomoyo Yahata and Kenji Hamaoka

Abbreviations

BAP	Biological antioxidant potential
CRP	C-reactive protein
EDRF	Endothelium-derived relaxing factor
eNOS	Endothelial NO synthase
%FMD	Endothelium-dependent vasodilation response
IL-6	Interleukin-6
iNOS	Inducible NO synthase
IVIG	Intravenous immunoglobulin
KD	Kawasaki disease
NADH/NADPH	Nicotinamide adenine dinucleotide/nicotinamide adenine dinucleotide phosphate
NE	Neutrophil elastase
NF-κB	Nuclear factor kappa B
NO	Nitric oxide
NOx	Nitrogen oxides
ROM	Reactive oxygen metabolites
ROS	Reactive oxygen species
TAT	Thrombin-antithrombin complex
TNF-α	Tumor necrosis factor-α

T. Yahata • K. Hamaoka, M.D., Ph.D. (✉)
Department of Pediatric Cardiology and Nephrology, Graduate School of Medical Science, Kyoto Prefectural University of Medicine, 465 Kajicho, Kawaramachi-Hirokoji, Kamigyo-ku, Kyoto 602-8566, Japan
e-mail: khamaoka@koto.kpu-m.ac.jp

H. Tsukahara and K. Kaneko (eds.), *Studies on Pediatric Disorders*, Oxidative Stress in Applied Basic Research and Clinical Practice, DOI 10.1007/978-1-4939-0679-6_16,

16.1 Introduction

Kawasaki disease (KD), a condition occurring most frequently in infants, is characterized by systemic vasculitis of small and medium arteries [1]. The underlying mechanisms by which KD triggers vasculitis remain unclear despite considerable research. However, certain stimuli, principally infections, which generate and release various inflammatory substances within the body, are known to trigger vasculitis, causing panangiitis through complex interactions.

When exposed to inflammatory substances, the vascular endothelial cells partially contract, forming intercellular gaps. Plasma proteins and other substances that are normally contained within the blood vessels exude from these gaps, forming chemotactic factors. Monocytes and macrophages then infiltrate the site of inflammation, are activated, and produce reactive oxygen species (ROS) [2].

The human body has evolved a redox system for maintaining homeostasis in response to various external stresses, and this ROS scavenging system counteracts the ROS production system. A healthy individual maintains a proper balance between generating and scavenging systems. If, however, the balance tilts toward generating, excess ROS are not scavenged but instead accumulate, leading to oxidative stress [3, 4] (Fig. 16.1). Excess ROS attack proteins, carbohydrates, lipids, nucleic acids, and other constituent substances of the body, resulting in cellular

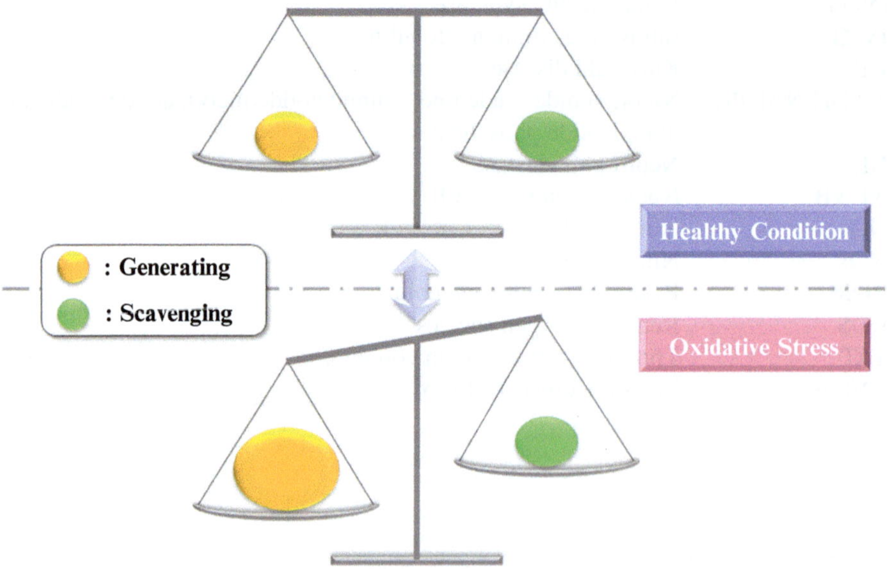

Fig. 16.1 What is oxidative stress? A healthy individual maintains a proper balance between generating and scavenging systems. If, however, the balance tilts toward generating, excess ROS are not scavenged but instead accumulate, leading to oxidative stress

damage that might cause various diseases. In KD specifically, infiltrating inflammatory cells produce excess ROS, and the resulting cytotoxicity is presumed to play a role in the pathogenesis of vasculitis.

16.2 Oxidative Stress and Vasculitis in Acute Phase

16.2.1 Phlogogenous Substances Involved in the Formation of Acute Vasculitis

It is no exaggeration to state that ROS are involved in all proinflammatory pathways. Infiltrating neutrophils and macrophages are the primary sources of ROS produced during inflammation. Activation of NADH/NADPH oxidase in these inflammatory cells causes rapid and abundant production of ROS that can lead to a state of oxidative stress if not removed by the ROS scavenging system. The arachidonic acid cascade, activation of xanthine oxidase in vascular endothelial cells [5], or inhibition of the electron transport system in intracellular mitochondria can also cause overproduction of ROS.

ROS are thought to affect cell function via two mechanisms. The first involves direct cellular impairment, while the other acts as a second messenger for other inflammatory substances such as tumor necrosis factor (TNF)-α.

The nuclear factor (NF)-κB-dependent cytokines, which include TNF-α, interleukin (IL)-6, and other inflammatory cytokines that increase during the acute phase of KD [6–8], are released by infiltrating monocytes/macrophages. NF-κB is a transcription factor activated by a specific inflammatory stimulus as well as by ROS themselves. Once released, TNF-α and IL-6 stimulate C-reactive protein (CRP) genes in the liver and promote inflammatory CRP production. IL-6 also acts independently on vascular smooth muscle to promote the production of CRP, inflammatory cytokines, and chemokines.

Neutrophil elastase (NE), a substance released from activated neutrophils, is also an important player in the acute inflammatory response [9]. NE is a powerful proteolytic enzyme with low substrate specificity. Infiltration of NE-positive neutrophils has been histologically proven by immunostaining in coronary artery lesions.

The neutrophils, macrophages, and vascular smooth muscle cells accumulating at the site of inflammation are also known to transiently produce large amounts of nitric oxide (NO) via inducible NO synthase (iNOS) [10]. NO is an unstable radical that forms peroxynitrite ($ONOO^-$) in the presence of excess ROS. NO initially acts as an endothelium-derived relaxing factor (EDRF). The resistant vessels constrict when the action of NO is lost, and the effect exacerbates the dynamics of blood circulation. In addition, $ONOO^-$ creates highly reactive radicals that cause significant tissue damage, and this substance is also thought to induce edema via promotion of vascular permeability.

In summary, various inflammatory substances, including ROS, are positively correlated, highly damaging to tissue, and thought to be involved in the etiology of acute vasculitis in KD.

16.2.2 Dynamics of Oxidative Stress in the Acute Phase

We measured blood levels of reactive oxygen metabolites (ROM), an index of the ROS generating system, and blood levels of the biological antioxidant potential (BAP), an index of the ROS scavenging system, in patients with acute KD [11] (Fig. 16.2).

The subjects included 19 acute patients with KD, of whom 13 responded favorably to the first course of IVIG treatment (2 g/kg in one dose). The remaining six patients did not respond favorably to the first course of IVIG. Blood ROM and BAP levels were measured in all cases immediately prior to IVIG treatment, immediately after IVIG treatment (24 h after the end of administration), and 2 weeks after the end of IVIG treatment. In the group who responded poorly to the first round of IVIG, treatment with either additional IVIG or steroids resulted in resolution of fever within 1 week. Although the blood ROM levels were clearly elevated immediately prior to IVIG in the group that responded favorably to treatment, the initial course of IVIG caused a favorable decline. In the group that responded poorly, however, no decline was observed after the initial course of IVIG treatment even though their ROM levels immediately before IVIG were similar to those in the group that responded favorably. However, there was a significant drop after 2 weeks. No clear fluctuation in BAP levels was observed before or after IVIG treatment in the group who responded favorably to IVIG, but levels tended to gradually increase 2 weeks after IVIG treatment. In the group who responded poorly, however, no significant change in BAP was noted throughout the entire course. The BAP levels immediately before IVIG treatment were significantly lower in the group that responded poorly vs. the group that responded favorably ($p < 0.01$).

These results suggest that ROS generation is significantly enhanced in acute KD, and drops immediately with IVIG treatment. This finding, along with an observed reduction of inflammation via an independent mechanism, may be useful for determining the therapeutic efficacy. In addition, the slight lag in the ROS scavenging system behind the generating system suggests that the ROS scavenging system has a gradual functional increase triggered by the stimulus of increased ROS. However, it is also possible that the capacity of the scavenging system might regulate responsiveness to IVIG treatment and the subsequent ability to recover from the damage sustained during the acute phase.

16.2.3 Future Prospects for Acute-Phase Treatments

Current acute-phase treatments focus on reducing inflammation, and there is a consensus that IVIG is the first line of defense [12]. The results of the recent large-scale RAISE STUDY trial have also brought the efficacy of steroid administration under review [13]. The growing diversity of acute-phase treatments is illustrated by increasing numbers of clinical reports on the effectiveness of anti-TNF-α antibodies which has no indication for KD [14, 15].

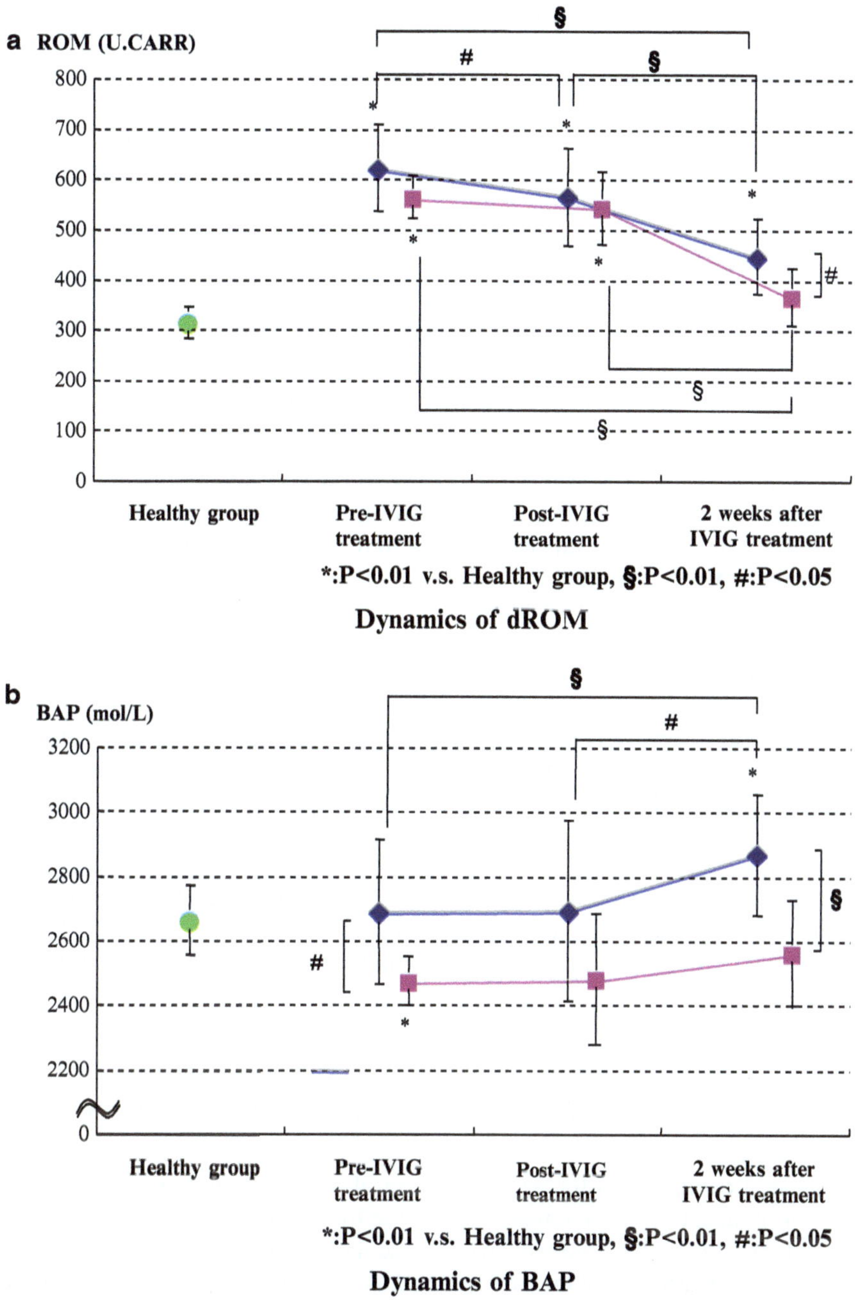

Fig. 16.2 Dynamics of reactive oxygen metabolites and biological antioxidant potential in acute stage. (**a**) Dynamics of dROM. Reproduced from [11] with permission from © 2011 by the Japanese Circulation Society. Changes in dROM level in both treatment groups (*blue*: favorable, *pink*: not favorable) in response to IVIG treatment. Data given as mean ± SD. (**b**) Dynamics of BAP. Reproduced from [11] with permission from © 2011 by the Japanese Circulation Society. Changes in BAP level in both treatment groups (*blue*: favorable, *pink*: not favorable) in response to IVIG treatment. Data given as mean ± SD

In acute KD, coronary artery involvement is regarded as the most serious complication. While this has followed a downward trend in association with the growing diversity of treatments, the problem has not yet been eliminated. The significant role of oxidative stress in the etiology of acute vasculitis, which our research documents, is an important consideration in the future planning of new treatment strategies. Control of oxidative stress, a factor closely related to inflammation, may enable quick alleviation of acute inflammation.

16.3 Oxidative Stress and Vascular Injury in Chronic Phase

16.3.1 Vascular Pathology Following the Acute Phase

Heavily impaired blood vessels do not immediately return to normal even after the acute phase is successfully overcome and severe acute inflammation is alleviated. The body's repair mechanisms work slowly to restore original conditions. The time needed for repair is impacted by various conditions such as the extent of damage, the capabilities of the individual's repair mechanism, and the presence of repair-inhibiting factors. During this repair period, each patient experiences growth, and some approach adulthood.

At present, no consensus exists on whether late-stage residual vascular involvement is present. However, advances in testing methods have revealed evidence of intimal thickening in blood vessels that would have been deemed healthy and normal by conventional testing [16]. In addition, vascular endothelial function testing has recently shown some dysfunction in patients who have a history of KD [17]. There are some reports that the vascular dysfunction may remain to be improved even in chronic phase. Atherosclerotic lesions have also been found in adult KD patients with residual coronary vessel involvement in chronic phase [18]. Taken together, these reports reflect a growing fear in recent years that KD vasculitis leads to arteriosclerosis.

16.3.2 Progression of Atherosclerosis and Oxidative Stress

The general explanations for the mechanism of onset and progression for atherosclerosis include Ross's inflammation response hypothesis [19] and Steinberg's oxidation hypothesis [20]. Namely, when vascular endothelial cells are damaged, various inflammatory cytokines are released. As a result, activated immunocompetent cells adhere to the endothelial cells, then migrate below the endothelium followed by the vascular smooth muscle cells, fibroblasts, and other cells, which results in further endothelial cell impairment. Atherosclerotic changes progress as this vicious circle repeats itself.

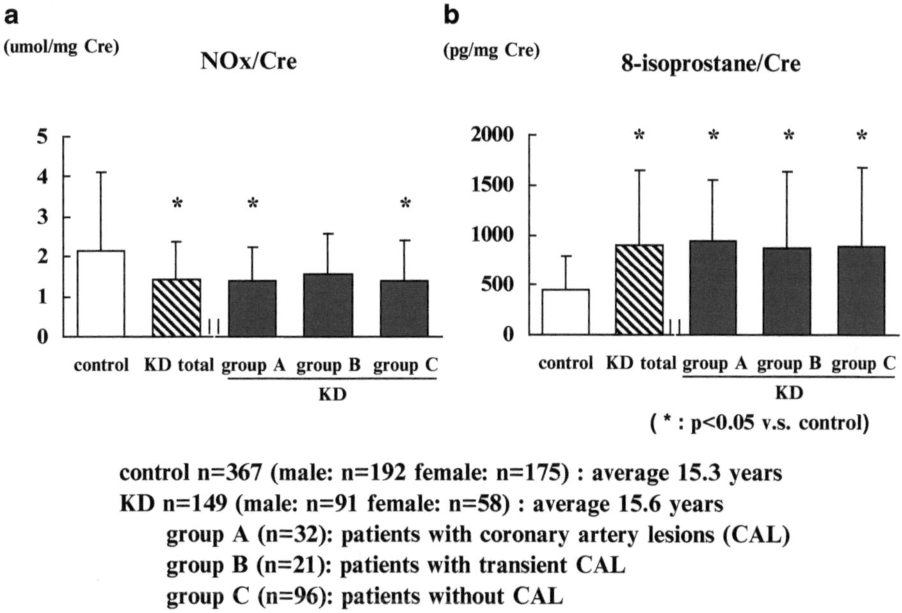

Fig. 16.3 NOx and 8-isoprostane of patients in chronic phase. Reproduced from [21] with permission from © 2006 by Ishiyaku Publishers, Inc., Japan. Data given as mean ± SD. The 8-isoprostane level was notably higher in patients with a history of KD, regardless of whether coronary lesions were present. KD patients also showed a significant reduction in NOx

Attention has recently been focused on the involvement of oxidative stress caused by ROS accumulation in the progression of these atherosclerotic changes. Oxidative stress triggers inflammation by inducing the expression of adhesion factors and chemokines via NF-κB activation in vascular endothelial and smooth muscle cells, promoting the formation of atherosclerosis. Superoxides also weaken the expression of vascular endothelial NO synthase (eNOS) and decrease NO production, thereby triggering not only contraction of the blood vessels but also enhanced platelet aggregation, thrombus formation, and intimal proliferation, factors believed to exacerbate the progression of vascular endothelial impairment.

16.3.3 Oxidative Stress and Impaired Endothelial Function in Chronic Phase

In order to study the relationship between oxidative stress and sustained vascular endothelial cell damage in chronic-phase KD, our group measured urine levels of NOx, a metabolite of nitric oxide, and urine 8-isoprostane levels, a sensitive marker of oxidative stress [21] (Fig. 16.3). The subjects included 149 patients with a history of KD who had reached puberty and a control group of 367 patients who had no

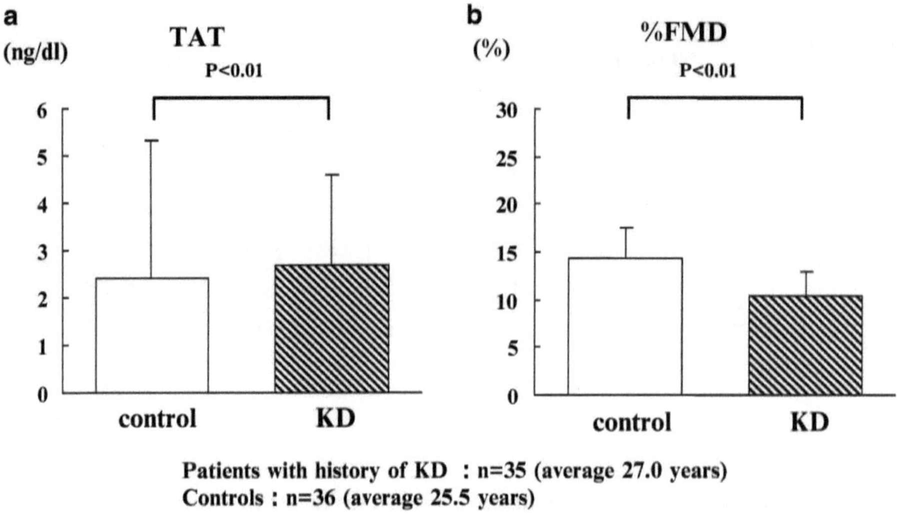

Fig. 16.4 TAT and %FMD of patients in chronic phase. Reproduced from [21] with permission from © 2006 by Ishiyaku Publishers, Inc., Japan. Data given as mean ± SD. The adults with KD showed a clear drop in %FMD compared with 36 patients in a control group. These patients also had a simultaneous clear rise in TAT

coronary risk factors. The 8-isoprostane level was notably higher in patients with a history of KD, regardless of whether coronary lesions were present. KD patients also showed a significant reduction in NOx, a vasodilation-related factor that originates from vascular endothelial cells. These results suggest the possibility that patients with a history of KD have some degree of oxidative stress status even in the chronic phase, which is involved in the appearance and progression of vascular endothelial cell impairment.

Thirty-five patients with a history of KD who had reached adulthood were also assessed for endothelial cell dysfunction [21] (Fig. 16.4). The adults with KD showed a clear drop in endothelium-dependent vasodilation response (%FMD) in forearm arteries compared with 36 patients in a control group. These patients also had a simultaneous clear rise in thrombin-antithrombin complex (TAT), a marker of vascular endothelial cell impairment.

16.3.4 Atherosclerotic Changes in Chronic Phase

Chronic-phase vascular endothelial dysfunction and oxidative stress are serious problems for children with a prior history of KD. As these two factors are positively correlated with each other, it is reasonable to hypothesize that they are both involved in the onset and progression of atherosclerosis.

Atherosclerosis, also called arteriosclerosis, varies depending on the tissue, and some question the role of atherosclerosis development, a risk for coronary events, in KD.

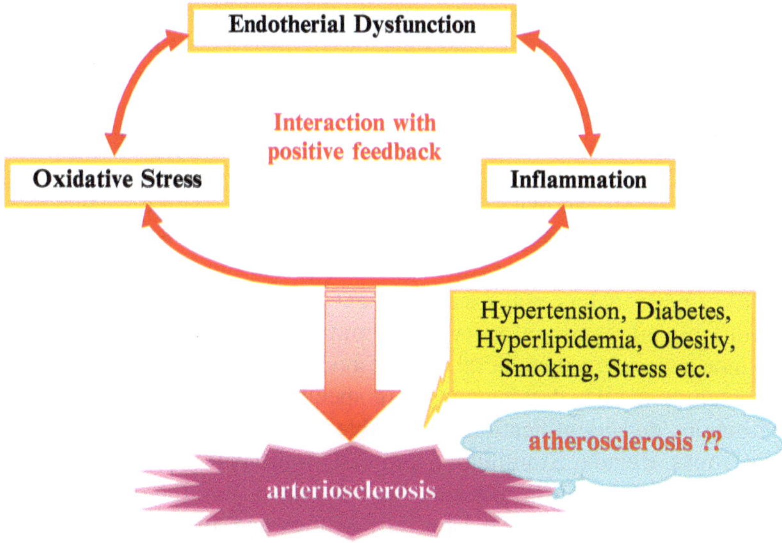

Fig. 16.5 Hypothetical mechanism of onset and progression for atherosclerosis in Kawasaki disease. Atherosclerosis is common in KD patients in chronic phase. Atherosclerosis may have an earlier onset in patients with a prior history of KD when other predisposing conditions linked to increased oxidative stress, such as hypertension, diabetes, hyperlipidemia, obesity, smoking, or stress coexists

The arteriosclerotic changes described in late-stage patients have often included the so-called post-inflammatory arteriosclerosis, which is rich in hyalinized fibrous tissue.

Using an allergic vasculitis model in weaning rabbits exhibiting KD-like vascular lesions, our group previously reported that vasculitis induction led to long-term vascular endothelial cell impairment [22]. In this KD-like model, the additional stress of a high-fat diet during the chronic phase in which vascular inflammation had already improved caused the atherosclerosis to worsen. This finding suggests that atherosclerosis is readily induced when hyperlipidemia and other risk factors are applied to vasculitis, which is itself a risk factor for atherosclerosis. That is, atherosclerosis may have an earlier onset in patients with a prior history of KD when other predisposing conditions linked to increased oxidative stress, such as hypertension, diabetes, hyperlipidemia, obesity, smoking, or stress coexist [23–27] (Fig. 16.5). Oxidative stress not only seems to trigger post-KD vascular impairment, as is the case for many diseases, but also appears to be a factor in deterioration after onset.

16.3.5 Antioxidant Therapy and Future Prospects

No clear evidence exists that administering antioxidants can mitigate vascular impairment and prevent the onset of arteriosclerosis. The incidence of coronary heart disease in France is low despite the high dietary intake of animal fat.

This "French paradox" brought attention to the antioxidant action of the polyphenols abundant in red wine [28]. Although subsequent animal research on various antioxidants substantiated their atherosclerosis-inhibiting effect [29], large-scale human clinical trials showed little evidence that they were effective in preventing atherosclerotic disease, and some studies in fact showed a tendency toward exacerbation [30, 31].

At present, we believe that administering drugs or taking late-stage corrective measures against oxidative stress in children with a prior history of KD may not necessarily immediately improve their vascular prognosis. Despite their cytotoxic effects, ROS nevertheless plays an important role in the biological defense and intracellular signal transduction systems, and their suppression by antioxidants drugs is not necessarily beneficial for the body. Important directions for future research include the development of more specifically targeted antioxidant treatment strategies and identification of suitable biomarkers for assessing vascular oxidative stress. Regular examination and early intervention such as diet and lifestyle guidance are critical for reducing the risk factors for atherosclerosis, in children with a prior history of KD, especially as they age, in order to improve their vascular prognosis.

References

1. Kawasaki T (1967) Acute febrile muco-cutaneous lymph node syndrome in young children with unique digital desquamation: clinical observation of 50 cases. Jpn J Allergol 16:178–222 (in Japanese)
2. Cave AC, Brewer AC, Narayanapanicker A, Ray R, Grieve DJ, Walker S, Shah AM (2006) NADPH oxidases in cardiovascular health and disease. Antioxid Redox Signal 8:691–728
3. Sies H (1985) Oxidative stress: inductor remarks. In: Sies H (ed) Oxidative stress. Academic, London, pp 1–8
4. Sies H (2000) What is oxidative stress? In: Kency JF Jr (ed) Oxidative stress and vascular disease. Kluwer Academic Publishers, Boston, pp 1–8
5. Phan SH, Gannon DE, Varani J, Ryan US, Ward PA (1989) Xanthine oxidase activity in rat pulmonary artery endothelial cells and its alteration by activated neutrophils. Am J Pathol 134:1201–1211
6. Furukawa S, Matsubara T, Jujoh K, Yone K, Sugawara T, Sasaki K, Kato H, Yabuta K (1988) Peripheral blood monocyte/macrophages and serum tumor necrosis factor in Kawasaki disease. Clin Immunol Immunopathol 48:247–251
7. Furukawa S, Matsubara T, Yone K, Hirano Y, Okumura K, Yabuta K (1992) Kawasaki disease differs from anaphylactoid purpura and measles with regard to tumour necrosis factor-alpha and interleukin 6 in serum. Eur J Pediatr 151:44–47
8. Lin CY, Lin CC, Hwang B, Chiang B (1992) Serial changes of serum interleukin-6, interleukin-8, and tumor necrosis factor alpha among patients with Kawasaki disease. J Pediatr 21:924–926
9. Imano Y, Harada K, Okuni M, Kimoto K, Takeuchi S, Sakurabayashi I (1987) Immunoreactive polymorphonuclear leukocyte elastase in complex with alpha 1-antitrypsin in Kawasaki disease. Acta Paediatr Jpn 29:202–205
10. Buttery LDK, Springall DR, Chester AH, Evans TJ, Standfield N, Parums DV, Yacoub MH, Polak JM (1996) Inducible nitric oxide synthase is present within human atherosclerotic lesions and promotes the formation and activity of peroxynitrite. Lab Invest 75:77–85

11. Yahata T, Suzuki C, Hamaoka A, Fujii M, Hamaoka K (2011) Dynamics of reactive oxygen metabolites and biological antioxidant potential in the acute stage of Kawasaki disease. Circ J 75:2453–2459
12. JCS Joint Working Group (2010) Guidelines for diagnosis and management of cardiovascular sequelae in Kawasaki disease (JCS 2008)—digest version. Circ J 74:1989–2020
13. Kobayashi T, Saji T, Otani T, Takeuchi K, Nakamura T, Arakawa H, Kato T, Hara T, Hamaoka K, Ogawa S, Miura M, Nomura Y, Fuse S, Ichida F, Seki M, Fukazawa R, Ogawa C, Furuno K, Tokunaga H, Takatsuki S, Hara S, Morikawa A (2012) RAISE study group investigators. Efficacy of immunoglobulin plus prednisolone for prevention of coronary artery abnormalities in severe Kawasaki disease (RAISE study): a randomized, open-label, blinded-endpoints trial. Lancet 28:1613–1620
14. Weiss JE, Eberhard BA, Chowdhury D, Gottlieb BS (2004) Infliximab as a novel therapy for refractory Kawasaki disease. J Rheumatol 31:808–810
15. Burns JC, Mason WH, Hauger SB, Janai H, Bastian JF, Wohrley JD, Balfour I, Shen CA, Michel ED, Shulman ST, Melish ME (2005) Infliximab treatment for refractory Kawasaki disease. J Pediatr 146:662–667
16. Sugimura T, Kato H, Inoue O, Fukuda T, Sato N, Ishii M, Takagi J, Akagi T, Maeno Y, Kawano T (1994) Intravascular ultrasound of coronary arteries in children. Assessment of the wall morphology and the lumen after Kawasaki disease. Circulation 89:258–265
17. Niboshi A, Hamaoka K, Sakata K, Yamaguchi N (2008) Endothelial dysfunction in adult patients with a history of Kawasaki disease. Eur J Pediatr 167:189–196
18. Takahashi K, Oharaseki T, Naoe S (2001) Pathological study of postcoronary arteritis in adolescents and young adults: with reference to the relationship between sequelae of Kawasaki disease and atherosclerosis. Pediatr Cardiol 22:138–142
19. Ross R (1999) Atherosclerosis—an inflammatory disease. N Engl J Med 340:115–126
20. Steinberg D, Parthasarathy S, Carew TE, Khoo JC, Witztum JL (1989) Beyond cholesterol. Modifications of low-density lipoprotein that increase its atherogenicity. N Engl J Med 320:915–924
21. Niboshi A, Hamaoka K (2006) Recent advances in free radical biology and medicine—oxidative stress and vascular injury of Kawasaki disease. J Clin Exp Med (Tokyo) 64:265–269 (in Japanese)
22. Liu Y, Onouchi Z, Sakata K, Ikuta K (1996) An experimental study on the role of smooth muscle cells in the pathogenesis of atherosclerosis of the coronary arteritis. J Jpn Pediatr Soc 100:1453–1458 (in Japanese)
23. Bruno RS, Pernomian L, Bendhack LM (2012) Contribution of oxidative stress to endothelial dysfunction in hypertension. Front Physiol 3:441
24. Stephens JW, Khanolkar MP, Bain SC (2009) The biological relevance and measurement of plasma markers of oxidative stress in diabetes and cardiovascular disease. Atherosclerosis 202:321–329
25. Araujo FB, Barbosa DS, Hsin CY, Maranhao RC, Abdalla DS (1995) Evaluation of oxidative stress in patients with hyperlipidemia. Atherosclerosis 117:61–71
26. Furukawa S, Fujita T, Shimabukuro M, Iwaki M, Yamada Y, Nakajima Y, Nakayama O, Makishima M, Matsuda M, Shimomura I (2004) Increased oxidative stress in obesity and its impact on metabolic syndrome. J Clin Invest 114:1752 1761
27. Morrow JD, Frei B, Longmire AW, Gaziano JM, Lynch SM, Shyr Y, Strauss WE, Oates JA, Roberts LJ 2nd (1995) Increased in circulating products of lipid peroxidation (F2-isoprostanes) in smokers. Smoking as a cause of oxidative damage. N Engl J Med 332:1198–1203
28. Ferrieres J (2004) The French paradox: lessons for other countries. Heart 90:170–211
29. Pratlco D, Tangirala RK, Rader DJ, Rokach J, FitzGerald GA (1998) Vitamin E suppresses isoprostane generation in vivo and reduces atherosclerosis in ApoE-deficient mice. Nat Med 4:1189–1192
30. Bjelakovic G, Nikolova D, Gluud LL, Simonetti RG, Gluud C (2007) Mortality in randomized trials of antioxidant supplements for primary and secondary prevention: systemic review and meta-analysis. JAMA 297:842–857

31. Baillie JK, Thompson AA, Irving JB, Bates MG, Sutherland AI, Macnee W, Maxwell SR, Webb DJ (2009) Oral antioxidant supplementation does not prevent acute mountain sickness: double blind, randomized placebo-controlled trial. QJM 102:341–348
32. Griendling KK, FitzGerald GA (2003) Oxidative stress and cardiovascular injury part 1: basic mechanisms and in vivo monitoring of ROS. Circulation 108:1912–1916

Chapter 17
Environmental Pollution and Health Consequences

Pavel Rossner Jr. and Radim J. Sram

Abbreviations

8-oxodG	8-Oxo-7,8-dihydro-2′-deoxyguanosine
15-F_{2t}-IsoP	15-F_{2t}-isoprostane
B[a]P	Benzo[a]pyrene
BMI	Body mass index
c-PAHs	Carcinogenic PAHs
DEPs	Diesel exhaust particles
ETS	Environmental tobacco smoke
IUGR	Intrauterine growth retardation
LBW	Low birth weight
MDA	Malondialdehyde
OR	Odds ratio
PAHs	Polycyclic aromatic hydrocarbons
PM	Particulate matter
PM1	PM of aerodynamic diameter <1 μm
PM2.5	PM of aerodynamic diameter <2.5 μm
PM10	PM of aerodynamic diameter <10 μm
RNS	Reactive nitrogen species
ROS	Reactive oxygen species
SNP	Single nucleotide polymorphism
VOC	Volatile organic compound

P. Rossner Jr. • R.J. Sram (✉)
The Institute of Experimental Medicine ASCR, Academy of Sciences of the Czech Republic, Videnska 1083, 142 20 Prague 4, Czech Republic
e-mail: sram@biomed.cas.cz

H. Tsukahara and K. Kaneko (eds.), *Studies on Pediatric Disorders*, Oxidative Stress in Applied Basic Research and Clinical Practice, DOI 10.1007/978-1-4939-0679-6_17,

17.1 Introduction

Air pollution is associated with many negative health effects including pulmonary and cardiovascular diseases or cancer as well as increased mortality [1]. Although underlying biological mechanisms of induction of negative health effects are mostly unknown, it has been demonstrated that pollutants cause damage to DNA and other important biomolecules (lipids, proteins) [2–4].

Air pollutants belong to diverse groups of chemical compounds that are generally grouped into four categories: gaseous pollutants (e.g., volatile organic compounds (VOCs), including benzene), persistent organic compounds (e.g., polycyclic aromatic hydrocarbons (PAHs), including benzo[a]pyrene (B[a]P)), heavy metals, and particulate matter (PM), onto which many chemicals are adsorbed [5]. As PM of various compositions is ubiquitous in the environment, health consequences of exposure to PM are of a great concern. Among a complex mixture of chemicals present in the ambient air, PAHs are particularly important because some of them are carcinogenic (c-PAHs) and may be directly responsible for the increased incidence of cancer due to PAH–DNA adduct formation and subsequent mutation induction [6]. PAHs, as well as other chemicals in the air, are adsorbed onto the surface of PM of various aerodynamic diameters. Inhalation of PM, particularly PM of aerodynamic diameter <2.5 μm (PM2.5) and smaller, leads to inflammation and subsequent production of reactive oxygen species (ROS) [7]. The production of ROS, that include, e.g., the hydroxyl radical, superoxide anion, or hydrogen peroxide, is caused by both the physical effects of PM (PM is phagocyted by macrophages that consequently produce ROS) and the presence of various chemicals on the surface of PM (e.g., metals, PAHs) with prooxidant properties. ROS may arise from exogenous or endogenous sources. Under normal physiologic conditions, a balance is maintained between endogenous oxidants and antioxidants. Excessive generation of oxidants or a decrease of antioxidants leads to an imbalance, and the abnormal oxidant system enters a state known as oxidative stress [8]. ROS can attack lipids, proteins, and nucleic acids simultaneously [9]. As a biomarker of oxidative DNA damage, 8-oxo-7,8-dihydro-2′-deoxyguanosine (8-oxodG) is most often studied. Urinary levels of 8-oxodG are believed to reflect the total DNA excision repair capacity of an organism [10]. Not only 8-oxodG is a biomarker of generalized, cellular oxidative stress, but it may also imply the risk of cancer, atherosclerosis, and diabetes [8, 9].

The attack of ROS on lipids leads to lipid peroxidation. Currently, isoprostanes are considered the most reliable markers of lipid peroxidation. These prostaglandin-like compounds are formed by free-radical-induced peroxidation of arachidonic acid, independent of cyclooxygenase enzymes. Lipid peroxidation products, including isoprostanes, play a role in the pathogenesis of many diseases [11].

Proteins may be damaged by ROS, reactive aldehydes, or by reactive sugars. These reactions generate carbonyl groups (aldehyde or ketone groups) mostly on side chains of protein molecules. These modifications result in alterations of protein structure and/or function. Damaged proteins are recognized by the proteolytic system of the cell and degraded by proteasomes [12]. However, this system may be inefficient and damaged proteins may accumulate in the organism. Carbonyl groups in damaged proteins may then be detected as a biomarker of protein oxidation.

17.2 The Effect of Air Pollution on Pregnancy Outcomes

The effect of various ambient air pollutants on adverse pregnancy outcomes has been analyzed in numerous studies with conflicting results [13]. Exposure to PM of aerodynamic diameter <10 μm (PM10) during the first month of pregnancy was associated with increased risk of intrauterine growth retardation (IUGR) [14]. Similarly, c-PAH exposure during the first gestational month significantly elevated IUGR [15]. Low birth weight (LBW) and the prevalence of premature births were associated with the concentrations of sulfur dioxide and total suspended particles during individual trimesters of pregnancy: LBW was associated with concentrations of both sulfur dioxide and total suspended particles in all three trimesters, while prematurity with sulfur dioxide in three trimesters and total suspended particles only in the first trimester of pregnancy [16]. LBW was increased in association with residence within 50 m of highways, but no direct effect of PM2.5 on LBW was found [17].

Oxidative stress during pregnancy is a normal physiological condition, and the fetus possesses mechanisms that help to minimize the deleterious effects of the production of ROS. These mechanisms include the expression of the antioxidant enzymes catalase, glutathione peroxidase, glutathione S-transferase, thiol/disulfide oxidoreductase, and superoxide dismutase as well as the accumulation of antioxidants (glutathione and vitamins C and E) [18]. However, in the case of excessive ROS production, apoptosis in the trophoblast is induced [19, 20], which in turn increases the risk of IUGR [18].

Oxidative stress, mediated, e.g., by PM [21], plays an important role in adverse pregnancy outcomes (Fig. 17.1, [22]).

It has been shown that increased oxidative stress, accompanied by reduced translation of proteins, including key signaling molecules, is present in the placentas of IUGR pregnancies [23]. Oxidative stress causes injury to chorionic villi, reducing the functional mass of the syncytiotrophoblast and limiting the capacity of the villi to mediate nutrient transport, thus contributing to IUGR induction [24]. Inflammation, also related to oxidative stress induction, is another key process associated with LBW syndromes. While normal pregnancy is a state of permanent mild inflammation, in adverse pregnancy outcomes, the inflammation is increased above normal levels [25]. The resulting inflammatory state is accompanied by maternal leukocyte activation, the release of cytokines from immune cells and uteroplacental tissues, endothelial cell activation, as well as immune/coagulation interactions. Activated immune cells express various molecules that are important in the formation of a dysfunctional placenta, resulting in excessive trophoblast apoptosis, shallow trophoblast invasion, and impaired spiral artery remodeling [25]. These processes cause adverse pregnancy outcomes.

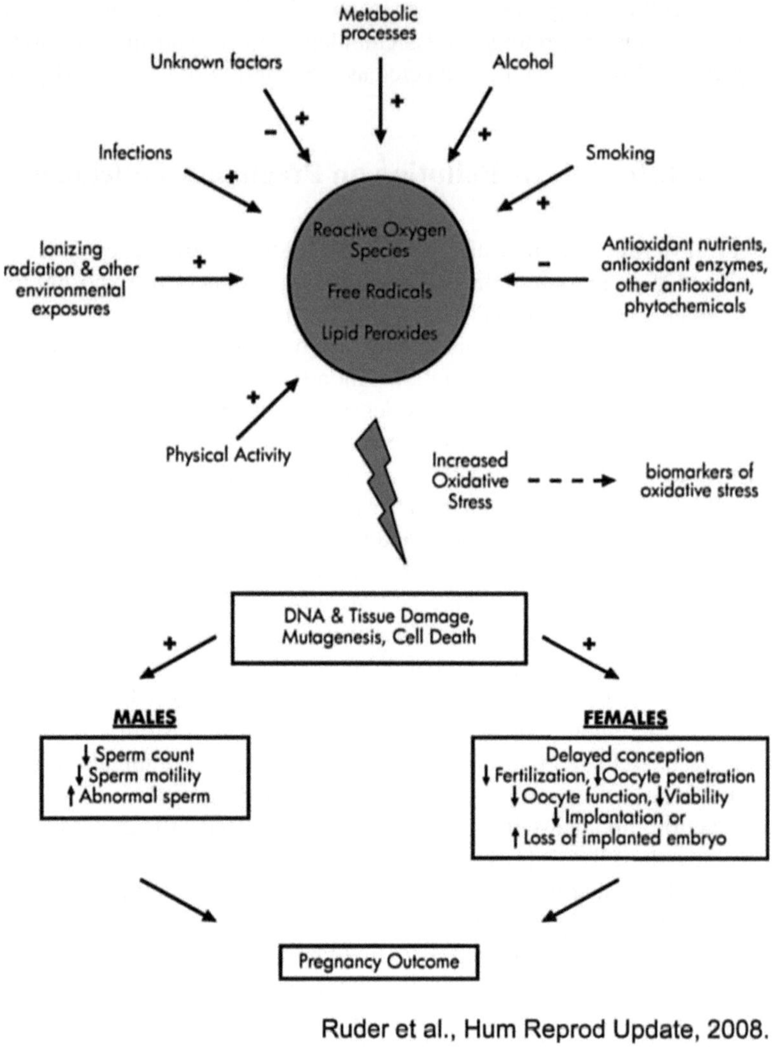

Fig. 17.1 Factors affecting oxidative stress-mediated adverse pregnancy outcomes

Several authors have reported increased levels of various oxidative stress markers during pregnancy and subsequent unfavorable pregnancy outcomes. The levels of malondialdehyde (MDA) and xanthine oxidase were elevated in maternal and umbilical cord plasma and placental tissues with IUGR when compared with controls [26]. Elevated urinary 8-oxodG levels during early pregnancy were predictors of LBW [27, 28]. Other authors reported associations of increased levels of oxidative stress markers measured in the third trimester of pregnancy with adverse pregnancy outcomes [29, 30].

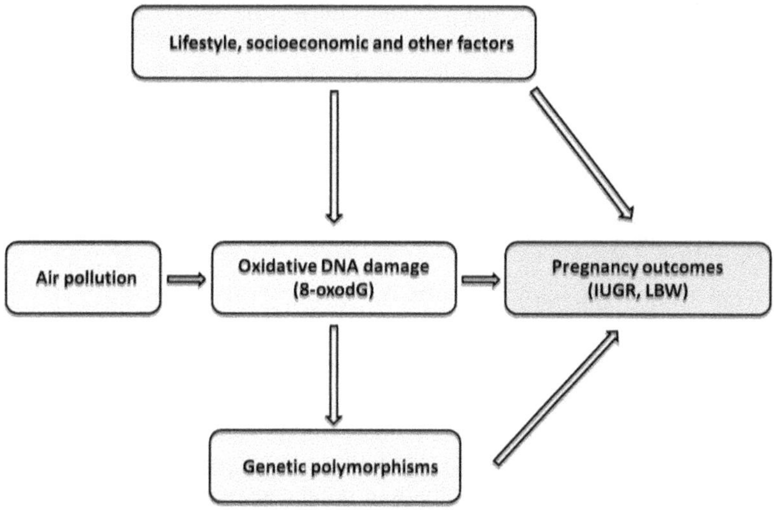

Fig. 17.2 Possible relationships between air pollution, oxidative DNA damage, genetic polymorphisms, lifestyle, socioeconomic and other factors, and pregnancy outcomes. 8-oxodG is an intermediate in the air pollution–pregnancy outcome pathway

Rossner et al. [31] investigated factors associated with LBW and IUGR risks in a group of 891 newborns born in two districts (Teplice and Prachatice) of the Czech Republic with different levels of air pollution. The analyzed factors included the levels of air pollutants ([B[a]P], PM2.5) during pregnancy, 8-oxodG levels in placental DNA, and a panel of 576 single nucleotide polymorphisms (SNPs) in 97 genes related to DNA repair, oxidative stress, xenobiotic metabolism, and immune functions. They hypothesized that air pollution induces oxidative DNA damage and this, in turn, increases the risk of adverse pregnancy outcomes. These effects may be modulated by SNPs in selected genes (Fig. 17.2). The principal aim was to determine whether oxidative stress was a major factor affecting LBW and IUGR in the cohort.

Out of 891 subjects, 674 pregnancies were normal, 88 newborns were IUGR- but not LBW-affected, 76 newborns were born with LBW only, and 53 subjects were affected by both IUGR and LBW. No difference in 8-oxodG between the districts was observed, despite higher levels of air pollutants in the Teplice region, which may induce oxidative DNA damage. Also, 8-oxodG levels did not differ between the two ethnic groups (Romani and European). In placental samples from IUGR subjects an increase in 8-oxodG levels was observed that was on the borderline of significance (median 8-oxodG/10^5 dG: 2.03, range: 0.29–6.17; median 8-oxodG/10^5 dG: 1.74, range: 0.20–6.50 for IUGR and non-IUGR samples, respectively; $p = 0.055$). LBW was associated with significantly higher oxidative DNA damage (median 8-oxodG levels/10^5 dG: 2.25, range 0.27–6.28; median 8-oxodG levels/10^5 dG: 1.75, range 0.20–6.50 for LBW and non-LBW samples, respectively; $p < 0.05$).

The exposure to environmental pollutants during the individual months of pregnancy for IUGR and non-IUGR, as well as LBW and non-LBW subjects was compared. While no significant differences were observed for IUGR and non-IUGR newborns exposed to both B[a]P and PM2.5, as well as for LBW and non-LBW children exposed to B[a]P, LBW-affected children were exposed to higher concentrations of PM2.5 during the first to fifth months of pregnancy (31.3 ± 11.4–33.6 ± 12.6 vs. 28.8 ± 12.8–$30.2 \pm 13.6\ \mu g/m^3$ PM2.5, for LBW and non-LBW-affected children, respectively).

In univariate analyses 8-oxodG levels in placental DNA were significantly associated with IUGR [odds ratio (OR); 95 % CI: 1.47; 1.02, 2.12; $p=0.037$], but not with LBW. Other factors associated with the IUGR risk included ethnicity, the mother's body mass index (BMI), maternal smoking, and the mother's and father's length of education. Exposure to PM2.5 had no effect on IUGR. District, ethnicity, the child's gender and gestational age, the mother's BMI, smoking, marital status and vitamin intake during pregnancy, the father's length of education and smoking, and exposure to PM2.5 in the first month of pregnancy were among variables associated with the risk of LBW. B[a]P exposure affected neither IUGR nor LBW.

The multivariate-adjusted logistic regression investigating the factors associated with IUGR indicates that 8-oxodG levels remained the only variable significantly associated with the risk of IUGR (OR; 95 % CI: 1.57; 1.06, 2.34; $p=0.026$). Factors associated with LBW after multivariate adjustment include 8-oxodG levels (OR; 95 % CI: 1.83; 1.12, 3.00; $p=0.017$), gender (OR; 95 % CI: 1.96; 1.21, 3.18; $p=0.006$), gestational age (OR; 95 % CI: 0.05; 0.03, 0.11; $p<0.001$), maternal smoking (OR; 95 % CI: 0.52; 0.29, 0.91; $p=0.023$), and the haplotype in the *MBL2* gene (OR; 95 % CI: 2.59; 1.59, 4.20; $p<0.001$). Thus, it seems that oxidative damage is one of the independent factors (a key factor in the case of IUGR) that collectively contribute to adverse pregnancy outcomes (Fig. 17.3).

All these data correspond to recent studies, proposing in vivo oxidative DNA damage (8-oxodG) and lipid peroxidation as biomarkers of oxidative stress in preterm LBW infants [32] and the relationship between 8-oxodG and lipid peroxidation to adverse birth outcomes due to exposure to PAHs [33].

17.3 Air Pollution and Oxidative Damage in Children

It is generally accepted that children are more sensitive to air pollution than adults because their immune and respiratory systems are not fully developed [34]. Bronchial asthma is the most common inflammatory disease of the respiratory system which impacts the immune response of the organism [35]. There is accumulating evidence that oxidative stress plays a critical role in the pathogenesis of this disease.

ROS not only cause damage to cell membranes which leads to cell death but also activate signaling pathways, which may cause pathological changes in the lungs. Oxidative stress may be further affected by individual susceptibility (e.g., genetic polymorphisms) as well as by diet, rich in antioxidants [36] (Fig. 17.4).

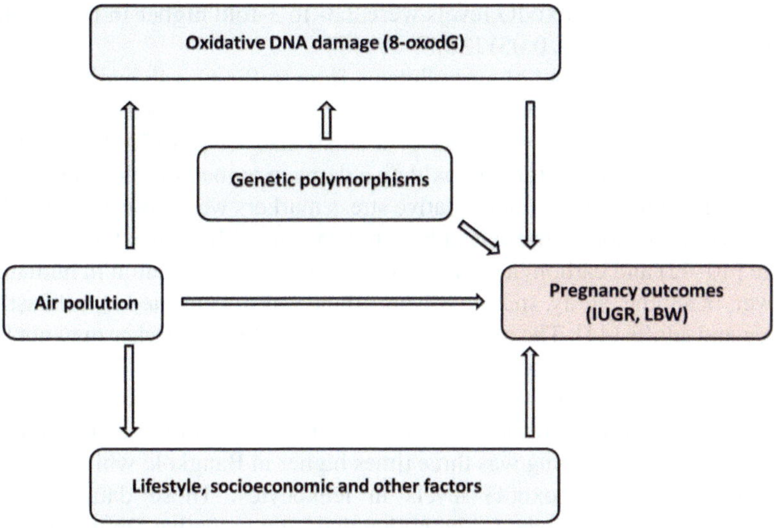

Fig. 17.3 Possible relationships between air pollution, oxidative DNA damage, genetic polymorphisms, lifestyle, socioeconomic and other factors, and pregnancy outcomes. Oxidative damage is an independent factor affecting IUGR and/or LBW

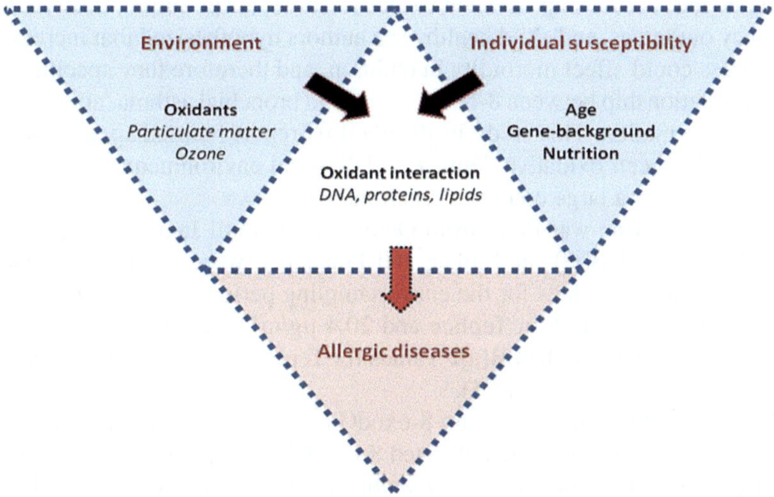

Fig. 17.4 Interactions between gene and environmental factors, oxidative stress, and susceptibility factors

Exposure to PM2.5 has been associated with impaired growth or lung function. Most studies reported increased prevalence of symptoms with increased exposure to fine PM [37]. A study in Mexico City showed that in children constantly exposed to ROS and reactive nitrogen species (RNS) from exogenous and endogenous sources, 8-oxodG levels were 2.3- to 3-fold higher in nasal epithelial cells than in controls ($p < 0.05$) [38].

Exposure to a high level of air pollutants from traffic was shown to increase the incidence of asthma in children [39]. The effect of traffic is mostly related to diesel exhaust particles (DEPs), which are of a small size <1 μm (PM1) and therefore increase airway inflammation and oxidative damage to macromolecules. Indeed, in several studies higher levels of oxidative stress markers were observed in asthmatic children including concentrations of 8-isoprostane and MDA in exhaled breath condensate [40–42] and carbonylated proteins in allergic inflammation in humans [43]. However, in another study, sputum 8-isoprostane was not elevated in mild asthmatic children and adults [44]. The authors concluded that this biomarker may not be sensitive in reflecting oxidant burden in mild asthma.

Oxidative DNA damage was also studied among urban and rural school children in Thailand, affected by traffic air pollution. The concentration of benzene measured by personal monitoring was three times higher in Bangkok, which was accompanied by increased 8-oxodG levels in leukocytes. Those data indicate the relationship between benzene in the polluted air and oxidative DNA damage, suggesting an increased health risk from traffic benzene emissions [45].

Svecova et al. [46] studied the effect of independent factors, including PM10, PM2.5, c-PAHs, and B[a]P, on urinary levels of 8-oxodG in children from the polluted district of Teplice and the control Prachatice district. Another objective was to investigate the relationship between oxidative stress, air pollution, lifestyle factors, pregnancy outcomes, and child health. The authors hypothesized that increased oxidative stress could affect morbidity in children, and therefore they specifically analyzed the relationship between 8-oxodG level and bronchial asthma, allergic rhinitis, and atopic dermatitis. This study is also the first report focused specifically on the association between oxidative damage to DNA and environmental tobacco smoke (ETS) exposure in a large cohort of children.

Sampling of urine was done from October 2004 until January 2005. Levels of PM10, PM2.5, and B[a]P in Teplice and Prachatice were obtained by stationary monitoring. Median values for the entire sampling period were 30 $\mu g/m^3$ of PM10 and 22.7 $\mu g/m^3$ of PM2.5 in Teplice and 20.4 $\mu g/m^3$ of PM10 and 16.8 $\mu g/m^3$ of PM2.5 in Prachatice. Median B[a]P values for Teplice and Prachatice were 1.8 and 0.9 ng/m^3, respectively (Fig. 17.5).

Although no difference in urinary 8-oxodG levels in children from both locations was observed, the authors noted elevated 8-oxodG excretion in subjects exposed to ETS. Because many constituents of ambient air pollution from manufacturing, motor vehicles, and home heating are also components of cigarette smoke, exposure to ETS places children at a greater risk for many negative health outcomes, including LBW, prenatal mortality, deficits in childhood growth, sudden infant death syndrome, middle-ear disease, bronchitis, pneumonia, cough, asthma, and wheezing [38, 47–51]. Further, a strong correlation was observed between urinary cotinine, a marker of ETS exposure, and 8-oxodG levels indicating oxidative DNA damage may be a starting point for respiratory and allergic morbidity [46].

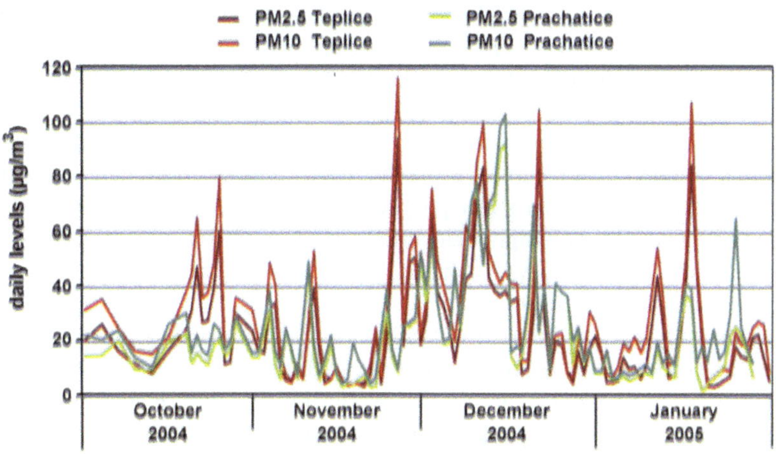

Fig. 17.5 Daily concentrations of PM10 and PM2.5 in Teplice and Prachatice districts during collection of urine samples (October 2004–January 2005)

Svecova et al. [46] also found a positive association of allergic rhinitis with 8-oxodG levels in subjects from Teplice, but a negative one in subjects from Prachatice. A different origin of nasal allergies in the two districts may be one of the reasons for this difference. In the greener district of Prachatice, there is a significantly higher proportion of children sensitized to pollen, whereas in Teplice air pollution is a major source of allergens.

Multivariate models of the effect of air pollutants, the child's age, ETS exposure, and allergic rhinitis on 8-oxodG levels indicate that oxidative damage is more pronounced in the district of Teplice, with higher levels of air pollution and ETS than in the district of Prachatice. Thus, urinary 8-oxodG levels seem to be a complex biomarker summing up the effect of air pollution and lifestyle.

17.4 Biomarker Studies on Oxidative Stress in Adults

17.4.1 Oxidative Damage to DNA

Although PAHs may induce oxidative stress during their metabolism by the formation of reactive quinones, the results of epidemiological studies on 8-oxodG levels after occupational, as well as environmental, PAH exposure are contradictory. Several authors reported a positive correlation between PAH exposure, measured by urinary levels of 1-hydroxypyrene, and 8-oxodG levels, either in lymphocytes or in urine [52–55]. Others did not see any correlation between 8-oxodG levels and urinary 1-hydroxypyrene and/or PAH levels in the air [56–58].

Similarly, Rossner et al. [59] did not find any effect of either personal or stationary c-PAH exposure on 8-oxodG levels in urine. These observations suggest that while c-PAHs may have the potential to induce oxidative DNA damage, other factors probably play a role in the response of the organism to environmental pollution.

VOCs, particularly benzene, are known to induce oxidative damage [60]. In occupational exposure studies, 8-oxodG levels in urine correlated with exposure to benzene [61, 62]; another study [63] reported a positive correlation between air benzene and 8-oxodG in lymphocytes. In the study by Rossner et al. [59], 8-oxodG levels were not affected by VOC, probably due to the relatively low environmental concentrations of these pollutants.

It is generally accepted that PM, particularly PM2.5 and ultrafine particles, plays a key role in oxidative damage caused by air pollution [62]. It has been shown that personal exposure to PM2.5 correlated well with 8-oxodG levels in lymphocyte DNA in a group of 50 Copenhagen students sampled repeatedly during a period of 1 year. No such correlation has been observed for stationary PM2.5 monitoring and 8-oxodG in lymphocyte DNA, though a weak association with urinary 8-oxodG was observed [64]. The authors conclude that for moderate exposure to PM, personal monitors are more informative than stationary monitors. It has been proposed that to show an association between PM exposure measured by stationary monitoring and 8-oxodG levels, high exposure levels are required [62]. In another study, personal exposure to PM2.5, black smoke, and nitrogen dioxide measured in 2-day periods in 30 subjects was compared with bedroom, front door, and background (stationary) levels of these pollutants. Personal PM2.5 concentrations were correlated only with bedroom and front door levels, but not with background concentrations [65]. However, Rossner et al. [66] identified both PM10 and PM2.5 stationary levels as independent factors increasing urinary 8-oxodG concentrations. Stationary PM2.5 exposure levels in Prague were almost threefold higher than in Copenhagen (median levels: 24.3 vs. 9.2 $\mu g/m^3$). Also the chemical composition of PM may differ between the two cities. Differences in the chemical composition of PM have been shown to be very important for the induction of DNA damage [67]. Thus, differences in PM quantities and composition may explain why the results of both studies [59] and [64] differ.

The originality of Rossner et al. [59] study lies in the fact that stationary monitoring was performed continuously during both years that the study was conducted. The collected air pollution data allowed them to correlate individual urinary 8-oxodG levels with the levels of pollutants measured at different times and for different periods before the collection of urine samples and thus identify the time frame that may be crucial for the induction of DNA oxidative damage. Their results indicate that exposure to PM2.5 and PM10 3 days before sampling positively affects 8-oxodG levels in urine. This result was also generally consistent for other 3-day intervals up to 42 days before urine collection.

Smoking was found to be a significant factor increasing 8-oxodG levels, though not all studies confirmed this observation (reviewed in [68]).

17.4.2 Oxidative Damage to Lipids and Proteins

After being attacked by ROS, lipids are peroxidized, forming reactive intermediates that may further react with other molecules and propagate oxidative stress [69]. Apart from this process, oxidized lipids also change the properties of cellular membranes, including their fluidity and the inactivation of membrane-bound proteins. As a result, normal cellular functions are impaired. To monitor lipid peroxidation in vivo, several biomarkers have been proposed, including MDA and 15-F_{2t}-isoprostane (15-F_{2t}-IsoP).

Because proteins are major components of most biological systems, they are an important target for ROS, scavenging 50–75 % of reactive intermediates [70]. Protein oxidation leads to the formation of a number of products, with protein carbonyls being the best characterized. Analyses of protein carbonyl levels are generally accepted methods for the measurement of protein oxidation [71].

Singh et al. [72] found significant correlations between PAHs exposure and lipid peroxidation, measured by MDA levels. However, measurements of MDA using a thiobarbituric acid assay do not represent a specific marker of lipid peroxidation [73]. In another study [74], urinary 15-F_{2t}-IsoP levels were not found to be increased in roofers applying asphalt. Thus, the effect of exposure to c-PAHs on lipid peroxidation is not conclusive. Unlike c-PAHs and B[a]P, Rossner et al. [75] did not observe any effect of stationary PM2.5 and PM10 on levels of 15-F_{2t}-IsoP when PM concentrations over a 3-day period immediately preceding urine collection were used in the statistical analysis. Only when the multivariate model was applied did the effect of PM2.5 exposure reach borderline significance. Interestingly, when concentrations of PM2.5 and PM10 over various 3-day periods before sampling were included in the statistical analysis, they found significant effects of PM exposure on 15-F_{2t}-IsoP levels. This observation suggests that 15-F_{2t}-IsoP levels, unlike 8-oxodG, are not markers of immediate effect; rather, they reflect the exposure to pollutants that occurred at least 3–4 weeks before sampling. The possible explanation may be the existence of DNA repair mechanisms that remove damage to DNA shortly after it occurs. Damage to lipids is not repaired; rather the lesions accumulate and thus allow the detection of exposure to pollutants longer period before sampling.

Several studies have described a correlation between PM exposure and lipid peroxidation. Significant results were observed in an animal system for MDA levels [76] and after wood smoke exposure for 15-F_{2t}-IsoP levels [77]. A study of Sorensen et al. [78] measured PM2.5 and carbon black exposure in 50 students in Copenhagen four times in 1 year and analyzed, among others, also MDA levels. The authors found a significant effect of personal PM2.5 exposure on lipid peroxidation in women only; they did not see any significant correlation in men or when background (stationary) PM2.5 exposure levels were used. The effect of c-PAHs and B[a]P exposure determined from both personal and stationary monitors on protein carbonyl levels was unexpectedly negative [75]. The effect of PM2.5 studied over 3-day periods before sampling showed the same result, while exposure to PM10 revealed a similar time-dependent pattern as observed for 15-F_{2t}-IsoP levels.

These findings, however, do not imply that increased exposure to c-PAHs and PM has a protective effect against protein oxidative damage. These observations are probably the result of the interaction of other pollutants (including ozone, as discussed below) with the organism. A few other studies have analyzed the association between air pollution exposure and protein oxidation [78–81]. In studies [78, 81], protein oxidation was measured by using the levels of 2-aminoadipic semialdehyde and γ-glutamyl semialdehyde in hemoglobin. One of these studies observed a negative association between γ-glutamyl semialdehyde and 1-hydroxypyrene, a urinary PAH metabolite, in a group of bus drivers and postal workers; the correlation with 2-aminoadipic semialdehyde was not significant. The authors suggested that a negative correlation may indicate the effect of additional pollutants that were not measured [81]. In the other study, no significant association between either marker of protein oxidation and personal or stationary exposure to PM2.5 was observed [78].

Oxidative stress may also be affected by other air pollutants and factors, including ozone. Rossner et al. [75] tested the correlation of ozone concentrations with 15-F_{2t}-IsoP and protein carbonyl levels and found that ozone levels were significantly positively associated with protein carbonyl levels in all the performed statistical analyses. Ozone is known to induce oxidative damage exposure to it correlated with oxidative damage to lipids [82]. It is possible that the negative correlation observed between protein carbonyl levels and c-PAHs and PM exposure by Rossner et al. [75] reflects the seasonal variability of ozone, c-PAHs, and PM and their effect on protein oxidation: the summer levels of c-PAHs and PM were low, but the ozone concentration was increased thus affecting protein carbonyl levels. However, there still may be other unidentified factors responsible for these surprising results.

Among other factors that seem to affect the levels of oxidative stress markers, a positive association between cotinine (considered as a marker of passive smoking) and 15-F_{2t}-IsoP levels was observed [83–85], although negative reports have also been published [86].

The increased oxidative damage to proteins and lipids may also be interpreted as a marker of increased risk of various diseases, including cancer [87, 88] and cardiovascular diseases [6]. The results of human studies consistently show that mechanisms of oxidative stress play a significant role in the cardiovascular effect of PM. Studies of the role of oxidative mechanisms associated with airborne particle toxicity are used as evidence of a mechanistic role of oxidative stress mediated by ROS as a mechanism of PM-induced inflammation and damage [89–92].

17.5 Conclusions

Ambient air pollution is an important exogenous factor inducing oxidative stress in the organism. Particulate matter and its components (transition metals, benzene and other VOCs, and PAHs) are mostly responsible for this induction. Although pregnancy is a state characterized by increased oxidative stress in the fetus and mother and the organism is adapted to it, excessive ROS formation during pregnancy is deleterious and has been associated with miscarriage and various complications, such as preeclampsia, IUGR, and diabetes. Maternal exposure to polluted air causes oxidative stress in the fetus, and may result in higher respiratory neonatal mortality, allergy incidence, and reduced lung function. In young children exposure to air pollution, usually accompanied by oxidative damage to macromolecules, increases the incidence of respiratory and other diseases, including bronchial asthma. Finally, in adults oxidative stress may lead to pulmonary and cardiovascular diseases, cancer, and increased mortality. It also speeds up the ageing process. All these facts supported by the results of hundreds of studies underline the importance of decreasing and observing emission limits of major industrial pollutants.

References

1. Yang W, Omaye ST (2009) Air pollutants, oxidative stress and human health. Mutat Res 674:45–54
2. Barbato DL, Tomei G, Tomei F, Sancini A (2010) Traffic air pollution and oxidatively generated DNA damage: can urinary 8-oxo-7,8-dihydro-2-deoxyguanosine be considered a good biomarker? A meta-analysis. Biomarkers 15:538–545
3. Moller P, Loft S (2010) Oxidative damage to DNA and lipids as biomarkers of exposure to air pollution. Environ Health Perspect 118:1126–1136
4. Bagryantseva Y, Novotna B, Rossner P Jr, Chvatalova I, Milcova A, Svecova V, Lnenickova Z, Solansky I, Sram RJ (2010) Oxidative damage to biological macromolecules in Prague bus drivers and garagemen: impact of air pollution and genetic polymorphisms. Toxicol Lett 199:60–68
5. Kampa M, Castanas E (2008) Human health effects of air pollution. Environ Pollut 151: 362–367
6. Lewtas J (2007) Air pollution combustion emissions: characterization of causative agents and mechanisms associated with cancer, reproductive, and cardiovascular effects. Mutat Res 636:95–133
7. Mazzoli-Rocha F, Fernandes S, Einicker-Lamas M, Zin WA (2010) Roles of oxidative stress in signaling and inflammation induced by particulate matter. Cell Biol Toxicol 26:481–498
8. Nair J, De Flora S, Izzotti A, Bartsch H (2007) Lipid peroxidation-derived etheno-DNA adducts in human atherosclerotic lesions. Mutat Res 621:95–105
9. Cooke MS, Evans MD, Dizdaroglu M, Lunec J (2003) Oxidative DNA damage: mechanisms, mutation, and disease. FASEB J 17:1195–1214
10. Sorensen M, Autrup H, Moller P, Hertel O, Jensen SS, Vinzents P, Knudsen LE, Loft S (2003) Linking exposure to environmental pollutants with biological effects. Mutat Res 544:255–271
11. Montuschi P, Barnes P, Roberts LJ 2nd (2007) Insights into oxidative stress: the isoprostanes. Curr Med Chem 14:703–717
12. Dunlop RA, Brunk UT, Rodgers KJ (2009) Oxidized proteins: mechanisms of removal and consequences of accumulation. IUBMB Life 61:522–527
13. Sram RJ, Binkova B, Dejmek J, Bobak M (2005) Ambient air pollution and pregnancy outcomes: a review of the literature. Environ Health Perspect 113:375–382
14. Dejmek J, Selevan SG, Benes I, Solansky I, Sram RJ (1999) Fetal growth and maternal exposure to particulate matter during pregnancy. Environ Health Perspect 107:475–480
15. Dejmek J, Solansky I, Benes I, Lenicek J, Sram RJ (2000) The impact of polycyclic aromatic hydrocarbons and fine particles on pregnancy outcome. Environ Health Perspect 108: 1159–1164
16. Bobak M (2000) Outdoor air pollution, low birth weight, and prematurity. Environ Health Perspect 108:173–176

17. Brauer M, Lencar C, Tamburic L, Koehoorn M, Demers P, Karr C (2008) A cohort study of traffic-related air pollution impacts on birth outcomes. Environ Health Perspect 116:680–686
18. Negre-Salvayre A, Auge N, Ayala V, Basaga H, Boada J, Brenke R, Chapple S, Cohen G, Feher J, Grune T, Lengyel G, Mann GE, Pamplona R, Poli G, Portero-Otin M, Riahi Y, Salvayre R, Sasson S, Serrano J, Shamni O, Siems W, Siow RC, Wiswedel I, Zarkovic K, Zarkovic N (2010) Pathological aspects of lipid peroxidation. Free Radic Res 44:1125–1171
19. Myatt L, Cui X (2004) Oxidative stress in the placenta. Histochem Cell Biol 122:369–382
20. Smith SC, Guilbert LJ, Yui J, Baker PN, Davidge ST (1999) The role of reactive nitrogen/oxygen intermediates in cytokine-induced trophoblast apoptosis. Placenta 20:309–315
21. Kannan S, Misra DP, Dvonch JT, Krishnakumar A (2006) Exposures to airborne particulate matter and adverse perinatal outcomes: a biologically plausible mechanistic framework for exploring potential effect modification by nutrition. Environ Health Perspect 114:1636–1642
22. Ruder EH, Hartman TJ, Blumberg J, Goldman MB (2008) Oxidative stress and antioxidants: exposure and impact on female fertility. Hum Reprod Update 14:345–357
23. Yung HW, Calabrese S, Hynx D, Hemmings BA, Cetin I, Charnock-Jones DS et al (2008) Evidence of placental translation inhibition and endoplasmic reticulum stress in the etiology of human intrauterine growth restriction. Am J Pathol 173:451–462
24. Scifres CM, Nelson DM (2009) Intrauterine growth restriction, human placental development and trophoblast cell death. J Physiol 587:3453–3458
25. Li M, Huang SJ (2009) Innate immunity, coagulation and placenta-related adverse pregnancy outcomes. Thromb Res 124:656–662
26. Biri A, Bozkurt N, Turp A, Kavutcu M, Himmetoglu O, Durak I (2007) Role of oxidative stress in intrauterine growth restriction. Gynecol Obstet Invest 64:187–192
27. Peter Stein T, Scholl TO, Schluter MD, Leskiw MJ, Chen X, Spur BW et al (2008) Oxidative stress early in pregnancy and pregnancy outcome. Free Radic Res 42:841–848
28. Potdar N, Singh R, Mistry V, Evans MD, Farmer PB, Konje JC et al (2009) First-trimester increase in oxidative stress and risk of small-for-gestational-age fetus. BJOG 116:637–642
29. Scholl TO, Stein TP (2001) Oxidant damage to DNA and pregnancy outcome. J Matern Fetal Med 10:182–185
30. Orhan H, Onderoglu L, Yucel A, Sahin G (2003) Circulating biomarkers of oxidative stress in complicated pregnancies. Arch Gynecol Obstet 267:189–195
31. Rossner P Jr, Tabashidze N, Dostal M, Novakova Z, Chvatalova I, Spatova M, Sram RJ (2011) Genetic, biochemical and environmental factors associated with pregnancy outcome in newborns from the Czech Republic. Environ Health Perspect 19:265–271
32. Negi R, Pande D, Kumar A, Khanna RS, Khanna HD (2012) Evaluation of biomarkers of oxidative stress and antioxidant capacity in the cord blood of preterm low birth weight neonates. J Matern Fetal Neonatal Med 25:1338–1341
33. Al-Saleh I, Alsabbahen A, Shinwari N, Billedo G, Mashhour A, Al-Sarraj Y, Mohamed Gel D, Rabbah A (2013) Polycyclic aromatic hydrocarbons (PAHs) as determinants of various anthropometric measures of birth outcome. Sci Total Environ 444:565–578
34. Jedrychowski WA, Perera FP, Maugeri U, Mroz E, Klimaszewska-Rembiasz M, Flak E, Edwards S, Spengler JD (2010) Effect of prenatal exposure to fine particulate matter on ventilatory lung function of preschool children of non-smoking mothers. Paediatr Perinat Epidemiol 24:492–501
35. Holgate ST (2008) Pathogenesis of asthma. Clin Exp Allergy 38:872–897
36. Byoung-Ju K, Soo-Jong H (2012) Ambient air pollution and allergic diseases in children. Korean J Pediatr 55:185–192
37. World Health Organisation (WHO) (2005) Effect of air pollution on children's health and development. World Health Organisation, Geneva
38. Calderon-Garciduenas L, Wang L, Zhang YJ, Rodriguez-Alcaraz A, Osnaya A, Villarreal-Calderon A, Santella RM (1999) 8-Hydroxy-2'-deoxyguanosine, a major mutagenic oxidative DNA lesion and DNA strand breaks in nasal respiratory epithelium of children exposed to urban pollution. Environ Health Perspect 107:469–474

39. Gehring U, Wijga AH, Brauer M, Fischer P, de Joungste JC, Kerkhof M et al (2010) Traffic-related air pollution and the development of asthma and allergies during the first 8 years of life. Am J Respir Crit Care Med 181:596–603
40. Caballero Balanza S, Martorell Aragones A, Cerda Mir JC, Remirez JB, Navarro Ivanez R, Navarro Soriano A, Felix Toledo R, Escribano Montaner A (2010) Leukotriene B4 and 8-isoprostane in exhaled breath condensate of children with episodic and persistent asthma. J Investig Allergol Clin Immunol 20:237–243
41. Hasan RA, Thomas J, Davidson B, Barnes J, Reddy R (2010) 8-Isoprostane in the exhaled breath condensate of children hospitalized for status asthmaticus. Pediatr Crit Care Med 12(1):e25–e28
42. Dut R, Dizdar EA, Birben E, Sackesen C, Soyer OU, Besler T, Kalayci O (2008) Oxidative stress and its determinants in the airways of children with asthma. Allergy 63:1605–1609
43. Nagai K, Betsuyaku T, Konno S, Ito Y, Nasuhara Y, Hizawa N, Kondo T, Nishimura M (2008) Diversity of protein carbonylation in allergic airway inflammation. Free Radic Res 42:921–929
44. Louhelainen N, Rytila P, Obase Y, Makela M, Haahtela T, Kinnula VL, Pelkonen A (2008) The value of sputum 8-isoprostane in detecting oxidative stress in mild asthma. J Asthma 45:149–154
45. Buthbumrung N, Mahidol C, Navasumrit P, Promvijit J, Hunsonti P, Autrup H, Ruchirawat M (2008) Oxidative DNA damage and influence of genetic polymorphisms among urban and rural schoolchildren exposed to benzene. Chem Biol Interact 172:185–194
46. Svecova V, Rossner P Jr, Dostal M, Topinka J, Solansky I, Sram RJ (2009) Urinary 8-oxodeoxyguanosine levels in children exposed to air pollutants. Mutat Res 662:37–43
47. Hertz-Picciotto I, Baker RJ, Yap PS, Dostal M, Joad JP, Lipsett M, Greenfield T, Herr CE, Benes I, Shumway RH, Pinkerton KE, Sram RJ (2007) Early childhood lower respiratory illness and air pollution. Environ Health Perspect 115:1510–1518
48. DiFranza JR, Lew RA (1995) Effect of maternal cigarette smoking on pregnancy complications and sudden infant death syndrome. J Fam Pract 40:385–394
49. DiFranza JR, Lew RA (1996) Morbidity and mortality in children associated with the use of tobacco products by other people. Pediatrics 97:560–568
50. Fox NL, Sexton M, Hebel JR (1990) Prenatal exposure to tobacco: I. Effects on physical growth at age three. Int J Epidemiol 19:66–71
51. Strachan DP, Cook DG (1997) Health effects of passive smoking. 1. Parental smoking and lower respiratory illness in infancy and early childhood. Thorax 52:905–914
52. Nilsson R, Nordlinder R, Moen BE, Ovrebo S, Bleie K, Skorve AH, Hollund BE, Tagesson C (2004) Increased urinary excretion of 8-hydroxydeoxyguanosine in engine room personnel exposed to polycyclic aromatic hydrocarbons. Occup Environ Med 61:692–696
53. Hu CW, Wu MT, Chao MR, Pan CH, Wang CJ, Swenberg JA, Wu KY (2004) Comparison of analyses of urinary 8-hydroxy-2'-deoxyguanosine by isotope-dilution liquid chromatography with electrospray tandem mass spectrometry and by enzyme-linked immunosorbent assay. Rapid Commun Mass Spectrom 18:505–510
54. Chuang CY, Lee CC, Chang YK, Sung FC (2003) Oxidative DNA damage estimated by urinary 8-hydroxydeoxyguanosine: influence of taxi driving, smoking and areca chewing. Chemosphere 52:1163–1171
55. Marczynski B, Rihs H, Rossbach B, Holzer J, Angerer J, Scherenberg M, Hoffmann G, Bruning T, Wilhelm M (2002) Analysis of 8-oxo-7,8-dihydro-2'-deoxyguanosine and DNA strand breaks in white blood cells of occupationally exposed workers: comparison with ambient monitoring, urinary metabolites and enzyme polymorphisms. Carcinogenesis 23:273–281
56. Marczynski B, Preuss R, Mensing T, Angerer J, Seidel A, El MA, Wilhelm M, Bruning T (2005) Genotoxic risk assessment in white blood cells of occupationally exposed workers before and after alteration of the polycyclic aromatic hydrocarbon (PAH) profile in the production material: comparison with PAH air and urinary metabolite levels. Int Arch Occup Environ Health 78:97–108

57. Zhang J, Ichiba M, Hanaoka T, Pan G, Yamano Y, Hara K, Takahashi K, Tomokuni K (2003) Leukocyte 8-hydroxydeoxyguanosine and aromatic DNA adduct in coke-oven workers with polycyclic aromatic hydrocarbon exposure. Int Arch Occup Environ Health 76:499–504
58. Casado A, De LN, Lopez-Fernandez E, Sanchez A, Jimenez JA (2006) Lipid peroxidation, occupational stress and aging in workers of a prehospital emergency service. Eur J Emerg Med 13:165–171
59. Rossner P Jr, Svecova V, Milcova A, Lnenickova Z, Solansky I, Sram RJ (2008) Seasonal variability of oxidative stress markers in city bus drivers—part I: oxidative damage to DNA. Mutat Res 642:14–20
60. Lagorio S, Tagesson C, Forastiere F, Iavarone I, Axelson O, Carere A (1994) Exposure to benzene and urinary concentrations of 8-hydroxydeoxyguanosine, a biological marker of oxidative damage to DNA. Occup Environ Med 51:739–743
61. Nilsson RI, Nordlinder RG, Tagesson C, Walles S, Jarvholm BG (1996) Genotoxic effects in workers exposed to low levels of benzene from gasoline. Am J Ind Med 30:317–324
62. Risom L, Moller P, Loft S (2005) Oxidative stress-induced DNA damage by particulate air pollution. Mutat Res 592:119–137
63. Liu L, Zhang Q, Feng J, Deng L, Zeng N, Yang A, Zhang W (1996) The study of DNA oxidative damage in benzene-exposed workers. Mutat Res 370:145–150
64. Sorensen M, Autrup H, Hertel O, Wallin H, Knudsen LE, Loft S (2003) Personal exposure to PM2.5 and biomarkers of DNA damage. Cancer Epidemiol Biomarkers Prev 12:191–196
65. Sorensen M, Loft S, Andersen HV, Raaschou-Nielsen O, Skovgaard LT, Knudsen LE, Nielsen IV, Hertel O (2005) Personal exposure to PM2.5, black smoke and NO2 in Copenhagen: relationship to bedroom and outdoor concentrations covering seasonal variation. J Expo Anal Environ Epidemiol 15:413–422
66. Rossner P Jr, Svecova V, Milcova A, Lnenickova Z, Solansky I, Santella RM, Sram RJ (2007) Oxidative and nitrosative stress markers in bus drivers. Mutat Res 617:23–32
67. Gutierrez-Castillo ME, Roubicek DA, Cebrian-Garcia ME, De Vizcaya-Ruiz A, Sordo-Cedeno M, Ostrosky-Wegman P (2006) Effect of chemical composition on the induction of DNA damage by urban airborne particulate matter. Environ Mol Mutagen 47:199–211
68. Pilger A, Rudiger HW (2006) 8-Hydroxy-2'-deoxyguanosine as a marker of oxidative DNA damage related to occupational and environmental exposures. Int Arch Occup Environ Health 80:1–15
69. Montuschi P, Barnes PJ, Roberts LJ (2004) Isoprostanes: markers and mediators of oxidative stress. FASEB J 18:1791–1800
70. Dalle-Donne I, Aldini G, Carini M, Colombo R, Rossi R, Milzani A (2006) Protein carbonylation, cellular dysfunction, and disease progression. J Cell Mol Med 10:389–406
71. Shacter E (2000) Quantification and significance of protein oxidation in biological samples. Drug Metab Rev 32:307–326
72. Singh VK, Patel DK, Singh J, Ram S, Mathur N, Siddiqui MK (2008) Blood levels of polycyclic aromatic hydrocarbons in children and their association with oxidative stress indices: an Indian perspective. Clin Biochem 41:152–161
73. Lykkesfeldt J (2007) Malondialdehyde as biomarker of oxidative damage to lipids caused by smoking. Clin Chim Acta 380:50–58
74. Toraason M, Hayden C, Marlow D, Rinehart R, Mathias P, Werren D, Olsen LD, Neumeister CE, Mathews ES, Cheever KL, Marlow KL, DeBord DG, Reid TM (2001) DNA strand breaks, oxidative damage, and 1-OH pyrene in roofers with coal-tar pitch dust and/or asphalt fume exposure. Int Arch Occup Environ Health 74:396–404
75. Rossner P Jr, Svecova V, Milcova A, Lnenickova Z, Solansky I, Sram RJ (2008) Seasonal variability of oxidative stress markers in city bus drivers. Part II. Oxidative damage to lipids and proteins. Mutat Res 642:21–27
76. Pereira CE, Heck TG, Saldiva PH, Rhoden CR (2007) Ambient particulate air pollution from vehicles promotes lipid peroxidation and inflammatory responses in rat lung. Braz J Med Biol Res 40:1353–1359

77. Barregard L, Sallsten G, Gustafson P, Andersson L, Johansson L, Basu S, Stigendal L (2006) Experimental exposure to wood-smoke particles in healthy humans: effects on markers of inflammation, coagulation, and lipid peroxidation. Inhal Toxicol 18:845–853
78. Sorensen M, Daneshvar B, Hansen M, Dragsted LO, Hertel O, Knudsen L, Loft S (2003) Personal PM2.5 exposure and markers of oxidative stress in blood. Environ Health Perspect 111:161–166
79. Rhoden CR, Lawrence J, Godleski JJ, Gonzales-Flecha B (2004) N-acetylcysteine prevents lung inflammation after short-term inhalation exposure to concentrated ambient particles. Toxicol Sci 79:296–303
80. Ceylan E, Kocyigit A, Gencer M, Aksoy N, Selek S (2006) Increased DNA damage in patients with chronic obstructive pulmonary disease who had once smoked or been exposed to biomass. Respir Med 100:1270–1276
81. Autrup H, Daneshvar B, Dragsted LO, Gamborg M, Hansen M, Loft S, Okkels H, Nielsen F, Nielsen PS, Raffn E, Wallin H, Knudsen LE (1999) Biomarkers for exposure to ambient air pollution—comparison of carcinogen-DNA adduct levels with other exposure markers and markers for oxidative stress. Environ Health Perspect 107:233–238
82. Chen C, Arjomandi M, Balmes J, Tager I, Holland N (2007) Effects of chronic and acute ozone exposure on lipid peroxidation and antioxidant capacity in healthy young adults. Environ Health Perspect 115:1732–1737
83. Block G, Dietrich M, Norkus EP, Morrow JD, Hudes M, Caan B, Packer L (2002) Factors associated with oxidative stress in human populations. Am J Epidemiol 156:274–285
84. Kato T, Inoue T, Morooka T, Yoshimoto N, Node K (2006) Short-term passive smoking causes endothelial dysfunction via oxidative stress in nonsmokers. Can J Physiol Pharmacol 84:523–529
85. Ahmadzadehfar H, Oguogho A, Efthimiou Y, Kritz H, Sinzinger H (2006) Passive cigarette smoking increases isoprostane formation. Life Sci 78:894–897
86. Kitano S, Hisatomi H, Hibi N, Kawano K, Harada S (2006) Improved method of plasma 8-isoprostane measurement and association analyses with habitual drinking and smoking. World J Gastroenterol 12:5846–5852
87. Rossner P Jr, Gammon MD, Terry MB, Agrawal M, Zhang FF, Teitelbaum SL, Eng SM, Gaudet MM, Neugut AI, Santella RM (2006) Relationship between urinary 15-F2t-isoprostane and 8-oxodeoxyguanosine levels and breast cancer risk. Cancer Epidemiol Biomarkers Prev 15:639–644
88. Rossner P Jr, Terry MB, Gammon MD, Agrawal M, Zhang FF, Ferris JS, Teitelbaum SL, Eng SM, Neugut AI, Santella RM (2007) Plasma protein carbonyl levels and breast cancer risk. J Cell Mol Med 11:1138–1148
89. Li N, Sioutas C, Cho A, Schmitz D, Misra C, Sempf J, Wang M, Oberley T, Froines J, Nel A (2003) Ultrafine particulate pollutants induce oxidative stress and mitochondrial damage. Environ Health Perspect 111:455–460
90. Hemminki K, Pershagen G (1994) Cancer risk of air pollution: epidemiological evidence. Environ Health Perspect 102(Suppl 4):187–192
91. Tao F, Gonzalez-Flecha B, Kobzik L (2003) Reactive oxygen species in pulmonary inflammation by ambient particulates. Free Radic Biol Med 35:327–340
92. Li N, Kim S, Wang M, Froines J, Sioutas C, Nel A (2002) Use of a stratified oxidative stress model to study the biological effects of ambient concentrated and diesel exhaust particulate matter. Inhal Toxicol 14:459–486

Chapter 18
Oxidative Stress in Inflammatory Bowel Disease

Tomohisa Takagi, Kazuhiko Uchiyama, and Yuji Naito

Abbreviations

5-ASA	5-Aminosalicylate
CD	Crohn's disease
Cu/Zn	Copper/zinc
DSS	Dextran sodium sulfate
EC	Extracellular
eNOS	Endothelial nitric oxide synthase
GPx	Glutathione peroxidases
GR	Glutathione reductase
GSH	Reduced glutathione
GSSG	Glutathione disulfide
H_2O_2	Hydrogen peroxide
HO	Heme oxygenase
IBD	Inflammatory bowel disease
IL	Interleukin
iNOS	Inducible nitric oxide synthase
LPS	Lipopolysaccharide
Mn	Manganese
MT	Metallothionein
nNOS	Neuronal nitric oxide synthase
NO	Nitric oxide
NOS	Nitric oxide synthase
Nox	NADPH oxidase

T. Takagi, M.D., Ph.D. • K. Uchiyama, M.D., Ph.D. • Y. Naito, M.D., Ph.D. (✉)
Department of Molecular Gastroenterology and Hepatology, Graduate School of Medical Science, Kyoto Prefectural University of Medicine, 465 Kajii-cho, Kawaramachi-Hirokoji, Kamigyo-ku, Kyoto 602-8566, Japan
e-mail: ynaito@koto.kpu-m.ac.jp

H. Tsukahara and K. Kaneko (eds.), *Studies on Pediatric Disorders*, Oxidative Stress in Applied Basic Research and Clinical Practice, DOI 10.1007/978-1-4939-0679-6_18,

NT	Nitrotyrosine
O_2^-	Superoxide
PC	Lecithinized
Prx	Peroxiredoxins
RNS	Reactive nitrogen species
ROS	Reactive oxygen species
Se	Selenium
SOD	Superoxide dismutase
TNBS	2,4,6-Trinitrobenzene sulfonic acid
TRX	Thioredoxin
UC	Ulcerative colitis

18.1 Introduction

Inflammatory bowel disease (IBD), which includes ulcerative colitis (UC) and Crohn's disease (CD), is a chronic and recurrent intestinal inflammatory disorder [1]. Although the precise etiology of IBD remains unknown, genetic susceptibility, environmental factors including the intestinal microbiota, and a disrupted immune system are believed to contribute to the pathogenesis of the disease [1, 2]. The key features of UC include diffuse mucosal inflammation that extends proximally from the rectum. Histopathological features include the presence of a considerable number of neutrophils in the lamina propria and the crypts, and it is well known that the disease activity of UC is linked to an influx of neutrophils into the mucosa and subsequently into the intestinal lumen, resulting in the formation of the so-called crypt abscesses (Fig. 18.1a, arrow head). On the other hand, CD is characterized by the aggregation of macrophages that frequently form noncaseating granulomas (Fig. 18.1b, arrow). Unlike UC, any site of the gastrointestinal tract may be affected in patients with CD. In addition, inflammatory lesions may be patchy, segmental, and transmural, and the involvement of the terminal ileum is common in CD.

The prevalence of IBD in Japan, which is still lower compared with that in Western countries, has been steadily increasing in recent decades [3]. According to the Japanese nationwide IBD registry, 10.6 % of CD and 5.9 % of UC were reported in patients <16 years old [4]. In comparison with adults, pediatric IBD patients tend to develop more severe disease activity and more extensive disease [4, 5]. Therefore, it is important to investigate the pathogenesis of IBD and to find new anti-inflammatory strategies.

Recent evidences from clinical and basic studies have revealed that the enhanced formation of reactive oxygen species (ROS) or reactive nitrogen species (RNS) contributes to the pathogenesis of IBD. In addition, the antioxidant defense systems that protect against excessive ROS and RNS are suppressed in the inflamed intestine of IBD patients. Here, we review the recent knowledge about the role of ROS, RNS, and the antioxidant defense system in IBD.

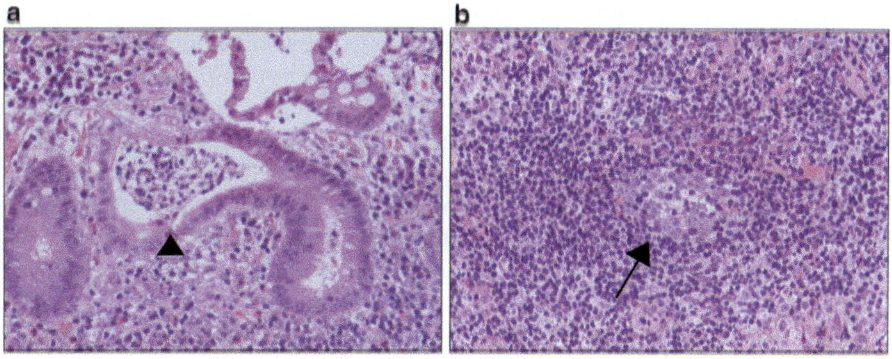

Fig. 18.1 Histologic hallmarks of inflammatory bowel disease. (**a**) Ulcerative colitis (resected colonic tissue from a 17-year-old female patient with active disease). The crypt abscess (*arrow head*) is composed of transmigrated neutrophils. (**b**) Crohn's disease (biopsy from a terminal ileum of a 29-year-old male patient with active disease). The figure shows a discrete granuloma composed of macrophages (*arrow*). There is marked infiltration of lymphoid cells surrounding the granuloma cells

18.2 ROS in IBD

Although IBD is a multifactorial disease, one of the most prominent histological features observed in IBD is infiltration of neutrophils into the inflamed mucosa. In addition, circulating activated neutrophils, a major source of inflammatory cytokines, are elevated in active IBD. Several studies have shown that granulocyte/monocyte adsorptive apheresis therapy can induce remission, especially in patients with active UC and CD that are refractory to conventional therapy [6–9]. Activated neutrophils are known to be potential sources of free radical production. On the other hand, monocytes/macrophages play an important role in free radical production in CD.

The presence of ROS has been extensively studied in patients with IBD. The chemiluminescence method was used to directly quantify the ROS level in colon biopsy specimens from patients with UC or CD, and this showed that the ROS level was markedly increased in these diseased mucosal samples as compared with normal mucosa. However, because ROS have short biological half-lives, their presence is generally measured indirectly by assessing the levels of oxidatively damaged molecules. This includes assessment of lipid peroxidation because unsaturated bonds in membrane phospholipids are the major targets for free radical reactions and oxidative DNA damage. Although there are some conflicting reports, the production of excessively oxidized molecules appears to be upregulated in patients with IBD as compared with healthy control subjects in a variety of organic systems, including serum, urine, and biopsy specimens from the colonic mucosa [10, 11].

It was previously thought that the increase in NADPH oxidase (Nox) due to neutrophil and macrophage invasion results in oxidative stress in the colon. Recent investigations have revealed the presence of six homologs of the cytochrome subunit of the phagocyte Nox: Nox1, Nox3, Nox4, Nox5, Duox1, and Duox2. Together with the phagocyte Nox (Nox2/gp91^{phox}), the homologs are now referred to as the Nox family of NADPH oxidases. All these enzymes can transport electrons across the plasma membrane and generate superoxide (O_2^-) and other downstream ROS. After the discovery of Nox homologs, it became clear that Nox1 is highly expressed in the distal colon epithelium [12, 13]. Although the function of Nox1 in the distal gastrointestinal tract is not yet understood, Nox1 may play a role in host defense. Nox1 expression is activated by bacterial lipopolysaccharide (LPS) [14] and flagellins [15], and is more highly expressed in the distal colon, as expected from the bacterial colonization pattern [16].

Furthermore, Nox1 overactivity could be involved in IBD pathogenesis. Interleukin (IL)-10-deficient mice are a well-established model of spontaneously developing colitis, and experiments with these mice showed that the absence of IL-10 significantly facilitated Nox1 expression in association with increased interferon-γ expression before the development of spontaneous colitis [17]. In situ hybridization studies of colon biopsy samples obtained from patients with CD or UC revealed that Nox1 expression is localized in lesional lymphocytes [16]. Therefore, Nox1 may be involved in the onset of IBD, although the role of Nox1 in lymphocyte action and the possible link with IBD pathophysiology remain unknown.

18.3 RNS in IBD

Accumulating evidence indicates that the inflamed mucosa of IBD patients contains increased levels of RNS such as nitric oxide (NO). The increased serum and urinary nitrate levels in patients with IBD promote the increase in NO production [18–20]. Increased NO production in the colon of patients with UC could be confirmed by the detection of gas during colonoscopy [21]. Moreover, high levels of NO metabolic end products such as nitrates and nitrites are found in the plasma, urine, and colonic lumen of IBD patients. However, because diet and the presence of bacteria can contribute to the increase in NO levels, these indirect measurements might not accurately reflect the changes in NO levels in the intestinal mucosa.

NO is produced by a group of enzymes known as NO synthase (NOS), which consists of three isoforms: the neuronal (nNOS), endothelial (eNOS), and inducible (iNOS) isoforms. iNOS expression is markedly increased in the inflamed colonic mucosa of patients with IBD, suggesting that iNOS is a major source of increased NO production [22–24]. High iNOS activity was also reported in experimental models of colitis. Although it is controversial whether the effects of iNOS-induced NO are beneficial or detrimental in experimental colitis [25], the majority of studies on selective iNOS inhibitors in murine colitis models have shown the effects to be beneficial. We reported that treatment with an iNOS-selective inhibitor, ONO-1714, ameliorated dextran sodium sulfate (DSS)-induced colitis in mice, suggesting that iNOS is involved in the progression of intestinal inflammation [26].

NO is inactivated by O_2^-, and the interaction of NO with O_2^- prevents O_2^--mediated hydroxylation reactions. Thus, the activity of NO may be dependent on the local O_2^- level and its scavenger systems [27]. Neutrophils can be induced to simultaneously produce NO and O_2^- in a concentrated localized manner by a variety of stimuli. NO and O_2^- then react to produce peroxynitrite, a potent and long-lived oxidant [28]. The peroxynitrite anion is cytotoxic because it inhibits mitochondrial electron transport, oxidizes protein sulfhydryl groups, initiates lipid peroxidation, and nitrates amino acids such as tyrosine, which affects many signal transduction pathways. The production of peroxynitrite can be indirectly inferred by the presence of nitrotyrosine (NT) residues [29]. NT is produced in the inflamed colonic mucosa of patients with IBD [24]. Keshavarzian et al. [30] demonstrated a correlation between the severity of colitis and NT levels, and reported that the actin cytoskeletal protein is a potential target of nitration. Ongoing studies in our laboratory include the analysis of oxidatively modified proteins such as NT-modified proteins. Our preliminary results identified NT-modified proteins in the inflamed colonic mucosa of the 2,4,6-trinitrobenzene sulfonic acid (TNBS)-induced colitis model, consistent with the report of Keshavarzian et al. [30]. This indicates that identifying specific nitrated proteins would provide both mechanistic information about nitrosative stress and insight into its downstream functional consequences.

18.4 Antioxidant Defense System in IBD

Oxidative stress is defined as an imbalance between the generation of ROS and the antioxidant defense systems. In patients with IBD, the disrupted antioxidant defense systems appear to be involved in the development of intestinal inflammation.

18.4.1 Superoxide Dismutases

Among the diverse endogenous mucosal antioxidant defense mechanisms, superoxide dismutases (SODs) are the primary defense against ROS. They convert O_2^- into hydrogen peroxide (H_2O_2), which is subsequently neutralized to oxygen and water by catalase or glutathione peroxidases (GPx). There are three SOD isoforms in humans: cytoplasmic copper/zinc (Cu/Zn)-SOD, mitochondrial manganese (Mn)-SOD, and extracellular (EC)-SOD. Accumulating evidence indicates that the protein levels and activities of Cu/Zn-SOD and EC-SOD are decreased in the inflamed mucosa of IBD patients. On the other hand, it was reported that Mn-SOD is increased in the inflamed mucosa of IBD patients compared with the noninflamed and normal control mucosa. However, an increased Mn-SOD level is not always associated with increased resistance to oxidative stress because most of the Mn-SOD exists in an enzymatically inactive form [31]. These observations provide a rationale for SOD-based intervention therapy in IBD. In our previous study, treatment with Mn-SOD inhibited DSS-induced colonic mucosal injury in mice [32]. On the other hand, Cu/Zn-SOD-overexpressing mice showed significantly lower colonic neutrophilic myeloperoxidase activity than their nontransgenic littermates [33]. The administration of Cu/Zn-SOD suppressed the development of IBD-related colitis in experimental animal models [34, 35]. However, subsequent Cu/Zn-SOD clinical trials have been unsuccessful because of its low stability in plasma [36]. Lecithinized SOD (PC-SOD) has improved plasma stability, and is a potentially beneficial clinical treatment for IBD. The administration of PC-SOD exerts an ameliorative effect against DSS-induced colitis by decreasing the colonic level of ROS [37]. Moreover, the efficacy of PC-SOD for the treatment of UC has already been demonstrated by a phase II clinical study [38]. In the future, the clinical application of PC-SOD for IBD therapy is anticipated.

18.4.2 Glutathione

The glutathione system (reduced glutathione [GSH], GPx, and glutathione reductase [GR]) is a key defense against H_2O_2 and other peroxides. In the presence of ROS, GSH is oxidized to glutathione disulfide (GSSG). This reaction is catalyzed by GPx and reversed by GR [39]. The reported changes in GSH levels vary across different studies. Iantomasi et al. [40] reported that GSH was decreased in the inflamed ileum of CD patients compared with the noninflamed ileum and controls. Tsunada et al. [41] also demonstrated that GSH levels were significantly depleted in UC patients with active colitis as compared with the normal controls. On the other hand, some investigators have reported that the GSH level in inflamed mucosal samples from both UC and CD patients was unchanged compared with normal subjects [42]. Sido et al. [43] identified increased GSSG levels in the inflamed mucosa of patients with both UC and CD, consistent with previous reports [41]. Holmes et al. [44] also reported decreased activity of γ-glutamylcysteine synthetase, the rate-limiting enzyme for glutathione biosynthesis, in the inflamed mucosa of patients with IBD. Most studies revealed high levels of GPx in the inflamed mucosa of both CD and UC patients, which appears to be a response against oxidative stress [41, 42, 44].

18.4.3 Thioredoxin

Thioredoxin (TRX) is one of the most important molecules that control the redox regulation system. It contains a redox-active disulfide/dithiol (two –SH groups that reside close to each other because of neighboring Cys residues) site: –Cys–Gly–Pro–Cys–. TRX plays a pivotal role in scavenging ROS with peroxiredoxins (Prx) and prevents apoptosis of various cell types such as lymphocytes, monocytes, and epithelial cells. Tamaki et al. [45] reported that serum TRX levels were significantly higher in patients with both active UC and active CD than in normal controls, and TRX levels correlated with disease activity. They also revealed that the administration of TRX significantly ameliorated DSS-induced colitis and colonic inflammation in IL-10-deficient mice. Interestingly, the expression of TRX interacting protein, which is a negative regulator of TRX, was significantly lower in the colonic mucosa of UC than in normal tissues [46]. These data indicate that TRX might be a new therapeutic molecule for IBD treatment.

18.4.4 Peroxiredoxin

Prx are recently characterized members of the selenium (Se)-independent peroxidase superfamily that detoxify ROS and RNS. There are six described members of the Prx family in mammalian tissues, and these Prx proteins were further classified on the basis of whether they contain one or two conserved Cys residues [47]. Our previous proteomic analysis of intestinal mucosa from mice with DSS-induced colitis demonstrated decreased expression of Prx6, the sole mammalian 1-Cys Prx [48]. Prx6 is unlike the other mammalian members of the Prx family because it uses glutathione instead of TRX as a reductant. Interestingly, Iizuka et al. [49] have reported that some CD patients have an antibody against Prx6-like protein, which may be involved in the pathogenesis of CD. Our preliminary study showed decreased colonic expression of Prx6 mRNA and protein in the inflamed mucosa of IBD patients compared with normal colon tissues [50]. The role of Prx6 in the intestinal inflammation has not yet been described; however, downregulation of Prx6 might be involved in the pathogenesis of IBD.

18.4.5 Heme Oxygenase

Heme oxygenase-1 (HO-1) is one of three mammalian HO isozymes and is a stress-responsive protein induced by various oxidative agents, including oxidative stress, heat shock, ultraviolet radiation, ischemia–reperfusion, heavy metals, LPS, cytokines, NO, and heme (its substrate). The strong adaptive response of HO-1 to various stimuli suggests an entirely new paradigm in which HO-1 may play a significant role in protecting against intestinal inflammation [51]. We showed that the expression of HO-1, which was mainly localized in mononuclear cells, was significantly increased in the colonic mucosa of patients with active UC as compared with normal mucosa [52]. In an animal colitis model, HO activity and HO-1 expression markedly increased after induction of murine experimental colitis, and administration of an HO inhibitor potentiated colonic damage and inflammation [53–55]. Recent investigation has shown that upregulation of HO-1 by several HO-1 inducers significantly reduces various types of colitis (Table 18.1) [56–73]. In addition, it has been demonstrated that HO-1 mainly localized in macrophages in intestinal lamina propria, and macrophages highly expressing HO-1 manifested M2 macrophage markers (Arginase-1, Fizz-1, Ym1, and MRC1) and acquired anti-inflammatory function [58]. These results indicate that HO-1 is an inducible protein responsible for host defense against intestinal inflammation and may be a novel therapeutic molecule for IBD [74].

In this respect, it has been shown that 5-aminosalicylate (5-ASA), which is a conventional agent for the treatment of IBD, can induce HO-1 [60, 75]. However, because 5-ASA exerts anti-inflammatory effects through various other functional activities, including by scavenging ROS and by inhibiting NF-κB as well as T-cell proliferation, its therapeutic importance as an HO-1 inducer remains unclear. Interestingly, recent investigations have demonstrated that treatment of UC patients

Table 18.1 Therapeutic efficacy of heme oxygenase-1 (HO-1) induction in using in vivo animal colitis model

Type of experimental models	HO-1 induction	Animal	References
DSS-induced colitis	Peracetylated (−)-epigallocatechin-3-gallate	Mouse	Chiou et al. [57]
	BTZO-15	Rat	Yukitake et al. [64]
	Hemin	Mouse	Zhong et al. [66]
	Tranilast	Mouse	Sun et al. [65]
	Cobalt protoporphyrin (CoPP)	Mouse	Paul et al. [68]
	Cobalt protoporphyrin (CoPP)	Mouse	Berberat et al. [56]
TNBS-induced colitis	BTB and CNC homolog 1 (Bach1) deficiency	Mouse	Harusato et al. [58]
	Oligosaccharides	Mouse	Higashimura et al. [59]
	5-Aminosalicylic acid	Rat	Horvath et al. [60]
	Heme, cadmium	Rat	Varga et al. [67]
	Chalcone	Mouse	Lee et al. [69]
	Glutamine	Rat	Giris et al. [70]
	Octreotide	Rat	Erbil et al. [71]
	Gliotoxin	Mouse	Jun et al. [72]
Acetic acid-induced colitis	Nadroparin	Rat	Yalniz et al. [63]
Interleukin (IL)-10 deficient (genetic colitis model)	Cobalt protoporphyrin (CoPP)	Mouse	Onyiah et al. [61]
	Carbon monoxide (CO)	Mouse	Hegazi et al. [73]
TCR-α deficient (genetic colitis model)	Carbon monoxide (CO), cobalt protoporphyrin (CoPP)	Mouse	Sheikh et al. [62]

DSS dextran sodium sulfate, *TNBS* 2,4,6-trinitrobenzene sulfonic acid, *TCR* T-cell receptor

with 5-ASA compounds can prevent UC-associated colon carcinogenesis [76]. This result suggests that at least one of the agents that are used for IBD treatment can provide excellent therapeutic effect by inducing HO-1 expression. In the near future, we anticipate the development of novel IBD therapeutic strategies that are based on the regulation of HO-1 expression.

18.4.6 Metallothionein

Metallothionein (MT) is a Cys-rich low-molecular-weight protein that can act as an ROS scavenger. The role of MT in IBD has not yet been clarified. Concerning MT expression in the intestine, there are conflicting reports on the role of MTs in intestinal inflammation. Although most investigations have demonstrated increased MT expression in patients with IBD [77, 78], other studies have demonstrated that MT expression was decreased in the colonic mucosa of patients with IBD [42, 79]. Experimental colitis model studies on MT-deficient mice revealed that MT does not protect against the development of colitis [80, 81]. In our recent study, colonic injury was significantly aggravated in the DSS-treated MT-deficient mice, as compared with the DSS-treated wild-type mice [82]. Interestingly, the expression of MT localized in F4/80-positive macrophages in murine colonic mucosa and MT contributed to anti-inflammatory function in macrophages, indicating that endogenous MTs play an important role in the protection of the intestinal mucosa through the regulation of macrophage function.

18.5 Conclusion

We have summarized the recent findings about the role of oxidative stress in IBD. A growing body of evidence indicates that oxidants such as ROS and RNS play a key role in the pathophysiology of IBD, particularly in the initiation and perpetuation of inflammation and in subsequent tissue damage. Furthermore, we focused on the depressed antioxidant defense system involved in IBD pathogenesis (Fig. 18.2). Regulating excessive oxidative stress and restoring the absent antioxidant defense system may be important therapeutic strategies for IBD treatment. In the near future, we anticipate the development of novel IBD therapeutic strategies that are based on the regulation of oxidative stress.

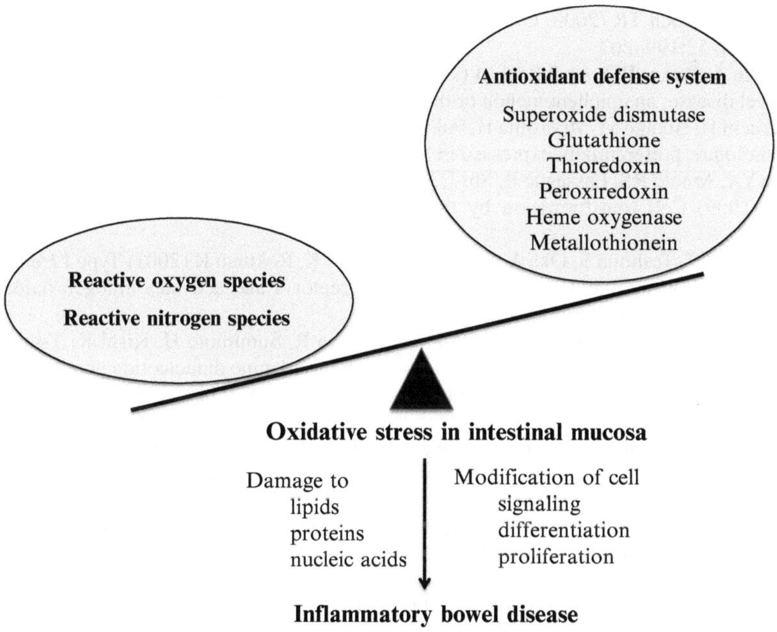

Fig. 18.2 The concept of oxidative stress in inflammatory bowel disease. An imbalance between the production of reactive oxygen species (ROS) and reactive nitrogen species (RNS) and protection against ROS and RNS by the antioxidant defense system in intestinal mucosa results in the occurrence of oxidative stress. This leads to the onset of inflammatory bowel disease through the damage of lipids, proteins, and nucleic acids and the modification of cell signaling, differentiation, and proliferation

References

1. Xavier RJ, Podolsky DK (2007) Unravelling the pathogenesis of inflammatory bowel disease. Nature 448:427–434
2. Podolsky DK (2002) Inflammatory bowel disease. N Engl J Med 347:417–429
3. Asakura K, Nishiwaki Y, Inoue N, Hibi T, Watanabe M, Takebayashi T (2009) Prevalence of ulcerative colitis and Crohn's disease in Japan. J Gastroenterol 44:659–665
4. Ishige T, Tomomasa T, Takebayashi T, Asakura K, Watanabe M, Suzuki T, Miyazawa R, Arakawa H (2010) Inflammatory bowel disease in children: epidemiological analysis of the nationwide IBD registry in Japan. J Gastroenterol 45:911–917
5. Biank V, Broeckel U, Kugathasan S (2007) Pediatric inflammatory bowel disease: clinical and molecular genetics. Inflamm Bowel Dis 13:1430–1438
6. Fukuda Y, Matsui T, Suzuki Y, Kanke K, Matsumoto T, Takazoe M, Matsumoto T, Motoya S, Honma T, Sawada K, Yao T, Shimoyama T, Hibi T (2004) Adsorptive granulocyte and monocyte apheresis for refractory Crohn's disease: an open multicenter prospective study. J Gastroenterol 39:1158–1164
7. Nagase K, Fukunaga K, Kashiwamura S, Kono T, Kamikozuru K, Yokoyama Y, Hida N, Ohda Y, Takeda N, Yoshida K, Iimuro M, Kikuyama R, Kato K, Miwa H, Matsumoto T (2011) Immunoregulatory effects of adsorptive granulocyte and monocyte apheresis in patients with drug refractory Crohn's disease. Ther Apher Dial 15:367–373
8. Naito Y, Takagi T, Yoshikawa T (2007) Neutrophil-dependent oxidative stress in ulcerative colitis. J Clin Biochem Nutr 41:18–26
9. Naito Y, Takagi T, Yoshikawa T (2007) Molecular fingerprints of neutrophil-dependent oxidative stress in inflammatory bowel disease. J Gastroenterol 42:787–798
10. Karp SM, Koch TR (2006) Oxidative stress and antioxidants in inflammatory bowel disease. Dis Mon 52:199–207
11. Rezaie A, Parker RD, Abdollahi M (2007) Oxidative stress and pathogenesis of inflammatory bowel disease: an epiphenomenon or the cause? Dig Dis Sci 52:2015–2021
12. Kikuchi H, Hikage M, Miyashita H, Fukumoto M (2000) NADPH oxidase subunit, gp91(phox) homologue, preferentially expressed in human colon epithelial cells. Gene 254:237–243
13. Suh YA, Arnold RS, Lassegue B, Shi J, Xu X, Sorescu D, Chung AB, Griendling KK, Lambeth JD (1999) Cell transformation by the superoxide-generating oxidase mox1. Nature 401: 79–82
14. Kawahara T, Teshima S, Oka A, Sugiyama T, Kishi K, Rokutan K (2001) Type I *Helicobacter pylori* lipopolysaccharide stimulates toll-like receptor 4 and activates mitogen oxidase 1 in gastric pit cells. Infect Immun 69:4382–4389
15. Kawahara T, Kuwano Y, Teshima-Kondo S, Takeya R, Sumimoto H, Kishi K, Tsunawaki S, Hirayama T, Rokutan K (2004) Role of nicotinamide adenine dinucleotide phosphate oxidase 1 in oxidative burst response to toll-like receptor 5 signaling in large intestinal epithelial cells. J Immunol 172:3051–3058
16. Szanto I, Rubbia-Brandt L, Kiss P, Steger K, Banfi B, Kovari E, Herrmann F, Hadengue A, Krause KH (2005) Expression of NOX1, a superoxide-generating NADPH oxidase, in colon cancer and inflammatory bowel disease. J Pathol 207:164–176
17. Kamizato M, Nishida K, Masuda K, Takeo K, Yamamoto Y, Kawai T, Teshima-Kondo S, Tanahashi T, Rokutan K (2009) Interleukin 10 inhibits interferon gamma- and tumor necrosis factor alpha-stimulated activation of NADPH oxidase 1 in human colonic epithelial cells and the mouse colon. J Gastroenterol 44:1172–1184
18. Melichar B, Karlicek R, Tichy M (1994) Increased urinary nitrate excretion in inflammatory bowel disease. Eur J Clin Chem Clin Biochem 32:3–4
19. Oudkerk Pool M, Bouma G, Visser JJ, Kolkman JJ, Tran DD, Meuwissen SG, Pena AS (1995) Serum nitrate levels in ulcerative colitis and Crohn's disease. Scand J Gastroenterol 30: 784–788

20. Sasajima K, Yoshida Y, Yamakado S, Sato J, Miyashita M, Okawa K, Matsutani T, Onda M, Kawano E (1996) Changes in urinary nitrate and nitrite during treatment of ulcerative colitis. Digestion 57:170–173
21. Perner A, Nordgaard I, Matzen P, Rask-Madsen J (2002) Colonic production of nitric oxide gas in ulcerative colitis, collagenous colitis and uninflamed bowel. Scand J Gastroenterol 37:183–188
22. Kimura H, Hokari R, Miura S, Shigematsu T, Hirokawa M, Akiba Y, Kurose I, Higuchi H, Fujimori H, Tsuzuki Y, Serizawa H, Ishii H (1998) Increased expression of an inducible isoform of nitric oxide synthase and the formation of peroxynitrite in colonic mucosa of patients with active ulcerative colitis. Gut 42:180–187
23. Rachmilewitz D, Stamler JS, Bachwich D, Karmeli F, Ackerman Z, Podolsky DK (1995) Enhanced colonic nitric oxide generation and nitric oxide synthase activity in ulcerative colitis and Crohn's disease. Gut 36:718–723
24. Singer II, Kawka DW, Scott S, Weidner JR, Mumford RA, Riehl TE, Stenson WF (1996) Expression of inducible nitric oxide synthase and nitrotyrosine in colonic epithelium in inflammatory bowel disease. Gastroenterology 111:871–885
25. Cross RK, Wilson KT (2003) Nitric oxide in inflammatory bowel disease. Inflamm Bowel Dis 9:179–189
26. Naito Y, Takagi T, Ishikawa T, Handa O, Matsumoto N, Yagi N, Matsuyama K, Yoshida N, Yoshikawa T (2001) The inducible nitric oxide synthase inhibitor ONO-1714 blunts dextran sulfate sodium colitis in mice. Eur J Pharmacol 412:91–99
27. Moncada S, Palmer RM, Higgs EA (1991) Nitric oxide: physiology, pathophysiology, and pharmacology. Pharmacol Rev 43:109–142
28. Beckman JS, Koppenol WH (1996) Nitric oxide, superoxide, and peroxynitrite: the good, the bad, and ugly. Am J Physiol 271:C1424–C1437
29. Ischiropoulos H (1998) Biological tyrosine nitration: a pathophysiological function of nitric oxide and reactive oxygen species. Arch Biochem Biophys 356:1–11
30. Keshavarzian A, Banan A, Farhadi A, Komanduri S, Mutlu E, Zhang Y, Fields JZ (2003) Increases in free radicals and cytoskeletal protein oxidation and nitration in the colon of patients with inflammatory bowel disease. Gut 52:720–728
31. Kruidenier L, Kuiper I, van Duijn W, Marklund SL, van Hogezand RA, Lamers CB, Verspaget HW (2003) Differential mucosal expression of three superoxide dismutase isoforms in inflammatory bowel disease. J Pathol 201:7–16
32. Naito Y, Takagi T, Handa O, Ishikawa T, Matsumoto N, Yagi N (2001) Role of superoxide and lipid peroxidation in the pathogenesis of dextran sulfate sodium-colitis in mice. ITE Lett 2:663–667
33. Kruidenier L, van Meeteren ME, Kuiper I, Jaarsma D, Lamers CB, Zijlstra FJ, Verspaget HW (2003) Attenuated mild colonic inflammation and improved survival from severe DSS-colitis of transgenic Cu/Zn-SOD mice. Free Radic Biol Med 34:753–765
34. Keshavarzian A, Morgan G, Sedghi S, Gordon JH, Doria M (1990) Role of reactive oxygen metabolites in experimental colitis. Gut 31:786–790
35. Segui J, Gironella M, Sans M, Granell S, Gil F, Gimeno M, Coronel P, Pique JM, Panes J (2004) Superoxide dismutase ameliorates TNBS-induced colitis by reducing oxidative stress, adhesion molecule expression, and leukocyte recruitment into the inflamed intestine. J Leukoc Biol 76:537–544
36. Greenwald RA (1990) Superoxide dismutase and catalase as therapeutic agents for human diseases. A critical review. Free Radic Biol Med 8:201–209
37. Ishihara T, Tanaka K, Tasaka Y, Namba T, Suzuki J, Ishihara T, Okamoto S, Hibi T, Takenaga M, Igarashi R, Sato K, Mizushima Y, Mizushima T (2009) Therapeutic effect of lecithinized superoxide dismutase against colitis. J Pharmacol Exp Ther 328:152–164
38. Suzuki Y, Matsumoto T, Okamoto S, Hibi T (2008) A lecithinized superoxide dismutase (PC-SOD) improves ulcerative colitis. Colorectal Dis 10:931–934
39. Meister A, Anderson ME (1983) Glutathione. Annu Rev Biochem 52:711–760

40. Iantomasi T, Marraccini P, Favilli F, Vincenzini MT, Ferretti P, Tonelli F (1994) Glutathione metabolism in Crohn's disease. Biochem Med Metab Biol 53:87–91
41. Tsunada S, Iwakiri R, Ootani H, Aw TY, Fujimoto K (2003) Redox imbalance in the colonic mucosa of ulcerative colitis. Scand J Gastroenterol 38:1002–1003
42. Kruidenier L, Kuiper I, Van Duijn W, Mieremet-Ooms MA, van Hogezand RA, Lamers CB, Verspaget HW (2003) Imbalanced secondary mucosal antioxidant response in inflammatory bowel disease. J Pathol 201:17–27
43. Sido B, Hack V, Hochlehnert A, Lipps H, Herfarth C, Droge W (1998) Impairment of intestinal glutathione synthesis in patients with inflammatory bowel disease. Gut 42:485–492
44. Holmes EW, Yong SL, Eiznhamer D, Keshavarzian A (1998) Glutathione content of colonic mucosa: evidence for oxidative damage in active ulcerative colitis. Dig Dis Sci 43: 1088–1095
45. Tamaki H, Nakamura H, Nishio A, Nakase H, Ueno S, Uza N, Kido M, Inoue S, Mikami S, Asada M, Kiriya K, Kitamura H, Ohashi S, Fukui T, Kawasaki K, Matsuura M, Ishii Y, Okazaki K, Yodoi J, Chiba T (2006) Human thioredoxin-1 ameliorates experimental murine colitis in association with suppressed macrophage inhibitory factor production. Gastroenterology 131:1110–1121
46. Takahashi Y, Masuda H, Ishii Y, Nishida Y, Kobayashi M, Asai S (2007) Decreased expression of thioredoxin interacting protein mRNA in inflamed colonic mucosa in patients with ulcerative colitis. Oncol Rep 18:531–535
47. Rhee SG, Chae HZ, Kim K (2005) Peroxiredoxins: a historical overview and speculative preview of novel mechanisms and emerging concepts in cell signaling. Free Radic Biol Med 38:1543–1552
48. Naito Y, Takagi T, Okada H, Omatsu T, Mizushima K, Handa O, Kokura S, Ichikawa H, Fujiwake H, Yoshikawa T (2010) Identification of inflammation-related proteins in a murine colitis model by 2D fluorescence difference gel electrophoresis and mass spectrometry. J Gastroenterol Hepatol 25(Suppl):144–148
49. Iizuka M, Nakagomi O, Nanjo H, Chiba M, Fukushima T, Sugita A, Sagara S, Horie Y, Watanabe S (2012) Molecular cloning reveals nearly half of patients with Crohn's disease have an antibody to peroxiredoxin 6-like protein. J Gastroenterol Hepatol 27:1388–1394
50. Takagi T, Naito Y, Yoshikawa T (2011) Free radicals in inflammatory bowel disease. In: Lerch MM (ed) Frontiers of gastrointestinal research, vol 29, Free radical biology in digestive diseases. S. Karger, Basel, pp 128–136
51. Naito Y, Takagi T, Yoshikawa T (2004) Heme oxygenase-1: a new therapeutic target for inflammatory bowel disease. Aliment Pharmacol Ther 20(Suppl 1):177–184
52. Takagi T, Naito Y, Mizushima K, Nukigi Y, Okada H, Suzuki T, Hirata I, Omatsu T, Okayama T, Handa O, Kokura S, Ichikawa H, Yoshikawa T (2008) Increased intestinal expression of heme oxygenase-1 and its localization in patients with ulcerative colitis. J Gastroenterol Hepatol 23(Suppl 2):S229–S233
53. Naito Y, Takagi T, Tomatsuri N (2003) Role of heme oxygenase-1 in dextran sulfate sodium-induced intestinal inflammation in mice. Gastroenterology 124(Suppl):A-490
54. Takagi T, Naito Y, Katada K (2004) Heme oxygenase regulates the balance of inflammatory cytokines in dextran sulfate sodium-induced colitis. Gastroenterology 126(Suppl):A-564
55. Wang WP, Guo X, Koo MW, Wong BC, Lam SK, Ye YN, Cho CH (2001) Protective role of heme oxygenase-1 on trinitrobenzene sulfonic acid-induced colitis in rats. Am J Physiol 281:G586–G594
56. Berberat PO, Yi AR, Yamashita K, Warny MM, Csizmadia E, Robson SC, Bach FH (2005) Heme oxygenase-1-generated biliverdin ameliorates experimental murine colitis. Inflamm Bowel Dis 11:350–359
57. Chiou YS, Ma NJ, Sang S, Ho CT, Wang YJ, Pan MH (2012) Peracetylated (-)-epigallocatechin-3-gallate (AcEGCG) potently suppresses dextran sulfate sodium-induced colitis and colon tumorigenesis in mice. J Agric Food Chem 60:3441–3451
58. Harusato A, Naito Y, Takagi T, Uchiyama K, Mizushima K, Hirai Y, Higashimura Y, Katada K, Handa O, Ishikawa T, Yagi N, Kokura S, Ichikawa H, Muto A, Igarashi K, Yoshikawa T

(2013) BTB and CNC homolog 1 (Bach1) deficiency ameliorates TNBS colitis in mice: role of M2 macrophages and heme oxygenase-1. Inflamm Bowel Dis 19:740–753
59. Higashimura Y, Naito Y, Takagi T, Mizushima K, Hirai Y, Harusato A, Ohnogi H, Yamaji R, Inui H, Nakano Y, Yoshikawa T (2013) Oligosaccharides from agar inhibit murine intestinal inflammation through the induction of heme oxygenase-1 expression. J Gastroenterol 48:897–909.
60. Horvath K, Varga C, Berko A, Posa A, Laszlo F, Whittle BJ (2008) The involvement of heme oxygenase-1 activity in the therapeutic actions of 5-aminosalicylic acid in rat colitis. Eur J Pharmacol 581:315–323
61. Onyiah JC, Sheikh SZ, Maharshak N, Steinbach EC, Russo SM, Kobayashi T, Mackey LC, Hansen JJ, Moeser AJ, Rawls JF, Borst LB, Otterbein LE, Plevy SE (2013) Carbon monoxide and heme oxygenase-1 prevent intestinal inflammation in mice by promoting bacterial clearance. Gastroenterology 144:789–798
62. Sheikh SZ, Hegazi RA, Kobayashi T, Onyiah JC, Russo SM, Matsuoka K, Sepulveda AR, Li F, Otterbein LE, Plevy SE (2011) An anti-inflammatory role for carbon monoxide and heme oxygenase-1 in chronic Th2-mediated murine colitis. J Immunol 186:5506–5513
63. Yalniz M, Demirel U, Orhan C, Bahcecioglu IH, Ozercan IH, Aygun C, Tuzcu M, Sahin K (2012) Nadroparin sodium activates Nrf2/HO-1 pathway in acetic acid-induced colitis in rats. Inflammation 35:1213–1221
64. Yukitake H, Kimura H, Suzuki H, Tajima Y, Sato Y, Imaeda T, Kajino M, Takizawa M (2011) Btzo-15, an ARE-activator, ameliorates DSS- and TNBS-induced colitis in rats. PLoS One 6:e23256
65. Sun X, Suzuki K, Nagata M, Kawauchi Y, Yano M, Ohkoshi S, Matsuda Y, Kawachi H, Watanabe K, Asakura H, Aoyagi Y (2010) Rectal administration of tranilast ameliorated acute colitis in mice through increased expression of heme oxygenase-1. Pathol Int 60:93–101
66. Zhong W, Xia Z, Hinrichs D, Rosenbaum JT, Wegmann KW, Meyrowitz J, Zhang Z (2010) Hemin exerts multiple protective mechanisms and attenuates dextran sulfate sodium-induced colitis. J Pediatr Gastroenterol Nutr 50:132–139
67. Varga C, Laszlo F, Fritz P, Cavicchi M, Lamarque D, Horvath K, Posa A, Berko A, Whittle BJ (2007) Modulation by heme and zinc protoporphyrin of colonic heme oxygenase-1 and experimental inflammatory bowel disease in the rat. Eur J Pharmacol 561:164–171
68. Paul G, Bataille F, Obermeier F, Bock J, Klebl F, Strauch U, Lochbaum D, Rummele P, Farkas S, Scholmerich J, Fleck M, Rogler G, Herfarth H (2005) Analysis of intestinal haem-oxygenase-1 (HO-1) in clinical and experimental colitis. Clin Exp Immunol 140:547–555
69. Lee SH, Sohn DH, Jin XY, Kim SW, Choi SC, Seo GS (2007) 2′,4′,6′-Tris(methoxymethoxy) chalcone protects against trinitrobenzene sulfonic acid-induced colitis and blocks tumor necrosis factor-alpha-induced intestinal epithelial inflammation via heme oxygenase 1-dependent and independent pathways. Biochem Pharmacol 74:870–880
70. Giris M, Erbil Y, Dogru-Abbasoglu S, Yanik BT, Alis H, Olgac V, Toker GA (2007) The effect of heme oxygenase-1 induction by glutamine on TNBS-induced colitis. The effect of glutamine on TNBS colitis. Int J Colorectal Dis 22:591–599
71. Erbil Y, Giris M, Abbasoglu SD, Barbaros U, Yanik BT, Necefli A, Olgac V, Toker GA (2007) Effect of heme oxygenase-1 induction by octreotide on TNBS-induced colitis. J Gastroenterol Hepatol 22:1852–1858
72. Jun CD, Kim Y, Choi EY, Kim M, Park B, Youn B, Yu K, Choi KS, Yoon KH, Choi SC, Lee MS, Park KI, Choi M, Chung Y, Oh J (2006) Gliotoxin reduces the severity of trinitrobenzene sulfonic acid-induced colitis in mice: evidence of the connection between heme oxygenase-1 and the nuclear factor-kappaB pathway in vitro and in vivo. Inflamm Bowel Dis 12:619–629
73. Hegazi RA, Rao KN, Mayle A, Sepulveda AR, Otterbein LE, Plevy SE (2005) Carbon monoxide ameliorates chronic murine colitis through a heme oxygenase 1-dependent pathway. J Exp Med 202:1703–1713
74. Takagi T, Naito Y, Uchiyama K, Yoshikawa T (2010) The role of heme oxygenase and carbon monoxide in inflammatory bowel disease. Redox Rep 15:193–201

75. Whittle BJ, Varga C (2010) New light on the anti-colitic actions of therapeutic aminosalicylates: the role of heme oxygenase. Pharmacol Rep 62:548–556
76. Croog VJ, Ullman TA, Itzkowitz SH (2003) Chemoprevention of colorectal cancer in ulcerative colitis. Int J Colorectal Dis 18:392–400
77. Bruwer M, Schmid KW, Metz KA, Krieglstein CF, Senninger N, Schurmann G (2001) Increased expression of metallothionein in inflammatory bowel disease. Inflamm Res 50:289–293
78. Dooley TP, Curto EV, Reddy SP, Davis RL, Lambert GW, Wilborn TW, Elson CO (2004) Regulation of gene expression in inflammatory bowel disease and correlation with IBD drugs: screening by DNA microarrays. Inflamm Bowel Dis 10:1–14
79. Ioachim E, Michael M, Katsanos C, Demou A, Tsianos EV (2003) The immunohistochemical expression of metallothionein in inflammatory bowel disease. Correlation with HLA-DR antigen expression, lymphocyte subpopulations and proliferation-associated indices. Histol Histopathol 18:75–82
80. Oz HS, Chen T, de Villiers WJ, McClain CJ (2005) Metallothionein overexpression does not protect against inflammatory bowel disease in a murine colitis model. Med Sci Monit 11: BR69–BR73
81. Tran CD, Ball JM, Sundar S, Coyle P, Howarth GS (2007) The role of zinc and metallothionein in the dextran sulfate sodium-induced colitis mouse model. Dig Dis Sci 52:2113–2121
82. Tsuji T, Naito Y, Takagi T, Kugai M, Yoriki H, Horie R, Fukui A, Mizushima K, Hirai Y, Katada K, Kamada K, Uchiyama K, Handa O, Konishi H, Yagi N, Ichikawa H, Yanagisawa R, Suzuki JS, Takano H, Satoh M, Yoshikawa T (2013) Role of metallothionein in murine experimental colitis. Int J Mol Med 31:1037–1046

Chapter 19
Genetic Disorders and Oxidative Stress (Especially Chronic Granulomatous Disease)

Shoji Tsuji and Kazunari Kaneko

Abbreviations

8-oxodG	8-Oxo-2′-deoxyguanosine
CAT	Catalase
CGD	Chronic granulomatous disease
DGC	Dystrophin–glycoprotein complex
DMD	Duchenne muscular dystrophy
fMLP	*N*-formyl-methionyl-leucyl-phenylalanine
gp91*phox*	Glycoprotein 91 kDa phagocyte oxidase
GPx	Glutathione peroxidase
GR	Glutathione reductase
LPS	Lipopolysaccharide
NADH	Nicotinamide adenine dinucleotide
nNOS	Neuronal nitric oxide synthase
NO	Nitric oxide
NO_2^-	Nitrite
NO_3^-	Nitrate
Nox	Nicotinamide adenine dinucleotide phosphate oxidase
p22*phox*	Protein 22 kDa phagocyte oxidase
p40*phox*	Protein 40 kDa phagocyte oxidase
p47*phox*	Protein 47 kDa phagocyte oxidase
p67*phox*	Protein 67 kDa phagocyte oxidase
rHuIFN-gamma	Recombinant human interferon-gamma
RNS	Reactive nitrogen species
ROS	Reactive oxygen species

S. Tsuji, M.D., Ph.D. (✉) • K. Kaneko, M.D., Ph.D.
Department of Pediatrics, Kansai Medical University,
2-5-1 Shin-machi, Hirakata-shi, Osaka 573 1010, Japan
e-mail: tsujis@hirakata.kmu.ac.jp

H. Tsukahara and K. Kaneko (eds.), *Studies on Pediatric Disorders*, Oxidative Stress in Applied Basic Research and Clinical Practice, DOI 10.1007/978-1-4939-0679-6_19,

SOD	Superoxide dismutase
SOD-1	Copper–zinc superoxide dismutase
SOD-2	Manganese superoxide dismutase
TBARS	Thiobarbituric acid reactant substances
TMP–SMX	Trimethoprim–sulfamethoxazole

19.1 Introduction

Energy production from biological activity depends mainly on oxidative phosphorylation. Although oxygen plays an important role in this energy production, some incorporated oxygen becomes reactive oxygen, such as superoxide, during the metabolic process [1]. Furthermore, leukocytes generate large amounts of reactive oxygen during inflammation, etc. Although these reactive oxygen species (ROS) are helpful in eliminating foreign objects such as bacteria, their high reactivity damages surrounding cells, thereby causing functional disorders. These ROS are usually eliminated by active-oxygen scavenging enzymes or low-molecular weight antioxidants, including superoxide dismutase (SOD), to protect organisms from damage. Failure to maintain this balance engenders disorders [2]. This review explains the hereditary disorders of Down syndrome and Duchenne muscular dystrophy (DMD), which are associated with oxidative stress, and which we often encounter in daily medical practice. Hereditary disorders such as these, for which an association with oxidative stress has become apparent, are presented in Table 19.1.

Table 19.1 Oxidative stress in genetic disorders

Disorders	Inheritance	Causes of oxidative stress	Influence of oxidative stress	References
Down syndrome		Increased activation of SOD-1	Premature aging Alzheimer's disease	[6–10]
Duchenne muscular dystrophy	X-linked	Dislocation of nNOS from DGC Increased activation of Nox	Disrupted DGC proteins Muscle wasting	[34, 46–48]
Ataxia telangiectasia	Autosomal recessive	Absence or inactivation of ataxia telangiectasia mutated (ATM)	Cerebellar ataxia Immunodeficiency Premature aging High incidence of lymphoma	[78–81]
Fanconi anemia		Absence or inactivation of Fanconi anemia proteins	Anemia Cancer	[82–85]

19.2 Down Syndrome

Down first reported Down syndrome in 1866 [3]. The disease site (Down syndrome critical region) encompasses the distal long arm of chromosome 21, chromosome band 21q22.3 [4] with a trisomic model having three chromosomes 21. Subsequently, as part of the human genome project, decoding of 3.355 million bases on the long arm of chromosome 21 was completed in 2000, thereby confirming the presence of genes of 225 types. The presence of 329 genes was suggested in 2003. Results showed that at least 16 genes are related to the energy metabolism of mitochondria, which is involved in oxidative stress. Moreover, 10 genes strongly affect the configuration of the central nerves, folic acid, and methyl metabolism [5].

19.2.1 Down Syndrome and Oxidative Stress

Some reports have described investigations of the relation between Down syndrome and oxidative stress. Increased activation of copper–zinc superoxide dismutase (SOD-1) caused by trisomy 21 chromosomal abnormality is described in many reports. Reports of other antioxidant enzymes including catalase (CAT), glutathione peroxidase (GPx), and glutathione reductase (GR) describe the presence or absence of increased activation. Tanabe et al. [6] compared 22 children with Down syndrome with 25 cases of healthy children, and reported that SOD-1 activation and protein contents are higher in children with Down syndrome by approximately 1.5 times than in healthy children, although manganese SOD (SOD-2) is lower. Moreover, they reported that CAT activation and red blood cell GPx are lower in children with Down syndrome. Pastor et al. [7] compared 72 children with Down syndrome with 72 healthy children, and reported that SOD-1, CAT, GPx, and GR activation in the red blood cells were greater by 37 %, 24 %, 8.4 %, and 37 %, respectively. Muchova et al. [8] compared 37 children with Down syndrome to their sibling controls. Although increased SOD-1 and GPx activation were observed in red blood cells of the patient group, no significant differences related to CAT and GR activation were found between the two groups.

No fixed trend exists in relation to CAT, GPx, and GR. However, other reports have clarified that SOD-1 increases [9, 10]. With regard to antioxidative substances, increased uric acid concentration in the plasma is widely observed in Down syndrome [11–15]. Campos et al. [15] recently reported no differences in uric acid concentrations in urine between 13 adults with Down syndrome and healthy adults, but the uric acid/creatinine ratio was higher in 19 children with Down syndrome. Increased plasma concentrations of allantoin, a metabolite of uric acid, were also observed in the Down syndrome group [12, 14]. Increased lipid oxide, an oxidative stress marker, was observed in children with Down syndrome: Jovanovic et al. [16] measured urinary 8-oxo-2′-deoxyguanosine (8-oxodG) and thiobarbituric acid reactant substances (TBARS) in 85 children with Down syndrome and 81 healthy siblings as controls, and reported that both oxidative stress markers were high in children with Down syndrome. Pratico et al. [17] measured 8-isoprostane in the urine of 33 children with Down syndrome and 33 healthy infants. Recently, 8-isoprostane has been used widely as a biomarker of lipid peroxidation [18]. The average 8-isoprostane concentration in the urine was higher in children with Down syndrome. A positive correlation was found between the patient age and 8-isoprostane concentration. As described above, exacerbation of oxidative stress is widely associated with Down syndrome. In at least one study, vitamin E administration was used in an attempt to alleviate oxidative stress in Down syndrome, but no efficacy has been reported to date [19].

19.3 Duchenne Muscular Dystrophy

DMD is a form of progressive muscular dystrophy with major histopathologies including degeneration, necrosis, and muscle fiber regeneration. The dystrophin gene, which causes the disease, is present in the short arm of chromosome X (Xp21.2) leading to a form of X-linked inheritance. Configuration of the dystrophin–glycoprotein complex (DGC) becomes impossible because of the impairment of dystrophin [20–23]. In fact, DMD, with incidence of one among 3,500 born male infants [24, 25], is marked by symptoms of muscle weakness such as falling and low capacity of running, since about 3 years of age. Children with DMD become unable to breathe and walk without assistance by about 12 years old [26, 27].

19.3.1 Dystrophin Deficiency and Oxidative Stress

Correlation between the severity of muscle degeneration in DMD and oxidative stress is becoming clear [28–35]. Oxidative stress occurs as a result of increased ROS and reactive nitrogen species (RNS), as well as decreased, insufficient, or imbalanced stress proteins such as antioxidant enzymes and heat shock proteins. Insufficient antioxidants and increased ROS or RNS cause cell dysfunction and damage [33]. These oxidative stress marker-related findings have been made consistently from studies of DMD patients [28, 36] and mdx mice [29, 37, 38], which are DMD model mice.

The protein carbonyl value is markedly higher in the quadriceps muscles of DMD patients than in those of normal controls [28]. Rodriguez and Tarnopolsky [36] reported significantly increased 8-oxodG, the oxidative damage marker of DNA. Ragusa et al. [29] clarified that TBARS in the skeletal muscles of mdx mice persistently exhibited a higher value than in control mice. The nicotinamide adenine dinucleotide (NADH) O_2 and cytochrome C value in the quadriceps muscle and gastrocnemius muscle of mdx mice are higher than those of the control mice [37]. Inflammatory cells, nicotinamide adenine dinucleotide phosphate oxidase (Nox), xanthine–xanthine oxidase, and mitochondria might be regarded as sources of ROS generation in respiratory muscles and muscles of the extremities of DMD patients [39–43]. Among these, Nox, which is a major source of oxidative stress, is attracting particular attention because of its involvement in the pathology of dystrophin-knockout mice [42, 44, 45]: reports suggest that neuronal nitric oxide synthase (nNOS), which is an element configuring DGC, is also involved in DMD pathology. Many reports in the literature describe that oxidative stress increases because of nNOS decline and dislocation from DGC, causing a collapse in DGC protein and muscle exhaustion [34, 46–48]. Reportedly, the dislocation of nNOS from the DGC increases Nox activity and enhances ubiquitin ligase activity, thereby promoting protein decomposition [34, 35, 47, 49].

In terms of treatment, corticosteroid can reduce oxidative stress and apoptosis in DMD patients, consequently prolonging pathological progress [50]. However, it cannot be used for long periods of time because of the occurrence of many side effects. The prevention of pathological progress is being attempted using various antioxidants against DMD. Green tea extracts [51–53] including epigallocatehin-3-gallate, low-iron diet [54], and *N*-acetylcysteine [42] were all observed with the effect of preventing muscle damage. In contrast, the effects of preventing myofunctional disorders and advancement of the patient condition were not found in clinical trials using nicotinamide (vitamin B), tocopherols (vitamin E), and penicillamine [55–58].

19.4 Chronic Granulomatous Disease

Chronic granulomatous disease (CGD), a disorder of primary immunodeficiency syndrome, is characterized by severe bacterial and fungal infections associated with the formation of granuloma in various organs, which occur repeatedly from infancy [59]. As a result of the disability to produce superoxide, patients with CGD become unable to sterilize non-hydrogen peroxide-productive/CAT-positive bacteria such as *Staphylococcus aureus*, *Klebsiella*, *Escherichia coli*, *Candida*, and *Aspergillus*. Therefore, patients systematically form suppurative lesions caused by these bacteria. Although the mechanism of granuloma formation remains unclear, the unsterilized pathogens remain within cells, maintaining the active state of cells, thereby causing the accumulation of inflammatory cells around cells because of the release of various cytokines and causing granuloma formation [59]. More than one gene is responsible for CGD, which develops because of genetic abnormalities in any gene that encodes a component protein of Nox which is necessary in the production of reactive oxygen, such as glycoprotein 91 kDa phagocyte oxidase (gp91^{phox}), protein 22 kDa phagocyte oxidase (p22^{phox}), protein 67 kDa phagocyte oxidase (p67^{phox}), protein 47 kDa phagocyte oxidase (p47^{phox}), or protein 40 kDa phagocyte oxidase (p40^{phox}) [59, 60] (Table 19.2). Clinical symptoms vary depending on the type of responsible gene, although X-linked gp91^{phox} deficiency type presents the most severe clinical symptoms.

Table 19.2 Characteristics and distribution of gene defects in chronic granulomatous disease (CGD): results from the US and European studies of a total of 122 families

Component	$gp91^{phox}$	$p22^{phox}$	$p47^{phox}$	$p67^{phox}$	$p40^{phox}$	p21rac2
Disease	X-linked	Autosomal recessive	Autosomal recessive	Autosomal recessive	Autosomal recessive	Autosomal dominant
	X91 CGD[a]	A22 CGD[a]	A47 CGD[a]	A67 CGD[a]	A40 CGD[a]	
Numbers of affected families, and incidence	$X91^{0}$69 (56 %) $X91^{-}$8 (7 %) $X91^{+}$2 (2 %)	$A22^{0}$7 (6 %) $A22^{+}$ (1 %)	$A47^{0}$28 (23 %)	$A67^{0}$7 (6 %)	Not described	
Genetic locus	CYBB	CYBA	NCF-1	NCF-2	NCF-4	
Chromosomal location	Xp21.1	16q24	7q11.23	1q25	22q13.1	22q12
Gene/mRNA size	30 kb/4.7 kb	8.5 kb/0.8 kb	15.2 kb/1.4 kb	37 kb/2.4 kb	18 kb/1.2 kb	18 kb/1.5 kb
Exons	13	6	11	16	10	7
Tissue specificity	Myeloid Low levels in mesangial cells, and some B lymphocytes Pulmonary neuroepithelial bodies	mRNA ubiquitous Protein expression only in presence of $gp91^{phox}$	Myeloid	Myeloid	Myeloid	p21rac2, myeloid

Adapted from Goldblatt and Thrasher, Table 1 [59]. Copyright 2002 John Wiley & Sons Inc. Reprinted with permission of John Wiley & Sons Inc.
[a]Accepted classification of CGD, in which A or X denotes inheritance patterns. This is followed by the molecular weight of the affected component in kDa. The superscript refers to the level of detectable immunoreactive protein: (0) denotes no proteins, (–) denotes diminished proteins, and (+) denotes normal levels of defective proteins

The clinical symptoms of CGD are bacterial infectious diseases that develop repeatedly from infancy, such as suppurative dermatitis, lymphadenitis, otitis media, and perianal abscess, but fungal infections such as *Aspergillus* increase with age. The mortality rate related to these infectious diseases remains high even today. Despite the use of newly developed antibiotics or antifungal drugs, it remains difficult to prolong life for anyone beyond 30 years of age [61]. Because the basic treatment policy is the control of infectious disease in CGD, prophylactic administration is conducted using drugs such as trimethoprim–sulfamethoxazole (TMP–SMX) and itraconazole [62]. In addition to antibacterial agents, recombinant human interferon-gamma (rHuIFN-gamma) might also be administered subcutaneously for infection prevention [63]. Although gene therapy has been used against CGD [64–66], the occurrence of abnormal hematopoietic systems such as leukemia, originating from transgenic cells [67, 68], caused only a temporary suspension. Development is underway to achieve safer gene therapy. The current mainstream radical treatment is hematopoietic stem cell transplantation. During the early 1990s, the survival rate was low even for HLA-matched family members, and bone marrow transplantation was not popular in those days. Recent advancement of management after transplantation has improved the success rate, increasing the number of transplant cases [69, 70].

19.4.1 Relationship Between RNS and CGD

Because CGD is incapable of producing superoxide, no report describes a relation between ROS and oxidative stress. However, some studies have examined the production of active RNS that cause oxidative stress, similar to ROS. Condino-Neto et al. [71] first reported that neutrophils of CGD patients indirectly produce nitric oxide (NO) in 1993. Subsequently, they measured NO produced by the neutrophils of CGD patients after administration of rHuIFN-gamma, as well as nitrite (NO_2^-) and nitrate (NO_3^-) in plasma 6 months after rHuIFN-gamma treatment. However, because the NO level in urine increased, they concluded that the NO level possibly increased from cells other than leukocytes because of the hHuIFN-gamma effect [72]. In contrast, hHuIFN-gamma reportedly enhances neutrophil NO production: Ahlin et al. [73] measured the NO produced by the neutrophils of CGD patients, taking advantage of the manner in which oxyhemoglobin changes to methemoglobin when reacted with NO. Results show that the level of NO produced by *N*-formyl-methionyl-leucyl-phenylalanine (fMLP)-stimulated neutrophils reached the maximum on the third day of administration of rHuIFN-gamma to CGD patients. Furthermore, they demonstrated the maximum inhibitory effect of neutrophils on bacterial growth on the third day of administration of rHuIFN-gamma by bactericidal assay against *Staphylococcus aureus*. These findings, which show that the production level of NO is increased after the administration of rHuIFN-gamma to CGD patients, were confirmed by Fernandez-Boyanapalli et al. [74] and Naderi Beni et al. [75]. Compared to that of healthy adults, the lower consumption rate of NO, which is associated with superoxide produced by Nox in children with CGD, was putatively associated with pathophysiology such as granuloma formation [76].

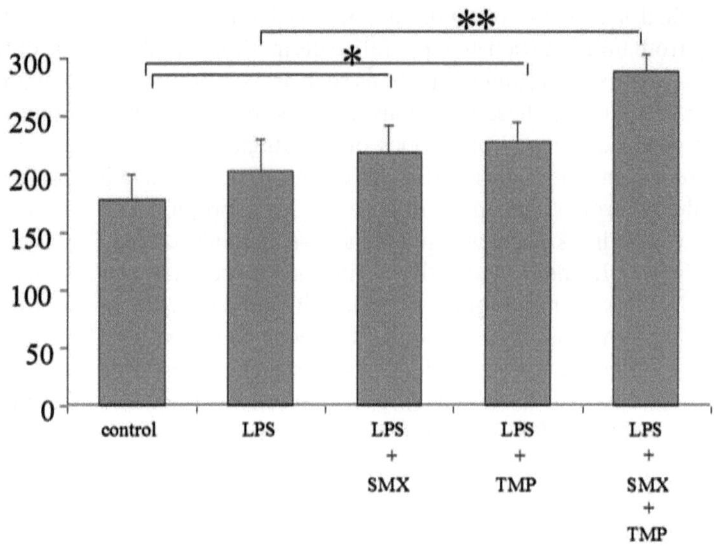

Fig. 19.1 Measurement of production of nitric oxide by neutrophils of chronic granulomatous disease (CGD) patients with trimethoprim–sulfamethoxazole (TMP–SMX). TMP–SMX with LPS in patients of CGD significantly increased the production of NO in comparison with the LPS stimulation ($n = 5$, $p < 0.05$). Ordinate, mean fluorescence intensity. *$p < 0.01$; **$p < 0.05$. Data are expressed as mean ± SE ($n = 5$). Adapted from Tsuji et al., Fig. 5 [77].

Based on the studies particularly addressing the role of NO produced by children with CGD, we examined the mechanism of TMP–SMX for preventing infection in children with CGD, which had long remained unknown. To clarify the role of TMP–SMX, we specifically examined the association of TMP–SMX with NO produced by neutrophils. Stimulation with lipopolysaccharide (LPS) upon adding TMP–SMX to the neutrophils in children with CGD increased NO production in the TMP–SMX-added group, compared to the group with single administration of LPS [77] (Fig. 19.1). These results suggest that TMP–SMX prevents infection in CGD through increased production of NO from neutrophils.

Although little is known about the association between CGD and RNS, future studies are expected to clarify important aspects of these and other relations between them.

19.5 Conclusion

This chapter reviewed genetic disorders and oxidative stress. Recent studies have clarified the involvement of oxidative stress in Down syndrome and DMD. Therefore, the establishment of treatments can be expected to reduce oxidative stress in each disorder and thereby improve patients' quality of life by delaying the progression of clinical conditions. Many aspects related to the association between CGD and oxidative stress remain unclear. Further studies in the future are expected to clarify the relation between CGD and RNS.

Acknowledgment This study was partly funded by grants from the Mami Mizutani Foundation.

References

1. Griendling KK, FitzGerald GA (2003) Oxidative stress and cardiovascular injury: Part I: basic mechanisms and in vivo monitoring of ROS. Circulation 108:1912–1916
2. Frei B (1994) Reactive oxygen species and antioxidant vitamins: mechanisms of action. Am J Med 97:5S–13S; discussion 22S–28S
3. Down JL (1866) Observations on an ethnic classification of idiots. London Hosp Rep 3:259–262
4. Delabar JM, Theophile D, Rahmani Z, Chettouh Z, Blouin JL, Prieur M, Noel B, Sinet PM (1993) Molecular mapping of twenty-four features of Down syndrome on chromosome 21. Eur J Hum Genet 1:114–124
5. Roizen NJ, Patterson D (2003) Down's syndrome. Lancet 361:1281–1289
6. Tanabe T, Kawamura N, Morinobu T, Murata T, Tamai H, Mino M, Takai T (1994) Antioxidant enzymes and vitamins in Down's syndrome. Pathophysiology 1:93–97
7. Pastor MC, Sierra C, Dolade M, Navarro E, Brandi N, Cabre E, Mira A, Seres A (1998) Antioxidant enzymes and fatty acid status in erythrocytes of Down's syndrome patients. Clin Chem 44:924–929
8. Muchova J, Sustrova M, Garaiova I, Liptakova A, Blazicek P, Kvasnicka P, Pueschel S, Durackova Z (2001) Influence of age on activities of antioxidant enzymes and lipid peroxidation products in erythrocytes and neutrophils of Down syndrome patients. Free Radic Biol Med 31:499–508
9. Brooksbank BW, Balazs R (1984) Superoxide dismutase, glutathione peroxidase and lipoperoxidation in Down's syndrome fetal brain. Brain Res 318:37–44
10. Kedziora J, Bartosz G, Gromadzinska J, Sklodowska M, Wesowicz W, Scianowski J (1986) Lipid peroxides in blood plasma and enzymatic antioxidative defence of erythrocytes in Down's syndrome. Clin Chim Acta 154:191–194
11. Pallardo FV, Degan P, d'Ischia M, Kelly FJ, Zatterale A, Calzone R, Castello G, Fernandez-Delgado R, Dunster C, Lloret A, Manini P, Pisanti MA, Vuttariello E, Pagano G (2006) Multiple evidence for an early age pro-oxidant state in Down Syndrome patients. Biogerontology 7:211–220
12. Tiano L, Padella L, Carnevali P, Gabrielli O, Bruge F, Principi F, Littarru GP (2008) Coenzyme Q10 and oxidative imbalance in Down syndrome: biochemical and clinical aspects. Biofactors 32:161–167
13. Nagyova A, Sustrova M, Raslova K (2000) Serum lipid resistance to oxidation and uric acid levels in subjects with Down's syndrome. Physiol Res 49:227–231
14. Zitnanova I, Korytar P, Aruoma OI, Sustrova M, Garaiova I, Muchova J, Kalnovicova T, Pueschel S, Durackova Z (2004) Uric acid and allantoin levels in Down syndrome: antioxidant and oxidative stress mechanisms? Clin Chim Acta 341:139–146
15. Campos C, Guzman R, Lopez-Fernandez E, Casado A (2010) Urinary uric acid and antioxidant capacity in children and adults with Down syndrome. Clin Biochem 43:228–233
16. Jovanovic SV, Clements D, MacLeod K (1998) Biomarkers of oxidative stress are significantly elevated in Down syndrome. Free Radic Biol Med 25:1044–1048

17. Pratico D, Iuliano L, Amerio G, Tang LX, Rokach J, Sabatino G, Violi F (2000) Down's syndrome is associated with increased 8,12-iso-iPF2alpha-VI levels: evidence for enhanced lipid peroxidation in vivo. Ann Neurol 48:795–798
18. Roberts LJ 2nd, Milne GL (2009) Isoprostanes. J Lipid Res 50(suppl):S219–S223
19. Ellis JM, Tan HK, Gilbert RE, Muller DP, Henley W, Moy R, Pumphrey R, Ani C, Davies S, Edwards V, Green H, Salt A, Logan S (2008) Supplementation with antioxidants and folinic acid for children with Down's syndrome: randomised controlled trial. BMJ 336:594–597
20. Matsumura K, Campbell KP (1994) Dystrophin-glycoprotein complex: its role in the molecular pathogenesis of muscular dystrophies. Muscle Nerve 17:2–15
21. Brenman JE, Chao DS, Xia H, Aldape K, Bredt DS (1995) Nitric oxide synthase complexed with dystrophin and absent from skeletal muscle sarcolemma in Duchenne muscular dystrophy. Cell 82:743–752
22. Chang WJ, Iannaccone ST, Lau KS, Masters BS, McCabe TJ, McMillan K, Padre RC, Spencer MJ, Tidball JG, Stull JT (1996) Neuronal nitric oxide synthase and dystrophin-deficient muscular dystrophy. Proc Natl Acad Sci U S A 93:9142–9147
23. Grady RM, Grange RW, Lau KS, Maimone MM, Nichol MC, Stull JT, Sanes JR (1999) Role for alpha-dystrobrevin in the pathogenesis of dystrophin-dependent muscular dystrophies. Nat Cell Biol 1:215–220
24. Nakamura A, Takeda S (2011) Mammalian models of Duchenne Muscular Dystrophy: pathological characteristics and therapeutic applications. J Biomed Biotechnol 2011:184393
25. Spencer MJ, Tidball JG (2001) Do immune cells promote the pathology of dystrophin-deficient myopathies? Neuromuscul Disord 11:556–564
26. Escolar DM, Scacheri CG (2001) Pharmacologic and genetic therapy for childhood muscular dystrophies. Curr Neurol Neurosci Rep 1:168–174
27. Lynch GS (2004) Role of contraction-induced injury in the mechanisms of muscle damage in muscular dystrophy. Clin Exp Pharmacol Physiol 31:557–561
28. Haycock JW, MacNeil S, Jones P, Harris JB, Mantle D (1996) Oxidative damage to muscle protein in Duchenne muscular dystrophy. Neuroreport 8:357–361
29. Ragusa RJ, Chow CK, Porter JD (1997) Oxidative stress as a potential pathogenic mechanism in an animal model of Duchenne muscular dystrophy. Neuromuscul Disord 7:379–386
30. Disatnik MH, Dhawan J, Yu Y, Beal MF, Whirl MM, Franco AA, Rando TA (1998) Evidence of oxidative stress in mdx mouse muscle: studies of the pre-necrotic state. J Neurol Sci 161:77–84
31. Rando TA, Disatnik MH, Yu Y, Franco A (1998) Muscle cells from mdx mice have an increased susceptibility to oxidative stress. Neuromuscul Disord 8:14–21
32. Rando TA (2001) Role of nitric oxide in the pathogenesis of muscular dystrophies: a "two hit" hypothesis of the cause of muscle necrosis. Microsc Res Tech 55:223–235
33. Rando TA (2002) Oxidative stress and the pathogenesis of muscular dystrophies. Am J Phys Med Rehabil 81:S175–S186
34. Tidball JG, Wehling-Henricks M (2004) Expression of a NOS transgene in dystrophin-deficient muscle reduces muscle membrane damage without increasing the expression of membrane-associated cytoskeletal proteins. Mol Genet Metab 82:312–320
35. Tidball JG, Wehling-Henricks M (2007) The role of free radicals in the pathophysiology of muscular dystrophy. J Appl Physiol 102:1677–1686
36. Rodriguez MC, Tarnopolsky MA (2003) Patients with dystrophinopathy show evidence of increased oxidative stress. Free Radic Biol Med 34:1217–1220
37. Hauser E, Hoger H, Bittner R, Widhalm K, Herkner K, Lubec G (1995) Oxyradical damage and mitochondrial enzyme activities in the mdx mouse. Neuropediatrics 26:260–262
38. Kaczor JJ, Hall JE, Payne E, Tarnopolsky MA (2007) Low intensity training decreases markers of oxidative stress in skeletal muscle of mdx mice. Free Radic Biol Med 43:145–154
39. Baker MS, Austin L (1989) The pathological damage in Duchenne muscular dystrophy may be due to increased intracellular OXY-radical generation caused by the absence of dystrophin and subsequent alterations in Ca_2+ metabolism. Med Hypotheses 29:187–193

40. Adams V, Jiang H, Yu J, Mobius-Winkler S, Fiehn E, Linke A, Weigl C, Schuler G, Hambrecht R (1999) Apoptosis in skeletal myocytes of patients with chronic heart failure is associated with exercise intolerance. J Am Coll Cardiol 33:959–965
41. Williams IA, Allen DG (2007) The role of reactive oxygen species in the hearts of dystrophin-deficient mdx mice. Am J Physiol Heart Circ Physiol 293:H1969–H1977
42. Whitehead NP, Pham C, Gervasio OL, Allen DG (2008) N-Acetylcysteine ameliorates skeletal muscle pathophysiology in mdx mice. J Physiol 586:2003–2014
43. Spurney CF, Knoblach S, Pistilli EE, Nagaraju K, Martin GR, Hoffman EP (2008) Dystrophin-deficient cardiomyopathy in mouse: expression of Nox4 and Lox are associated with fibrosis and altered functional parameters in the heart. Neuromuscul Disord 18:371–381
44. Whitehead NP, Streamer M, Lusambili LI, Sachs F, Allen DG (2006) Streptomycin reduces stretch-induced membrane permeability in muscles from mdx mice. Neuromuscul Disord 16:845–854
45. Whitehead NP, Yeung EW, Froehner SC, Allen DG (2010) Skeletal muscle NADPH oxidase is increased and triggers stretch-induced damage in the mdx mouse. PLoS One 5:e15354
46. Wehling M, Spencer MJ, Tidball JG (2001) A nitric oxide synthase transgene ameliorates muscular dystrophy in mdx mice. J Cell Biol 155:123–131
47. Nguyen HX, Tidball JG (2003) Null mutation of gp91phox reduces muscle membrane lysis during muscle inflammation in mice. J Physiol 553:833–841
48. Shiao T, Fond A, Deng B, Wehling-Henricks M, Adams ME, Froehner SC, Tidball JG (2004) Defects in neuromuscular junction structure in dystrophic muscle are corrected by expression of a NOS transgene in dystrophin-deficient muscles, but not in muscles lacking alpha- and beta1-syntrophins. Hum Mol Genet 13:1873–1884
49. Suzuki N, Motohashi N, Uezumi A, Fukada S, Yoshimura T, Itoyama Y, Aoki M, Miyagoe-Suzuki Y, Takeda S (2007) NO production results in suspension-induced muscle atrophy through dislocation of neuronal NOS. J Clin Invest 117:2468–2476
50. Lim JH, Kim DY, Bang MS (2004) Effects of exercise and steroid on skeletal muscle apoptosis in the mdx mouse. Muscle Nerve 30:456–462
51. Buetler TM, Renard M, Offord EA, Schneider H, Ruegg UT (2002) Green tea extract decreases muscle necrosis in mdx mice and protects against reactive oxygen species. Am J Clin Nutr 75:749–753
52. Nakae Y, Hirasaka K, Goto J, Nikawa T, Shono M, Yoshida M, Stoward PJ (2008) Subcutaneous injection, from birth, of epigallocatechin-3-gallate, a component of green tea, limits the onset of muscular dystrophy in mdx mice: a quantitative histological, immunohistochemical and electrophysiological study. Histochem Cell Biol 129:489–501
53. Dorchies OM, Wagner S, Vuadens O, Waldhauser K, Buetler TM, Kucera P, Ruegg UT (2006) Green tea extract and its major polyphenol (-)-epigallocatechin gallate improve muscle function in a mouse model for Duchenne muscular dystrophy. Am J Physiol Cell Physiol 290:C616–C625
54. Bornman L, Rossouw H, Gericke GS, Polla BS (1998) Effects of iron deprivation on the pathology and stress protein expression in murine X-linked muscular dystrophy. Biochem Pharmacol 56:751–757
55. Roelofs RI, de Arango GS, Law PK, Kinsman D, Buchanan DC, Park JH (1979) Treatment of Duchenne's muscular dystrophy with penicillamine. Results of a double-blind trial. Arch Neurol 36:266–268
56. Fenichel GM, Brooke MH, Griggs RC, Mendell JR, Miller JP, Moxley RT 3rd, Park JH, Provine MA, Florence J, Kaiser KK et al (1988) Clinical investigation in Duchenne muscular dystrophy: penicillamine and vitamin E. Muscle Nerve 11:1164–1168
57. Stern LZ, Ringel SP, Ziter FA, Menander-Huber KB, Ionasescu V, Pellegrino RJ, Snyder RD (1982) Drug trial of superoxide dismutase in Duchenne's muscular dystrophy. Arch Neurol 39:342–346
58. Walton JN, Nattrass FJ (1954) On the classification, natural history and treatment of the myopathies. Brain 77:169–231

59. Goldblatt D, Thrasher AJ (2000) Chronic granulomatous disease. Clin Exp Immunol 122:1–9
60. Matute JD, Arias AA, Wright NA, Wrobel I, Waterhouse CC, Li XJ, Marchal CC, Stull ND, Lewis DB, Steele M, Kellner JD, Yu W, Meroueh SO, Nauseef WM, Dinauer MC (2009) A new genetic subgroup of chronic granulomatous disease with autosomal recessive mutations in p40 phox and selective defects in neutrophil NADPH oxidase activity. Blood 114:3309–3315
61. Hasui M (1999) Chronic granulomatous disease in Japan: incidence and natural history. The Study Group of Phagocyte Disorders of Japan. Pediatr Int 41:589–593
62. Kobayashi Y, Amano D, Ueda K, Kagosaki Y, Usui T (1978) Treatment of seven cases of chronic granulomatous disease with sulfamethoxazole-trimethoprim (SMX-TMP). Eur J Pediatr 127:247–254
63. Ezekowitz RA, Dinauer MC, Jaffe HS, Orkin SH, Newburger PE (1988) Partial correction of the phagocyte defect in patients with X-linked chronic granulomatous disease by subcutaneous interferon gamma. N Engl J Med 319:146–151
64. Sekhsaria S, Gallin JI, Linton GF, Mallory RM, Mulligan RC, Malech HL (1993) Peripheral blood progenitors as a target for genetic correction of p47phox-deficient chronic granulomatous disease. Proc Natl Acad Sci U S A 90:7446–7450
65. Ding C, Kume A, Bjorgvinsdottir H, Hawley RG, Pech N, Dinauer MC (1996) High-level reconstitution of respiratory burst activity in a human X-linked chronic granulomatous disease (X-CGD) cell line and correction of murine X-CGD bone marrow cells by retroviral-mediated gene transfer of human gp91phox. Blood 88:1834–1840
66. Roesler J, Brenner S, Bukovsky AA, Whiting-Theobald N, Dull T, Kelly M, Civin CI, Malech HL (2002) Third-generation, self-inactivating gp91(phox) lentivector corrects the oxidase defect in NOD/SCID mouse-repopulating peripheral blood-mobilized CD34+ cells from patients with X-linked chronic granulomatous disease. Blood 100:4381–4390
67. Hacein-Bey-Abina S, Hauer J, Lim A, Picard C, Wang GP, Berry CC, Martinache C, Rieux-Laucat F, Latour S, Belohradsky BH, Leiva L, Sorensen R, Debre M, Casanova JL, Blanche S, Durandy A, Bushman FD, Fischer A, Cavazzana-Calvo M (2010) Efficacy of gene therapy for X-linked severe combined immunodeficiency. N Engl J Med 363:355–364
68. Stein S, Ott MG, Schultze-Strasser S, Jauch A, Burwinkel B, Kinner A, Schmidt M, Kramer A, Schwable J, Glimm H, Koehl U, Preiss C, Ball C, Martin H, Gohring G, Schwarzwaelder K, Hofmann WK, Karakaya K, Tchatchou S, Yang R, Reinecke P, Kuhlcke K, Schlegelberger B, Thrasher AJ, Hoelzer D, Seger R, von Kalle C, Grez M (2010) Genomic instability and myelodysplasia with monosomy 7 consequent to EVI1 activation after gene therapy for chronic granulomatous disease. Nat Med 16:198–204
69. Seger RA, Gungor T, Belohradsky BH, Blanche S, Bordigoni P, Di Bartolomeo P, Flood T, Landais P, Muller S, Ozsahin H, Passwell JH, Porta F, Slavin S, Wulffraat N, Zintl F, Nagler A, Cant A, Fischer A (2002) Treatment of chronic granulomatous disease with myeloablative conditioning and an unmodified hemopoietic allograft: a survey of the European experience, 1985–2000. Blood 100:4344–4350
70. Horwitz ME, Barrett AJ, Brown MR, Carter CS, Childs R, Gallin JI, Holland SM, Linton GF, Miller JA, Leitman SF, Read EJ, Malech HL (2001) Treatment of chronic granulomatous disease with nonmyeloablative conditioning and a T-cell-depleted hematopoietic allograft. N Engl J Med 344:881–888
71. Condino-Neto A, Muscara MN, Grumach AS, Carneiro-Sampaio MM, De Nucci G (1993) Neutrophils and mononuclear cells from patients with chronic granulomatous disease release nitric oxide. Br J Clin Pharmacol 35:485–490
72. Condino-neto A, Muscara MN, Bellinati-Pires R, Carneiro-Sampaio MM, Brandao AC, Grumach AS, De Nucci G (1996) Effect of therapy with recombinant human interferon-gamma on the release of nitric oxide by neutrophils and mononuclear cells from patients with chronic granulomatous disease. J Interferon Cytokine Res 16:357–364
73. Ahlin A, Larfars G, Elinder G, Palmblad J, Gyllenhammar H (1999) Gamma interferon treatment of patients with chronic granulomatous disease is associated with augmented production of nitric oxide by polymorphonuclear neutrophils. Clin Diagn Lab Immunol 6:420–424

74. Fernandez-Boyanapalli R, McPhillips KA, Frasch SC, Janssen WJ, Dinauer MC, Riches DW, Henson PM, Byrne A, Bratton DL (2010) Impaired phagocytosis of apoptotic cells by macrophages in chronic granulomatous disease is reversed by IFN-gamma in a nitric oxide-dependent manner. J Immunol 185:4030–4041
75. Naderi Beni F, Fattahi F, Mirshafiey A, Ansari M, Mohsenzadegan M, Movahedi M, Pourpak Z, Moin M (2012) Increased production of nitric oxide by neutrophils from patients with chronic granulomatous disease on interferon-gamma treatment. Int Immunopharmacol 12:689–693
76. Clark SR, Coffey MJ, Maclean RM, Collins PW, Lewis MJ, Cross AR, O'Donnell VB (2002) Characterization of nitric oxide consumption pathways by normal, chronic granulomatous disease and myeloperoxidase-deficient human neutrophils. J Immunol 169:5889–5896
77. Tsuji S, Taniuchi S, Hasui M, Yamamoto A, Kobayashi Y (2002) Increased nitric oxide production by neutrophils from patients with chronic granulomatous disease on trimethoprim-sulfamethoxazole. Nitric Oxide 7:283–288
78. Bagley J, Singh G, Iacomini J (2007) Regulation of oxidative stress responses by ataxia–telangiectasia mutated is required for T cell proliferation. J Immunol 178:4757–4763
79. Guo Z, Kozlov S, Lavin MF, Person MD, Paull TT (2010) ATM activation by oxidative stress. Science 330:517–521
80. Chen P, Peng C, Luff J, Spring K, Watters D, Bottle S, Furuya S, Lavin MF (2003) Oxidative stress is responsible for deficient survival and dendritogenesis in Purkinje neurons from ataxia–telangiectasia mutated mutant mice. J Neurosci 23:11453–11460
81. Okuno Y, Nakamura-Ishizu A, Otsu K, Suda T, Kubota Y (2012) Pathological neoangiogenesis depends on oxidative stress regulation by ATM. Nat Med 18:1208–1216
82. Pagano G, Manini P, Bagchi D (2003) Oxidative stress-related mechanisms are associated with xenobiotics exerting excess toxicity to Fanconi anemia cells. Environ Health Perspect 111:1699–1703
83. Park SJ, Ciccone SL, Beck BD, Hwang B, Freie B, Clapp DW, Lee SH (2004) Oxidative stress/damage induces multimerization and interaction of Fanconi anemia proteins. J Biol Chem 279:30053–30059
84. Zhang X, Li J, Sejas DP, Pang Q (2005) Hypoxia-reoxygenation induces premature senescence in FA bone marrow hematopoietic cells. Blood 106:75–85
85. Mukhopadhyay SS, Leung KS, Hicks MJ, Hastings PJ, Youssoufian H, Plon SE (2006) Defective mitochondrial peroxiredoxin-3 results in sensitivity to oxidative stress in Fanconi anemia. J Cell Biol 175:225–235

Chapter 20
Growth, Puberty, and Nutritional Disturbances

Valentina Chiavaroli, Aurelia Guarini, Francesco Chiarelli, and Angelika Mohn

Abbreviations

8-oxodG	8-Oxo-7,8-dihydro-2′-deoxyguanosine
AGA	Appropriate for gestational age
AGE/RAGE	Advanced glycation end products/receptor for advanced glycation end products
AN	Anorexia nervosa
AOPP	Advanced oxidation protein products
BMI	Body mass index
cGMP	Cyclic guanosine monophosphate
CRP	C-reactive protein
DHEA-S	Dehydroepiandrosterone-sulfate
ER	Endoplasmic reticulum
FCH	Familial combined hyperlipidemia
FDA	Food and Drug Administration
FH	Familial hypercholesterolemia
FSH	Follicle stimulating hormone
GH	Growth hormone
GHD	Growth hormone deficiency
GnRH	Gonadotropin releasing hormone
GnRHa	GnRH analog
Hcy	Homocysteine
IGF-1	Insulin-like growth factor 1
IL	Interleukin

V. Chiavaroli • A. Guarini • F. Chiarelli • A. Mohn (✉)
Department of Pediatrics, University of Chieti, Via dei Vestini, 5, 66100 Chieti, Italy
e-mail: amohn@unich.it

H. Tsukahara and K. Kaneko (eds.), *Studies on Pediatric Disorders*, Oxidative Stress in Applied Basic Research and Clinical Practice, DOI 10.1007/978-1-4939-0679-6_20,

IR	Insulin resistance
KS	Klinefelter syndrome
LDL-C	Low-density lipoprotein cholesterol
LGA	Large for gestational age
LH	Luteinizing hormone
LPO	Lipoperoxidation
MC4R	Melanocortin-4 receptor
MFS	Marfan syndrome
NO	Nitric oxide
NOS	Nitric oxide synthase
PAI-1	Plasminogen activator inhibitor-1
PCOS	Polycystic ovary syndrome
PUFAs	Polyunsaturated fatty acids
ROS	Reactive oxygen species
SD	Standard deviations
SGA	Small for gestational age
SOD	Superoxide dismutase
sRAGE/esRAGE	Soluble RAGE/endogenous secretory RAGE
SS	Short stature

20.1 Introduction

Oxidative stress represents an impairment between oxidant and antioxidant status, which occurs in many physiological and pathological conditions. It has been demonstrated that the impairment of the oxidant–antioxidant system is responsible for many diseases in adulthood as well as in childhood. The aim of this chapter is to summarize the most important and recent findings regarding the role of oxidative stress during *growth*, *puberty*, and *nutritional disturbances*. The content is divided into three sections describing the physiological processes of growth, puberty, and nutrition and the pathological conditions which can occur during them related to oxidative stress.

20.2 Growth

Human growth is a continuous process characterized by the development of physical, mental, social, and emotional abilities of the child. This crucial period of life is influenced by genetic as well as environmental factors and their interaction.

Table 20.1 The most important phases of child growth and development

I Phase (until 2 years of age)	Nutrition
II Phase (from 2 years until puberty)	GH–IGF-1 axis Hypothalamus–pituitary–thyroid axis
III Phase (puberty)	GH–IGF-1 axis Hypothalamus–pituitary–gonadal axis

20.2.1 Physiology of Growth

Human growth can be divided into three different periods (see Table 20.1). The first phase lasts up to 2 years of age and is controlled by nutrition [1]. In fact, thanks to substrates like amino acids and glucose that the baby receives through nutrition, insulin-like growth factor 1 (IGF-1) is synthesized and acts on tissues. Breast milk represents the best source of nutrients for infants. In addition, it furnishes a variety of antioxidant components. In this respect, Aycicek et al. demonstrated that breast milk provides better antioxidant power than does formula [2]. Infants fed solely on breast milk and infants fed commercial cow's milk-modified formulas differ considerably in the amount and composition of consumed fatty acids. A major difference is the content of long chain polyunsaturated fatty acids (PUFAs). In fact, long chain PUFAs with 20–22 carbons, of both the w-3 and w-6 classes constituting 2 % of total fatty acids in human milk, are undetectable in the more commonly used commercial formulas. It has been proven a protective role of PUFAs in preventing the cytotoxic effects caused by enhanced reactive oxygen species (ROS) formation [2].

Proteins, iron, carbohydrates, and minerals are also indispensable in the diet of the child. Usually infants are fed on breast milk until the age of 6 months, when the weaning occurs [2, 3]. Cereals are a source of iron and they are usually introduced at first, and then vegetables and fruits are followed by meat and eggs.

The second phase of human growth is controlled by growth hormone (GH)/IGF-1 axis. After the first 2 years of life, GH is secreted thanks to the activation of the hypothalamus–pituitary axis. GH stimulates cellular metabolism and acts on bones, muscles, and other tissues. In fact, GH and IGF-1 control remodeling and mineralization of the skeleton, in part via direct actions on bone. GH plays the main role because it mechanistically acts on bone inducing skeletal IGF-1 synthesis, proliferation on prechondrocytes, and hypertrophy of osteoblasts [3]. In this second phase of growth, also thyroxin plays an important role. After being secreted by the thyroid gland, thyroxin exerts a major influence on growth and development of brain and bones as well as metabolism throughout adult life [1].

The last third phase, named puberty, is regulated by both GH and sexual hormones through the activation of the pituitary–gonadal axis [1]. This phase requires the interaction and the coordination between GH/IGF-1 and gonadotropin releasing hormone (GnRH)/luteinizing hormone (LH)/sexual hormones. In fact, IGF-1 enhances hypothalamic GnRH, potentiates GnRH-stimulated LH release, and synergizes with LH and follicle stimulating hormone (FSH) in stimulating ovarian and testicular steroidogenesis. In addition, endogenous gonadal sex steroids amplify the synthesis of GH and IGF-1 and regulate the availability of IGF binding proteins. Other functions include the influence that sexual hormones have with some insulin-mimetic peptides in regulating appetite, behavior, and energy expenditure via central and peripheral pathways [3].

However, not only nutrition and hormones are responsible for human growth. In fact, also environmental factors are implicated in this process. In particular, the relationship between environment and oxidant–antioxidant system can have important effects in healthy subjects. It has been suggested that oxidative stress in healthy children is caused by pollution or passive smoking. Kosecik et al. demonstrated that plasma total peroxide levels were higher in children exposed to passive smoking than in those not exposed. Chronic oxidant exposure promotes oxidative injury in tissues and organs. The oxidative injury also induces production of inflammatory cytokines, such as interleukin (IL)-6, tumor necrosis factor-alpha, and IL-1, and may compromise important metabolic processes that influence both endothelial and vascular smooth muscle cell function, key components of atherogenesis [4].

20.2.2 Growth Disorders

Many disorders can occur during children growth. They belong to two different groups: short and tall stature, respectively, caused by several causes.

20.2.2.1 Short Stature

Short stature (SS) is considered present when height is more than 2 standard deviations (SD) below the corresponding mean height for a given age, sex, and population group [5]. According to the classification of the European Society for Paediatric Endocrinology, 20 % of SS is due to primary and secondary growth disorders. Primary growth disorders include syndromes (such as Turner syndrome and Down syndrome), being born small for gestational age (SGA) with failure of catch-up growth, and skeletal dysplasias (including defects in the short stature homeobox gene (*SHOX*)). Secondary growth disorders include malnutrition, organic disorders, GH deficiency (GHD) and disorders of the GH/IGF-1 axis, other endocrine disorders (such as Cushing syndrome and hypothyroidism), metabolic diseases, and psychosocial and iatrogenic causes. Of note, in the 80 % of subjects the pathogenesis of SS remains unknown and the term "idiopathic short stature" has been coined [5].

Short Stature and Oxidative Stress

Emerging evidences suggest that numerous conditions characterized by SS are linked to an impaired oxidant–antioxidant status [6].

GHD represents the secondary growth disorders and is one of the most common cause of SS. Adult GHD is known to be associated with oxidative stress determined by the underlying GH/IGF-1 axis alteration. Furthermore, GHD in adulthood is associated with many diseases, such as insulin resistance (IR) and dyslipidemia [5]. The oxidant–antioxidant status in children with GHD has not been widely studied. We demonstrated an increased oxidative stress in prepubertal children with GHD, which reduced after 12 months of GH replacement therapy (Fig. 20.1). We also found a close correlation between indices of oxidative stress and both IGF-1 and IGF binding protein-3 levels. These results suggest that impaired levels of IGF-1 could play a major role in the induction of oxidative stress even during infancy [7].

Being born SGA, defined as neonates whose birth weight and/or length are at least 2 SD below the mean for gestational age, is one of the most common risk factors for SS [5]. In fact, SS is detectable in about 10 % of subjects born SGA. In addition, these children are known to be at increased risk of adult degenerative diseases, such as cardiovascular dysfunction and other metabolic diseases including obesity and metabolic syndrome. One of the common features of these pathologies is IR, which seems to increase free radical production via an increased plasma concentration of free fatty acids. We found an impaired oxidant–antioxidant status in a group of prepubertal SGA children compared with subjects born appropriate for gestational age (AGA). In particular, decreased plasma vitamin E levels were found in SGA children than in those born AGA [8]. Afterwards, our study group evaluated the effect of birth weight and obesity on oxidative stress in a group of prepubertal children born SGA and large for gestational age (LGA) compared to AGA subjects [9]. We found increased IR and oxidative stress (urinary isoprostanes) in normal-weight SGA and LGA children, with a continuous alteration related to obesity (Fig. 20.2). These data suggest that birth weight represents an independent risk factor for degenerative diseases.

Other genetic conditions responsible for SS include syndromes. In subjects with Down syndrome, chronic oxidative stress has been suggested to play a role in the development of Alzheimer-like brain changes and in the accelerated overall aging process. In this respect, malonildialdehyde levels have been found higher in patients with Down syndrome and oxidative stress has been attributed to the enhanced activity of superoxide dismutase (SOD). In fact, this antioxidant enzyme is coded on chromosome 21, present in excess in these individuals [6].

20.2.2.2 Tall Stature and Oxidative Stress

Tall stature is defined as a height greater than 2 SD above the mean for a given age, sex, and population group. There are several causes of tall stature:

- *Familial or genetic tall stature*: It is the most common cause. These children are tall from early childhood with bone age usually compatible with their chronologic age, and have tall parents [10].
- *Constitutional tall stature*: These children have normal length at birth, while the growth velocity accelerates in early childhood and tall stature is evident at the

Fig. 20.1 Oxidative stress parameters in GH-deficient children before (T_0) and after (T_{12}) therapy (reproduced with permission from Mohn et al., Alterations in the oxidant-antioxidant status in prepubertal children with growth hormone deficiency: effect of growth hormone replacement therapy, Clin Endocrinol, Wiley publisher [7])

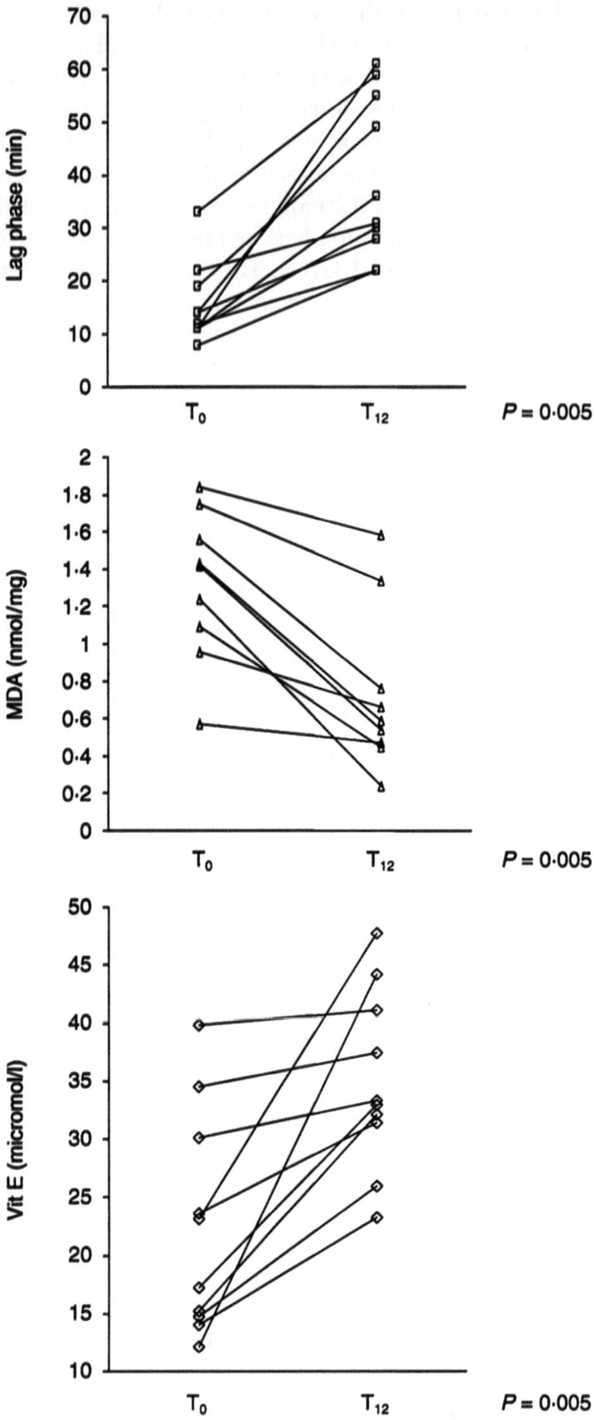

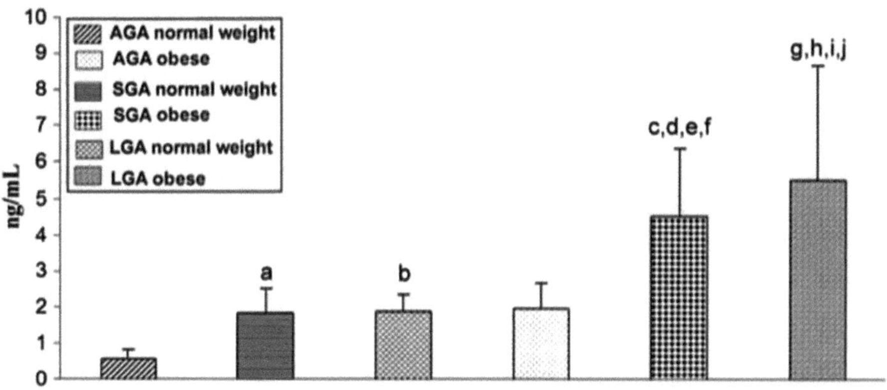

Fig. 20.2 Oxidant–antioxidant status of the study population ($P=0.0001$, by one-way analysis of variance). Significant values by post hoc analysis—a: normal-weight SGA versus normal-weight AGA; b: normal-weight LGA versus normal-weight AGA; c: obese SGA versus obese AGA; d: obese SGA versus normal-weight AGA; e: obese SGA versus normal-weight SGA; f: obese SGA versus normal-weight LGA; g: obese LGA versus obese AGA; h: obese LGA versus normal-weight AGA; i: obese LGA versus normal-weight SGA; and j: obese LGA versus normal-weight LGA (reproduced with permission from Pediatrics, Vol. 124(2), Pages 695-702, Copyright 2013 by the AAP [9])

age of 3–4 years. Growth velocity slows down after the child reaches 4 or 5 years. Thereafter, the growth curve is parallel to and above normal curve. These children have advanced bone age [10].

- *GH hypersecretion*: In children it is usually caused by a GH adenoma of the pituitary gland. Before the epiphyses close, an excess of GH produces accelerated linear growth and results in gigantism. Bone age is often advanced. Pituitary tumors are associated with headache, decreased visual acuity, visual field defects, and symptoms or signs suggesting increased intracranial pressure [10]. There are no studies that assess oxidative stress in children with GH excess. However, Nishizawa et al. found increased IGF-1 levels in subjects with acromegaly. Of note, the enhancement of IGF-1 levels was related to increased serum levels of oxidative stress markers, such as 8-oxo-7,8-dihydro-2′-deoxyguanosine (8-oxodG) [11] that indicates oxidization of DNA [12].
- *Marfan syndrome* (*MFS*): It is an inherited connective tissue disorder characterized by tall stature, arachnodactyly, superior lens subluxation or dislocation, and aortic or mitral regurgitation. The arm span is greater than height, and the upper to lower body segment ratio is diminished. Other clinical features are hypertension, pectus excavatum, scoliosis, megacornea, and myopia [10]. Fiorillo et al. demonstrated that markers of oxidative stress were higher in MFS patients compared to healthy controls. The cause is still unknown but it has recently been demonstrated in an MFS mouse model a reduced expression of SOD and an increased expression of inducible nitric oxide synthase (NOS), nicotinamide adenine dinucleotide (P) hydrogen oxidase, and xanthine oxidase, which represent the main source of ROS in the aortic tissue. As a matter of fact, one of

the most important disorders of patients with MFS is the myxomatous degeneration of aortic valve [13].

- *Homocystinuria*: It is an autosomal recessive disorder caused by a deficiency of cystathionine-beta-synthase, with high levels of homocysteine (Hcy) and methionine in biological fluids. Patients have same clinical features of those with MFS. Additional features are mental retardation, malar flush, inferior lens dislocation and increased incidence of seizures, arterial and venous thrombosis, osteoporosis, and vertebral collapse [10]. Vanzin et al. found that tissue antioxidant defenses were altered in these patients. These data indicate that oxidative stress may represent an important underlying mechanism in the pathogenesis of homocystinuria, probably caused by increased tissue Hcy levels [14].
- *Klinefelter syndrome* (*KS*): It has a chromosomic aetiology and the most common karyotype is 47, XXY. Affected children are tall with long arm span and legs, and present hypogonadism, cryptorchidism, gynecomastia, and, in some cases, mental retardation and antisocial behavior.

20.3 Puberty

Puberty represents an important period of life characterized by the acquisition of sexual characteristics and reproductive capability.

20.3.1 Physiology of Puberty

The study of mechanisms of puberty is still incomplete, although it is known that pubertal onset is controlled by different neuroendocrine systems [1, 15]. It has been hypothesized that neurotransmitters such as neuropeptide Y, gamma-aminobutyric acid, leptin, and other paracrine and autocrine factors modulate this axis. Furthermore, it is clear that genetic and environmental factors can change pubertal onset.

Gonadotropins are secreted from the anterior pituitary gland in response to pulsatile secretion of GnRH from the hypothalamus [15, 16]. In a prepubertal status the hypothalamus is quiescent and low LH levels are secreted during sleep. When it activates, GnRH is secreted by GnRH neurons and activates the gonadotropic axis to drive complete gonadal maturation [1]. The onset of secondary sexual characteristics represents the hypothalamus activation even better than GnRH neurons [16]. FSH stimulates Sertoli cells in male and follicular development in female, and the production of LH receptors in Leydig cells. LH regulates gonadal steroid production by Leydig cells of the testes and by ovarian follicles through the LH receptor [15, 16]. The age of puberty onset is variable. Tanner stages identify the five stages of maturation of secondary sexual characteristics [17, 18]. Breast development (thelarche), that occurs approximately at the age of 11 years, follows the secretion of ovarian estrogen, whereas the development of female pubic hair (pubarche) is

Table 20.2 Classification of the most common causes of precocious puberty

Gonadotropin-dependent precocious puberty
Idiopathic
Cerebral injuries
Gonadotropin-independent precocious puberty
McCune–Albright syndrome
Ovarian cysts
Ovarian tumors
Leydig cell tumors
Adrenal tumors
hCG secreting tumors
Hypothyroidism
"*Partial*" *or* "*incomplete*" *precocious puberty*
Premature adrenarche
Premature thelarche
Premature menarche

caused by androgens from the adrenal glands or ovaries [16]. Testicular enlargement (3–4 mL in volume) usually precedes pubic hair development and the onset is between the age of 11 and 13 years [15]. Adrenal androgens also play an important role in puberty. Dehydroepiandrosterone and dehydroepiandrosterone-sulfate (DHEA-S) levels increase before LH is secreted [15]. They are produced by the adrenal cortex and play an important role in the adrenarche, acting on the development of pubic hair in female and secondary sexual characteristics in male [1, 15].

Oxidative stress markers have been measured in both prepubertal and pubertal age. Pérez-Navero et al. detected higher lipoperoxidation (LPO) products in healthy pubertal subjects compared to prepubertal children, while no difference was detected between the two groups in terms of catalase activity and glutathione. The enhancement of LPO is possibly linked to the biochemical changes which occur during puberty. Nevertheless, more studies are needed to confirm these results and to understand the role of oxidative stress in puberty [19].

20.3.2 Disorders of Pubertal Development

Precocious puberty and delayed puberty represent the two important groups in which the disorders of pubertal development can be well summarized.

20.3.2.1 Precocious Puberty

Precocious puberty refers to the appearance of physical and hormonal signs of puberty before the age of 8 years in girls and before the age of 9 years in boys. It can be gonadotropin dependent or gonadotropin independent (see Table 20.2). The first condition is idiopathic in 75 % of cases and is characterized by early activation of the hypothalamus–pituitary–gonadal axis resulting in an increased pulsatile LH secretion. However, gonadotropin-dependent precocious puberty can also be related to occult intracranial tumors (15–20 % of cases) [18] or be secondary to other disorders of central nervous system (5–10 % of cases). Gonadotropin-independent precocious puberty can be caused by several diseases, such as ovarian and testicular tumors, adrenal tumors or adrenogenital syndrome, McCune–Albright syndrome, and hypothyroidism.

It has to be reported also the presence of normal variants of puberty. They include premature thelarche, premature menarche, and premature adrenarche. Premature thelarche is a common disorder, in which breast development occurs with no other signs of puberty. There are two forms of premature thelarche: the classical and the atypical. The classical is self-limiting, while the atypical is associated with occasional menstrual withdrawal bleeding. Isolated premature menarche is described as a condition in which girls have repeated vaginal bleeding with no other secondary sexual characteristics. In this case it is important to exclude lesions of the genital tract, McCune–Albright syndrome, exogenous administration of estrogens, and child abuse. Premature adrenarche or pubarche is defined as the appearing of sexual hair (pubic hair, axillary hair, or both) before 8 years in girls and 9 years in boys. The aetiology is possibly related to a dysfunction of the hormones secreted by the adrenal cortex, due to heterozygosis or homozygosis of the enzyme 22-hydroxylase. In this condition plasma levels of DHEA-S and androstenedione are elevated for age [18]. It has been reported that girls with history of premature adrenarche are at elevated risk of developing polycystic ovary syndrome (PCOS) and features of the metabolic syndrome later in life. Other diseases that cause oxidative stress, such as coagulation disorders and inflammatory status, have been reported. Livadas et al. found that C-reactive protein (CRP) and plasminogen activator inhibitor-1 (PAI-1) were higher in adolescents with history of premature adrenarche than healthy controls. In this study the authors considered premature adrenarche as a prodromal stage of PCOS development. Another interesting factor in this study was that girls with premature adrenarche had coagulation disorders justified by higher values of PAI-1 and lower tissue plasminogen activator than controls [20]. In addition PCOS, known to be a risk factor for developing disorders linked to oxidative stress, could be induced by GnRH analog (GnRHa) therapy in girls with early puberty. In fact, we demonstrated an increased prevalence of PCOS in adolescents previously treated with GnRHa for early puberty. However, more studies are necessary to establish if GnRHa therapy might be administered just in few selected cases to avoid PCOS in the early post-menarchal years [21].

20.3.2.2 Delayed Puberty

Delayed puberty is defined by the absence of testicular enlargement (testicular volume lower than 4 mL) in boys beyond 14-year-old or by the absence of breast development in girls beyond 13-year-old [22]. Many conditions can cause delayed puberty:

- *Hypogonadotropic hypogonadism*: It is characterized by low levels of sex hormones and gonadotropin secretion and is caused by mutations in several genes, such as *KAL* gene (Kallmann syndrome), *DAXI* (X-linked adrenal hypoplasia congenita), GnRH receptor, FSH, and LH [22].

- *Hypergonadotropic hypogonadism*: It can be caused by several conditions such as Turner syndrome, the commonest gonadal dysgenesis in girls (45, XO karyotype or mosaic). In boys, KS is the commonest disease responsible for seminiferous tubular dysgenesis. Esteves et al. found that individuals with KS had higher levels of oxidative stress markers than healthy individuals. Moreover, KS men previously treated with testosterone had a better sperm retrieval rate because testosterone reduced oxidative stress [23]. Another disorder is the androgen insensitivity syndrome, an X-linked disorder in 46, XY individuals with normal androgen production and metabolism. Hypo/hypergonadotrophic hypogonadism can also have secondary causes, such as chemotherapy responsible for gonadal damage. Radiotherapy induces dysfunction by direct injury to the gonads and, with less frequency, by neuroendocrine injury to the hypothalamic–pituitary axis [22].

20.4 Nutritional Disturbances

Eating disorders are complex syndromes affecting children and adolescents with increasing frequency. The main nutritional disturbances include obesity, anorexia, and hypercholesterolemia.

20.4.1 Obesity

Obesity is a public health problem in both developed and developing countries [24–26]. According to the National Health and Nutrition Examination Survey, 16 % of children is overweight while 31 % has the risk of becoming overweight. Childhood overweight and obesity are defined as conditions in which body mass index (BMI) values are respectively between the 85th and 95th percentiles and over the 95th percentiles than in the reference population. In 97 % of cases obesity is essential and related to several environmental and genetic factors. The increasing phenomenon of childhood obesity is due to decreased physical activity and increased consumption of snacks and high-calorie food. In addition, a major risk factor for obesity is high birth weight, possibly related to maternal diabetes or obesity [24, 25]. Interestingly, over the past years many discoveries regarding gene mutations influencing obesity have been made [26]. The Fat Mass- and Obesity-Associated Gene, called *FTO*, is located in the chromosome 16 and is expressed in the brain, specifically in hypothalamic nuclei. Individuals homozygous for this condition gain weight easily than other people. These subjects have reduced feelings of satiety, prefer energy-dense food, and consume more fats and calories. The mutations of the melanocortin-4 receptor (*MC4R*) have also been investigated and described in genetic disorders, such as Bardet–Biedl syndrome. Individuals that present the *MC4R* mutation prefer saturated fats and have reduced feelings of satiety [26]. In 3 % of cases obesity is secondary to other diseases, such as genetic (Beckwith–Wiedemann's syndrome, Prader–Willi syndrome) or endocrine diseases (including Cushing syndrome).

Childhood obesity is associated with several consequences during both childhood and adult life. Numerous comorbid conditions can complicate obesity, such as IR, type 2 diabetes, dyslipidemia, hypertension, atherosclerosis, and nonalcoholic fatty liver disease [25, 27].

20.4.1.1 Obesity and Inflammatory Stress

Obesity is a state of chronic inflammation: the adipose tissue is an endocrine tissue and produces different molecules having several effects [24]. Chronic nutrient overload leads to an energy imbalance, which manifests as an increase in the size of adipose tissue with both adipocyte size and number (hypertrophy and hyperplasia). Preadipocytes differentiate into adult adipocyte which secretes a large number of cytokines, called adipokines. Adipokines act as paracrine and endocrine molecules, and their purpose is maintaining the body's energy balance in distant organs. The enlargement of adipocytes induces secretion of inflammatory cytokines. The most important cytokines are leptin, IL-6, and resistin. Adipocyte hypoxia favors cell necrosis, which is a process that recruits macrophages and other phagocytic cells and creates a milieu favoring the perpetuation of inflammation within the adipose tissue. Increased amounts of visceral adipose tissue lead to an increased delivery of inflammatory adipocytokines and free fatty acids to the liver via the portal venous system. It is important to stress the concept that different anatomic depots of adipose tissue manifest distinct metabolic properties. Visceral fat seems to produce several adipokines more actively than does subcutaneous adipose tissue.

The production of ROS can occur in response to the metabolic overload caused by increased macronutrients in the diet. When caloric intake exceeds energy expenditure, there is an imbalance and free radicals are generated. A high-fat, high-carbohydrate meal induces an intense oxidative and inflammatory stress response with higher ROS secretion in obese patients than normal-weight subjects. The mechanisms leading to ROS generation include specific metabolic systems, such as the nitrogen oxides system and other oxidases including lipoxygenase and xanthine oxidase. Mechanisms responsible for increased nitrogen oxides system-mediated ROS production in adipocytes are unclear. It has been suggested that local adipose tissue hypoxia could induce the upregulation of inducible NOS messenger RNA. Of note, the mitochondrial respiratory chain produces the highest quantity of ROS. Obesity is associated with an increase in oxidative stress at the level of adipocyte mitochondria because the processing of excess free fatty acids causes mitochondrial uncoupling with an increased release of ROS. Indeed, it has been suggested that mitochondrial ROS work as signaling molecules for the production of pro-inflammatory cytokines [24]. The endoplasmic reticulum (ER) controls the regulation of both lipid and protein metabolism. The ER also controls the protein synthesis, an energy consuming process that requires a lot of energy leading to ROS production.

Such "ER stress" can start due to the increasing demand for more proteins and their synthesis or by the accumulation of misfolded proteins promoted by inflammatory stimuli. In addition, the ER stress leads to the activation of inflammatory signaling pathways with increased mitochondrial ROS generation. In this sense, excessive generation of ROS in adipose tissue occurs by several mechanisms, such as nutrient metabolic overload, mitochondrial dysfunction, and ER stress. ROS generation is perpetuated by an inflammatory response feeding a vicious cycle. Also inducible or inflammatory NOS production is stimulated by adipocyte inflammatory stimuli [24], and when superoxide and nitric oxide (NO) are produced simultaneously both peroxynitrite and hydroxyl radicals are produced.

Emerging evidences suggest an impairment of oxidant–antioxidant system in obese children. In fact, we found that the oxidative stress markers urinary 8-isoprostane F-2 alpha were elevated in obese children rather than in controls [27]. Advanced oxidation protein products (AOPP) are considered important markers to assess the degree of oxidant-mediated protein damage, and they have several advantages compared to other biomarkers because of their early formation, stability, reliability, and longer life-span. Atabek et al. demonstrated that AOPP levels were higher in children affected by obesity and IR [28]. The study group of Kaneko et al. found that urinary levels of 8-oxodG were higher in younger patients with rapid growth and sustained immune activation. They also found that oxidative stress markers were higher in children with heart diseases that received cardiac catheterization and radiations [29]. Unfortunately, obese children have also reduced antioxidant capacity compared to normal-weight healthy children. In this respect, reduced vitamin E levels have been detected in obese children caused by the sequestration by the adipose tissue with alteration in assimilation and metabolism [24].

Advanced glycation end products/receptor for advanced glycation end products (AGE/RAGE) system is a new pathway identified in the field of vascular biology and is a new marker linked to the onset and progression of atherosclerosis [30]. RAGE is expressed on the surface of numerous cells implicated in plaque formation and progression, such as endothelial cells. AGE and its ligands lead to the induction of oxidative stress. The activation of these pathways promotes the migration of circulating monocyte/macrophages and lymphocytes into arterial wall and induces release of other cytokines. Recent studies suggest that the obese status is related to low levels of soluble RAGE (sRAGE) and endogenous secretory RAGE (esRAGE). In fact, we found that obese children had lower levels of sRAGE and esRAGE than healthy controls [30]. In addition, we detected reduced sRAGE and esRAGE levels also in a group of children born SGA and LGA, particularly in those showing excessive body weight during childhood [31].

20.4.1.2 Obesity, Oxidative Stress, and Endothelial Dysfunction

Endothelial dysfunction is defined as the loss of the endothelium to serve its normal physiologic and protective function. The role of endothelial cells is to produce different substances with a vasoactive function. In addition, they secrete several mediators that regulate platelet aggregation, coagulation, and fibrinolysis. Endothelial dysfunction is characterized by a reduction in the bioavailability of vasodilators, in particular NO, and by an increased production of many factors that enhance the contraction of the smooth muscle favoring inflammation and thrombosis. All these processes lead to the formation of the arterial plaque. The excessive ROS production determines the reduction of NO levels and perpetuates the inflammation through the production of arachidonic acid derivates, such as isoprostanes which in turn determine platelet activation. The inflammatory status also determines an increase of nitrosamine due to the action of ROS, and these products of NO play a main role in the endothelial dysfunction. Many studies have demonstrated that obese children have high levels of CRP, which represents an important marker of inflammation. We demonstrated in a group of prepubertal children that the exposure to metabolic alterations typical of the obesity status contributed to increase in carotid intima media thickness. It is known that the increase of adipose tissue promotes the initial abnormalities in glucose metabolism and the development of IR in obesity. Furthermore, increased values of high sensitivity-CRP and prostaglandin F-2 alpha, considered indices of chronic inflammatory status and oxidative stress [32], have been detected in obese children compared with controls [27, 33].

20.4.1.3 Treatment of Obesity

Obesity and its complications can be prevented and treated through preventive measures started early in life, such as lifestyle modifications consisting of reduction in caloric intake by performing a diet focused on healthy micro- and macronutrient in association to regular exercise programs [24]. Lifestyle intervention can reduce rates of weight gain and fat deposition in children and may delay or prevent some long-term risks, such as type 2 diabetes. However, lifestyle modification needs to be intensive (i.e., calories restriction, individual and family counseling, daily exercise) and continued to be effective in children and adolescents. We demonstrated that a dietary restriction-weight loss program reduced the oxidant status in prepubertal obese children. It is important to stress that during the period of the study, five patients refused to adhere to the diet and when they regained weight with a hypercaloric diet, oxidative stress markers increased again. These data suggest an important correlation between obesity and hypercaloric unbalanced diet and oxidative stress [34]. Llorente-Cantarero et al. demonstrated that physical inactivity caused enhancement of oxidative stress biomarkers also in healthy children. They demonstrated that acute aerobic exercise induced oxidative stress, whereas regular aerobic exercise decreased oxidant markers and increased antioxidant enzyme activities. In addition, the authors found that regular exercise increased antioxidant defenses, while sedentary children showed higher levels of glutathione and oxidized glutathione and lower reduced glutathione/oxidized glutathione ratio [17].

Pharmacotherapy may be considered as an adjunct if children or adolescents are not reaching weight loss goals with lifestyle modification or have significant comorbidities. Pharmacotherapy is an option approved by the Food and Drug Administration (FDA) for extremely obese (i.e., BMI ≥2 units above the 95th percentile) children older than 12 years of age who have not responded to 1-year dietary and lifestyle treatments, and for those with impaired glucose tolerance or IR, steatohepatitis, ovarian hyperandrogenism, or a strong family history of diabetes, myocardial infarction, or stroke. It has been demonstrated that a combination of medication and lifestyle modification decreases weight more than lifestyle change alone [35]. Among medications, orlistat prevents absorption of fats from the diet reducing caloric intake. Chanoine et al. demonstrated that BMI decreased more in the group of obese children treated with orlistat (–0.55 kg/m^2) than in the placebo group (–0.31 kg/m^2) [36].

20.4.2 Anorexia Nervosa

Anorexia nervosa (AN) is a serious psychiatric illness associated with significant morbidity and mortality rate, which is among the highest for psychiatric illness. An important clinical feature of this disease is the decreased food intake. The cause is still unknown although biological feature and psychological as well as sociocultural factors seem to play an important role. Eating disorders are increased especially among adolescents and, in particular, among young women. In Brazil 1.4 % of individuals 7–14 year-old are at risk of AN and bulimia nervosa. Death in subjects affected of AN is attributed to cardiac and pulmonary events ranging from electrical abnormalities [37].

20.4.2.1 Anorexia Nervosa and Oxidative Stress

The first cause of death in subjects with AN is heart failure, related to oxidative stress induced by this disease. Impaired NO bioavailability plays an important role in cardiovascular disorders, perhaps reflecting endothelial dysfunction and platelet activation [37]. Oxidative stress and production of free radicals create an important damage to cell structure. NO reacts with superoxide forming peroxynitrite, which is harmful. In addition, in AN the inadequate intake of protein energy and micronutrients might cause an increase in oxidative stress, which aggravates this severe nutritional disorder. Rodrigues Pereira et al. evaluated the different steps of the L-arginine-NO-cyclic guanosine monophosphate (cGMP) pathway activity in adolescents with AN and its association with platelet expression of NOS and arginase isoforms, platelet aggregation, oxidative stress, and plasma levels of L-arginine [37]. The study established that NO production decreased in AN patients. This result could be related to diminished L-arginine transport capacity via the y^+ L system, leading to the decreased basal NOS activity and cGMP in platelets. Another factor that can limit L-arginine transport is the low plasma L-arginine concentration. It is possible that low intracellular Na^+ content and/or a small quantity of intraplatelet amino acids lead to a reduced exchange between extracellular L-arginine and intracellular neutral amino acids plus Na^+. Another possible explanation for decreased NO secretion is the activation of arginase which catalyzes the hydrolysis of L-arginine to L-ornithine and urea. Arginase competes with NOS for L-arginine and treatment with an arginase inhibitor in animal models of hypertension can increase NO production. Other results demonstrated an overexpression of arginase II in human platelets of patients with AN. This result confirms recent data in the literature showing that an increase in arginase II mRNA and arginase activity limits intracellular L-arginine bioavailability for platelet NO synthesis in pathological conditions. Rodrigues Pereira et al. also demonstrated that NOS activity was diminished in the presence of unchanged expression of endothelial and inducible NOS. NOS requires several cofactors and prosthetic groups for activity, such as thiolate-bound heme, calmodulin and Ca^{2+} (endothelial NOS), and (6*R*)-5,6,7,8-tetrahydro-L-biopterin (tetrahydrobiopterin; BH4). Because of the malnourishment of AN patients, they probably lack the right concentration of cofactors necessary for NOS activation. Despite this decreased NO production, platelet aggregation was unaffected. In this study authors did not demonstrate that platelet aggregation induced by collagen was affected in AN, even though NO produced by platelets and/or by endothelium could be enough to maintain normal levels of aggregability during adolescence with a further impairment in adulthood [37]. In subjects affected by AN already presenting comorbidity, such as renal failure, reduced activity of L-arginine-NO pathway has been found in platelets. In conclusion, the unchanged platelet aggregation in the presence of diminished NO synthesis suggests an NO-cGMP-independent anti-aggregatory mechanism in adolescents with AN [37, 38].

20.4.3 Dyslipidemia

Dyslipidemia, established risk factor for coronary heart disease [39], includes disorders of lipoprotein metabolism that result in excess of total cholesterol, low-density lipoprotein cholesterol (LDL-C), and triglyceride or deficiency of high-density lipoprotein cholesterol. The definition of hypercholesterolemia in childhood is different from the one used in adults [39, 40] because in children lipid profile is influenced by age and sex [41]. For this reason, during childhood dyslipidemias are defined according to strict criteria. An elevated LDL-C level is the most common clinically significant marker of dyslipidemia in children. The majority of children have idiopathic dyslipidemias (polygenic, risk factor-associated, or multifactorial), whereas a minority have monogenic or secondary dyslipidemias. The more common genetic dyslipidemias include familial hypercholesterolemia (FH), familial combined hyperlipidemia (FCH), familial defective apoprotein-B, and familial hypertriglyceridemia [39].

20.4.3.1 Hypercholesterolemia and Oxidative Stress

LDL-C oxidation is one of the primary steps that finally leads to atherogenesis. The oxidative stress related to a condition of hypercholesterolemia leads to free radical production and, in particular, to the isoprostanes production. Isoprostanes are free radical-catalyzed products of arachidonic acid. A very important isoprostane, 8-iso-PGF-2 alpha, is secreted in monocytes/macrophages and vascular smooth muscle cell in human atherosclerotic plaque reflecting lipid peroxidation. 8-iso-PGF-2 alpha is a vasoconstrictor and mitogen and may modulate platelet function in vitro [39]. Cracowski et al. found that urinary levels of 8-iso-PGF-2 alpha did not differ in children with type IIa hypercholesterolemia from those of the control group. This finding suggests that hypercholesterolemia is not associated with an increased lipid peroxidation in childhood [40]. Although endothelial dysfunction has been described in hypercholesterolemic children, the clinical manifestations of coronary artery disease are rare in heterozygous familial and polygenic hypercholesterolemic children, even in the second decade of life, and children develop atherosclerosis by middle age. As a consequence the elevated 8-iso-PGF-2 alpha levels observed in hypercholesterolemic adults compared with controls may be due to early atherosclerosis development rather than a direct effect of LDL-C elevation [40].

20.4.3.2 Treatment of Hypercholesterolemia

Most treatment recommendations advise a low-fat, low-cholesterol diet [42], such as the American Heart Association Step I diet, for children with dyslipidemia beginning at the age of 2 years or older. Children younger than 2 years should not perform a low-fat, low-cholesterol diet, because their rapid growth and development require adequate fat and cholesterol intake. Children and adolescents with FH or FCH are the only non-adults for whom trials of drug therapy are available and drugs are approved by FDA. Bile-acid-binding resins are the only medications approved for treatment of dyslipidemia for children younger than 8 years of age. 3-Hydroxy-3-methylglutaryl coenzyme A reductase inhibitors (statins) are approved for use in older children with heterozygous FH. Other medications used in adults for treatment of hyperlipidemia, such as niacin, either are not recommended for children or have not been evaluated for safety and efficacy in childhood. Additional interventions for children include dietary supplements (fiber, sterol or stanol margarines, and omega-3 fatty acids), exercise, weight loss for overweight and obese children, and identification and treatment of diabetes or other causes of secondary dyslipidemia.

20.5 Conclusions

Emerging evidences suggest that children affected by growth, puberty, as well as nutritional disturbances are highly exposed to oxidative stress. Further studies are needed to establish possible prophylactic actions or treatment protocols limiting the oxidative damage in childhood.

References

1. Styne DM (2003) The regulation of pubertal growth. Horm Res 60(1):22–26
2. Aycicek A, Erel O, Kocyigit A, Selek S, Demirkol MR (2006) Breast milk provides better antioxidant power than does formula. Nutrition 22(6):616–619
3. Veldhuis JD, Roemmich JN, Richmond EJ, Rogol AD, Lovejoy JC, Sheffield-Moore M, Mauras N, Bowers CY (2005) Endocrine control of body composition in infancy, childhood and puberty. Endocr Rev 26(1):114–146
4. Kosecik M, Erel O, Sevinc E, Selek S (2005) Increased oxidative stress in children exposed to passive smoking. Int J Cardiol 100(1):61–64
5. Wit JM, Clayton PE, Rogol AD, Savage MO, Saenger PH, Cohen P (2008) Idiopathic short stature: definition, epidemiology, and diagnostic evaluation. Growth Horm IGF Res 18(2): 89–110
6. Granot E, Kohen R (2004) Oxidative stress in childhood—in health and disease states. Clin Nutr 23(1):3–11
7. Mohn A, Marzio D, Giannini C, Capanna R, Marcovecchio M, Chiarelli F (2005) Alterations in the oxidant-antioxidant status in prepubertal children with growth hormone deficiency: effect of growth hormone replacement therapy. Clin Endocrinol (Oxf) 63(5):537–542
8. Mohn A, Chiavaroli V, Cerruto M, Blasetti A, Giannini C, Bucciarelli T, Chiarelli F (2007) Increased oxidative stress in prepubertal children born small for gestational age. J Clin Endocrinol Metab 92(4):1372–1378
9. Chiavaroli V, Giannini C, D'Adamo E, de Giorgis T, Chiarelli F, Mohn A (2009) Insulin resistance and oxidative stress in children born small and large for gestational age. Pediatrics 124(2):695–702
10. Leung AK, Robson WL (1995) Evaluating tall children. Can Fam Physician 41:457–458, 461–462, 465–468
11. European Standards Committee on Urinary (DNA) Lesion Analysis, Evans MD, Olinski R, Loft S, Cooke MS (2010) Toward consensus in the analysis of urinary 8-oxo-7,8-dihydro-2′-deoxyguanosine as a noninvasive biomarker of oxidative stress. FASEB J 24(4):1249–1260
12. Nishizawa H, Handayaningsih AE, Iguchi G, Cho Y, Takahashi M, Yamamoto M, Suda K, Kasahara K, Hakuno F, Yamanouchi K, Nishihara M, Seino S, Takahashi S, Takahashi Y (2012) Enhanced oxidative stress in GH-transgenic rat and acromegaly in humans. Growth Horm IGF Res 22(2):64–68
13. Fiorillo C, Becatti M, Attanasio M, Lucarini L, Nassi N, Evangelisti L, Porciani MC, Nassi P, Gensini GF, Abbate R, Pepe G (2010) Evidence for oxidative stress in plasma of patients with Marfan syndrome. Int J Cardiol 145(3):544–546
14. Vanzin CS, Biancini GB, Sitta A, Wayhs CA, Pereira IN, Rockenbach F, Garcia SC, Wyse AT, Schwartz IV, Wajner M, Vargas CR (2011) Experimental evidence of oxidative stress in plasma of homocystinuric patients: a possible role for homocysteine. Mol Genet Metab 104(1–2): 112–117
15. Traggiai C, Stanhope R (2003) Disorders of pubertal development. Best Pract Res Clin Obstet Gynaecol 17(1):41–56
16. Tena-Sempere M (2012) Deciphering puberty: novel partners, novel mechanisms. Eur J Endocrinol 167(6):733–747
17. Llorente-Cantarero FJ, Gil-Campos M, Benitez-Sillero JD, Muñoz-Villanueva MC, Túnez I, Pérez-Navero JL (2012) Prepubertal children with suitable fitness and physical activity present reduced risk of oxidative stress. Free Radic Biol Med 53(3):415–420
18. Berberoğlu M (2009) Precocious puberty and normal variant puberty: definition, etiology, diagnosis and current management. J Clin Res Pediatr Endocrinol 1(4):164–174
19. Pérez-Navero JL, Benítez-Sillero JD, Gil-Campos M, Guillén-del Castillo M, Tasset I, Túnez I (2009) Changes in oxidative stress biomarkers induced by puberty. An Pediatr (Barc) 70(5): 424–428

20. Livadas S, Dracopoulou M, Vasileiadi K, Lazaropoulou C, Magiakou MA, Xekouki P, Voutetakis A, Kanaka-Gantenbein C, Papassotiriou I, Stefanadis C, Chrousos GP, Dacou-Voutetakis C (2009) Elevated coagulation and inflammatory markers in adolescents with a history of premature adrenarche. Metabolism 58(4):576–581
21. Chiavaroli V, Liberati M, D'Antonio F, Masuccio F, Capanna R, Verrotti A, Chiarelli F, Mohn A (2010) GNRH analog therapy in girls with early puberty is associated with the achievement of predicted final height but also with increased risk of polycystic ovary syndrome. Eur J Endocrinol 163(1):55–62
22. Traggiai C, Stanhope R (2002) Delayed puberty. Best Pract Res Clin Endocrinol Metab 16(1):139–151
23. Esteves SC, Agarwal A (2011) Novel concepts in male infertility. Int Braz J Urol 37(1):5–15
24. Codoñer-Franch P, Valls-Bellés V, Arilla-Codoñer A, Alonso-Iglesias E (2011) Oxidant mechanisms in childhood obesity: the link between inflammation and oxidative stress. Transl Res 158(6):369–384
25. Flegal KM, Ogden CL (2011) Childhood obesity: are we all speaking the same language? Adv Nutr 2(2):159S–166S
26. Rhee KE, Phelan S, McCaffery J (2012) Early determinants of obesity: genetic, epigenetic, and in utero influences. Int J Pediatr 2012:463850
27. Giannini C, de Giorgis T, Scarinci A, Ciampani M, Marcovecchio ML, Chiarelli F, Mohn A (2008) Obese related effects of inflammatory markers and insulin resistance on increased carotid intima media thickness in pre-pubertal children. Atherosclerosis 197(1):448–456
28. Atabek ME, Keskin M, Yazici C, Kendirci M, Hatipoglu N, Koklu E, Kurtoglu S (2006) Protein oxidation in obesity and insulin resistance. Eur J Pediatr 165(11):753–756
29. Kaneko K, Kimata T, Tsuji S, Ohashi A, Imai Y, Sudo H, Kitamura N (2012) Measurement of urinary 8-oxo-7,8-dihydro-2-deoxyguanosine in a novel point-of-care testing device to assess oxidative stress in children. Clin Chim Acta 413(23–24):1822–1826
30. de Giorgis T, D'Adamo E, Giannini C, Chiavaroli V, Scarinci A, Verrotti A, Chiarelli F, Mohn A (2012) Could receptors for advanced glycation end products be considered cardiovascular risk markers in obese children? Antioxid Redox Signal 17(2):187–191
31. Chiavaroli V, D'Adamo E, Giannini C, de Giorgis T, De Marco S, Chiarelli F, Mohn A (2012) Serum levels of receptors for advanced glycation end products in normal-weight and obese children born small and large for gestational age. Diabetes Care 35(6):1361–1363
32. Chiavaroli V, Giannini C, De Marco S, Chiarelli F, Mohn A (2011) Unbalanced oxidant-antioxidant status and its effects in pediatric diseases. Redox Rep 16(3):101–107
33. Giannini C, de Giorgis T, Scarinci A, Cataldo I, Marcovecchio ML, Chiarelli F, Mohn A (2009) Increased carotid intima-media thickness in pre-pubertal children with constitutional leanness and severe obesity: the speculative role of insulin sensitivity, oxidant status, and chronic inflammation. Eur J Endocrinol 161(1):73–80
34. Mohn A, Catino M, Capanna R, Giannini C, Marcovecchio M, Chiarelli F (2005) Increased oxidative stress in prepubertal severely obese children: effect of a dietary restriction-weight loss program. J Clin Endocrinol Metab 90(5):2653–2658
35. Matson KL, Fallon RM (2012) Treatment of obesity in children and adolescents. J Pediatr Pharmacol Ther 17(1):45–57
36. Chanoine JP, Richard M (2011) Early weight loss and outcome at one year in obese adolescents treated with orlistat or placebo. Int J Pediatr Obes 6(2):95–101
37. Rodrigues Pereira N, Bandeira Moss M, Assumpção CR, Cardoso CB, Mann GE, Brunini TM, Mendes-Ribeiro AC (2010) Oxidative stress, l-arginine-nitric oxide and arginase pathways in platelets from adolescents with anorexia nervosa. Blood Cells Mol Dis 44(3):164–168
38. Muñoz MT, Argente J (2002) Anorexia nervosa in female adolescents: endocrine and bone mineral density disturbances. Eur J Endocrinol 147(3):275–286
39. Reilly MP, Praticò D, Delanty N, DiMinno G, Tremoli E, Rader D, Kapoor S, Rokach J, Lawson J, FitzGerald GA (1998) Increased formation of distinct F2 isoprostanes in hypercholesterolemia. Circulation 98(25):2822–2828

40. Cracowski JL, Ploin D, Bessard J, Baguet JP, Stanke-Labesque F, Mallion JM, Bost M, Bessard G (2001) Formation of isoprostanes in children with type IIa hypercholesterolemia. J Cardiovasc Pharmacol 38(2):228–231
41. Skinner AC, Steiner MJ, Chung A, Perrin EM (2012) Cholesterol curves to identify population norms by age and sex in healthy weight children. Clin Pediatr (Phila) 51(3):233–237
42. Kavey RE, Allada V, Daniels SR, Hayman LL, McCrindle BW, Newburger JW, Parekh RS, Steinberger J, American Heart Association Expert Panel on Population and Prevention Science, American Heart Association Council on Cardiovascular Disease in the Young, American Heart Association Council on Epidemiology and Prevention, American Heart Association Council on Nutrition, Physical Activity and Metabolism, American Heart Association Council on High Blood Pressure Research, American Heart Association Council on Cardiovascular Nursing, American Heart Association Council on the Kidney in Heart Disease, Interdisciplinary Working Group on Quality of Care and Outcomes Research (2006) Cardiovascular risk reduction in high-risk pediatric patients: a scientific statement from the American Heart Association Expert Panel on Population and Prevention Science; the Councils on Cardiovascular Disease in the Young, Epidemiology and Prevention, Nutrition, Physical Activity and Metabolism, High Blood Pressure Research, Cardiovascular Nursing, and the Kidney in Heart Disease; and the Interdisciplinary Working Group on Quality of Care and Outcomes Research: endorsed by the American Academy of Pediatrics. Circulation 114(24):2710–2738

Chapter 21
Hematologic Disorders

Stephen Rogers, Manuel Silva, and Allan Doctor

Abbreviations

2,3-BPG	2,3-Bisphosphoglycerate
ATP	Adenosine triphosphate
Cat	Catalase
cdB3	Cytoplasmic domain of Band 3
CO_2	Carbon dioxide
DHA	Dehydroascorbic acid
EMP	Embden Meyerhof pathway
G6P	Glucose-6-phosphate
G6PD	Glucose-6-phosphate dehydrogenase
GAPDH	Glyceraldehyde phosphate dehydrogenase
GLUT-1	Glucose transporter 1
GR	Glutathione reductase
GSH	L-γ-Glutamyl-L-cysteinylglycine
GSHPx	Glutathione peroxidase
GSSG	Glutathione disulfide
H_2O_2	Hydrogen peroxide
Hb	Hemoglobin
HbS	Hemoglobin S
HMP	Hexose monophosphate pathway
$HO^•$	Hydroxyl radical

S. Rogers, Ph.D. • M. Silva, M.D. • A. Doctor, M.D. (✉)
Departments of Pediatrics and Biochemistry, Washington University in Saint Louis, School of Medicine, McDonnel Pediatric Research Building,
5th Floor/Campus Box 8208, 660 South Euclid Avenue, St. Louis, MO 63110, USA
e-mail: Rogers_S@wustl.kids.edu; Silva_M@wustl.kids.edu; Doctor_A@wustl.kids.edu; doctor@wustl.edu

H. Tsukahara and K. Kaneko (eds.), *Studies on Pediatric Disorders*, Oxidative Stress in Applied Basic Research and Clinical Practice, DOI 10.1007/978-1-4939-0679-6_21,

LDH	Lactate dehydrogenase
metHb	Methemoglobin
metHbR	Methemoglobin reductase
NADH	Nicotinamide adenine dinucleotide
NADPH	Nicotinamide adenine dinucleotide phosphate
$NO^{\bullet}$	Nitric oxide
O_2	Oxygen
O_2^-	Superoxide
PK	Pyruvate kinase
PMOR	Plasma membrane oxidoreductases
Prx	Periredoxin
RBC	Red blood cell
RNS	Reactive nitrogen species
ROS	Reactive oxygen species
SCD	Sickle cell disease
SOD	Superoxide dismutase
TrxR	Thioredoxin reductase

21.1 Introduction

Erythrocytes, the most abundant cells in the human body, are well known for the essential part they play in oxygen (O_2) and carbon dioxide (CO_2) transport. Less appreciated are their more subtle functions including context responsive vascular signaling (regulating vascular smooth muscle tone as a function of O_2-sensitive processing of reactive nitrogen species (RNS) [1, 2], reactive oxygen species (ROS) [3], and metabolites of adenosine [4]), and of most relevance to this review—the detoxification of damaging oxidants [5, 6]. A number of factors allow erythrocytes to fulfill these essential functions: their ubiquitous distribution, high turnover (a desirable attribute for a detoxification unit [6]), highly evolved structure/composition, and perhaps most important their metabolic specialization, which is dedicated to maintaining reversible O_2 binding capacity and redox homeostasis in blood [7]. Herein, we will review features of erythrocyte metabolism relevant to antioxidant systems as well as perturbations of these systems in congenital and acquired disease that affect erythrocyte function.

21.1.1 Erythrocyte Development/Life-span

Initially, as multipotent hematopoietic stems cells in bone marrow, erythrocytes undergo a series of differentiations (erythropoiesis) prior to entry into circulation as reticulocytes. After circulating for 1–2 days, reticulocytes ("immature" red blood cells—RBCs) mature into erythrocytes. During this final developmental phase, having lost their nucleus just prior to entry into the circulation, reticulocytes also lose ribosomes, mitochondria, and endoplasmic reticulum, and therefore no longer possess capacity for cell division, protein synthesis, or mitochondrial-based energy metabolism [8]. This leaves the mature erythrocyte to rely on cellular constituents developed prior to maturation for protection and life-span preservation [9].

In healthy humans, erythrocytes circulate for 100–120 days, during which time they may travel up to 400 km in the course of circulatory transit [7]. At the end of their life-span, aged/damaged erythrocytes are phagocytosed by macrophages, predominantly in the spleen [7]. The molecular mechanism determining erythrocyte removal is incompletely understood; however, it is thought to involve recognition of senescence antigens [8, 10] (e.g., the clustering of Band-3 membrane protein [11]).

21.1.2 Erythrocyte Membrane Structure

Mature erythrocytes have evolved a membrane structure ideally matched to functional requirements. The biconcave disc enhances gas exchange and enables the distortion required for microcirculatory traversal. The membrane is anchored to the cytoskeleton by special transmembrane proteins [10], which also form highly organized networks of specialized domains built around the integral proteins which span the bilayer [12]. These domains not only control cell shape but also, via the attachment of substrates/cytoplasmic enzymes, play an essential role in the regulation of erythrocyte energy metabolism [13] and antioxidant defense [14].

21.1.3 Erythrocyte Cytoplasmic Composition

The metalloprotein hemoglobin (Hb) dominates cytoplasmic composition of the erythrocyte, accounting for between 95 and 97 % of all proteins. Hb reversibly binds O_2 at a heme group carried by each globin chain (of which there are four). In order to bind O_2, the heme group must be maintained in its reduced ferrous (+2) oxidation state; once oxidized to a ferric (+3) state forming methemoglobin (metHb) (consequent to oxygen release as superoxide O_2^-, rather than as O_2), Hb is no longer able to fulfill this function [15].

Recent proteomic analyses of erythrocytes have revealed over 500 other cytosolic proteins, many of which have redox regulation capabilities [16], necessary to address the substantive oxidative stress associated with O_2 transport. Most of these are involved in the process of metabolizing glucose, synthesizing adenosine triphosphate (ATP), carrying O_2 and CO_2, and neutralizing the oxidants arising from these processes and as a consequence of baseline O_2 release from heme as superoxide [17, 18].

21.2 Erythrocyte Energetics

The mature erythrocyte, devoid of the organelles and enzymes required for oxidative energy production, relies on the relatively inefficient anaerobic glycolytic pathway for ATP generation [7]. While glucose is the principle source of energy, other substrates including inosine, fructose, mannose, and galactose may also be metabolized. Under normal circumstances, plasma glucose is consumed at a rate of approximately 1.25 mmol/L of blood each hour [19]. In this metabolic process, several essential by-products are formed, including allosteric effectors and reducing equivalents. These play important roles in regulating O_2 binding capacity and delivery, as well as in redox balance, and vascular homeostasis.

21.2.1 Metabolic Pathways

Glucose enters erythrocytes by facilitated diffusion (which does not require energy expenditure) via the insulin-independent glucose transporter GLUT-1 [20]. Upon entry, it is immediately phosphorylated to glucose-6-phosphate (G6P), thus preventing its escape from the cell. G6P is then shared as the initial substrate for both metabolic pathways in the cell; (1) the Embden Meyerhof pathway (EMP; glycolytic pathway) and (2) the hexose monophosphate pathway (HMP; or pentose shunt). These two pathways fulfill distinct roles in the requirements of the erythrocyte.

21.2.1.1 Embden Meyerhof Pathway (EMP)

The main function of the EMP (direct glycolysis) is the generation of cellular energy in the form of ATP. In this process, glucose is anaerobically catabolized to pyruvate or lactate in three defined phases (Fig. 21.1). First is an initial "investment phase" during which 2 mol of ATP are utilized to prepare glucose for further metabolism. Next, a "splitting phase" in which fructose 1,6-bisphosphate is cleaved to two three-carbon triose phosphates. Finally comes the "yield phase" (or "payoff phase"), resulting in the formation of four molecules of ATP and the production of pyruvate or lactate—the overall net gain being 2 mol of ATP per mole of glucose.

The allosteric effector 2,3-bisphosphoglycerate (2,3-BPG) is also produced via the EMP, by the subsidiary Luebering–Rapoport shunt. 2,3-BPG stabilizes the deoxygenated Hb "t-state" conformation, thereby decreasing Hb affinity for O_2 and facilitating O_2 release to tissue (shifting the oxyhemoglobin dissociation curve to the right). Flux through the 2,3-BPG pathway is O_2- and pH dependent; at an intracellular pH of 7.2 (physiological pH in erythrocytes), this pathway accounts for approximately 20 % of glycolytic flux. However, greater flux is observed in response to hypoxia and increases in pH.

The glyceraldehyde phosphate dehydrogenase (GAPDH) reaction of the EMP also furnishes the cells with nicotinamide adenine dinucleotide (NADH), an important reducing equivalent (Fig. 21.2a). NADH is utilized by the enzyme

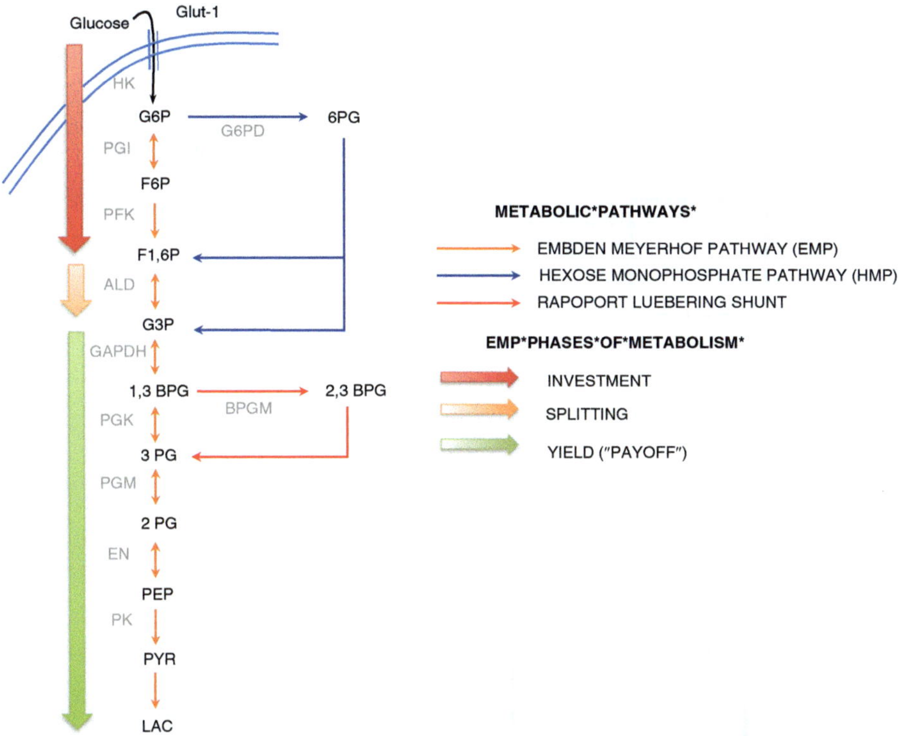

Fig. 21.1 There are three metabolic "loops" that comprise glucose metabolism in erythrocytes: the Embden Meyerhof pathway (EMP), hexose monophosphate pathway (HMP), and the Rapoport–Leubering shunt (RLS). Metabolism through the EMP generates energy in the form of ATP, during which glucose is anaerobically catabolized to pyruvate or lactate in three defined phases. First is an initial "investment phase" during which 2 mol of ATP are utilized to prepare glucose for further metabolism. Next, a "splitting phase" in which fructose 1,6-bisphosphate is cleaved to two three-carbon triose phosphates. Finally comes the "yield phase" (or "payoff phase"), resulting in the formation of four molecules of ATP and the production of pyruvate or lactate—the overall net gain being 2 mol of ATP per mole of glucose

methemoglobin reductase (metHbR; also known as NADH diaphorase or cytochrome b5 reductase) to reduce the heme group of Hb from its oxidized ferric state (Fe^{3+}) to its reduced ferrous state (Fe^{2+}) [21], thus maintaining the ability of Hb to reversibly bind O_2. This system maintains metHb concentrations in healthy individuals at less than 1 % of total Hb [7].

21.2.1.2 Hexose Monophosphate Pathway (HMP)

The HMP, which shares G6P (with the EMP) as its initial substrate, does not generate any high-energy phosphate bonds. Instead, its primary function is the reduction of nicotinamide adenine dinucleotide phosphate ($NADP^+$) to its reduced form

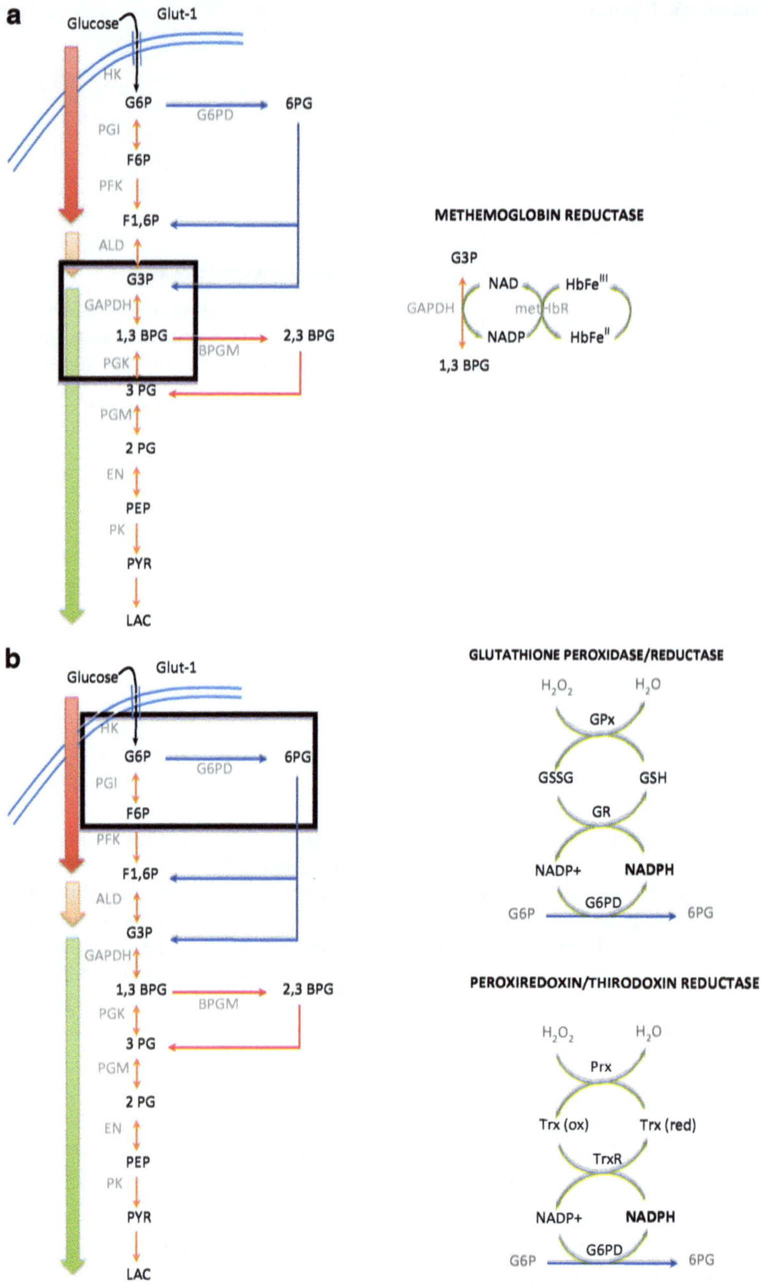

Fig. 21.2 (**a**) The glyceraldehyde phosphate dehydrogenase (GAPDH) reaction of the Embden Meyerhof pathway furnishes the cells with NADH, an important reducing equivalent. NADH is utilized by the methemoglobin reductase (*inset*) to reduce the heme group of Hb from its oxidized ferric state (Fe^{3+}) to its reduced ferrous state (Fe^{2+}), thus maintaining the ability of Hb to reversibly bind O_2. (**b**) This is achieved via the glucose 6-phosphate dehydrogenase (G6PD) reaction of the hexose monophosphate pathway, which shares G6P (with the EMP) as its initial substrate and does not generate any high-energy phosphate bonds. Rather, the primary function of this step is the reduction of nicotinamide adenine dinucleotide phosphate ($NADP^+$) to its reduced form NADPH, which is utilized by multiple antioxidant systems in the cell (such as the glutathione and periredoxin systems)

NADPH (Fig. 21.2b). This is achieved via the glucose-6-phosphate dehydrogenase (G6PD) reaction. The reducing equivalent NADPH is highly relevant in terms of cellular antioxidant capacity, as it is utilized by several enzymes (including glutathione reductase (GR), catalase, and thioredoxin reductase (TrxR) [22]) to maintain functionality of antioxidant systems.

21.2.2 Erythrocytic Energy Consumption

Erythrocytic ATP is used almost exclusively in and on the plasma membrane to maintain electrochemical and ion gradients, phosphorylation reactions, and turnover of phospholipid phosphates (particularly di- and tri-phosphatidylinositol [19]). Cytosolic processes account for less than 10 % of total ATP utilization [7, 19]. These minor components include the maintenance of glycolysis, and reduction reactions involving Hb (i.e., the maintenance of the heme group in its active reduced state—allowing the reversible binding of O_2 [23]) and enzyme and membrane protein sulfhydryl groups [7, 24].

21.2.3 Regulation of Erythrocyte Energetics

Regulation of erythrocyte energetics was originally thought to be under the control of covalent modifications and/or feedback inhibition/activation of key allosteric enzymes. A new hypothesis of metabolic regulation has since emerged, based upon an O_2-sensitive reciprocal binding relationship between key EMP enzymes and hemoglobin for the cytoplasmic domain of Band 3 (cdB3) membrane protein [25–29]; EMP binding to cdB3 is inhibitory, decelerating this pathway in oxygenated erythrocytes [30, 31]. Specifically, it has been demonstrated that (1) deoxygenated Hb displays a higher affinity for the cdB3 than oxygenated Hb [32], (2) phosphorylation of specific tyrosine residues (positions 8 and 21) on the cdB3 displaces EMP enzymes from the membrane [13, 25], and that (3) tyrosine phosphorylation up-regulates the EMP and down-regulates the HMP [33].

O_2-sensitive regulation of glucose flux perfectly fits the cellular requirements of the erythrocyte. Under fully oxygenated conditions, the inhibition of glycolytic enzymes (due to their binding to cdB3) releases substrate constraint on G6P, allowing greater HMP glucose flux. This ensures efficient NADPH recycling, necessary to protect the erythrocyte from oxidative stress derived from the high erythrocyte O_2 load/environment. Conversely, the displacement and activation of EMP enzymes from the cdB3 under low O_2 conditions up-regulate EMP activity, furnishing erythrocytes with ATP, 2,3-BPG (aiding in O_2 unloading), and NADH (necessary to combat the formation of metHb formation), which occurs upon O_2 unloading [27].

21.2.4 Metabolic Flux

In unstressed conditions, 90–95 % of total erythrocytic glucose is metabolized via the EMP [34, 35]. Under such conditions, the G6PD reaction operates at less than 1 % of its capacity [36], largely due to inhibition of the enzyme by NADPH [37]. Exposure of erythrocytes to oxidant stress results in the oxidation of NADPH (to $NADP^+$), releasing the inhibition on G6PD. In such situations, the HMP can account for 80–100 % of total glucose consumed [38, 39]. This ability to up-regulate the HMP serves to maintain levels of the reducing equivalent NADPH, which protects the cell by preserving the ongoing functionality of several antioxidant systems, including catalase, glutathione reductase, and TrxR.

21.3 Free Radicals and Reactive Oxygen Species

During circulation, erythrocytes encounter a variety of naturally produced free radicals. These chemical species, which contain one or more unpaired electrons, play a number of important signaling roles (e.g., superoxide, O_2^-; hydroxyl radical, $HO^{\bullet}$; and nitric oxide, $NO^{\bullet}$) [1, 40]. Erythrocytes also encounter various ROS, including non-radical derivatives of O_2 such as hydrogen peroxide (H_2O_2). While low concentrations of these species are essential for normal function, uncontrolled accumulation has the potential to be extremely damaging.

The reactivity of the various ROS is subject to reaction rates and is dependent upon the concentration of reactants and the rate of competing reactions, including the respective concentration of their reactants. This complexity has been simplified into a concept known as the reaction target area principle [41], which states that the relative amount of a compound that reacts with respective targets can be determined by multiplying the reaction rate and the concentration of the target. As such, ROS compartmentalization plays a significant role in determining reactivity. For example, uncharged molecules such as H_2O_2 freely cross the erythrocyte membrane, whereas charged molecules such as O_2^- enter only via transmembrane anion channels. For this reason, O_2^- is often concentrated in the intracellular compartment. Additionally, H_2O_2 and O_2^- are more selective in terms of their reactivity with biological molecules (leaving most targets unscathed). For example, the main reactions of O_2^- are with itself (dismutation), with other radical species, or with transition metals. On the other hand, $HO^{\bullet}$ is a much more aggressive oxidating species and reacts with numerous biological targets, making it potentially much more damaging.

21.3.1 RBC Oxidant Exposure

By virtue of the environment to which they are exposed, which is rich in O_2 and heme-iron (in the form of the oxygen carrying protein Hb), erythrocytes are continuously exposed to oxidant stress [42]. During reversible O_2 binding, occasional spontaneous conformational fluctuations in the heme pocket of oxygenated Hb allow water or a small anion to enter, resulting in transfer of an electron from the iron to O_2 to produce metHb and the O_2^- radical [43]. In fact, approximately 2–3 % of total Hb undergoes auto-oxidation daily [44]. In addition to scavenging endogenously produced oxidants, erythrocytes are also ideally positioned (due to ubiquitous distribution) to act as circulating detoxification agents, neutralizing other oxidants produced throughout the circulation [45, 46].

21.3.2 RBC Antioxidant Defenses

Erythrocytes contain a robust antioxidant system, with ability to rapidly counter the detrimental effects of oxidative insult. The efficacy of this system is reflected by the fact that erythrocytes generally demonstrate little evidence of oxidant damage. Oxidant damage occurring to erythrocytes may result from either an increase in ROS production or a reduction in antioxidant capacity. Generally, such situations are mainly observed under pathological conditions, leading to disruption of membrane transport systems, enzymes, proteins, and lipids and the premature removal of the erythrocyte from the circulation.

Classically, the erythrocyte antioxidant system is thought of as just comprising systems to scavenge and detoxify ROS and prevent the generation of radical chain reactions [15]. However, additional secondary mechanisms are also evident, which selectively rid the erythrocyte of damaged proteins [47, 48], thereby prolonging circulating cell life-span.

21.3.2.1 Enzymatic Antioxidants

Several enzymatic antioxidants are found in erythrocytes. These include superoxide dismutase (SOD), catalase (Cat), glutathione peroxidase (GSHPx), glutathione reductase (GR), periredoxin (Prx), TrxR, plasma membrane oxidoreductases (PMOR), and metHb reductase (metHbR). Ongoing functionality of these enzymatic antioxidant systems requires the recycling of the reducing equivalents NADPH and NADH, which are furnished in the erythrocyte via glucose metabolism through the HMP and EMP, respectively (Fig. 21.3). This enzyme family is described in detail elsewhere in this text.

Methemoglobin Reductase

In healthy erythrocytes, the majority of metHb reduction is achieved through an NADH-linked system [49]. NADH, furnished via the EMP (glycolytic) pathway in the G3PD reaction, is utilized by the enzyme metHbR. NADH reduces cytochrome

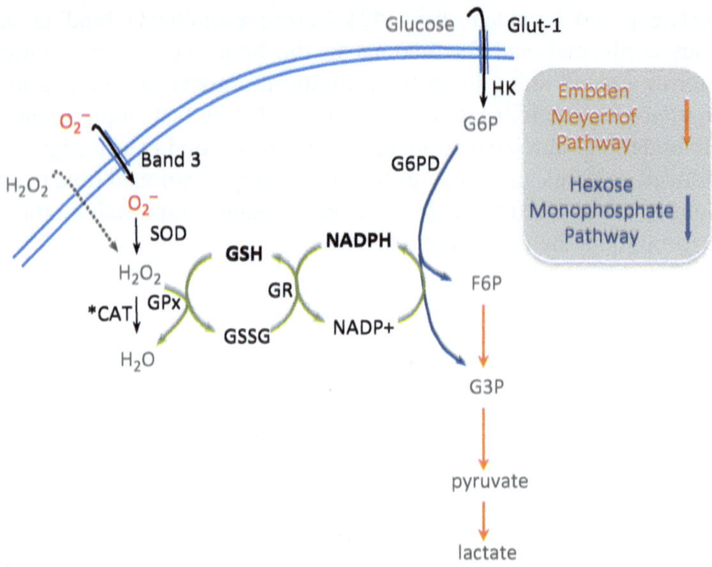

Fig. 21.3 Cartoon illustrating relationship amongst glycolytic pathways, glutathione reductase (GR), catalase (CAT), and superoxide dismutase (SOD), as this system addresses commonly encountered ROS (e.g., superoxide: O_2^- or hydrogen peroxide: H_2O_2)

b5, which in turn reduces the iron of metHb from its ferric trivalent (+3) state to its ferrous divalent (+2) state. GSH and ascorbic acid also play a role in metHb reduction, although their contribution is minimal. Ongoing functionality of metHbR requires metabolic flux via the EMP, which is the main route via which erythrocytes recycle NAD to its reduced form (NADH).

21.3.2.2 Nonenzymatic Antioxidants

In addition to the enzymatic antioxidant systems, erythrocytes also contain numerous nonenzymatic antioxidants, including glutathione (GSH), the vitamins (A, C—ascorbate, and E—alpha tocopherol), and other less familiar species such as uric acid and melatonin.

Glutathione

GSH (L-y-glutamyl-L-cysteinylglycine) is the most abundant low molecular weight thiol found in erythrocytes (concentration ~ 2 mM). GSH can act as a direct scavenger of free radicals, or as a substrate for GSHPx1 during the detoxification of H_2O_2 and lipid hydroperoxides [50]. GSH also directly reduces hemoglobin sulfhydryls, generating glutathione disulfide (GSSG) which is recycled by GR. Steady state GSH levels reflect a balance between synthesis (including both de novo synthesis and GR-mediated regeneration of GSH from GSSG) and loss (due to the efflux of GSSG from the cell; the erythrocyte membrane is permeable to GSSG, but not GSH). GSSG removal from the cell occurs as a result of active transport, in a process that is temperature sensitive and dependent upon metabolic energy [51, 52]. When the cell is no longer able to maintain its content of GSH, irreversible damage occurs.

Vitamins A, C, and E

Three vitamins (A, C, and E) provide important defense against oxidant stress in erythrocytes. Carotenoids, the precursors to vitamin A, also exert antioxidant effects in erythrocytes [53]. In total, six carotenoid species have been detected in erythrocytes including lutenin, zeaxanthin, β-cryptoxanthin, α-carotene, β-carotene, and lycopene [54].

Vitamin C (ascorbic acid or ascorbate) readily donates one or two electrons to a variety of oxidants, including oxygen free radicals, peroxides, and superoxide [55]. Additionally vitamin C functions to recycle vitamin E (alpha tocopherol) in a non-enzymatic pathway involving dihydrolipoic acid [56]. Each stage of vitamin C oxidation is reversible, thus permitting recycling [57].

Erythrocytes lack an active transporter for vitamin C. Uptake occurs by simple diffusion and is consequently very slow, with a half-time of hours [58]. Dehydroascorbic acid (DHA), however, is taken up by erythrocytes by facilitated diffusion through the glucose transporter (GLUT-1) [59], after which it is rapidly converted to ascorbate and trapped within the cells [60]. Erythrocytes also have enzymes that can facilitate GSH-dependent reduction of DHA, including the thiol-transferase glutaredoxin.

Vitamins A and E are both lipophilic and reside in the erythrocyte membrane. Vitamin E acts as the major radical chain-breaking antioxidant. In the process of scavenging lipid and non-lipid radical species, vitamin E is oxidized to a tocopheroxyl radical. This radical can be recycled back to vitamin E in an NADH-cytochrome b5-dependent reaction or via a nonenzymatic pathway involving vitamin C [56].

Melatonin

Melatonin (*N*-acetyl-5-methoxytryptamine), the chief secretory product of the pineal gland, is also synthesized in erythrocytes [61]. This synthetic product functions as a direct ROS scavenger, including O_2^- [62], H_2O_2 [63], and $HO^{\bullet}$ [64], in addition to lipid peroxides [65]. Furthermore, melatonin has also been demonstrated to stimulate other antioxidant enzymes, including Cu-Zn-SOD, catalase, GSHPx, glutathione reductase, and G6PD [66].

21.4 Disorders Resulting in Erythrocyte Oxidative Stress

21.4.1 Enzymopathies

21.4.1.1 Glucose-6-Phosphate Dehydrogenase Deficiency

G6PD deficiency is the most common enzyme defect in erythrocytes, with a worldwide prevalence of approximately 400 million. The geographical distribution of people with G6PD deficiency coincides with the location of malaria, suggesting protection. The most common pathologic manifestation of this defect is hemolytic anemia, provoked by an exposure to drug or infection [67].

G6PD is the first enzyme in the HMP (AKA pentose shunt, see Sect. 21.2.1.2) and in erythrocytes, the HMP is the sole source of the key reducing equivalent, nicotinamide adenine dinucleotide phosphate (NADPH). Mutations in the G6PD gene located in the Xq28 region cause this X-linked hereditary disease. There are more than 100 mutations or combinations of mutations that have been described, leading to a wide array of phenotypes. G6PD has 515 amino acids and most mutations cause amino acid substitutions [68].

Most people affected with G6PD deficiency are asymptomatic and undiagnosed. This enzyme deficiency is usually discovered only after development of acute hemolytic anemia following exposure to an infectious agent and ingestion of fava beans or a xenobiotic. Severe infections or systemic disorders associated with oxidative stress such as diabetes or congestive heart failure may also trigger hemolytic anemia in patients with G6PD deficiency [69]. Jaundice and hemolysis with the presence of Heinz bodies develop within 72 h of exposure to an offending drug or event. Patients develop back pain and hemoglobinuria if hemolysis is severe, although more frequently hemolysis is self-limited and the anemia is relatively mild. Several drugs have been shown to cause substantial hemolysis in patients with G6PD deficiency, such as primaquine, dapsone, nitrofurantoin, and sulfomethoxazole [67].

Studies have linked G6PD deficiency and jaundice in newborns [70]. The mechanism by which neonatal jaundice is caused in G6PD-deficient newborns is not well elucidated. Neonates that develop hyperbilirubinemia within the first day of life or those who have a sibling history of severe neonatal jaundice should be evaluated for G6PD deficiency [67].

Some patients with G6PD deficiency develop chronic hemolysis; these variants are mostly caused by non-inherited sporadic mutations. Patients with these variants can present at birth with severe jaundice, chronic anemia, gallstones, and splenomegaly. Patients with this phenotype may have all the markers of chronic hemolysis; the degree of hemolysis varies greatly, with some patients requiring chronic transfusion therapy [71].

Preventing oxidant exposure in patients with G6PD deficiency is the most effective means to limit hemolysis. Once hemolysis has begun, however, removal of the offending agent should be prioritized or treatment of infection should be initiated with careful monitoring for anemia and consequences of hemolysis, such as hemoglobinuria and renal injury. Blood transfusions may be required if the anemia is severe, especially after ingestion of fava beans. Significant neonatal jaundice with unconjugated hyperbilirubinemia should be treated with phototherapy.

21.4.1.2 Pyruvate Kinase Deficiency

Pyruvate kinase (PK) is a 200 kDa tetramer that is a key component of the EMP; PK transfers a phosphate group from phosphoenolpyruvate to ADP, producing pyruvate and a molecule of ATP in glycolysis. The PK erythrocyte gene is located in chromosome 1 [72]. Missense mutations are the most frequent causes of this defect [24], with autosomal recessive inheritance and prevalence of 50 per million. In addition to decreased ATP production, PK deficiency results in increased 2,3-DPG levels (due to increased flux through the Rapoport–Leubering shunt). Erythrocytes are dependent on ATP generated by glycolysis for membrane integrity; consequently, amongst the EMP defects, PK deficiency is the most common cause of hereditary non-spherocytic anemia.

Most of the clinical manifestations of PK deficiency are limited to erythrocytes. Disease severity varies from mild anemia to severe neonatal anemia presenting with hydrops fetalis requiring exchange transfusion. The phenotype may also manifest as chronic hemolysis that intensifies during an infection or other metabolic stress. Of note, increased production of 2,3-DPG (consequence of diminished PK activity) attenuates the impact of anemia since diminished O_2 affinity increases O_2 offloading efficiency.

The hemolysis found in PK deficiency has typical laboratory findings, such as anemia, reticulocytosis, elevated lactate dehydrogenase (LDH), and increased unconjugated bilirubin. Hemoglobin is usually 7 g/dL; splenectomy improves the anemia, with a median hemoglobin of 9.8 g/dL [72]. Infrequent complications of this mutation are pancreatitis, gallbladder disease, kernicterus, thromboembolic events, and extramedullary hematopoietic production [73]. The diagnosis is made by assessment of the PK activity and specific DNA testing.

There has been an effort to match specific PK gene mutations to disease phenotype, dividing patients in three groups [72, 74]. The severe phenotype has a median age of 4 years at diagnosis, though it frequently presents with neonatal jaundice requiring exchange transfusion (median hemoglobin was 6.8 g/dL). Splenectomy for these patients usually increases hemoglobin by 1.8 g/dL and diminishes the transfusion requirement. Studies at the molecular level revealed that this phenotype is associated with missense mutations that affect protein stability and active site function. The moderate phenotype has a median age of 25 years at diagnosis, with a median hemoglobin of 9 g/dL. The most common mutation related to this phenotype is (homozygous) 1529A, with a very low residual PK activity. The mild phenotype is diagnosed at adulthood (mutation: 1456T); transfusion need is rare and is usually precipitated by concomitant disease or pregnancy [72].

Hemoglobinopathies

.2.1 Sickle Cell Disease

ıckle cell disease (SCD) is a hemoglobinopathy caused by a mutation in β-globin gene (val6glu), which results in abnormally shaped, rigid, and unstable erythrocytes with diminished half-life (10 vs. 50 days, in normals) and a phenotype characterized by chronic hemolytic anemia, recurrent vaso-occlusive crises, acute chest syndrome, pulmonary hypertension, stroke, chronic pain, and ischemia–reperfusion injury.

SCD is frequently accompanied by significant ROS production and chronic oxidative stress that contributes to phenotype severity [75]. However, we lack clear mechanistic understanding of the significant oxidative stress complicating SCD, which is also a key feature of phenotype variation [75]. Of note, even non-polymerized, solution-phase hemoglobin S (HbS) may promote oxidative stress; specifically, the low redox potential for heme in HbS [76] and avid binding affinity of HbS for the cdB3 regulatory domain [77, 78] strongly impact RBC energetics and antioxidant systems [14, 26, 27, 79] and notably, do so as a function of RBC O_2 content. As such, both the genesis and the disposal of ROS are abnormal in SCD, creating a baseline state of oxidative stress, which worsens in hypoxia [79]. Moreover, vaso-occlusive crises lead to ischemia and hypoxia in affected tissues. After reperfusion, superoxide, hydroxyl, and other ROS are produced. Xanthine oxidase is also released to the circulation increasing ROS and causing vascular dysfunction [80].

ROS create erythrocyte membrane instability and promote hemolysis. Additionally, there is an association between oxidative stress, phospholipase A2 activation, and the production of phosphatidylserine, marking erythrocytes as senescent and further accelerating hemolysis. Accelerated hemolysis overwhelms plasma haptoglobin resulting in free hemoglobin in plasma, which diminishes nitric oxide (NO) bioavailability directly (by Hb-based quenching of NO equivalents in plasma); this effect is amplified by release of arginase from erythrocytes, which depletes substrate for nitric oxide synthase [81]. Both adversely impact NO-dependent vasodilation and O_2 delivery. Moreover, SCD is also characterized by blunted antioxidant systems. The enzymatic mechanisms such as SOD, catalase, and heme oxygenase-1 are decreased; moreover, the free radical scavengers vitamin C, glutathione, and vitamin E are depleted in sickle erythrocytes [75]. Due to the clear and mounting evidence of oxidative stress in SCD, there are efforts to create therapeutics that enhance antioxidant systems. For example, early evidence suggests decreased sickling after a period of vitamins E and C supplementation [82, 83].

21.4.2.2 Thalassemia

β-Thalassemia is a hemoglobinopathy that arises from reduced production of the β-globin chain and an imbalance in α/β globin chain ratio, leading to a significant increase in ROS genesis and chronic hemolytic anemia. As noted above, ROS damage the erythrocyte membrane, further increasing hemolysis and releasing free iron, which deplete antioxidant systems and exacerbate hemolysis in a vicious cycle [84]. Though the mechanism is incompletely understood, increased ROS production in thalassemia is thought to arise from the precipitation of excess, unpaired alpha-globin chains and release of free iron [85]. Current treatment for thalassemia is supportive with blood transfusions and iron chelation. As with SCD, results of antioxidant therapy are controversial, though no definitive trial has yet been attempted [86].

21.4.3 Acquired Disorders

21.4.3.1 Sepsis

The pathophysiology of sepsis is complicated and includes systemic inflammation, coagulation, and endothelial dysfunction, as well as perfusion abnormalities that impair O_2 delivery and lead to organ dysfunction. Moreover, oxidative stress plays a significant role in its pathophysiology and is implicated in the progression of organ failure by a variety of mechanisms, including direct cellular injury (damage to lipids, DNA, and structural proteins and enzymes) and dysfunctional activation of gene expression involved in the inflammation pathways such as nuclear factor κβ and cytokines that mediate inflammation [87, 88].

There is evidence that RBCs play a key role in the maintenance of antioxidant defenses against oxidative stress, as a mobile ROS scavenger in the microcirculation [6]. Notably, outcomes for sepsis correlate with markers of oxidative stress in erythrocytes and non-survivors demonstrate evidence of erythrocyte injury as well as evidence of antioxidant system dysfunction, such as glutathione depletion and depressed catalase activity [89].

21.4.3.2 Diabetes Mellitus

There are several mechanisms by which hyperglycemia in diabetes produces an increase in ROS and oxidative stress contributes to vascular complications such as atherosclerosis and endothelial dysfunction [90]. It has been suggested that inappropriate protein glycosylation disrupts normal regulatory interactions between Hb, cdB3, and key EMP enzymes, constraining HMP flux and NADPH recycling capacity. Several lines of indirect evidence indicate such a disturbance. Notably (1) Hb glycosylation (HbA1C) affects the quaternary structure, allostery, and O_2 binding properties of Hb in a manner that likely promotes Hb–cdB3 association, explaining the observation that (2) HbA1C binds avidly to the RBC membrane [91]; moreover, (3) diabetic RBCs demonstrate abnormally glycated and dysfunctional Band 3 [92]. Imbalanced EMP/HMP is further suggested by the observation that (4) diabetic RBCs have limited NADPH [93] and GSH recycling capacity [94–96], and (5) diabetic RBC membranes show signs of oxidative damage/dysregulation [97, 98]. It is likely that chronic hyperglycemia disrupts EMP/HMP balance in RBCs and further impairs diabetic RBC antioxidant status, in addition to the known abnormalities in glucose G6PD [99] and GSH reductase [100]. Consequently, this results in chronic oxidation of the RBC and plasma thiol pool, thereby disturbing NO exchange between RBCs and endothelium vital to regulation of blood vessel tone. This hypothesis suggests a novel mechanism for the initiation and progression of diabetic vascular disease, which is associated with HbA1C levels [101, 102]. There is abundant literature on the benefits of the effects of antioxidant therapies in the pathogenesis of vascular disease in diabetes. The greatest number of studies have been done with vitamin E and vitamin C [103, 104].

21.4.3.3 Metabolic Syndrome

Metabolic syndrome is a highly prevalent condition comprised by central obesity, insulin resistance, dyslipidemia, and hypertension; together, metabolic syndrome components increase the chances of cardiovascular disease greater than the collective likelihood of its separate components [105]. Oxidative stress plays a key role in metabolic syndrome and its pathogenesis with increased activity of phagocytic NADPH oxidases, oxidized LDL, and nitrotyrosine levels has been found in large arteries of patients with metabolic syndrome [106, 107]. Notably, those with metabolic syndrome have been found to have lower concentrations of vitamins C and E [106]. Where examined, oxidative injury to erythrocytes and impaired hemorheology are ubiquitous in patients with metabolic syndrome [108].

21.4.3.4 Uremia

Oxidative stress is nearly uniformly present in patients with end-stage renal disease [109]. In uremia, there is a substantial increase of ROS, which, combined with uremic toxins, leads to a ubiquitous dysfunctional protein modification known as "carbonyl stress" [110, 111]. There is also reduced concentration of glutathione-dependent antioxidant systems in uremic patients. Patients with renal failure have an increased ROS production, decreased renal clearance of some of these agents, and a reduced enzymatic antioxidant system [112]. As in other chronic conditions with oxidative stress, erythrocytes suffer significant injury in uremia, with adverse impact upon O_2 bonding/release, antioxidant systems, rheology, and signaling (NO processing)—all of which contribute to disease progression by contributing to tissue dysoxia [113].

21.4.3.5 Blood Storage

Mounting evidence suggests negative outcomes associated with RBC transfusions; this has led to attention to the quality of stored blood used for transfusion [114, 115]. Many groups have demonstrated storage time-dependent oxidative damage to erythrocytes, collectively described as the "storage lesion." Glycolysis is impaired in erythrocytes after approximately 2 weeks of storage (and therefore ATP, NADPH, and NADH generation), with resulting reduction in erythrocyte glutathione content and glutathione reductase [116]. Moreover, the high glucose concentration in storage media (required to counteract the metabolic disturbance above) leads to nonenzymatic glycosylation of proteins and formation of advanced glycosylation end products, which impair normal erythrocyte function [116].

References

1. Singel DJ, Stamler JS (2005) Chemical physiology of blood flow regulation by red blood cells: the role of nitric oxide and S-nitrosohemoglobin. Annu Rev Physiol 67:99–145
2. Doctor A, Stamler JS (2011) Nitric oxide transport in blood: a third gas in the respiratory cycle. Compr Physiol 1:541–568
3. Buehler PW, Alayash AI (2004) Oxygen sensing in the circulation: "cross talk" between red blood cells and the vasculature. Antioxid Redox Signal 6:1000–1010
4. Ellsworth ML et al (2009) Erythrocytes: oxygen sensors and modulators of vascular tone. Physiology (Bethesda) 24:107–116
5. Buehler PW, Alayash AI (2005) Redox biology of blood revisited: the role of red blood cells in maintaining circulatory reductive capacity. Antioxid Redox Signal 7:1755–1760
6. Richards RS, Roberts TK, McGregor NR, Dunstan RH, Butt HL (1998) The role of erythrocytes in the inactivation of free radicals. Med Hypotheses 50:363–367
7. Siems WG, Sommerburg O, Grune T (2000) Erythrocyte free radical and energy metabolism. Clin Nephrol 53:S9–S17
8. Volpe EP (1993) Blood and circulation. McGraw-Hill College, Columbus
9. Hattangadi SM, Lodish HF (2007) Regulation of erythrocyte lifespan: do reactive oxygen species set the clock? J Clin Invest 117:2075–2077
10. Seda Artis SAS (2012) Carnosine and its role on the erythrocyte rheology. In: Seda Artis A (ed) Hemodynamics—new diagnostic and therapeutic approaches. InTech Europe, Croatia
11. Bosman GJ, Willekens FL, Werre JM (2005) Erythrocyte aging: a more than superficial resemblance to apoptosis? Cell Physiol Biochem 16:1–8
12. Smith C, Marks AD, Lieberman M (2005) Mark's basic medical biochemistry. Lippincott Williams & Wilkins, Philadelphia
13. Campanella ME, Chu H, Low PS (2005) Assembly and regulation of a glycolytic enzyme complex on the human erythrocyte membrane. Proc Natl Acad Sci U S A 102:2402–2407
14. Rogers SC et al (2009) Hypoxia limits antioxidant capacity in red blood cells by altering glycolytic pathway dominance. FASEB J 9:3159–3170
15. Cimen MY (2008) Free radical metabolism in human erythrocytes. Clin Chim Acta 390:1–11
16. Chakrabarti A et al (2011) Differential expression of red cell proteins in hemoglobinopathy. Proteomics Clin Appl 5:98–108
17. Rifkind JM, Nagababu E (2013) Hemoglobin redox reactions and red blood cell aging. Antioxid Redox Signal 18:2274–2283
18. Bhattacharya D, Mukhopadhyay D, Chakrabarti A (2007) Hemoglobin depletion from red blood cell cytosol reveals new proteins in 2-D gel-based proteomics study. Proteomics Clin Appl 1:561–564
19. Rapoport SM, Dubiel W, Maretzki D, Siems W (1985) In: Proceedings of the 16th FEBS meeting, Part A. VNU Science Press, Utrecht, pp 165–176
20. Baldwin SA, Lienhard GE (1989) Purification and reconstitution of glucose transporter from human erythrocytes. Methods Enzymol 174:39–50

21. Prchal JT et al (1990) Congenital methemoglobinemia due to methemoglobin reductase deficiency in two unrelated American black families. Am J Med 89:516–522
22. Mustacich D, Powis G (2000) Thioredoxin reductase. Biochem J 346(Pt 1):1–8
23. Telen MJ, Kaufman RE (1999) The mature erythrocyte. In: Greer JP, Foerster J (eds) Clinical hematology. Lippincott Williams & Wilkins, Philadelphia, pp 217–247
24. van Wijk R, van Solinge WW (2005) The energy-less red blood cell is lost: erythrocyte enzyme abnormalities of glycolysis. Blood 106:4034–4042
25. Harrison ML, Rathinavelu P, Arese P, Geahlen RL, Low PS (1991) Role of band 3 tyrosine phosphorylation in the regulation of erythrocyte glycolysis. J Biol Chem 266:4106–4111
26. Low PS, Rathinavelu P, Harrison ML (1993) Regulation of glycolysis via reversible enzyme binding to the membrane protein, band 3. J Biol Chem 268:14627–14631
27. Messana I et al (1996) Human erythrocyte metabolism is modulated by the O2-linked transition of hemoglobin. FEBS Lett 390:25–28
28. Castagnola M, Messana I, Sanna MT, Giardina B (2010) Oxygen-linked modulation of erythrocyte metabolism: state of the art. Blood Transfus 8(suppl 3):s53–s58
29. De Rosa MC, Alinovi CC, Galtieri A, Russo A, Giardina B (2008) Allosteric properties of hemoglobin and the plasma membrane of the erythrocyte: new insights in gas transport and metabolic modulation. IUBMB Life 60:87–93
30. Tsai IH, Murthy SN, Steck TL (1982) Effect of red cell membrane binding on the catalytic activity of glyceraldehyde-3-phosphate dehydrogenase. J Biol Chem 257:1438–1442
31. Solti M, Friedrich P (1976) Partial reversible inactivation of enzymes due to binding to the human erythrocyte membrane. Mol Cell Biochem 10:145–152
32. Walder JA et al (1984) The interaction of hemoglobin with the cytoplasmic domain of band 3 of the human erythrocyte membrane. J Biol Chem 259:10238–10246
33. Lewis IA, Campanella ME, Markley JL, Low PS (2009) Role of band 3 in regulating metabolic flux of red blood cells. Proc Natl Acad Sci U S A 106:18515–18520
34. Albrecht V, Roigas H, Schultze M, Jacobasch G, Rapoport S (1971) The influence of pH and methylene blue on the pathways of glucose utilization and lactate formation in erythrocytes of man. Eur J Biochem 20:44–50
35. Gaetani GD, Parker JC, Kirkman HN (1974) Intracellular restraint: a new basis for the limitation in response to oxidative stress in human erythrocytes containing low-activity variants of glucose-6-phosphate dehydrogenase. Proc Natl Acad Sci U S A 71:3584–3587
36. Thorburn DR, Kuchel PW (1985) Regulation of the human-erythrocyte hexose-monophosphate shunt under conditions of oxidative stress. A study using NMR spectroscopy, a kinetic isotope effect, a reconstituted system and computer simulation. Eur J Biochem 150:371–386
37. Galiano S, Mareni C, Gaetani GF (1978) Effect of haemolysis on the hexose monophosphate pathway in normal and in glucose-6-phosphate dehydrogenase-deficient erythrocytes. Biochim Biophys Acta 501:1–9
38. Morelli A et al (1979) In vitro correction of erythrocyte glucose 6-phosphate dehydrogenase (G6PD) deficiency. Arch Biochem Biophys 197:543–550
39. Roigas H, Zoellner E, Jacobasch G, Schultze M, Rapoport S (1970) Regulatory factors in methylene blue catalysis in erythrocytes. Eur J Biochem 12:24–30 (in German)
40. Forman HJ, Fukuto JM, Torres M (2004) Redox signaling: thiol chemistry defines which reactive oxygen and nitrogen species can act as second messengers. Am J Physiol Cell Physiol 287:C246–C256
41. Beckman JS, Crow JP (1993) Pathological implications of nitric oxide, superoxide and peroxynitrite formation. Biochem Soc Trans 21:330–334
42. Low FM, Hampton MB, Winterbourn CC (2008) Peroxiredoxin 2 and peroxide metabolism in the erythrocyte. Antioxid Redox Signal 10:1621–1630
43. Winterbourn CC (1990) Oxidative denaturation in congenital hemolytic anemias: the unstable hemoglobins. Semin Hematol 27:41–50
44. Rifkind JM, Ramasamy S, Manoharan PT, Nagababu E, Mohanty JG (2004) Redox reactions of hemoglobin. Antioxid Redox Signal 6:657–666

45. Winterbourn CC, Stern A (1987) Human red cells scavenge extracellular hydrogen peroxide and inhibit formation of hypochlorous acid and hydroxyl radical. J Clin Invest 80:1486–1491
46. van Asbeck BS et al (1985) Protection against lethal hyperoxia by tracheal insufflation of erythrocytes: role of red cell glutathione. Science 227:756–759
47. Fujino T, Tada T, Hosaka T, Beppu M, Kikugawa K (2000) Presence of oxidized protein hydrolase in human cell lines, rat tissues, and human/rat plasma. J Biochem 127:307–313
48. Fujino T, Tada T, Beppu M, Kikugawa K (1998) Purification and characterization of a serine protease in erythrocyte cytosol that is adherent to oxidized membranes and preferentially degrades proteins modified by oxidation and glycation. J Biochem 124:1077–1085
49. Elahian F, Sepehrizadeh Z, Moghimi B, Mirzaei SA (2012) Human cytochrome b5 reductase: structure, function, and potential applications. Crit Rev Biotechnol; Early online 1–11
50. Masella R, Di Benedetto R, Vari R, Filesi C, Giovannini C (2005) Novel mechanisms of natural antioxidant compounds in biological systems: involvement of glutathione and glutathione-related enzymes. J Nutr Biochem 16:577–586
51. Lunn G, Dale GL, Beutler E (1979) Transport accounts for glutathione turnover in human erythrocytes. Blood 54:238–244
52. Srivastava SK, Beutler E (1969) The transport of oxidized glutathione from human erythrocytes. J Biol Chem 244:9–16
53. Miller NJ, Sampson J, Candeias LP, Bramley PM, Rice-Evans CA (1996) Antioxidant activities of carotenes and xanthophylls. FEBS Lett 384:240–242
54. Miyazawa T, Nakagawa K, Miyazawa T (2012) Liquid chromatography-based assay for carotenoids in human blood. In: Preedy VR (ed) Vitamin A and carotenoids: chemistry, analysis, function and effects. RSC Publishing, Cambridge
55. Buettner GR (1993) The pecking order of free radicals and antioxidants: lipid peroxidation, alpha-tocopherol, and ascorbate. Arch Biochem Biophys 300:535–543
56. Constantinescu A, Han D, Packer L (1993) Vitamin E recycling in human erythrocyte membranes. J Biol Chem 268:10906–10913
57. Mendiratta S, Qu ZC, May JM (1998) Erythrocyte ascorbate recycling: antioxidant effects in blood. Free Radic Biol Med 24:789–797
58. Hughes RE, Maton SC (1968) The passage of vitamin C across the erythrocyte membrane. Br J Haematol 14:247–253
59. Bianchi J, Rose RC (1986) Glucose-independent transport of dehydroascorbic acid in human erythrocytes. Proc Soc Exp Biol Med 181:333–337
60. Wagner ES, White W, Jennings M, Bennett K (1987) The entrapment of [14C]ascorbic acid in human erythrocytes. Biochim Biophys Acta 902:133–136
61. Ebadi M (1993) Multiple pineal receptors in regulating melatonin synthesis. In: Yu HS, Reiter RJ (eds) Melatonin: biosynthesis, physiological effects, and clinical applications. CRC Press, Boca Raton
62. Marshall KA, Reiter RJ, Poeggeler B, Aruoma OI, Halliwell B (1996) Evaluation of the antioxidant activity of melatonin in vitro. Free Radic Biol Med 21:307–315
63. Tan DX et al (2000) Melatonin directly scavenges hydrogen peroxide: a potentially new metabolic pathway of melatonin biotransformation. Free Radic Biol Med 29:1177–1185
64. Poeggeler B et al (1994) Melatonin—a highly potent endogenous radical scavenger and electron donor: new aspects of the oxidation chemistry of this indole accessed in vitro. Ann N Y Acad Sci 738:419–420
65. Reiter RJ (1998) Oxidative damage in the central nervous system: protection by melatonin. Prog Neurobiol 56:359–384
66. Reiter RJ et al (2003) Melatonin as an antioxidant: biochemical mechanisms and pathophysiological implications in humans. Acta Biochim Pol 50:1129–1146
67. Cappellini MD, Fiorelli G (2008) Glucose-6-phosphate dehydrogenase deficiency. Lancet 371:64–74
68. Ho HY, Cheng ML, Chiu DT (2007) Glucose-6-phosphate dehydrogenase—from oxidative stress to cellular functions and degenerative diseases. Redox Rep 12:109–118

69. Hecker PA, Leopold JA, Gupte SA, Recchia FA, Stanley WC (2013) Impact of glucose-6-phosphate dehydrogenase deficiency on the pathophysiology of cardiovascular disease. Am J Physiol Heart Circ Physiol 304:H491–H500
70. Watchko JF, Lin Z (2010) Exploring the genetic architecture of neonatal hyperbilirubinemia. Semin Fetal Neonatal Med 15:169–175
71. Mason PJ, Bautista JM, Gilsanz F (2007) G6PD deficiency: the genotype-phenotype association. Blood Rev 21:267–283
72. Zanella A, Fermo E, Bianchi P, Chiarelli LR, Valentini G (2007) Pyruvate kinase deficiency: the genotype-phenotype association. Blood Rev 21:217–231
73. Zanella A, Fermo E, Bianchi P, Valentini G (2005) Red cell pyruvate kinase deficiency: molecular and clinical aspects. Br J Haematol 130:11–25
74. Zanella A, Bianchi P (2000) Red cell pyruvate kinase deficiency: from genetics to clinical manifestations. Baillieres Best Pract Res Clin Haematol 13:57–81
75. Chirico EN, Pialoux V (2012) Role of oxidative stress in the pathogenesis of sickle cell disease. IUBMB Life 64:72–80
76. Hebbel RP, Morgan WT, Eaton JW, Hedlund BE (1988) Accelerated autoxidation and heme loss due to instability of sickle hemoglobin. Proc Natl Acad Sci U S A 85:237–241
77. Platt OS, Falcone JF (1995) Membrane protein interactions in sickle red blood cells: evidence of abnormal protein 3 function. Blood 86:1992–1998
78. Shaklai N, Sharma VS (1980) Kinetic study of the interaction of oxy- and deoxyhemoglobins with the erythrocyte membrane. Proc Natl Acad Sci U S A 77:7147–7151
79. Rogers SC et al (2013) Sickle hemoglobin disturbs normal coupling among erythrocyte O2 content, glycolysis, and antioxidant capacity. Blood 121:1651–1662
80. Nur E et al (2011) Oxidative stress in sickle cell disease; pathophysiology and potential implications for disease management. Am J Hematol 86:484–489
81. Kato GJ, Gladwin MT, Steinberg MH (2007) Deconstructing sickle cell disease: reappraisal of the role of hemolysis in the development of clinical subphenotypes. Blood Rev 21:37–47
82. Natta CL, Machlin LJ, Brin M (1980) A decrease in irreversibly sickled erythrocytes in sickle cell anemia patients given vitamin E. Am J Clin Nutr 33:968–971
83. Amer J et al (2006) Red blood cells, platelets and polymorphonuclear neutrophils of patients with sickle cell disease exhibit oxidative stress that can be ameliorated by antioxidants. Br J Haematol 132:108–113
84. Pfeifer WP et al (2008) Vitamin E supplementation reduces oxidative stress in beta thalassaemia intermedia. Acta Haematol 120:225–231
85. Scott MD et al (1993) Effect of excess alpha-hemoglobin chains on cellular and membrane oxidation in model beta-thalassemic erythrocytes. J Clin Invest 91:1706–1712
86. Dhawan V, Kumar KR, Marwaha RK, Ganguly NK (2005) Antioxidant status in children with homozygous thalassemia. Indian Pediatr 42:1141–1145
87. Huet O et al (2007) Plasma-induced endothelial oxidative stress is related to the severity of septic shock. Crit Care Med 35:821–826
88. Wheeler DS (2011) Oxidative stress in critically ill children with sepsis. Open Inflamm J 4:74–81
89. Dyson A et al (2011) An integrated approach to assessing nitroso-redox balance in systemic inflammation. Free Radic Biol Med 51:1137–1145
90. Droge W (2002) Free radicals in the physiological control of cell function. Physiol Rev 82:47–95
91. Petropoulos IK, Margetis PI, Antonelou MH, Koliopoulos JX, Gartaganis SP, Margaritis LH, Papassideri IS (2007) Structural alterations of the erythrocyte membrane proteins in diabetic retinopathy. Graefes Arch Clin Exp Ophthalmol 245:1179–1188
92. Gaczynska M, Judkiewicz L, Szosland K (1993) Abnormal degradation of red cell membrane proteins in diabetes. Cytobios 75:7–11
93. Carroll J et al (2006) An altered oxidant defense system in red blood cells affects their ability to release nitric oxide-stimulating ATP. Mol Biosyst 2:305–311

94. Bono A, Caimi G, Catania A, Sarno A, Pandolfo L (1987) Red cell peroxide metabolism in diabetes mellitus. Horm Metab Res 19:264–266
95. Dincer Y, Akcay T, Alademir Z, Ilkova H (2002) Effect of oxidative stress on glutathione pathway in red blood cells from patients with insulin-dependent diabetes mellitus. Metabolism 51:1360–1362
96. Thornalley PJ, McLellan AC, Lo TW, Benn J, Sonksen PH (1996) Negative association between erythrocyte reduced glutathione concentration and diabetic complications. Clin Sci 91:575–582
97. Jiang M et al (2003) Protein disregulation in red blood cell membranes of type 2 diabetic patients. Biochem Biophys Res Commun 309:196–200
98. Jain SK, McVie R, Duett J, Herbst JJ (1989) Erythrocyte membrane lipid peroxidation and glycosylated hemoglobin in diabetes. Diabetes 38:1539–1543
99. Xu Y, Osborne BW, Stanton RC (2005) Diabetes causes inhibition of glucose-6-phosphate dehydrogenase via activation of PKA, which contributes to oxidative stress in rat kidney cortex. Am J Physiol Renal Physiol 289:F1040–F1047
100. Blakytny R, Harding JJ (1992) Glycation (non-enzymic glycosylation) inactivates glutathione reductase. Biochem J 288(Pt 1):303–307
101. The relationship of glycemic exposure (HbA1c) to the risk of development and progression of retinopathy in the diabetes control and complications trial (1995). Diabetes 44:968–983
102. Effect of intensive diabetes management on macrovascular events and risk factors in the Diabetes Control and Complications Trial (1995). Am J Cardiol 75:894–903
103. Ceriello A et al (1991) Vitamin E reduction of protein glycosylation in diabetes. New prospect for prevention of diabetic complications? Diabetes Care 14:68–72
104. Varvarovska J et al (2004) Aspects of oxidative stress in children with type 1 diabetes mellitus. Biomed Pharmacother 58:539–545
105. Lee L, Sanders RA (2012) Metabolic syndrome. Pediatr Rev 33:459–466; quiz 467–458
106. Hutcheson R, Rocic P (2012) The metabolic syndrome, oxidative stress, environment, and cardiovascular disease: the great exploration. Exp Diabetes Res 2012:271028
107. Goodwill AG, Frisbee JC (2012) Oxidant stress and skeletal muscle microvasculopathy in the metabolic syndrome. Vasc Pharmacol 57:150–159
108. Ziobro A, Duchnowicz P, Mulik A, Koter-Michalak M, Broncel M (2013) Oxidative damages in erythrocytes of patients with metabolic syndrome. Mol Cell Biochem 378:267–273
109. Himmelfarb J, Hakim RM (2003) Oxidative stress in uremia. Current Opin Nephrol Hypertens 12:593–598
110. Rutkowski P et al (2006) Interrelationship between uremic toxicity and oxidative stress. J Ren Nutr 16:190–193
111. Suzuki D, Miyata T, Kurokawa K (2001) Carbonyl stress. Contrib Nephrol 134:36–45
112. Floccari F et al (2005) Oxidative stress and uremia. Med Res Rev 25:473–486
113. Yilmaz MI et al (2009) Hemoglobin is inversely related to flow-mediated dilatation in chronic kidney disease. Kidney Int 75:1316–1321
114. Doctor A, Spinella P (2012) Effect of processing and storage on red blood cell function in vivo. Semin Perinatol 36:248–259
115. Spinella PC, Doctor A, Blumberg N, Holcomb JB (2011) Does the storage duration of blood products affect outcomes in critically ill patients? Transfusion 51:1644–1650
116. Kanias T, Acker JP (2010) Biopreservation of red blood cells—the struggle with hemoglobin oxidation. FEBS J 277:343–356

Chapter 22
Infectious and Inflammatory Disorders

Hirokazu Tsukahara, Masato Yashiro, Yoshiharu Nagaoka, and Tsuneo Morishima

Abbreviations

BOM	Bilirubin oxidative metabolites
CNS	Central nervous system
CSF	Cerebrospinal fluid
ELISA	Enzyme-linked immunosorbent assay
GPX	Glutathione peroxidase
HIV	Human immunodeficiency virus
HO	Heme oxygenase
IAE	Influenza-associated acute encephalopathy
MDA	Malonaldehyde
MPO	Myeloperoxidase
NO	Nitric oxide
NOS	Nitric oxide synthase
8-OHdG	8-hydroxy-2′-deoxyguanosine
ROS	Reactive oxygen species
RSV	Respiratory syncytial virus
SOD	Superoxide dismutase
TAC	Total antioxidant capacity
TH	Total hydroperoxides

H. Tsukahara, M.D., Ph.D. (✉) • M. Yashiro, M.D., Ph.D.
Y. Nagaoka, M.D. • T. Morishima, M.D., Ph.D.
Department of Pediatrics, Okayama University Graduate School of Medicine, Dentistry and Pharmaceutical Sciences, Okayama 700-8558, Japan
e-mail: hirokazu010661@yahoo.co.jp

H. Tsukahara and K. Kaneko (eds.), *Studies on Pediatric Disorders*, Oxidative Stress in Applied Basic Research and Clinical Practice, DOI 10.1007/978-1-4939-0679-6_22,

22.1 Introduction

Reactive species with nitrogen-based or oxygen-based unpaired electrons appear to play diverse roles in many aspects of physiological and pathological events in the pediatric field [1]. This is readily evident in infectious and inflammatory disorders such as sepsis, meningitis, encephalopathy, pneumonia, gastritis, enterocolitis, urinary infection, skin infection, burn injury, and immune activation syndromes.

Nitric oxide (NO), produced endogenously in cells and tissues of various types [2, 3], is a biological messenger molecule involved in numerous homeostatic processes. Constitutive isoenzymes of nitric oxide synthase (NOS) contribute to important physiological processes such as vasorelaxation and neurotransmission. Inducible NOS, which is expressed in various cells including macrophages, neutrophils, epithelial cells, endothelial cells, and hepatocytes, produces excessive NO in infectious and inflammatory conditions. Furthermore, reactive oxygen species (ROS) including superoxide (O_2^-), hydrogen peroxide (H_2O_2), and hydroxyl radical (OH) are produced through enzymes such as xanthine oxidoreductase and NADPH oxidases by infiltrating phagocytic cells and resident cells in inflamed tissues. Many effects of NO and O_2^- might be mediated by their reaction product peroxynitrite ($ONOO^-$). These reactive species, which have antimicrobial actions via their cytotoxic or cytostatic effects, contribute to innate (nonspecific) and immunological host defenses.

It is necessary to mention that NO and ROS act as a double-edged sword, having both beneficial and deleterious effects in biological systems [1–3]. They can attack various substrates in the body including lipids, proteins, nucleic acids, and carbohydrates. Oxidation of any of these substances, if unchecked, can contribute to the development of acute and chronic disorders. Biomarker research in this field can help guide the selection, efficiency, and sufficiency of medical interventions for pathological states that are characterized by the enhanced generation of NO and ROS [1].

This chapter presents a review of the medical literature and discusses the pathophysiological roles of NO and ROS in infectious and inflammatory disorders of children. It provides a detailed explanation of some devastating conditions affecting the brain (i.e., purulent meningitis and encephalopathy). We hope that this chapter will foster further investigation into the mechanisms by which oxidative stress influences the progression of these disorders, and that it will suggest potential therapeutic measures.

22.2 Oxidative Stress Biomarkers

Results of recent studies implicate oxidative stress in various acute and chronic disorders of children [1]. Accordingly, it is important to evaluate the stress conditions objectively and non-invasively in this population.

Any biomolecule can be damaged by NO and ROS. Direct measurement of these species is difficult because of their short half-life. Therefore, stress conditions are evaluated by measuring stable products of the oxidative modification processes

Table 22.1 Clinically important oxidative stress biomarkers of the two categories

(a) *Formation of modified molecules by nitric oxide and reactive oxygen species*
Lipid peroxidation: malondialdehyde,[a] 4-hydroxy-2,3-nonenal,[a] acrolein,[a] hexanoyl-lysine,[a] F2-isoprostane,[a] hydroperoxides[a]
Oxidative DNA damage: 8-hydroxy-2′-deoxyguanosine[a]
Glyco-oxidation: carboxymethyl-lysine, pentosidine, methylglyoxal
Nitro-oxidation: nitrotyrosine, nitrite/nitrate[a]
Others: hydrogen peroxide, o,o′-dityrosine, carbonyl protein,[a] bilirubin oxidative metabolites,[a] oxidized glutathione[a]
(b) *Antioxidative enzymes and molecules*
Enzymes: superoxide dismutase,[a] catalase,[a] glutathione peroxidase,[a] glutathione reductase, glutathione-S-transferase, thioredoxin reductase, heme oxygenase[a]
Proteins: albumin, transferrin, ceruloplasmin, thioredoxin, L-type fatty acid binding protein
Low molecular weight molecules: bilirubin,[a] retinol,[a] ascorbate,[a] tocopherols,[a] ubiquinol/ubiquinone (coenzyme Q_{10}),[a] reduced glutathione,[a] nitrite/nitrate,[a] selenium,[a] zinc[a]

[a]These markers are explained in this chapter

Only biomarkers that are determined in samples of blood or urine are shown. Nitric oxide behaves either as a prooxidant or as an antioxidant according to its environment. Therefore, its stable metabolites, nitrite/nitrate, are listed in both categories (a) and (b)

in vivo [1]. Oxidative stress is also evaluated by measuring the consumption or induction of certain enzymes or antioxidants. Measurement of specific biomarkers in blood or urine enables repeated monitoring of systemic oxidative stress status, which is not possible with invasive tests. Clinically applicable and reliable biomarkers are presented in Table 22.1.

The first category (a) includes molecules that are generated in reactions with NO and ROS. They are subjected to either scission, cross-linking, or covalent modification in these reactions. Accordingly, the amount of these molecules is increased when ROS is generated excessively. Some are removed or repaired rapidly, but others remain in the body for a long time. Major targets of NO and ROS in the molecular components of the cells are membrane lipids, proteins, nucleic acids, and carbohydrates.

The second category (b) includes antioxidative enzymes and molecules that are associated with NO and ROS metabolism. In most cases, these molecules are destroyed or modified and exhibit decreased activity or quantity. Conversely, they often show an overshooting response for a period of hours, days, or weeks.

22.3 Infectious and Inflammatory Disorders Associated with Oxidative Stress

Numerous infectious and inflammatory disorders appear to be linked to oxidative damage attributable to NO and ROS in their pathogenesis and progression [1]. In most studies, oxidative stress biomarkers were determined in samples of blood (such as whole blood serum, plasma, erythrocytes, and leukocytes) or urine [4–36] (Table 22.2). In other studies, the parameters were measured using tissues or

Table 22.2 Infectious and inflammatory diseases associated with enhanced oxidative stress

Condition	Primary findings
Bacterial or fungal infection:	
Purulent meningitis	Higher urinary nitrite/nitrate [4]; higher serum MDA, ascorbate, lower serum-reduced glutathione [16]; higher serum nitrite/nitrate, lipid hydroperoxides, SOD [29]; higher serum MDA, MPO, lower serum TAC [31]
Sepsis	Lower whole blood-reduced glutathione [6]
Cholera	Higher plasma nitrite/nitrate, MPO [8]
Purulent skin infection	Higher urinary acrolein, 8-OHdG, bilirubin oxidative metabolites [11]
Chronic tonsillitis	Higher plasma MDA [12]
Chronic nail candidiasis	Higher serum MDA, 4-hydroxy-2,3-nonenal [15]
Chronic otitis media	Higher plasma MDA, lower plasma-reduced glutathione [18]
Acute otitis media, acute tonsillitis	Higher whole blood MDA, lower whole blood reduced glutathione [20]
Lower urinary tract infection	Higher urinary MDA, TAC [21]
Acute pneumonia	Higher whole blood MDA, lower whole blood reduced glutathione, GPX, SOD [22]; higher plasma MDA [23]
Viral infection:	
Human immunodeficiency virus infection	Lower serum TAC [5]; higher serum MDA [9]; lower plasma ascorbate, lower serum tocopherol [28]; lower serum ascorbate [34]
Hepatitis C virus infection	Higher leukocyte 8-OHdG [7]
Influenza encephalopathy	Higher serum nitrite/nitrate [10]
Viral bronchiolitis	Higher serum MDA, lower serum selenium [14]
Rotavirus-associated convulsion	Higher serum nitrite/nitrate [17]
Measles	Higher whole blood MDA, lower whole blood-reduced glutathione, lower serum retinol, ascorbate, tocopherol [25]
Human herpes virus 6 encephalopathy	Higher urinary 8-OHdG [26]
Pandemic influenza (H1N1)	Lower serum TAC, ubiquinol/ubiquinone, zinc [35]
Protozoan infection:	
Cutaneous leishmaniasis	Higher erythrocyte MDA, SOD [13]; higher serum nitrite/nitrate, MDA, lower serum SOD, GPX [19]
Schistosoma mansoni infection	Higher serum hydroperoxides, lower serum retinol, tocopherol [24]
Fasciola hepatica infection	Higher serum "total oxidant status," lower plasma TAC [32]
Others:	
Acute rheumatic fever	Lower serum TAC [27]
Burn	Higher plasma MDA [30]
Kawasaki disease	Higher plasma hydroperoxides [33]; higher serum hydroperoxides [36]

GPX glutathione peroxidase, *MDA* malondialdehyde, *MPO* myeloperoxidase, *8-OHdG* 8-hydroxy-2′-deoxyguanosine, *SOD* superoxide dismutase, *TAC* total antioxidant capacity

Only studies in which systemic oxidative stress status was evaluated are listed here. In these studies, specific biomarkers were determined in blood or urine samples from the patients. Conditions associated with bacterial or fungal infection are described in the early part of the table

Numbers in square brackets indicate cited references

different body fluids, either alone or in combination with samples of blood or urine. The results are the following: higher levels of myeloperoxidase (MPO) and inducible NO synthase in rectal tissue in cholera [8]; higher leukocyte levels and lower urinary levels of 8-hydroxy-2′-deoxyguanosine (8-OHdG) in *Helicobacter pylori* infection [37]; higher levels of malondialdehyde (MDA) and catalase in tonsillar and adenoid tissue in chronic adenotonsillitis [38]; lower "total antioxidant capacity (TAC)" in saliva in human immunodeficiency virus (HIV) infection [39]; higher 8-isoprostane, carbonyl protein, and glutathione peroxidase (GPX) in bronchoalveolar lavage fluid in postinfectious bronchiolitis obliterans [40]; and higher MDA and 8-isoprostane in nasopharyngeal secretion in hypoxic respiratory syncytial virus (RSV) bronchiolitis [41]. The results for cerebrospinal fluid (CSF) examination are presented in the next section.

Recently, enzyme-linked immunosorbent assay (ELISA) systems have been constructed for various stress biomarkers [11, 26, 40, 41]. Onerous pretreatments and expensive apparatus are virtually obviated by ELISA, which is a labor-saving, cost-saving method [42]. Examination using a few sample aliquots is possible. Furthermore, compact machines by which serum/plasma total hydroperoxides (TH) and "TAC" [24, 33, 36, 43] or urinary 8-OHdG [44] can be measured are available. These machines provide highly reproducible results quickly.

It is noteworthy that oxidative damage might take place in a selective manner. For instance, lipid peroxidation and oxidative DNA damage are not always accompanied by overproduction of NO. Detection of more than one marker is an important key because a single marker might yield misleading results. It might also be crucial to determine which particular markers, alone or in combination with others, can serve as a precise indicator of the contribution of oxidative stress to a disease, thereby allowing the success (or failure) of the treatment to be monitored.

22.4 Biomarkers in CSF

Purulent meningitis and acute encephalopathy are severe central nervous system (CNS) disorders that can result in sudden death or development of neurological sequelae [45, 46]. These diseases, which are not rare in children, persist as an important public health problem worldwide.

Computed tomography and magnetic resonance imaging might be useful to evaluate the severity of brain injury, but it is often difficult to perform such radiological examinations during the critical period when key therapeutic decisions are made. Therefore, assessing ongoing brain injury and predicting outcomes using CSF samples is extremely valuable. These markers might be of various kinds, including cytokines/chemokines (such as tumor necrosis factor-α, interleukins-1β, -6, -8, -10, -13, granulocyte colony-stimulating factor, monocyte chemotactic protein-1, macrophage inflammatory protein-1, transforming growth factor-β, interferon-γ, interferon-γ-inducible protein-10), brain injury marker (S-100B protein, neuron-specific enolase, glial fibrillary acidic protein, neurofilaments, tau protein), tissue degradation enzymes (matrix metalloproteinases-8, -9), and oxidative stress markers (as described below).

The brain is regarded as vulnerable to free radical damage because of high oxygen consumption, in addition to its consequent generation of high levels of NO and ROS, high contents of unsaturated lipids and cellular iron, and weakened antioxidant defense system [47]. Oxidative stress is a predisposing factor for neuronal destruction in meningitis and encephalopathy, and for neuronal degeneration in developmental brain disorders [48].

Measurement of oxidative stress biomarkers in CSF samples has been introduced into the pediatric field. These markers presumably permit sequential biological monitoring of the CNS of the patients. Clinically applicable CSF markers are the following: lipid hydroperoxides, MDA, acrolein, TH, 8-OHdG, nitrite/nitrate, bilirubin oxidative metabolites (BOM), and TAC [17, 29, 49–62]. All these markers have also been determined in blood or urine samples for evaluation of systemic oxidative stress status (Table 22.2). ELISA [53, 54, 62] and rapid analysis methods [50, 57, 58, 60, 61] are applied to practice in CSF analyses.

22.5 Acute Purulent Meningitis

Acute purulent meningitis is a severe bacterial infection of the CNS that occurs especially in children younger than 5 years of age. Although the introduction of antibiotics has made it curable, mortality and morbidity from the disease remain high [46, 63]. The mortality rate is about 5 %. Long-term morbidity, consisting mainly of persistent neurological sequelae, is about 20 %.

It is apparent that the excessive host immune response is incapable of controlling infection within the CNS, particularly the CSF within the subarachnoid space, and that this host inflammatory response contributes to many adverse events that can occur during purulent meningitis [63]. A complex series of events involving host cytokines/chemokines, proteolytic enzymes, and NO/ROS is responsible for meningitis-induced brain damage, at least during the early phase of the disease (Fig. 22.1).

In this section, we highlight the ROS-mediated brain damage and antioxidant defenses in pediatric patients with acute purulent meningitis. Previous results for the CSF analyses from the patients are presented in Table 22.3 [29, 31, 49–54, 58]. They show enhanced production of NO and ROS in the CNS of patients with purulent meningitis compared with those with aseptic meningitis as well as non-meningitis subjects. The pathogenic bacteria were *Streptococcus pneumoniae*, *Haemophilus influenzae*, Group B streptococcus, and others. These findings also suggest the contribution of elevated oxidative stress to disease severity and occurrence of neurological complications [29, 31, 49, 51, 52, 58]. Our earlier observations [53, 54], which are presented in Fig. 22.2, suggest that clinical and laboratory improvement is linked closely to the decrease in oxidant activation in CNS.

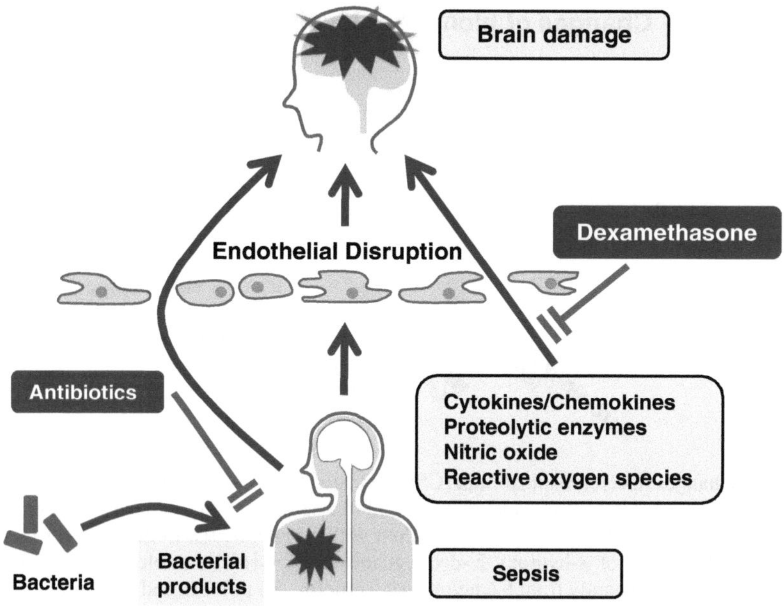

Fig. 22.1 Mechanisms of brain damage in purulent meningitis. Meningitis-associated brain damage and neuronal death is not mediated simply by the presence of viable bacterial components. Widely diverse inflammatory host factors (cytokines/chemokines, nitric oxide/reactive oxygen species, proteolytic enzymes) are now known to be involved in the complex pathophysiological cascade of purulent meningitis. Further exploration of these pathways is expected to contribute to the development of therapeutic adjunctive strategies in purulent meningitis

Table 22.3 CSF levels of oxidative stress biomarkers in acute purulent meningitis

Authors	Primary findings
Hamed et al. [29]	Higher nitrite/nitrate, lipid hydroperoxides
Miric et al. [31]	Higher malondialdehyde, myeloperoxidase, lower "total antioxidant capacity"
van Furth et al. [49]	Higher nitrite/nitrate
Tsukahara et al. [50]	Higher nitrite (levels were also determined using "urine reagent strips")[a]
Ray et al. [51]	Higher superoxide, malondialdehyde, superoxide dismutase
Murawska-Ciałowicz et al. [52]	Higher nitrite/nitrate
Tsukahara et al. [53]	Higher 8-hydroxy-2′-deoxyguanosine
Tsukahara et al. [54]	Higher acrolein-lysine, nitrite, bilirubin oxidative metabolites
Yamanaka et al. [58]	Higher hydroperoxides (levels were determined using the free radical analytical system; Diacron International, Italy)[a]

Numbers in square brackets indicate cited references
[a]Rapid analytical methods

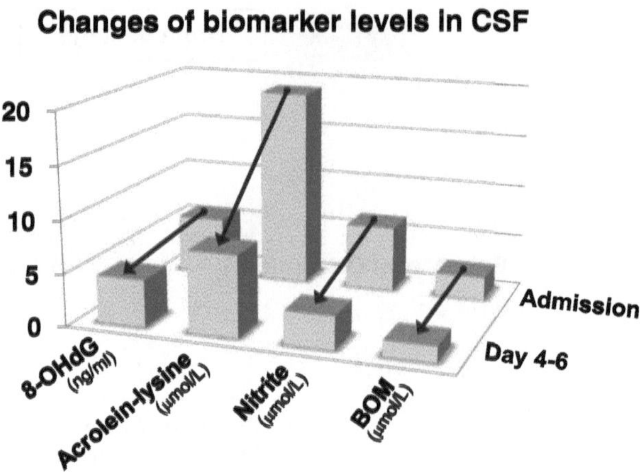

Fig. 22.2 Changes in cerebrospinal fluid (CSF) oxidative stress biomarker levels in children with purulent meningitis. All these markers were several times higher in children during the early phase of purulent meningitis than they were in children without meningitis. In the purulent meningitis group, the CSF levels of 8-hydroxy-2′-deoxyguanosine (8-OHdG), acrolein-lysine, nitrite, and BOM decreased significantly from 5.6 ng/mL, 19.0 μmol/L, 6.3 μmol/L, and 2.3 μmol/L on admission to 4.5 ng/mL, 7.8 μmol/L, 3.2 μmol/L, and 1.5 μmol/L at 4–6 hospital days, respectively, as patients responded to intravenous β-lactam antibiotics and dexamethasone administration [53, 54]. Presented data are mean values of the markers

The pathological hallmark of purulent meningitis is the dense inflammatory infiltrates in the subarachnoid and ventricular spaces, predominantly composed of neutrophils. Activated neutrophils secrete proteolytic enzymes and liberate NO and ROS. Excessive NO and ROS can exert various cytotoxic effects through lipid peroxidation, DNA strand breakage, mitochondrial damage, or potassium channel activity alterations [63, 64]. Oxidative alterations to vital macromolecules are observed in brain samples (most prominently in blood vessels and inflammatory cells) of patients who have died of meningitis [64, 65].

The presence of increased CSF levels of BOM in children with purulent meningitis is noteworthy here [54]. The levels were nearly eight times higher in these patients than in individuals of the non-meningitis group (Fig. 22.2). That significant increase of BOM is assumed to be attributable to local or systemic high production of bilirubin, mainly reflecting the overexpression of inducible heme oxygenase (HO) [66]. The HO induction and the subsequent bilirubin biosynthesis might be part of a common defense mechanism aimed at reducing oxidative brain injury in several neurological conditions [67].

Corticosteroids are beneficial for the early treatment of both human and experimental purulent meningitis [68]. Dexamethasone attenuates neuronal tissue damage by intervening at various levels of meningeal inflammatory cascade (Fig. 22.1). The proposed neuroprotective mechanisms include inhibition of ROS production, lipid peroxidation, leukocyte-endothelial interaction, and of transcription of various proinflammatory cytokines and inducible NOS. Dexamethasone treatment significantly lowered the nitrite concentration in the CSF of purulent meningitis children compared with those of non-treated children on the second day of the disease [52]. An additional piece of evidence for the beneficial effects of dexamethasone on purulent meningitis might be reduction in the levels of 8-OHdG, acrolein, and nitrite as the patients started to recover from the disease [53, 54] (Fig. 22.2).

Studies in rats have provided substantial evidence that antioxidant therapy (such as *N*-acetyl-L-cysteine, phenyl tert-butyl nitrone, and Mn(III) tetrakis(4-benzoic acid) porphyrin) is protective against vascular, cortical, and cochlear injury in pneumococcal meningitis [63, 64]. The adjunctive use of these antioxidants appears to offer promising future treatment options for purulent meningitis in humans.

22.6 Acute Encephalopathy

Acute encephalopathy is a severe CNS complication of common infections (such as influenza, exanthem subitum, and acute viral gastroenteritis) with high mortality and neurological sequelae [69]. As explained in previous sections, assessing brain injury and predicting outcomes using bedside point-of-care testing is valuable in these patients.

Here, we briefly highlight the ROS-mediated brain damage and antioxidant defenses in children with acute encephalopathy. Previous results for the CSF analyses from the patients are described in Table 22.4 [55–57, 59–62]. They indicate enhanced production of NO and ROS in the CNS of children with acute encephalopathy compared with those of non-encephalopathy subjects. These reports described the pathogenic viruses of influenza, enterovirus, RSV, and other viruses. Pleocytosis and high protein levels in CSF were found only occasionally in the patients. It remains to be clarified whether the NO/ROS production levels are related to the development of sequelae because available data in this research area remain limited.

Table 22.4 CSF levels of oxidative stress biomarkers in acute encephalopathy

Authors	Primary findings
Kawashima et al. [55]	Higher nitrite/nitrate in influenza encephalopathy
Kawashima et al. [56]	Higher nitrite/nitrate in influenza encephalopathy
Yamanaka et al. [57]	Higher hydroperoxides (free radical analytical system)[a] in influenza encephalopathy
Morichi et al. [59]	Higher nitrite/nitrate in respiratory syncytial virus encephalopathy
Kawashima et al. [60]	Higher hydroperoxides in enterovirus encephalopathy (free radical analytical system)[a]
Kawashima et al. [61]	Higher nitrite/nitrate, hydroperoxides (free radical analytical system)[a] in influenza encephalopathy
Miyata et al. [62]	Higher 8-hydroxy-2′-deoxyguanosine, hexanoyl-lysine in "clinically mild encephalopathy with a reversible splenial lesion"

Numbers in square brackets indicate cited references
[a]Rapid analytical method

It is noteworthy that in influenza encephalopathy patients, the increase of CSF NO levels was linked to the low serum zinc levels [56]. Trace elements are micronutrients that are present in small amounts in the body and which are necessary for the normal functioning of immune and antioxidant systems. The role of micronutrients in the pathogenesis, course, and outcome of encephalopathy warrants further study.

Stratification of acute encephalopathy is important clinically. Strategies for severe encephalopathy should be established in the near future. Prevention (or modulation) of excessive host inflammatory response including NO/ROS release, a possible therapeutic approach, is explained in the following section.

22.7 Redox Modulation Strategy for Influenza Encephalopathy

Influenza-associated acute encephalopathy (IAE) is an abrupt disorder of the CNS triggered by influenza virus infection, often engendering severe sequelae or death [70]. The Centers for Disease Control and Prevention have regarded IAE as an important public health problem at least since 2003, citing our report on IAE [71]. The 2009 pandemic influenza A (H1N1) virus emerged in Mexico in April 2009, thereafter spreading rapidly worldwide. In June 2009, The World Health Organization declared that the spreading novel influenza virus constituted a global pandemic. IAE of severe type also occurred following the 2009 pandemic influenza [72, 73].

Because influenza virus infection occurs predominantly in younger generations, great concern has arisen in relation to the severity of complications, such as IAE or severe pneumonia, among children [70]. Pathological findings, including the lack of viral antigen and sparse, if any, inflammatory infiltrates in the brain, imply that direct viral invasion and subsequent inflammation are unlikely to cause this encephalopathy. A prevailing theory is that excessive host inflammatory response characterized by massive production of pro-inflammatory cytokines/chemokines and NO/ROS and excessive apoptosis exacerbate IAE [74–79]. As a consequence, widespread vascular endothelial activation, dysfunction, and damage occur, ultimately resulting in multiple organ failure and death (Fig. 22.3).

In Japan, the guideline for diagnosis and management of pediatric IAE was formulated in 2005 by the collaborating study group on IAE, which was organized by the Japanese Ministry of Health, Labour, and Welfare. Dr. Morishima (an author of this chapter) has been the supervisor of the study since its inception. The guideline has been used widely among general and pediatric hospitals in our country. In addition to neuraminidase inhibitors, pulse steroid therapy, high-dose immunoglobulin, antioxidative agent (edaravone), coagulation modifying agent (thrombomodulin), plasma exchange, and hypothermia are listed as selectable treatments for severe IAE. These treatments are expected to impede excessive inflammatory host response and enhanced oxidative stress in the patients [57, 61, 78, 79] (Fig. 22.3).

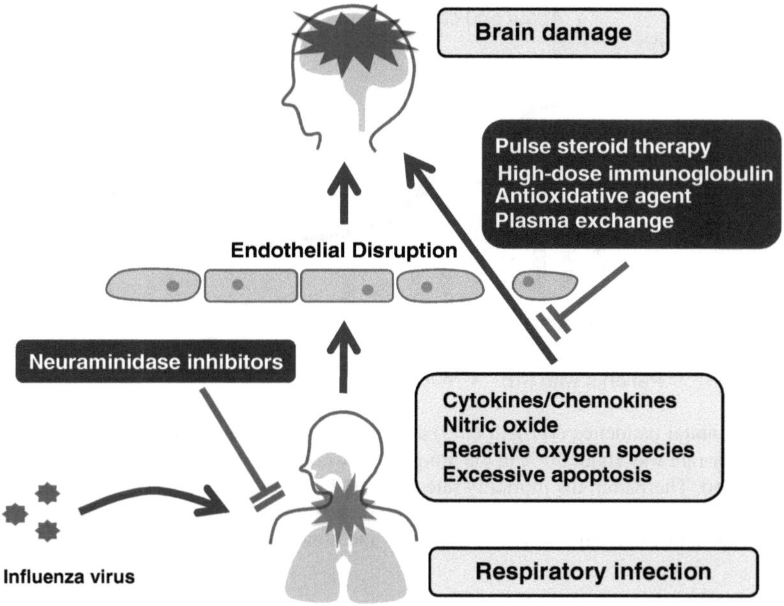

Fig. 22.3 Mechanisms of brain damage in influenza-associated acute encephalopathy (IAE). The IAE pathogenesis remains to be clarified. Viral RNA has not been detected in the CSF of most patients with IAE. The findings presented in recent reports suggest that, in cases of severe IAE, either seasonal or 2009 pandemic, pathological manifestations similarly result from complex biological phenomena including overproduction of cytokines/chemokines and nitric oxide/reactive oxygen species, apoptosis induction, and vascular endothelial disruption. Additional exploration of these pathways is expected to contribute to the development of more effective adjunctive strategies in IAE

For pediatric patients with IAE, the mortality rate was about 30 % in the pre-guideline era in Japan, when no efficient strategy had been proposed. Thereafter, the mortality rate decreased to about the one-fourth along with the nationwide distribution of this practical guideline (Fig. 22.4). Nevertheless, the incidence of poor outcomes of pediatric IAE has not been ideally low: 7 % for death and about 20 % for neurological sequelae. More global studies using edaravone and/or other antioxidants with sequential monitoring of oxidative stress biomarkers must be conducted to identify more effective strategies for IAE of severe type.

22.8 Summary and Conclusions

Infectious and inflammatory disorders are common and often severe in young generations. Severe forms of these disorders are occasionally fatal or leave severe sequelae, for which effective treatment is currently either insufficient or

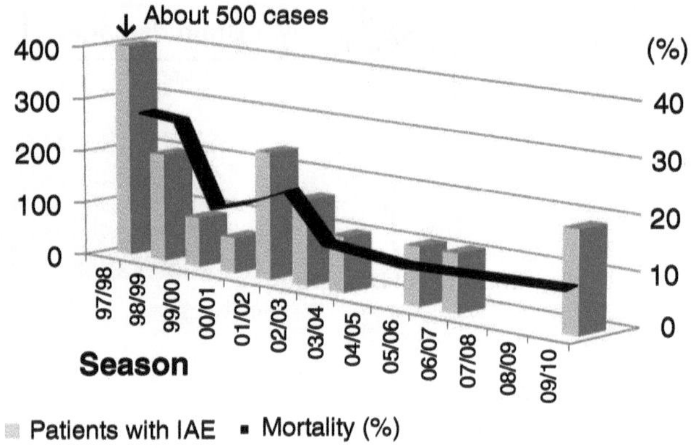

Fig. 22.4 Annual incidence (*vertical bars*) and mortality (*line graph*) of pediatric IAE in Japan. The mortality rate was about 30 % in the pre-guideline era in Japan, when no efficient strategy had been proposed. Thereafter, the mortality rate decreased to about one-fourth along with the nationwide distribution of this practical guideline. Nevertheless, the incidence of poor outcomes of pediatric IAE has not been ideally low: 7 % for death and about 20 % for neurological sequelae

unavailable. The excessive host response and enhanced oxidative stress are inferred to play an important role in the progression and deterioration of these disorders.

Nitric oxide and ROS are unstable molecules produced by various cells and tissues. Using biomarkers will be of great importance for the diagnosis and management of infectious and inflammatory disorders. Those levels are expected to reflect the severity of clinical illness and to predict long-term consequences (if appropriately efficient strategies are not selected). Measurement of oxidative stress biomarkers in CSF samples is especially valuable in patients with purulent meningitis and acute encephalopathy.

Anti-oxidative strategies represent a potential adjunctive and an effective approach to ameliorate these disorders in the pediatric field. Determination of these parameters might identify the disease state and enhance the timing of therapeutic approaches in future medical applications.

References

1. Tsukahara H (2007) Biomarkers for oxidative stress: clinical application in pediatric medicine. Curr Med Chem 14:339–351
2. Akaike T, Maeda H (2000) Nitric oxide and virus infection. Immunology 101:300–308
3. Goligorsky MS, Brodsky SV, Noiri E (2004) NO bioavailability, endothelial dysfunction, and acute renal failure: new insights into pathophysiology. Semin Nephrol 24:316–323
4. Nishimura M, Tsukahara H, Hiraoka M, Osaka Y, Ohshima Y, Tanizawa A, Mayumi M (2000) Systemic inflammatory response syndrome and acute renal failure associated with *Hemophilus influenzae* septic meningitis. Am J Nephrol 20:208–211
5. de Martino M, Chiarelli F, Moriondo M, Torello M, Azzari C, Galli L (2001) Restored antioxidant capacity parallels the immunologic and virologic improvement in children with perinatal

human immunodeficiency virus infection receiving highly active antiretroviral therapy. Clin Immunol 100:82–86
6. Lyons J, Rauh-Pfeiffer A, Ming-Yu Y, Lu XM, Zurakowski D, Curley M, Collier S, Duggan C, Nurko S, Thompson J, Ajami A, Borgonha S, Young VR, Castillo L (2001) Cysteine metabolism and whole blood glutathione synthesis in septic pediatric patients. Crit Care Med 29:870–877
7. Cardin R, Saccoccio G, Masutti F, Bellentani S, Farinati F, Tiribelli C (2001) DNA oxidative damage in leukocytes correlates with the severity of HCV-related liver disease: validation in an open population study. J Hepatol 34:587–592
8. Qadri F, Raqib R, Ahmed F, Rahman T, Wenneras C, Das SK, Alam NH, Mathan MM, Svennerholm AM (2002) Increased levels of inflammatory mediators in children and adults infected with *Vibrio cholerae* O1 and O139. Clin Diagn Lab Immunol 9:221–229
9. Jareño EJ, Romá J, Romero B, Marín N, Muriach M, Johnsen S, Bosch-Morell F, Marselou L, Romero FJ (2002) Serum malondialdehyde correlates with therapeutic efficiency of high activity antiretroviral therapies (HAART) in HIV-1 infected children. Free Radic Res 36:341–344
10. Kawashima H, Watanabe Y, Ichiyama T, Mizuguchi M, Yamada N, Kashiwagi Y, Takekuma K, Hoshika A, Mori T (2002) High concentration of serum nitrite/nitrate obtained from patients with influenza-associated encephalopathy. Pediatr Int 44:705–707
11. Tsukahara H, Shibata R, Ohshima Y, Todoroki Y, Sato S, Ohta N, Hiraoka M, Yoshida A, Nishima S, Mayumi M (2003) Oxidative stress and altered antioxidant defenses in children with acute exacerbation of atopic dermatitis. Life Sci 72:2509–2516
12. Kaygusuz I, Ilhan N, Karlidag T, Keles E, Yalçin S, Cetiner H (2003) Free radicals and scavenging enzymes in chronic tonsillitis. Otolaryngol Head Neck Surg 129:265–268
13. Kocyigit A, Gurel M, Ulukanligil M (2003) Erythrocyte antioxidative enzyme activities and lipid peroxidation levels in patients with cutaneous leishmaniasis. Parasite 10:277–281
14. Gurkan F, Atamer Y, Ece A, Kocyigit Y, Tuzun H, Mete M (2004) Relationship among serum selenium levels, lipid peroxidation, and acute bronchiolitis in infancy. Biol Trace Elem Res 100:97–104
15. Gangemi S, Saija A, Minciullo PL, Tomaino A, Cimino F, Bisignano G, Briuglia S, Merlino MV, Dallapiccola B, Salpietro DC (2004) Serum levels of malondialdehyde and 4-hydroxy-2,3-nonenal in patients affected by familial chronic nail candidiasis. Inflamm Res 53:601–603
16. Caksen H, Cemek M, Dede S, Dulger H, Cemek F (2004) Brief clinical study: lipid peroxidation and antioxidant status in children with acute purulent meningitis and encephalitis. Int J Neurosci 114:105–111
17. Kawashima H, Inage Y, Ogihara M, Kashiwagi Y, Takekuma K, Hoshika A, Mori T, Watanabe Y (2004) Serum and cerebrospinal fluid nitrite/nitrate levels in patients with rotavirus gastroenteritis induced convulsion. Life Sci 74:1397–1405
18. Yilmaz T, Koçan EG, Besler HT, Yilmaz G, Gürsel B (2004) The role of oxidants and antioxidants in otitis media with effusion in children. Otolaryngol Head Neck Surg 131:797–803
19. Serarslan G, Yilmaz HR, Söğüt S (2005) Serum antioxidant activities, malondialdehyde and nitric oxide levels in human cutaneous leishmaniasis. Clin Exp Dermatol 30:267–271
20. Cemek M, Dede S, Bayiroğlu F, Caksen H, Cemek F, Yuca K (2005) Oxidant and antioxidant levels in children with acute otitis media and tonsillitis: a comparative study. Int J Pediatr Otorhinolaryngol 69:823–827
21. Pavlova EL, Lilova MI, Savov VM (2005) Oxidative stress in children with kidney disease. Pediatr Nephrol 20:1599–1604
22. Cemek M, Caksen H, Bayiroğlu F, Cemek F, Dede S (2006) Oxidative stress and enzymic—non-enzymic antioxidant responses in children with acute pneumonia. Cell Biochem Funct 24:269–273
23. Mahalanabis D, Basak M, Paul D, Gupta S, Shaikh S, Wahed MA, Khaled MA (2006) Antioxidant vitamins E and C as adjunct therapy of severe acute lower-respiratory infection in infants and young children: a randomized controlled trial. Eur J Clin Nutr 60:673–680

24. Berhe N, Halvorsen BL, Gundersen TE, Myrvang B, Gundersen SG, Blomhoff R (2007) Reduced serum concentrations of retinol and alpha-tocopherol and high concentrations of hydroperoxides are associated with community levels of *S. mansoni* infection and schistosomal periportal fibrosis in Ethiopian school children. Am J Trop Med Hyg 76:943–949
25. Cemek M, Dede S, Bayiroglu F, Caksen H, Cemek F, Mert N (2007) Oxidant and non-enzymatic antioxidant status in measles. J Trop Pediatr 53:83–86
26. Fukuda M, Yamauchi H, Yamamoto H, Aminaka M, Murakami H, Kamiyama N, Miyamoto Y, Koitabashi Y (2008) The evaluation of oxidative DNA damage in children with brain damage using 8-hydroxydeoxyguanosine levels. Brain Dev 30:131–136
27. Kurban S, Mehmetoglu I, Oran B, Kiyici A (2008) Homocysteine levels and total antioxidant capacity in children with acute rheumatic fever. Clin Biochem 41:26–29
28. Srinivas A, Dias BF (2008) Antioxidants in HIV positive children. Indian J Pediatr 75:347–350
29. Hamed SA, Hamed EA, Zakary MM (2009) Oxidative stress and S-100B protein in children with bacterial meningitis. BMC Neurol 9:51
30. Barbosa E, Faintuch J, Machado Moreira EA, Gonçalves da Silva VR, Lopes Pereima MJ, Martins Fagundes RL, Filho DW (2009) Supplementation of vitamin E, vitamin C, and zinc attenuates oxidative stress in burned children: a randomized, double-blind, placebo-controlled pilot study. J Burn Care Res 30:859–866
31. Miric D, Katanic R, Kisic B, Zoric L, Miric B, Mitic R, Dragojevic I (2010) Oxidative stress and myeloperoxidase activity during bacterial meningitis: effects of febrile episodes and the BBB permeability. Clin Biochem 43:246–252
32. Karsen H, Sunnetcioglu M, Ceylan RM, Bayraktar M, Taskin A, Aksoy N, Erten R (2011) Evaluation of oxidative status in patients with *Fasciola hepatica* infection. Afr Health Sci 11:S14–S18
33. Yahata T, Suzuki C, Hamaoka A, Fujii M, Hamaoka K (2011) Dynamics of reactive oxygen metabolites and biological antioxidant potential in the acute stage of Kawasaki disease. Circ J 75:2453–2459
34. Oliveira KF, Cunha DF, Weffort VR (2011) Analysis of serum and supplemented vitamin C and oxidative stress in HIV-infected children and adolescents. J Pediatr (Rio J) 87:517–522
35. Kelekçi S, Evliyaoğlu O, Sen V, Yolbaş I, Uluca U, Tan I, Gürkan MF (2012) The relationships between clinical outcome and the levels of total antioxidant capacity (TAC) and coenzyme Q (CoQ10) in children with pandemic influenza (H1N1) and seasonal flu. Eur Rev Med Pharmacol Sci 8:1033–1038
36. Kaneko K, Takahashi M, Yoshimura K, Kitao T, Yamanouchi S, Kimata T, Tsuji S (2012) Intravenous immunoglobulin counteracts oxidative stress in Kawasaki disease. Pediatr Cardiol 33:1086–1088
37. Siomek A, Rytarowska A, Szaflarska-Poplawska A, Gackowski D, Rozalski R, Dziaman T, Czcrwionka-Szaflarska M, Olinski R (2006) *Helicobacter pylori* infection is associated with oxidatively damaged DNA in human leukocytes and decreased level of urinary 8-oxo-7,8-dihydroguanine. Carcinogenesis 27:405–408
38. Kiroglu AF, Noyan T, Oger M, Kara T (2006) Oxidants and antioxidants in tonsillar and adenoidal tissue in chronic adenotonsillitis and adenotonsillar hypertrophy in children. Int J Pediatr Otorhinolaryngol 70:35–38
39. Padmanabhan V, Rai K, Hegde AM, Shetty S (2010) Total antioxidant capacity of saliva in children with HIV. J Clin Pediatr Dent 34:347–350
40. Mallol J, Aguirre V, Espinosa V (2011) Increased oxidative stress in children with post infectious bronchiolitis obliterans. Allergol Immunopathol (Madr) 39:253–258
41. Hosakote YM, Jantzi PD, Esham DL, Spratt H, Kurosky A, Casola A, Garofalo RP (2011) Viral-mediated inhibition of antioxidant enzymes contributes to the pathogenesis of severe respiratory syncytial virus bronchiolitis. Am J Respir Crit Care Med 183:1550–1560
42. Noiri E, Tsukahara H (2005) Parameters for measurement of oxidative stress in diabetes mellitus: applicability of enzyme-linked immunosorbent assay for clinical evaluation. J Investig Med 53:167–175

43. Nakatsukasa Y, Tsukahara H, Tabuchi K, Tabuchi M, Magami T, Yamada M, Fujii Y, Yashiro M, Tsuge M, Morishima T (2013) Thioredoxin-1 and oxidative stress status in pregnant women at early third trimester of pregnancy: relation to maternal and neonatal characteristics. J Clin Biochem Nutr 52:27–31
44. Kaneko K, Kimata T, Tsuji S, Ohashi A, Imai Y, Sudo H, Kitamura N (2012) Measurement of urinary 8-oxo-7,8-dihydro-2-deoxyguanosine in a novel point-of-care testing device to assess oxidative stress in children. Clin Chim Acta 413:1822–1826
45. Ichiyama T (2010) Acute encephalopathy/encephalitis in childhood: a relatively common and potentially devastating clinical syndrome. Brain Dev 32:433–434
46. Woehrl B, Klein M, Grandgirard D, Koedel U, Leib S (2011) Bacterial meningitis: current therapy and possible future treatment options. Expert Rev Anti Infect Ther 9:1053–1065
47. Tsukahara H, Fujii Y, Hayashi Y, Morishima T (2012) Multiple biomarkers for oxidative stress in patients with brain disorders. J Neurol Sci 313:196
48. Hayashi M, Miyata R, Tanuma N (2012) Oxidative stress in developmental brain disorders. Adv Exp Med Biol 724:278–290
49. van Furth AM, Seijmonsbergen EM, Groeneveld PH, van Furth R, Langermans JA (1996) Levels of nitric oxide correlate with high levels of tumor necrosis factor alpha in cerebrospinal fluid samples from children with bacterial meningitis. Clin Infect Dis 22:876–878
50. Tsukahara H, Haruta T, Hori C, Matsuda M, Ono N, Hiraoka M, Shigematsu Y, Mayumi M (1999) Evaluation of a rapid reagent strip test for the diagnosis of childhood meningitis. Pediatr Int 41:443–446
51. Ray G, Aneja S, Jain M, Batra S (2000) Evaluation of free radical status in CSF in childhood meningitis. Ann Trop Paediatr 20:115–120
52. Murawska-Ciałowicz E, Szychowska Z, Tr busiewicz B (2000) Nitric oxide production during bacterial and viral meningitis in children. Int J Clin Lab Res 30:127–131
53. Tsukahara H, Haruta T, Ono N, Kobata R, Fukumoto Y, Hiraoka M, Mayumi M (2000) Oxidative stress in childhood meningitis: measurement of 8-hydroxy-2′-deoxyguanosine concentration in cerebrospinal fluid. Redox Rep 5:295–298
54. Tsukahara H, Haruta T, Todoroki Y, Hiraoka M, Noiri E, Maeda M, Mayumi M (2002) Oxidant and antioxidant activities in childhood meningitis. Life Sci 71:2797–2806
55. Kawashima H, Watanabe Y, Morishima T, Togashi T, Yamada N, Kashiwagi Y, Takekuma K, Hoshika A, Mori T (2003) NOx (nitrite/nitrate) in cerebral spinal fluids obtained from patients with influenza-associated encephalopathy. Neuropediatrics 34:137–140
56. Kawashima H, Amaha M, Ioi H, Yamanaka G, Kashiwagi Y, Sasamoto M, Takekuma K, Hoshika A, Watanabe Y (2005) Nitrite/nitrate (NOx) and zinc concentrations in influenza-associated encephalopathy in children with different sequela. Neurochem Res 30:311–314
57. Yamanaka G, Kawashima H, Suganami Y, Watanabe C, Watanabe Y, Miyajima T, Takekuma K, Oguchi S, Hoshika A (2006) Diagnostic and predictive value of CSF d-ROM level in influenza virus-associated encephalopathy. J Neurol Sci 243:71–75
58. Yamanaka G, Ishii C, Kawashima H, Oana S, Miyajima T, Hoshika A (2008) Cerebrospinal fluid diacron-reactive oxygen metabolite levels in pediatric patients with central nervous system diseases. Pediatr Neurol 39:80–84
59. Morichi S, Kawashima H, Ioi H, Ushio M, Yamanaka G, Kashiwagi Y, Takekuma K, Hoshika A, Watanabe Y (2009) Cerebrospinal fluid NOx (nitrite/nitrate) in RSV-infected children with CNS symptoms. J Infect 59:299–301
60. Kawashima H, Suzuki K, Yamanaka G, Kashiwagi Y, Takekuma K, Amaha M, Takahashi Y (2010) Anti-glutamate receptor antibodies in pediatric enteroviral encephalitis. Int J Neurosci 120:99–103
61. Kawashima H, Yamanaka G, Ishii C, Kashiwagi Y, Takekuma K, Watanabe Y (2010) NOx as new targets of treatment in influenza-associated encephalopathy. J Pediatr Infect Dis 5:171–176
62. Miyata R, Tanuma N, Hayashi M, Imamura T, Takanashi J, Nagata R, Okumura A, Kashii H, Tomita S, Kumada S, Kubota M (2012) Oxidative stress in patients with clinically mild encephalitis/encephalopathy with a reversible splenial lesion (MERS). Brain Dev 34:124–127

63. Klein M, Koedel U, Kastenbauer S, Pfister HW (2008) Nitrogen and oxygen molecules in meningitis-associated labyrinthitis and hearing impairment. Infection 36:2–14
64. Koedel U, Klein M, Pfister HW (2010) Modulation of brain injury as a target of adjunctive therapy in bacterial meningitis. Curr Infect Dis Rep 12:266–273
65. Kastenbauer S, Koedel U, Becker BF, Pfister HW (2002) Oxidative stress in bacterial meningitis in humans. Neurology 58:186–191
66. Balla J, Vercellotti GM, Jeney V, Yachie A, Varga Z, Jacob HS, Eaton JW, Balla G (2007) Heme, heme oxygenase, and ferritin: how the vascular endothelium survives (and dies) in an iron-rich environment. Antioxid Redox Signal 9:2119–2137
67. Syapin PJ (2008) Regulation of haeme oxygenase-1 for treatment of neuroinflammation and brain disorders. Br J Pharmacol 155:623–640
68. Borchorst S, Møller K (2012) The role of dexamethasone in the treatment of bacterial meningitis: a systematic review. Acta Anaesthesiol Scand 56:1210–1221
69. Hoshino A, Saitoh M, Oka A, Okumura A, Kubota M, Saito Y, Takanashi J, Hirose S, Yamagata T, Yamanouchi H, Mizuguchi M (2012) Epidemiology of acute encephalopathy in Japan, with emphasis on the association of viruses and syndromes. Brain Dev 34:337–343
70. Wang GF, Li W, Li K (2010) Acute encephalopathy and encephalitis caused by influenza virus infection. Curr Opin Neurol 23:305–311
71. Morishima T, Togashi T, Yokota S, Okuno Y, Miyazaki C, Tashiro M, Okabe N, Collaborative Study Group on Influenza-Associated Encephalopathy in Japan (2002) Encephalitis and encephalopathy associated with an influenza epidemic in Japan. Clin Infect Dis 35:512–517
72. Kawashima H, Morichi S, Okumara A, Nakagawa S, Morishima T, Collaborating study group on influenza-associated encephalopathy in Japan (2012) National survey of pandemic influenza A (H1N1) 2009-associated encephalopathy in Japanese children. J Med Virol 84:1151–1156
73. Okumura A, Nakagawa S, Kawashima H, Morichi S, Muguruma T, Saito O, Fujimoto J, Toida C, Kuga S, Imamura T, Shimizu T, Kondo N, Morishima T (2013) Severe form of encephalopathy associated with 2009 pandemic influenza A (H1N1) in Japan. J Clin Virol 56:25–30
74. Kawada J, Kimura H, Ito Y, Hara S, Iriyama M, Yoshikawa T, Morishima T (2003) Systemic cytokine responses in patients with influenza-associated encephalopathy. J Infect Dis 188:690–869
75. Ichiyama T, Isumi H, Ozawa H, Matsubara T, Morishima T, Furukawa S (2003) Cerebrospinal fluid and serum levels of cytokines and soluble tumor necrosis factor receptor in influenza virus-associated encephalopathy. Scand J Infect Dis 35:59–61
76. Hosoya M, Nunoi H, Aoyama M, Kawasaki Y, Suzuki H (2005) Cytochrome c and tumor necrosis factor-alpha values in serum and cerebrospinal fluid of patients with influenza-associated encephalopathy. Pediatr Infect Dis J 24:467–470
77. Hasegawa S, Matsushige T, Inoue H, Shirabe K, Fukano R, Ichiyama T (2011) Serum and cerebrospinal fluid cytokine profile of patients with 2009 pandemic H1N1 influenza virus-associated encephalopathy. Cytokine 54:167–172
78. Kawashima H, Togashi T, Yamanaka G, Nakajima M, Nagai M, Aritaki K, Kashiwagi Y, Takekuma K, Hoshika A (2005) Efficacy of plasma exchange and methylprednisolone pulse therapy on influenza-associated encephalopathy. J Infect 51:E53–E56
79. Kawashima H, Morichi S, Okumara A, Nakagawa S, Morishima T, The Collaborating Study Group on Influenza-Associated Encephalopathy in Japan (2012) Treatment of pandemic influenza A (H1N1) 2009-associated encephalopathy in children. Scand J Infect Dis 44:941–947

Chapter 23
Oxidative Stress in Inherited Metabolic Diseases

Hironori Nagasaka, Hirokazu Tsukahara, and Takashi Miida

Abbreviations

ADMA	Asymmetric dimethylarginine
AGC2	Aspartate/glutamate carrier isoform 2
CoQ_{10}	Coenzyme Q_{10}
CTLN2	Adult-onset type II citrullinemia
DTNB	5, 5-Dithibis [2-nitrobenzoic] acid
ELISA	Enzyme-linked immunosorbent assay
GPx	Glutathione peroxidase
GSH	Reduced glutathione
GSSG	Oxidized glutathione
MDA-LDL	Malondialdehyde-modified low-density lipoprotein
NICCD	Neonatal intrahepatic cholestasis caused by citrin deficiency
NO	Nitric oxide
NOx	Nitrite/nitrate
8-OHdG	8-Hydroxy-2′-deoxyguanosine
PKU	Phenylketonuria
PPAR	Peroxisome proliferator-activated receptor

H. Nagasaka, M.D. (✉)
Department of Pediatrics, Takarazuka City Hospital,
4-5-1 Kohama-cho, Takarazuka 665-0827, Japan
e-mail: nagasa-hirono@k2.dion.ne.jp

H. Tsukahara, M.D., Ph.D.
Department of Pediatrics, Okayama University Graduate School of Medicine,
Dentistry and Pharmaceutical Sciences, 2-5-1 Shikata-cho, Kita-ku,
Okayama 700-8558, Japan

T. Miida, M.D., Ph.D.
Department of Clinical Laboratory Medicine, Juntendo University School of Medicine,
2-1-1 Hongo, Bunkyo-ku, Tokyo 113-8421, Japan

H. Tsukahara and K. Kaneko (eds.), *Studies on Pediatric Disorders*, Oxidative Stress in Applied Basic Research and Clinical Practice, DOI 10.1007/978-1-4939-0679-6_23,

SOD	Superoxide-dismutase
TAR	Total antioxidant activity
TBARS	Thiobarbituric acid-reactive species
WD	Wilson's disease

23.1 Introduction

Oxidative stress and antioxidant defense systems in various diseases have been studied extensively. Deep associations of oxidative stress with inflammation, fibrosis, and apoptosis of hepatic tissues have been shown [1–3].

Regarding metabolic diseases, the contributions of enhanced oxidative stress to the developments of cardiovascular diseases such as atherosclerosis and hypertension, and liver steatosis have been shown [2, 4, 5]. However, information related to the oxidative stress in congenital metabolic diseases remains limited. Considering that large amounts of the disease-specific toxic substances and the metabolites accumulate in diseases of this kind, substantial amounts of oxidants may emerge in the affected subjects.

From this consideration, we studied oxidative stress for inherited metabolic diseases. First, we examined the status of oxidative stress in respective diseases using biomarkers for oxidative stress. Secondly, we analyzed metabolic changes in terms of the oxidative stress. Thirdly, based on the results, we tried to evaluate the metabolic control of affected subjects.

The oxidative stress can constitute or modify the clinical and biological features of phenylketonuria (PKU), citrin deficiency, and Wilson's disease (WD).

23.2 Materials and Methods

23.2.1 *Blood and Urinary Oxidative Stress Markers*

As blood markers for oxidative stress, plasma levels of thiobarbituric acid-reactive species (TBARS) and total antioxidant reactivity (TAR), and serum level of malondialdehyde-modified low-density lipoprotein (MDA-LDL) as oxidized LDL were determined [6–8].

As the urinary markers for oxidative stress, urinary acrolein-lysine and 8-hydroxy-2′-deoxyguanosine (8-OHdG), also known as 8-oxo-7,8-dihydro-2-deoxyguanosine, were examined [6].

As antioxidative enzymes in erythrocytes, catalase, superoxide-dismutase (SOD), and glutathione peroxidase (GPx) were examined.

As antioxidant substances, β-carotene and α-tocophenol in plasma were examined [6]. Coenzyme Q_{10} (CoQ_{10}), an electron carrier transporting electrons derived from mitochondrial respiratory chain complexes I and II to complex III, in plasma was also examined because CoQ_{10} serves as an antioxidant and its production is closely linked with phenylalanine metabolism [6–8].

Nitric oxide (NO) production is influenced considerably by oxidative stress [9, 10]. To evaluate NO production, we determined serum levels of nitrite/nitrate (NOx) as stable metabolites of NO and asymmetric dimethylarginine (ADMA), a competitor of arginine for NO synthase, together with plasma levels of arginine and citrulline constituting NO-citrulline cycle [9, 11, 12].

Blood was drawn from a peripheral vein the morning after overnight fasting. Serum and plasma were obtained for the determinations of TBARS, TAR, MDA-LDL, α-tocophenol, β-carotene, CoQ_{10}, NOx, ADMA, and amino acids. Erythrocytes were washed in cold 0.9 % NaCl solution. Lysates for the determinations of antioxidative enzyme activities were prepared by adding 100 μL of washed erythrocytes to 1 mL of distilled water, which was then frozen at −80 °C until further analyses. Urine samples (5–15 mL) for oxidative stress markers were collected 0.5–2 h before blood sample collection.

This study protocol was approved by the relevant institutional review boards. The parents of all patients provided written informed consent before the start of the study.

23.2.2 Assays for Determinations of Blood and Urinary Oxidative Stress Markers

Plasma TBARS level was determined using a method using fluorometric assay [13, 14]. Plasma TAR level, which represents the ability to attenuate oxidants, was determined by luminol chemiluminescence intensity induced by z,z-azo-bis-2-amidinopropane according to the method described by Lissi et al. [13, 14].

Serum MDA-LDL level was determined using a sensitive enzyme-linked immunosorbent assay (ELISA) with a monoclonal antibody interacting with MDA-apo B [15]. Plasma β-carotene, α-tocophenol, and CoQ_{10} (ubiquinol-10 plus ubiquinone-10) levels were measured using high-performance liquid chromatography [16].

Urinary acrolein-lysine and 8-OHdG levels were determined, respectively, using competitive ELISA kits: ACR-Lysine Adduct ELISA (NOF Corporation, Tokyo, Japan) and 8-OHdG Check (Institute for the Control of Aging, Shizuoka, Japan) [17]. These values were presented as acrolein-lysine and 8-OHdG to creatinine ratios.

23.2.3 Assays for Determinations of Antioxidative Enzyme Activities in Erythrocytes

The SOD activity was determined using spectrophotometry at 505 nm (RANSOD kit; Randox Laboratories Limited, Antrim, UK) [13, 14, 16]. Catalase activity was determined using the method described by Aebi [13, 14, 16]. In brief, we monitored the decrease in absorbance at 240 nm in a reaction medium containing 20 mM H_2O_2, 10 mM potassium phosphate buffer, pH 7.0, and 0.1–0.3 mg protein/mL. The GPx activity was determined by the method described by Wendel [13, 16]. The activity was determined monitoring the disappearance of NADPH at 340 nm in a medium containing 2 mM glutathione, 0.15 U/mL glutathione reductase, 0.4 mM azide, 0.5 mM tert-butyl-hydroperoxidase, and 0.1 mM NADPH.

23.2.4 Assays for Determination of NOx and ADMA

Serum NOx levels were measured using the Griess method, with a nitrate/nitrite colorimetric assay kit (Cayman Chemical, Ann Arbor, MI, USA). Serum ADMA levels were determined using an ELISA kit (DLD Diagnostika GmbH, Hamburg, Germany) [18].

23.2.5 Oxidative Stress Markers for the Liver Tissues

The magnitude of oxidative stress loaded on the liver was evaluated by reduced glutathione (GSH) concentration to oxidized glutathione (GSSG) concentration, together with liver TBARS in the liver [5, 19]. Antioxidative enzymes in the liver were also examined [19].

23.2.6 Assays for Liver TBARS and Glutathiones

Each liver sample was homogenized and sonicated in 10 mM phosphate buffer of pH 7.80. After precipitation of proteins with perchloric acid/EDTA and centrifugation at 8,000 × *g* for 2 min, the resultant supernatant was adjusted to pH 5–6 and was centrifuged further at 12,000 × *g* for 5 min.

The amount of total glutathione (GSH + GSSG) in the supernatant was determined according to the method described by Tietze [19, 20]. In brief, the supernatant was dissolved in phosphate-EDTA buffer of pH 7.5 and was then added to a solution containing 5, 5′-dithibis [2-nitrobenzoic] acid (DTNB) and glutathione reductase. GSH-dependent DTNB reduction initiated by addition of triphosphopyridine nucleotide was monitored using spectrophotometer at 412 nm (UV-160; Shimizu Corporation); the final volume (1 mL) of this reaction mixture contained 0.6 μmol DTNB, 1U glutathione reductase, and 0.2 μmol triphosphopyridine nucleotide.

GSSG in the supernatant was determined according to the method described by Sies et al. [19, 21]. In brief, GSSG-dependent NADH reduction initiated by the addition of 0.6 U glutathione reductase to 1 mL of the mixture containing 20 mM triethanolamine and 5 μM NADPH was monitored with spectrometer at 340 nm (Hitachi 181).

The concentration of TBARS in the liver, which was homogenated and sonicated as above, was determined using fluorometric assay according to the method described previously [19, 22, 23].

23.2.7 Assays for Liver Antioxidant Enzyme Activities

To determine enzyme activities, frozen liver samples were homogenized in 10 mM phosphate buffer pH 7.80 containing 1 mM EDTA; they were then sonicated for 1 min with icing.

Total-SOD activity was measured at pH 7.8 using a commercial kit according to the method described by McCord et al. [24]. The CuZn-SOD activity was measured at pH 10.2 using the same kit. The Mn-SOD activity was calculated by subtraction of CuZn-SOD from the total SOD.

The GPx activity was measured using a commercial kit according to the method described by Paglia et al. [25]. Catalase activity was measured using a spectrophotometric method using H_2O_2 as a substrate [26].

23.2.8 Statistical Analysis

Differences between values of patients and those of controls were estimated using Student's *t*-test. The relation between each pair of parameters was estimated using Pearson's correlation test. All $p \leq 0.05$ were considered statistically significant.

23.3 Phenylketonuria and Oxidative Stress

PKU presenting prominent hyperphenylalaninemia is an autosomal recessive disorder caused by deficiency of hepatic phenylalanine hydroxylase. This disorder causes mental retardation, seizure, behavioral difficulty, and other neurological symptoms unless the affected child is maintained on a strict low phenylalanine diet [27].

After the introduction of newborn mass screening in the 1960s–1970s, affected children can achieve normal development by early treatment of phenylalanine-restricted diet following to the early diagnosis.

Initially, the low phenylalanine diet prescribed in classical PKU was discontinued by the end of the first decade of life. However, thereafter, evidence that high serum phenylalanine level engenders the development of attention deficit, recognition dysfunction, and emotional fluctuation in diet-off adult patients has been accumulating [27]. Actually, those symptoms ameliorate after patients resume phenylalanine-restricted diet [28, 29]. Therefore, PKU patients should continue receiving phenylalanine-restricted diet therapy for life. However, optimal serum phenylalanine levels in later life remain to be established.

Several studies have drawn that oxidative stress status is enhanced in PKU patients [13, 14, 16, 30]. However, the available data are still limited, and more information must be obtained on this point.

In our latest study, we examined blood and urinary surrogate markers for oxidative stress, blood antioxidant substances, and erythrocyte antioxidant enzymes in 42 PKU patients aged 21–40 years (female/male, 22/20) and 30 age-matched healthy controls aged 20–38 (female/male, 15/15). We also investigated NO production in terms of oxidative stress status because NO is associated with vascular tone, neurological function, apoptosis, and anti-inflammation [9, 31, 32].

Plasma TBARS level was significantly increased in PKU patients, showing a strong positive correlation with the serum phenylalanine level ($p<0.001$) (Fig. 23.1 upper panel) (Table 23.1). Plasma MDA-LDL was also increased significantly ($p<0.001$). In contrast, plasma TAR level was significantly decreased in them with a significant negative correlation to the phenylalanine level ($p<0.001$) (Fig. 23.1 upper panel) (Table 23.1). β-carotene and CoQ_{10} were decreased significantly ($p<0.001$). Urinary acrolein-lysine and urinary 8-OHdG were increased (Table 23.1).

Among antioxidant enzymes, erythrocyte GPx activity was significantly decreased with a strong negative correlation to the phenylalanine level ($p<0.001$) (Fig. 23.1 lower panel) (Table 23.1). In contrast, erythrocyte SOD and catalase activities were significantly increased with positive significant correlations to the phenylalanine level ($p<0.01$).

Regarding NO metabolism, serum NOx was increased concomitantly with decreased serum ADMA (Table 23.1).

Thus, the status of oxidative stress is, to a considerable degree, determined by serum phenylalanine levels in PKU. The oxidative stress markers changed drastically as a boundary of 700–800 μmol/L of serum phenylalanine. The brain has a high content of lipids that are vulnerable to oxidants, in particular unsaturated fatty acids, but its antioxidant defense system is relatively weak [1–3]. In this context, strongly suppressed GPx activity together with enhanced oxidative stress and altered NO production may be considerably disadvantageous for the central nervous system in PKU patients.

23.4 Citrin Deficiency and Oxidative Stress

Citrin is a liver-type aspartate/glutamate carrier isoform 2 (AGC2) of mitochondria encoded by the SLC25A13 gene on chromosome 7q21.3. AGC2 plays an important role in the malate-aspartate NADH shuttle and urea synthesis (Fig. 23.2) [33–35]. Impairment of AGC2 function can induce an increased $NADH/NAD^+$ ratio in cytosol and can cause failure of the aspartate supply from the mitochondria to the cytoplasm for synthesis of argininosuccinate, thereby giving rise to hypercitrullinemia and hyperammonemia.

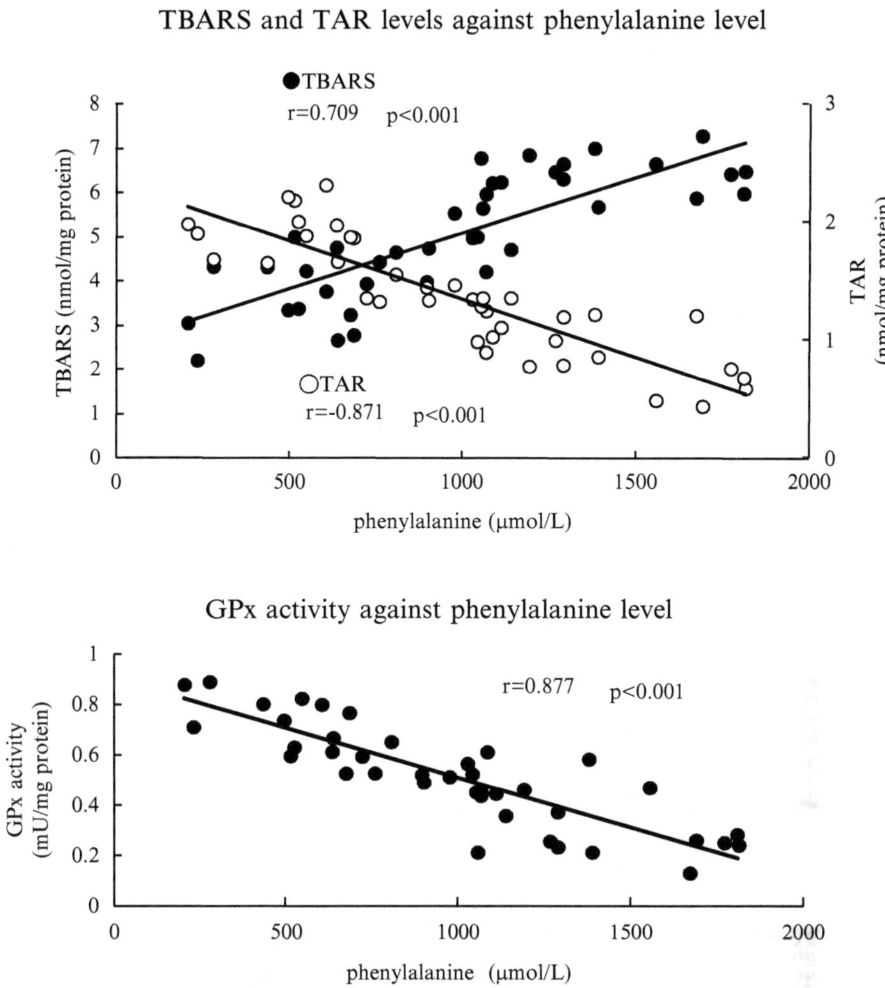

Fig. 23.1 Correlations of serum phenylalanine with blood oxidative stress parameters (upper panel) and erythrocyte GPx activity (lower panel). *TBARS* thiobarbituric acid-reactive species, *TAR* total antioxidant reactivity, *GPx* glutathione peroxidase

Clinical pictures of citrin deficiency vary dramatically by age [36–38]. Affected children of early life present with diverse clinical manifestations, namely neonatal intrahepatic cholestasis caused by citrin deficiency (NICCD), such as considerable liver dysfunction along with cholestasis, citrullinemia, mild hyperammonemia, galactosemia, and hypoglycemia. These clinical presentations of NICCD resolve during the first 6 months to 1 year of life. However, among patients who have manifested NICCD, only one fifth of patients develop adult-onset type II citrullinemia (CTLN2) manifested by frequent bouts of hyperammonemia, liver

Table 23.1 Oxidative stress markers, erythrocyte antioxidant enzymes, and nitric oxide production markers in adult phenylketonuria

	PKU patients ($n=42$)	Healthy controls ($n=30$)
TBARS in blood (nmol/mg protein)	5.05±1.16**	3.79±0.46
TAR in blood (nmol/mg protein)	1.36±0.40***	2.15±0.38
MDA-LDL in blood (U/L)	61±18***	39±11
α-tocophenol in blood (mg/dL)	0.74±0.13	0.85±0.14
β-carotene in blood (mg/dL)	28.2±15.0***	46.5±13.5
Coenzyme Q_{10} (ng/mL)	482±102***	970±237
8-OHdG in urine (ng/mgCr)	7.68±1.65	8.23±1.49
Acrolein-lysine in urine (nmol/mgCr)	279±142**	199±99
SOD activity in erythrocytes (U/mg protein)	1.54±0.25**	1.12±0.21
Cat activity in erythrocytes (U/mg protein)	3.49±0.47**	2.79±0.34
GPx activity in erythrocyte (mU/mg protein)	0.511±0.168***	0.756±0.122
NOx in blood (μmol/L)	47.2±17.2**	30.9±10.7
ADMA in blood (μmol/L)	0.44±0.10*	0.61±0.10
ADMA/NOx	0.014±0.008**	0.020±0.007
Arginine in blood (μmol/L)	66±22**	95±22
Citrulline in blood (μmol/L)	32±9*	26±6

TBARS thiobarbituric acid-reactive species, *TAR* total antioxidant reactivity, *MDA-LDL* malondialdehyde-modified LDL, *8-OHdG* 8-hydroxy-2′-deoxyguanosine, *Cr* creatinine, *SOD* superoxide dismutase, *Cat* catalase, *GPx* glutathione peroxidase

*$p<0.05$, **$p<0.01$, ***$p<0.001$ versus controls (Student's t-test)

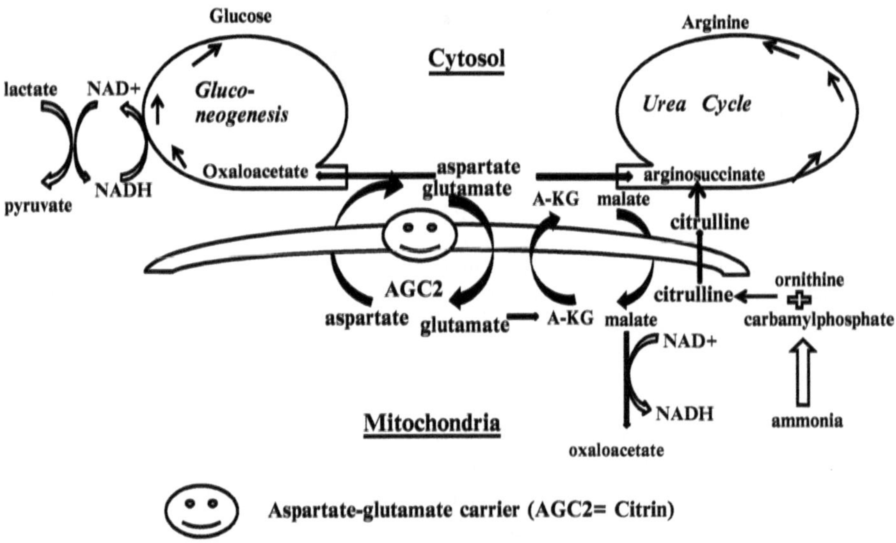

Fig. 23.2 Metabolic roles of aspartate clutamate carrier 2, namely citrin. Liver-type aspartate/glutamate carrier isoform 2 (AGC2) of mitochondria, namely citrin, plays an important role in the malate-aspartate NADH shuttle and urea synthesis. Impairment of AGC2 function can engender an increased NADH/NAD⁺ ratio in cytosol, in addition to failure of the aspartate supply from the mitochondria to the cytoplasm for synthesis of argininosuccinate, resulting in hypercitrullinemia and hyperammonemia

Table 23.2 Abnormalities in carbohydrate, amino acids and lipids, together with oxidative stress markers, in patients with citrin deficiency in silent stage

	Pyruvate (mg/dL)	Lactate (mg/dL)	L/P	Citrulline (μmol/L)	Ornithine (μmol/L)
Patients ($n=20$)	0.8 (0.3)	12 (3)	15 (2)*	40.8 (6.3)**	105.1 (24.2)***
Ranges	0.2–1.7	6–26	9–18	25.3–56.4	65.0–193.4
Controls ($n=32$)	0.8 (0.2)	10 (4)	11 (1)	28.2 (6.3)	61.3 (13.6)
Ranges	0.3–1.1	7–19	7–13	14.4–41.4	40.1–90.0
	LDL-C (mg/dL)	HDL-C (mg/dL)	Ox LDL (U/L)	8-OHdG (ng/mgCr)	Acrolein-lysine (nmol/mgCr)
Patients ($n=20$)	116(23)**	79(7)***	82 (24)***	67(21)***	481(125)**
Ranges	76–196	54–108	39–156	39–156	220–686
Controls ($n=32$)	85(13)	54(11)	25 (7)	19 (5)	272 (90)
Ranges	42–106	39–77	5–50	11–29	70–424

TC total cholesterol; *LDL-C* low-density lipoprotein cholesterol, *HDL-C* high-density lipoprotein cholesterol, *Ox LDL* oxidized LDL, *8-OHdG* 8-hydroxy-2′-deoxyguanosine, *Cr* creatinine
Presented data are mean (SD) values and the ranges
*$p<0.05$, **$p<0.01$, ***$p<0.001$ versus controls (Student's *t*-test)

steatosis, mental derangement, sudden attacks of unconsciousness, and ultimately death within a few years of onset.

This study enrolled 20 children with citrin deficiency aged 1–10 years (female/male, 10/10) in the silent period after NICCD and 32 age-matched healthy children aged 2–9 years (female/male, 16/16 boys) as the controls. The affected children's blood levels of transaminase, gamma-glutamyl transpeptide, total bile acids, and total bilirubin at presentation were entirely normal.

Urinary acrolein-lysine and urinary 8-OHdG in the affected children were significantly higher than those in the age-matched controls (acrolein-lysine, $p<0.01$; 8-OHdG, $p<0.001$) (Table 23.2). In contrast, blood vitamin E levels in the affected patients were significantly lower than those in the controls. Erythrocyte SOD and catalase activities in the affected patients were significantly higher than those in the age-matched controls ($p<0.05$). Collectively, these findings suggest persistently enhanced oxidative stress in citrin deficiency.

Three amino acids, ornithine, citrulline, and arginine, constitute urea cycle that is coupled with NO-citrulline cycle [9]. Serum ornithine and citrulline levels of the affected children were, respectively, 1.7 times ($p<0.001$) and 1.4 times ($p<0.01$) higher than those of the controls (Table 23.2). In contrast, their arginine level was 0.87 times as high as the controls' level, although no significant difference was found between these two groups. However, plasma ammonia level was not increased in the affected children. Serum NOx and ADMA levels were also not different between the two groups (Table 23.2).

Regarding the carbohydrate metabolism, no significant difference was found between these two groups' blood glucose, galactose, lactate, or pyruvate levels at fasting. However, the lactate-to-pyruvate ratio in the affected children was significantly higher than that in the controls ($p < 0.05$), suggesting a high ratio of NADH to NAD^+ and suppressed mitochondrial functions in the affected children (Table 23.2).

Serum low-density lipoprotein-cholesterol and high-density lipoprotein cholesterol levels in the affected children were higher than those in the age-matched controls. Surprisingly, MDA-LDL levels were much higher in the affected patients (Table 23.2).

Collectively, the results of this study showed multiple metabolic abnormalities and enhanced oxidative stress in children with citrin deficiency even during the silent period. Increased cholesterol and enhanced oxidative stress over a long period might engender the developments of liver steatosis and steatohepatitis [1, 4, 5].

23.5 Wilson's Disease and Oxidative Stress

WD is an autosomal recessive disorder. The gene responsible is *ATP7B1* encoding a copper-transporting P-type ATPase [39]. In this disorder, excess copper accumulates in various organs, particularly in the liver and brain, resulting in various clinical manifestations (Fig. 23.3) [40, 41].

Many reports have described hepatic presentations in WD precisely [40, 41]. Free radical generation by excess copper and the subsequent oxidative changes in hepatocyte organellar lipids and thiol-containing proteins putatively account for the development and progression of liver disease [42, 43]. Experiments using HepG2 cells have shown that proteins involved in antioxidant defenses in this cell line are

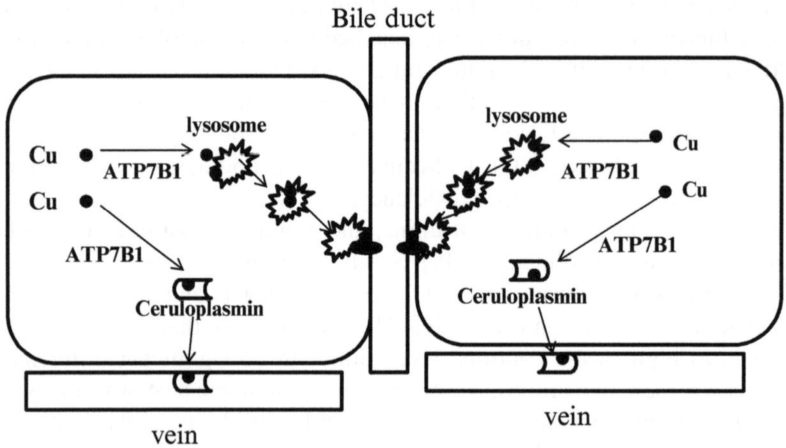

Fig. 23.3 Metabolic roles of ATP7B1 in copper transport. ATP7B1 transports copper to lysosome and ceruloplasmin, which excrete copper, respectively, into bile canaliculi and circulation. In Wilson's disease caused by ATP7B1 deficiency, excess copper accumulates in various organs, particularly in the liver and brain, engendering various clinical manifestations

Table 23.3 Liver glutathione and thiobarbituric acid-reactive species in patients with Wilson's disease

Group	GSH (nmol/mg protein)	GSSG (nmol/mg protein)	GSH/GSSG	TBARS (pmol/mg protein)
H	30 (2)***	1.5 (0.3)***	20 (2)	19 (2)
I	15 (5)	3.1 (0.6)***	5.5(1.5)***	47 (7)***
II	6(4)***	2.7 (0.3)***	1.7(1.0)***	185 (43)***
III	2(1)***	0.9 (0.2)*	1.1 (0.2)***	388 (59)***
Controls	16 (4)	0.5 (0.3)	20 (4)	14 (5)

GSH reduced glutathione, *GSSG* oxidized glutathione, *TBARS* thiobarbituric acid-reactive species
Group H consists of four heterozygotes lacking liver damage
Group I, II, and III are patient groups: group I, 5 patients with mild liver damage; group II, 10 patients with moderate or more liver damage; group III, 5 patients with fulminant hepatic failure
Control group consists of 14 healthy subjects
Presented data are mean (SD) values
$^*p<0.05$, $^{**}p<0.01$, $^{***}p<0.001$ versus control group (Student's t-test)

altered dramatically by chronic copper exposure [26, 44]. However, in humans, copper-derived oxidative stress and the state of antioxidant defense system in the liver have been studied only rarely [5, 19].

We evaluated the magnitude of oxidative stress in the livers of 20 WD patients aged 3–16 years (female/male, 10/10) with various liver manifestations and in the livers of four carriers (all male) with no clinical manifestations.

Patients were classified into three groups, I, II, and III according to their respective liver functions and liver histological findings. As a histological estimation of liver injury, the stage and grade scores that, respectively, reflect the fibrosis and inflammatory activity of the liver were determined according to the well-established METAVIR scoring system of stages F0–F4 and of grades A0–A3: group I, F1/A1 (mild damage, $n=5$); group II, F2–F3/A1–A2 (moderate or more damage, $n=10$); and group III, F1–F3/A3 (fulminant hepatic failure, $n=5$) [45].

Four related donors (group H) were confirmed to be heterozygous for *ATP7B* mutations. Their serum ceruloplasmin levels were below the normal range but were higher than those of patients. Copper contents of their livers were markedly higher than those of control livers, but were considerably lower than those of patients. Their liver functions and histology were entirely normal.

Fragments of livers were obtained from 12 non-related healthy donors for liver transplantation (female/male, 3/9) with ages of 28–42 years and were used as controls.

Both GSH and GSSG concentrations in the carriers (Group H) were apparently higher than the respective control levels, but the GSH/GSSG ratio was at the control level (Table 23.3). In group I, on the other hand, the GSSG concentration was high, whereas GSH concentrations were within the normal range, resulting in a GSH/GSSG ratio that was lower than those in the controls. In group II, the GSSG concentration was as high as that in group I, whereas GSH concentrations were apparently below the normal range, rendering the GSH/GSSG ratio much lower than the control level. In group III, the GSSG concentration showed a level at around the upper limit of the normal range, but the GSH concentration exhibited an extremely low value, rendering an extremely low GSH/GSSG ratio (Table 23.3).

The TBARS concentrations in the carriers were within the control range (Table 23.3). The TBARS concentration in group I was higher than the control level. Those in groups II and III were extremely high.

CuZn-SOD, Mn-SOD, and catalase activities in group H were increased substantially, whereas those in groups I, II, and III were decreased. Especially, the decreases in these three enzyme activities in group III were marked. On the other hand, the GPx activity was lower only in group III. Those in the other groups were maintained at the control level (Table 23.4).

These results suggest that the magnitude of oxidative stress is increased according to the progression of the liver damage and that copper-derived oxidants contribute to the development and progression of liver disease in WD.

Recently, peroxisome proliferator-activated receptors (PPARs) -α and -γ, nuclear receptors, have been studied in non-alcoholic fatty liver disease and non-alcoholic steatohepatitis [46, 47]. The effects of these PPARs on regulating the fat content of the liver have been shown. The PPARs have also been suggested as regulating the antioxidant enzymes and the anti-inflammation system [47, 48].

PPAR-α expression was increased over the control levels in groups H and I exhibiting mild or no liver damage, but was decreased in groups II and III exhibiting moderate or more liver damage (Table 23.4). On the contrary, PPAR-γ expression was increased as the liver damage progressed. Mn-dependent superoxide dismutase (Mn-SOD), CuZn-SOD, and catalase activities were decreased in the affected three groups and were increased in group H (Table 23.4). Among group II exhibiting substantial inter-individual variances in parameters, the severity of steatosis showed a significant positive correlation with PPAR-γ expression ($p<0.001$), but not PPAR-α expression (Fig. 23.4). CuZn-SOD activity was positively correlated with PPAR-α expression ($p<0.05$) but not PPAR-γ expression (Fig. 23.5).

These results suggest that changes of PPAR-γ and PPAR-α are associated with the steatosis and the regulation of antioxidant system in the liver of WD.

23.6 Other Metabolic Diseases and Oxidative Stress

For many other inherited metabolic diseases, oxidative stress and the associated metabolic and clinical consequences have been studied (Fig. 23.6) [49–55]. Nevertheless, the information related to many important areas remains insufficient. Comprehensive studies must be conducted.

Mitochondrial diseases, including respiratory chain complex deficiencies, are known to enhance oxidative stress strongly in many circumstances [49–51]. However, details of the resultant biological and clinical changes remain to be elucidated. In these diseases, the balance between vasomediators is also altered considerably, thereby complicating the clinical pictures and the biology [49–51].

Table 23.4 Liver antioxidant enzymes and peroxisome proliferator-activated receptors in patients with Wilson's disease

Group	CuZn-SOD (U/mg protein)	Mn-SOD (U/mg protein)	GPx (mU/mg protein)	catalase (U/mg protein)	PPAR-α (PPAR-α/β-actin)	PPAR-γ (PPAR-γ/β-actin)
H	34±4*	20±3*	552±50	103±17*	1.66±0.26**	0.24±0.05*
I	14.4±3.3**	3.9±1.1***	609±125	26.8±6.1****	1.84±0.29**	0.58±0.17**
II	5.8±2.1	2.5±0.3	600±133	19.6±5.1#	0.75±0.17#	1.50±0.37##
III	4.8±0.7	2.1±0.3	280±54###	6.4±1.7	0.37±0.06	2.26±0.69
Controls	15–28	6–16	350–720	30–70	0.8–1.5	0.1–0.3

CuZn-SOD, CuZn-dependent superoxide dismutase; Mn-SOD, Mn-dependent superoxide dismutase; GPx, glutathione peroxidase

PPAR-α, peroxisome proliferator-activated receptor-α; PPAR-γ, peroxisome proliferator-activated receptor-γ

Group H consists of five heterozygotes lacking liver damage

Group I, II, and III are patient groups: group I, 5 patients with mild liver damage; group II, 10 patients with moderate or more liver damage; group III, 5 patients with fulminant hepatic failure

The control ranges were obtained from livers of 16 controls

*$p<0.001$ versus groups I, II, and III, **$p<0.001$ versus groups II and III, ***$p<0.01$ versus groups II and III, ****$p<0.05$ versus group II and $p<0.001$ versus group III, #$p<0.001$ versus group III, ##$p<0.01$ versus group III, ###$p<0.001$ versus groups H, I, and II

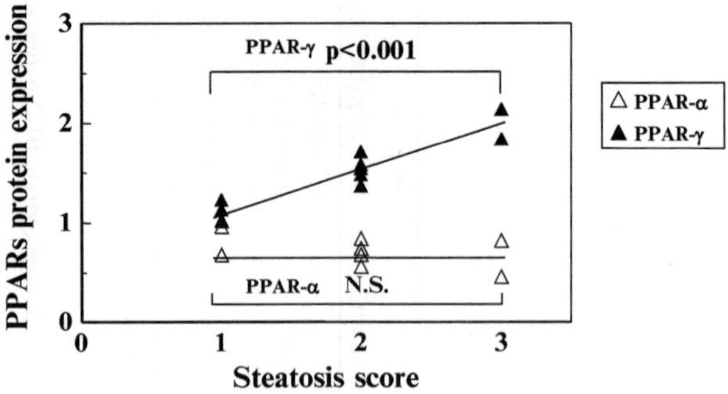

Fig. 23.4 Steatosis score and peroxisome-proliferative activator receptors in patients with Wilson's disease belonging to group II exhibiting moderate or more liver damage. PPAR, peroxisome-proliferative activator receptor

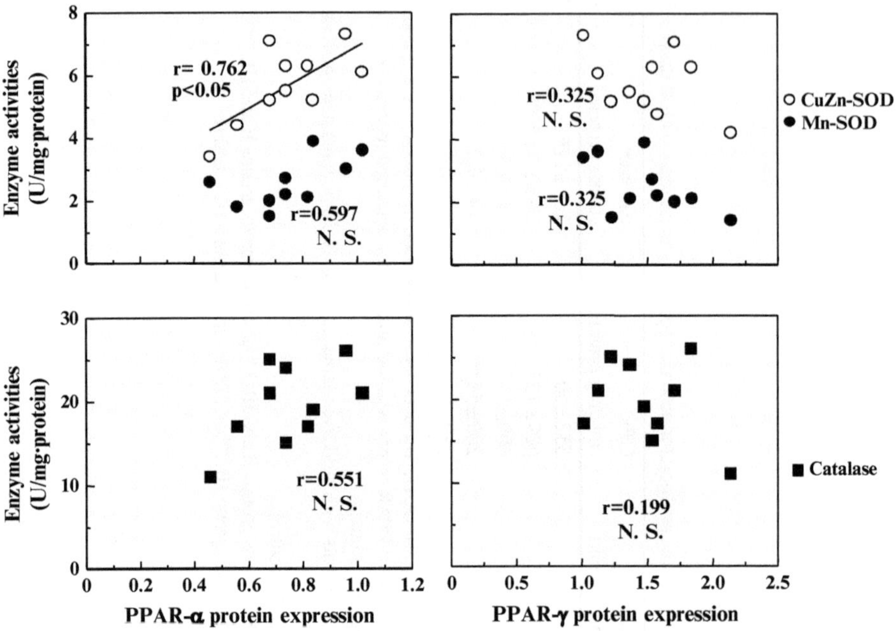

Fig. 23.5 Liver antioxidant enzyme activities and peroxisome-proliferative activator receptors in patients with Wilson's disease belonging to group II exhibiting moderate or more liver damage. *PPAR* peroxisome-proliferative activator receptor, *SOD* superoxide dismutase, *N.S.* not significant

Methylmalonic acidemia and propionic acidemia produce large amounts of oxidants and cause profound metabolic acidosis [52–54]. The magnitude of oxidative stress and the amounts of oxidants are closely correlated with methylmalonate and propionate production [52, 53]. Considerable impairments in mitochondrial functions by the oxidative stress in these diseases have also been described in earlier reports [52–54]. Especially, it is emphasized that these toxic organic acids and the increased oxidants injure cardiac muscles and kidneys.

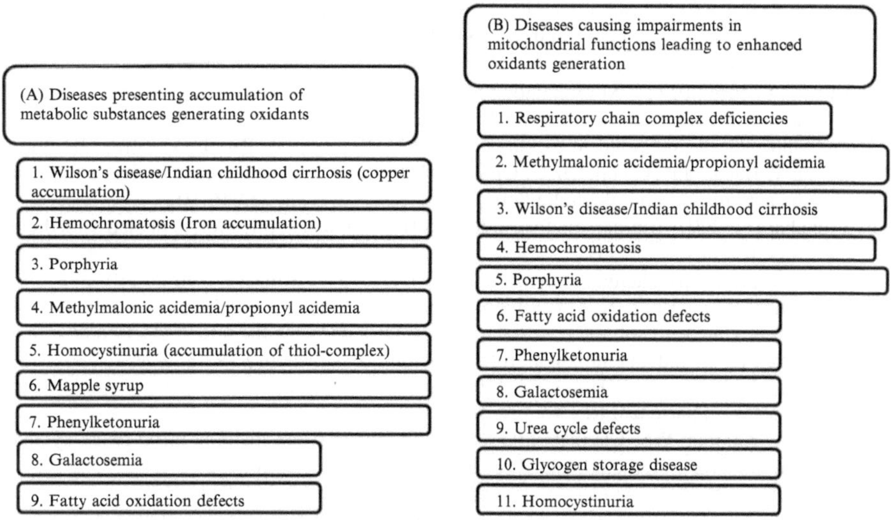

Fig. 23.6 Representative inherited metabolic diseases to enhance oxidative stress

Oxidative stress is enhanced in homocystinuria, which presents disturbances of the cardiovascular system, mainly attributable to thrombo-embolism and atherosclerosis [55]. The enhanced oxidative stress, at least in part, appears to contribute to the development and progression of cardiovascular disease. Reduced NO production might also exert deleterious effects on the cardiovascular system [55].

Thus, oxidative stress is likely to be a common and crucial factor among the clinical and biological features of inherited metabolic diseases.

23.7 Conclusions

Inherited metabolic diseases of many kinds enhance oxidative stress. The consequences of that enhanced stress might differ greatly among diseases. The clinical changes and metabolic derangements are often explained by enhanced oxidative stress. The attenuation of oxidative stress might be important for the treatment of metabolic diseases. Oxidative stress markers should be used more widely and more intensively for the management of inherited metabolic diseases.

References

1. Halliwell B (1994) Free radicals, antioxidants, and human disease: curiosity, cause, or consequence? Lancet 344:721–724
2. Reznick AZ, Packer L (1993) Free radicals and antioxidants in muscular neurological diseases and disorders. In: Poli G, Albano B, Dianzani MU (eds) Free radical: from basic science to medicine. Birkhauser, Basel, pp 425–437

3. Przedborski S, Donaldson D, Jakowec M, Kish SJ, Guttman M, Rosoklija G, Hays AP (1996) Brain superoxide dismutase, catalase, and glutathione peroxidase activities in amyotrophic lateral sclerosis. Ann Neurol 39:158–165
4. Khan S, O'Brien PJ (1995) Modulating hypoxia-induced hepatocyte injury by affecting intracellular redox state. Biochim Biophys Acta 1269:153–161
5. Perlemuter G, Davit-Spraul A, Cosson C, Conti M, Bigorgne A, Paradis V, Corre MP, Prat L, Kuoch V, Basdevant A, Pelletier G, Oppert JM, Buffet C (2005) Increase in liver antioxidant enzyme activities in non-alcoholic fatty liver disease. Liver Int 25:946–953
6. Tsukahara H (2007) Biomarkers for oxidative stress: clinical application in pediatric medicine. Curr Med Chem 14:339–351
7. Crane FL (2001) Biochemical functions of coenzyme Q_{10}. J Am Coll Nutr 20:591–598
8. Takayanagi R, Takeshige K, Minakimi S (1980) NADH- and NADPH-dependent lipid peroxidation in bovine heart submitchondrial particles. Dependent on the rate of electron flow in the respiratory chain and an antioxidant role for ubiquinol. Biochem J 192:853–860
9. Moncada S, Palmer RM, Higgs EA (1991) Nitric oxide: physiology, pathophysiology, and pharmacology. Pharmacol Rev 43:109–142
10. Karaa A, Kamoun WS, Clemens MG (2005) Oxidative stress disrupts nitric oxide synthase activation in liver endothelial cells. Free Radic Biol Med 39:1320–1331
11. Cooke JP (2005) ADMA: its role in vascular disease. Vasc Med 10:S11–S17
12. Sydow K, Munzel T (2003) ADMA and oxidative stress. Atheroscler Suppl 4:41–51
13. Sirtori LR, Dutra-Filho CS, Fitarelli D, Sitta A, Haeser A, Barschak AG, Wajner M, Coelho DM, Llesuy S, Belló-Klein A, Giugliani R, Deon M, Vargas CR (2005) Oxidative stress in patients with phenylketonuria. Biochim Biophys Acta 1740:68–73
14. Sitta A, Barschak AG, Deon M, de Mari JF, Barden AT, Vanzin CS, Biancini GB, Schwartz IV, Wajner M, Vargas CR (2009) L-carnitine blood level and oxidant stress in treated phenylketonuric patients. Cell Mol Neurobiol 29:211–218
15. Kotani K, Maekawa M, Kanno T, Kondo A, Toda N, Manabe M (1994) Distribution of immunoreactive malondialdehyde-modified low-density lipoprotein in human serum. Biochim Biophys Acta 1215:121–125
16. Artuch R, Colomé C, Sierra C, Brandi N, Lambruschini N, Campistol J, Ugarte D, Vilaseca MA (2004) A longitudinal study of antioxidant status in phenylketonuric patients. Clin Biochem 37:198–203
17. Tamura S, Tsukahara H, Ueno M, Maeda M, Kawakami H, Sekine K, Mayumi M (2006) Evaluation of a urinary multi-parameter biomarker set for oxidative stress in children, adolescents and young adults. Free Radic Res 40:1198–1205
18. Schulze F, Wesemann R, Schwedhelm E, Sydow K, Albsmeier J, Cooke JP, Böger RH (2004) Determination of asymmetric dimethylarginine (ADMA) using a novel ELISA assay. Clin Chem Lab Med 42:1377–1383
19. Summer KH, Eisenburg J (1985) Low content of hepatic reduced glutathione in patients with Wilson's disease. Biochem Med 34:107–111
20. Tietze F (1969) Enzymic method for quantitative determination of nanogram amounts of total and oxidized glutathione: applications to mammalian blood and other tissues. Anal Biochem 27:502–522
21. Sies H, Summer KH (1975) Hydroperoxide-metabolizing systems in rat liver. Eur J Biochem 57:503–512
22. Gutteridge JM, Halliwell B (1990) The measurement and mechanism of lipid peroxidation in biological systems. Trend Biochem Sci 15:129–135
23. Will ED (1969) Lipid peroxide formation in microsomes. Biochem J 113:315–341
24. McCord JM, Fridovich I (1969) Superoxide dismutase. An enzymic function for erythrocuprein (hemocuprein). J Biol Chem 244:6049–6055
25. Paglia DE, Valentine WN (1967) Studies on the quantitative and qualitative characterization of erythrocyte glutathione peroxidase. J Lab Clin Med 70:158–169
26. Jiménez I, Aracena P, Letelier ME, Navarro P, Speisky H (2002) Chronic exposure of HepG2 cells to excess copper results in depletion of glutathione and induction of metallothionein. Toxicol In Vitro 16:167–175

27. Scriver CR, Kaufman S (2001) Hyperphenylalaninemia: phenylalanine hydroxylase deficiency. In: Scriver CR, Beaudet al, Sly WS, Sly D (eds) The metabolic and molecular bases of inherited disease, 8th edn. McGraw-Hill, New York, pp 1667–1724
28. Hanley WB (2004) Adult phenylketonuria. Am J Med 117:590–595
29. Lee PJ, Amos A, Robertson L, Fitzgerald B, Hoskin R, Lilburn M, Weetch E, Murphy G (2009) Adults with late diagnosed PKU and severe challenging behaviour: a randomised placebo-controlled trial of a phenylalanine-restricted diet. J Neurol Neurosurg Psychiatry 80:631–635
30. Hargreaves IP (2007) Coenzyme Q_{10} in phenylketonuria and mevalonic aciduria. Mitochondrion 7(Suppl):S175–S180
31. Förstermann U, Münzel T (2006) Endothelial nitric oxide in vascular disease from marvel to menace. Circulation 113:1708–1714
32. Mori M, Gotoh T (2004) Arginine metabolic enzymes, nitric oxide and infection. J Nutr 134:2820–2825
33. Palmieri F (2008) Diseases caused by mitochondria transporters. Biochim Biophys Acta 1777:564–568
34. Yasuda T, Yamaguchi N, Kobayashi K, Nishi I, Horinouchi H, Jalil MA, Li MX, Ushikai M, Iijima M, Kondo I, Saheki T (2000) Identification of two novel mutations in the SLC25A13 gene and detection of seven mutations in 102 patients with adult-onset type II citrullinemia. Hum Genet 107:537–545
35. Yamaguchi N, Kobayashi K, Yasuda T, Nishi I, Iijima M, Nakagawa M, Osame M, Kondo I, Saheki T (2002) Screening of SLC25A13 mutations in early and late onset patients with citrin deficiency and in the Japanese population: identification of two novel mutations and establishment of multiple DNA diagnosis method for the nine mutations. Hum Mutat 19:122–130
36. Saheki T, Kobayashi K (2002) Mitochondrial aspartate glutamate carrier (citrin) deficiency as the cause of adult-onset type II citrullinemia (CTLN2) and idiopathic neonatal hepatitis (NICCD). J Hum Genet 47:333–341
37. Ohura T, Kobayashi K, Tazawa Y, Abukawa D, Sakamoto O, Tsuchiya S, Saheki T (2007) Clinical pictures of 75 patients with neonatal intrahepatic cholestasis caused by citrin deficiency. J Inherit Metab Dis 30:139–144
38. Dimmock D, Kobayashi K, Iijima M, Tabata A, Wong LJ, Saheki T, Lee B, Scaglia F (2007) Citrin deficiency: a novel cause of failure to thrive that responds to a high-protein, low-carbohydrate diet. Pediatrics 119:773–777
39. Tanzi RE, Petrukhin K, Chernov I, Pellequer JL, Wasco W, Ross B, Romano DM, Parano E, Pavone L, Brzustowicz LM, Devote M, Peppercorn J, Bush AI, Sternlieb I, Pirastu M, Gusella JF, Evgrafov O, Penchaszadeh GK, Honig B, Edelman IS, Soares MB, Sheinberg IH, Gilliam TC (1993) The Wilson disease gene is a copper transporting ATPase with homology to the Menkes disease gene. Nat Genet 5:344–350
40. Scheinberg IH, Sternlieb I (1984) Wilson disease. In: Smith LH (ed) Major problems in internal medicine. Saunders, Philadelphia, PA, pp 1–179
41. Sokol RJ, Narkewicz MR (2001) Copper and iron disorders. In: Suchy FJ, Sokol RJ, Balistreli WF (eds) Liver disease in children. Lippincott Williams & Wilkins, Philadelphia, PA, pp 595–640
42. Hochstein P, Kumar KS, Forman SJ (1980) Lipid peroxidation and the cytotoxicity of copper. Ann N Y Acad Sci 355:240–248
43. Sokol RJ, Twedt D, McKim JM Jr, Devereaux MW, Karrer FM, Kam I, von Steigman G, Narkewicz MR, Bacon BR, Britton RS (1994) Oxidant injury to hepatic mitochondria in patients with Wilson's disease and Bedlington terriers with copper toxicosis. Gastroenterology 107:1788–1798
44. Strand S, Hofmann WJ, Grambihler A, Hug H, Volkmann M, Otto G, Wesch H, Marian SM, Hack V, Stremmel W, Krammer PH, Galle PR (1998) Hepatic failure and liver cell damage in acute Wilson's disease involve CD95(APO-1/Fas) mediated apoptosis. Nat Med 4:588–593
45. Bedossa P, Poynard T (1996) An algorithm for the grading of activity in chronic hepatitis C. The METAVIR Cooperative Study Group. Hepatology 24:289–293

46. Tailleux A, Wouters K, Staels B (2012) Roles of PPARs in NAFLD: potential therapeutic targets. Biochim Biophys Acta 1821:809–818
47. Nagasawa T, Inada Y, Nakano S, Tamura T, Takahashi T, Maruyama K, Yamazaki Y, Kuroda J, Shibata N (2006) Effects of bezafibrate, PPAR pan-agonist, and GW501516, PPAR delta agonist, on development of steatohepatitis in mice fed a methionine- and choline-deficient diet. Eur J Pharmacol 536:182–191
48. Lutchman G, Modi A, Kleiner DE, Promrat K, Heller T, Ghany M, Borg B, Loomba R, Liang TJ, Premkumar A, Hoofnagle JH (2007) The effects of discontinuing pioglitazone in patients with nonalcoholic steatohepatitis. Hepatology 46:424–429
49. Enns GM, Kinsman SL, Perlman SL, Spicer KM, Abdenur JE, Cohen BH, Amagata A, Barnes A, Kheifets V, Shrader WD, Thoolen M, Blankenberg F, Miller G (2012) Initial experience in the treatment of inherited mitochondrial disease with EPI-743. Mol Genet Metab 105:91–102
50. Murphy MP (2009) How mitochondria produce reactive oxygen species. Biochem J 417: 1–13
51. Ryter SW, Kim HP, Hoetzel A, Park JW, Nakahira K, Wang X, Choi AM (2007) Mechanism of cell death in oxidative stress. Antioxid Redox Signal 9:49–89
52. Fernandes CG, Borges CG, Seminotti B, Amaral AU, Knebel LA, Eichler P, de Oliveira AB, Leipnitz G, Wajner M (2011) Experimental evidence that methylmalonic acid provokes oxidative damage and compromises antioxidant defenses in nerve terminal and striatum of young rats. Cell Mol Neurobiol 31:775–785
53. Kanaumi T, Takashima S, Hirose S, Kodama T, Iwasaki H (2006) Neuropathology of methylmalonic academia in a child. Pediatr Neurol 34:156–159
54. de Keyzer Y, Valayannopoulos V, Benoist JF, Batteux F, Lacaille F, Hubert L, Chrétien D, Chadefeaux-Vekemans B, Niaudet P, Touati G, Munnich A, de Lonlay P (2009) Multiple OXPHOS deficiency in the liver, kidney, heart, and skeletal muscle of patients with methylmalonic aciduria and propionic aciduria. Pediatr Res 66:91–95
55. Biancini GB, Sitta A, Wayhs CA, Pereira IN, Rockenbach F, Garcia SC, Wyse AT, Schwartz IV, Wajner M, Vargas CR (2011) Experimental evidence of oxidative stress in plasma of homocystinuric patients: a possible role for homocysteine. Mol Genet Metab 104:112–117

Chapter 24
Neurological Disorders (Especially Developmental Brain Disorders)

Masaharu Hayashi, Rie Miyata, and Naoyuki Tanuma

Abbreviations

4-HNE	4-hydroxynonenal
8-OHdG	8-hydroxy-2′-deoxyguanosine
AGE	Advanced glycation end product
ASD	Autism spectrum disorder
CP	Cerebral palsy
Cre	Creatinine
CSF	Cerebrospinal fluid
DS	Down syndrome
ELISA	Enzyme-linked immunosorbent assay
GSH	Glutathione
GSSG	Oxidized GSH
HEL	Hexanoyl-lysine adduct
HIE	Hypoxic ischemic encephalopathy
LGS	Lennox–Gastaut syndrome
NO	Nitric oxide
PAO	Potential antioxidant
PME	Progressive myoclonic epilepsy
RNS	Reactive nitrogen species
ROS	Reactive oxygen species
RTT	Rett syndrome

M. Hayashi, M.D. (✉) • R. Miyata, M.D. • N. Tanuma, M.D.
Department of Brain Development and Neural Regeneration, Tokyo Metropolitan Institute of Medical Science, 2-1-6 Kamikitazawa, Setagaya-ku, Tokyo 156-8506, Japan
e-mail: hayashi-ms@igakuken.or.jp

H. Tsukahara and K. Kaneko (eds.), *Studies on Pediatric Disorders*, Oxidative Stress in Applied Basic Research and Clinical Practice, DOI 10.1007/978-1-4939-0679-6_24,

SMID	Severe motor and intellectual disabilities
SOD	Superoxide dismutase
TAP	Total antioxidant power
TG	Thymidine glycol
WS	West syndrome

24.1 Introduction

Oxidative stress has been delineated in detail in other chapters, and it is briefly addressed in this chapter with respect to descriptions of pediatric neurological disorders. Oxygen is metabolized to generate energy in the form of ATP through a series of reductive steps at the inner membrane of the mitochondria, and reactive oxygen species (ROS) and reactive nitrogen species (RNS) are formed. Although the production of free radicals has a role in the regulation of biological function, ROS and RNS can cause functional disruptions in lipids, proteins, and nucleic acids, resulting in tissue damage in a process called oxidative stress [1]. ROS include the superoxide anion (O^{2-}), hydroxyl radicals (–OH), and hydrogen peroxide (H_2O_2), while nitric oxide (NO) and peroxynitrite ($ONOO^-$) are known as RNS. Oxidative stress is a condition where an imbalance exists between the production of ROS and RNS and the body's ability to neutralize these intermediates. The antioxidant defense system consists of large molecules, such as albumin, ferritin, and ceruloplasmin, and small molecules, including metal-chelating vitamins C and E, in addition to radical scavenging enzymes, such as catalase, superoxide dismutase (SOD), and glutathione (GSH) peroxidase [2]. Excess ROS/RNS production over detoxification results in a shift in the balance towards oxidative damage. The brain is susceptible to oxidative stress because it utilizes a large amount of oxygen, contains a high concentration of polyunsaturated fatty acids, is prone to lipid peroxidation, and is rich in iron and low in catalase activity. Hence, oxidative stress is responsible for neuronal and glial dysfunction, which contributes to disease pathogenesis in adult neurodegenerative diseases, such as Alzheimer's disease, Parkinson's disease, and amyotrophic lateral sclerosis [3].

The oxidative damage of DNA produces 8-hydroxy-2′-deoxyguanosine (8-OHdG), which has been used as a marker of oxidative nucleoside damage [4]. The oxidation of thymidine in DNA but not in RNA by hydroxyl radicals leads to the formation of thymidine glycol (TG), which is not degraded as easily as 8-OHdG, and TG is a stable oxidative marker specific for DNA [5]. Lipid peroxidation in the membranes of brain cells and myelin can form various aldehydes, including hexanoyl-lysine adduct (HEL) and 4-hydroxynonenal (4-HNE) [1]. Advanced glycation end products (AGEs) are formed by the glycation of proteins, lipids, and nucleic acids, and increasing oxidative stress, in addition to other mechanisms, may accelerate the formation of AGEs and cause multisystem tissue damage [6]. In our laboratory, markers of oxidative DNA damage, lipid peroxidation, and protein glycation have been examined in autopsy brains with immunohistochemistry and in urine and cerebrospinal fluid (CSF) with enzyme-linked immunosorbent assays (ELISA) of patients with childhood-onset neurological disorders [7, 8]. SOD is a reliable marker of the anti-oxidant system in autopsy brain [9]. The potential antioxidant (PAO) is a marker of antioxidant capacity in various biologic fluids, in which Cu^{2+} is reduced by antioxidants to Cu^{+}, and PAO evaluates both hydrophilic and hydrophobic antioxidants [10]. PAO enables the evaluation of not only hydrophilic antioxidants, such as vitamin C and GSH, but also hydrophobic antioxidants, such as vitamin E. Total antioxidant power (TAP) is also measured quantitatively with a spectrophotometric method of cupric-reducing antioxidant capacity [11].

24.2 Methods

The ethical committee of the Tokyo Metropolitan Institute of Medical Science approved the studies, and the families of the subjects gave consent for immunohistochemical examinations in autopsy brains and for experimental measurements in the CSF and urine. For the immunohistochemistry, serial sections were cut from paraffin-embedded blocks from the examined brain regions. Following microwave antigen retrieval, sections were treated with mouse monoclonal antibodies against 8-OHdG, TG, 4-HNE (dilution 1:2,000; Japan Institute for the Aging, Shizuoka, Japan), AGE (dilution 1:2,000; Trans Genic, Inc., Kumamoto, Japan), and rabbit polyclonal antibodies against Cu/ZnSOD and MnSOD (dilution 1:2,000; Stressgen Biotechnologies Corporation, Victoria, BC, Canada). Immune reactions were visualized with the streptavidin-biotin-complex/3,3′-diaminobenzidine tetrahydrochloride staining kit (Nichirei Corporation, Tokyo, Japan). Levels of 8-OHdG, HEL, PAO, and TAP in the urine and/or CSF were determined with commercial ELISA kits (8-OHdG and HEL: Japan Institute for the Aging, Shizuoka, Japan), and colorimetric microplate assay kits (PAO: Japan Institute for the Aging, Shizuoka, Japan; TAP: Oxford Biomedical Research, Oxford, MI, USA) [1, 12]. Samples were obtained from subjects and immediately stored at −80 °C until analysis. In the CSF samples, 1-h dialysis with a YM-10 centrifugal filter (EMD Millipore Corporation, Billerica, MA, USA) was performed in order to remove macromolecules. Urinary creatinine (Cre) levels were examined by a calorimetric assay, and the results in the analysis of the urinary biomarkers were expressed relative to urinary Cre levels (mg/dL) in order to adjust for muscle mass. The levels of oxidative stress markers and antioxidant capacities were determined with a Model 680 plate reader (Bio-Rad Laboratories, Inc., Hercules, CA, USA) and a Thermo Multiskan FC microplate photometer (Thermo Fisher Scientific, Inc., Waltham, MA, USA). We calculated the mean plus the value of 2 standard deviations (SD) in controls in order to determine a cutoff index for each oxidative marker. The differences in the averaged levels between patients and controls were statistically evaluated with multiple analyses.

24.3 Autism Spectrum Disorder

Autism spectrum disorders (ASDs) are neurodevelopmental disorders that are characterized by impairments in communication and social abilities and repetitive and stereotypic behaviors. Etiological hypotheses for ASDs include genetic susceptibilities, immunologic alterations, oxidative stress, and environmental exposures [13]. According to a review of research literatures on ASDs [14], percentage of publications, which indicated an association between ASDs and each pathomechanism, is in the following: immune abnormalities/inflammation (95 %), oxidative stress (100 %), mitochondrial dysfunction (95 %), and toxicant exposures (89 %). Several attempts have been made to validate biomarkers for screening purposes, but, as yet, none has been found to univocally predict ASDs.

One meta-analysis has shown that ASD patients, relative to controls, show decreased blood levels of GSH, GSH peroxidase, methionine, and cysteine and increased concentrations of oxidized GSH (GSSG), whereas SOD, homocysteine, and cystathionine levels were not associated with patients with ASDs [15]. A recent analysis has demonstrated increases in intraerythrocyte and plasma levels of 4-HNE, in addition to decreased erythrocyte levels of GSH, in 20 ASD children relative to 18 age-matched controls, suggesting the possibility of lipid peroxidation that is partly due to a GSH redox imbalance [16]. The intracellular GSH redox balance (GSH/GSSG) has been shown to be reduced in resting peripheral blood mononuclear cells, activated monocytes, and CD4 T cells in 43 ASD children [17]. The same research group has examined frozen samples of the cerebellum and temporal cortex from 15 ASD patients and 12 controls. Consistent with the results of the blood cell analysis, the cerebellum and temporal cortex in the ASD subjects were shown to demonstrate a decrease in GSH and GSH/GSSG, whereas biomarkers of oxidative protein damage (3-nitrotyrosine) and oxidative DNA damage (8-OHdG) were increased, and the latter was inversely correlated with GSH/GSSG in the cerebellum [18]. Chauhan et al. have examined the GSH redox balance in postmortem frozen brain specimens in ten each of ASD children and controls and found decreases in GSH and GSH/GSSG and increases in GSSG in the cerebellum and temporal cortex [19]. These data may indicate the involvement of a GSH redox imbalance in ASD neurological disorders. Regarding treatments, viral vitamin/mineral supplementation has been shown to be beneficial for the treatment of the metabolic statuses of ASD children, and these benefits include improvements in methylation, GSH, oxidative stress, and sulfation [20]. Some investigators have used hyperbaric oxygen treatment to treat ASD children, and this treatment has been shown to improve cerebral perfusion, decrease markers of inflammation, and not worsen the levels of oxidative stress markers [21]. However, most of the studies on treatments have a number of limitations due to their inconsistent findings across studies. Accordingly, a link between oxidative stress and ASDs remains controversial due to opposing views on its role in the pathogenesis.

We performed a preliminary analysis of 27 urine samples from 16 ASD patients who were aged less than 10 years and 9 age-matched controls who did not show

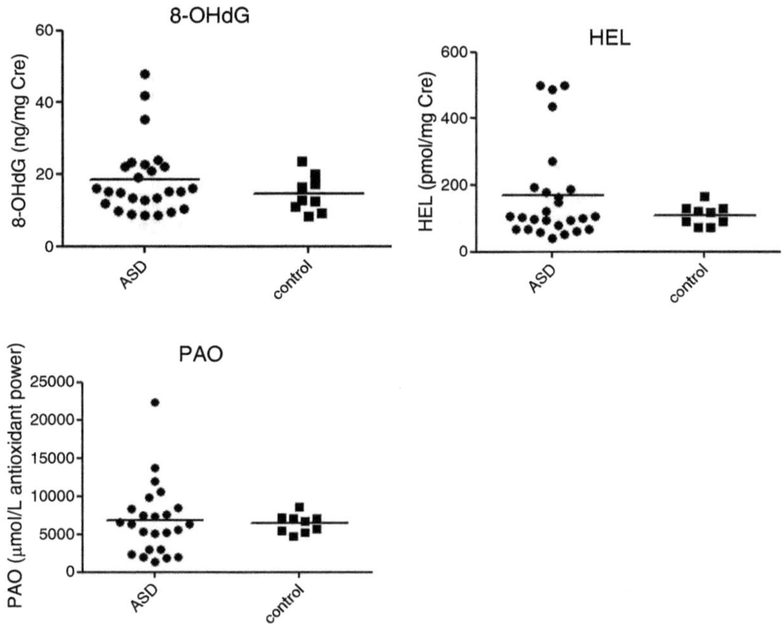

Fig. 24.1 Urinary level of 8-hydroxy-2′-deoxyguanosine (8-OHdG), hexanoyl lysine (HEL), and potential antioxidant (PAO) in patients with autism spectrum disorder (ASD) and controls

either neurological or developmental abnormalities. Seven ASD patients had a past history of epilepsy, including neonatal seizures, West syndrome (WS), and Lennox–Gastaut syndrome (LGS). There were no significant differences in the urinary levels of 8-OHdG, HEL, or PAO between the ASD patients and controls, although a few ASD patients demonstrated increased urinary levels of 8-OHdG or HEL (Fig. 24.1). We examined the effects of a past history of epileptic seizures and failed to find a difference in the urinary levels of 8-OHdG, HEL, and PAO between the ASD patients with and without epilepsy (Fig. 24.2). Because ASD is a multifactorial disease, the number of patients that are examined should be increased to at least over 100, taking age and etiology into consideration, in order to make a definite conclusion.

24.4 Rett Syndrome

Rett syndrome (RTT) is a childhood-onset developmental disorder that is characterized by a number of features that are observed in other disorders ranging from ASD to Parkinson's disease and dystonia. The disorder, which affects 1 in 10,000 females, is most often caused by mutations in gene coding that are mainly caused by

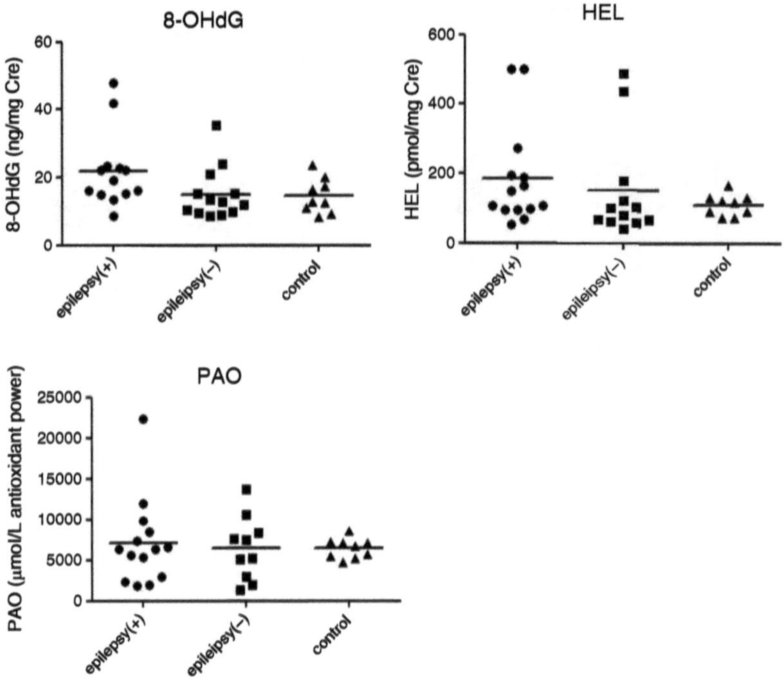

Fig. 24.2 Comparison of urinary level of 8-OHdG, HEL, and PAO among patients with ASD and epilepsy, those with ASD but not epilepsy and controls

X-chromosomal methyl-CpG-binding protein 2, which is a transcriptional regulatory protein [22]. Several potential treatments for RTT have been proposed [23]. However, its pathogenesis remains to be investigated, and no effective therapy is available to date. Data from several laboratories have indicated a possible involvement of oxidative stress [24]. First, reduced GSH levels have been found in the brain in one autopsy case of RTT [25]. Fourteen years after that study, RTT patients were reported to have increased plasma levels of lipid peroxidation markers and reduced SOD activities in erythrocytes [26]. In addition, patients with typical RTT have shown hypoxia-related increases in intraerythrocyte and plasma levels of nonprotein-bound iron and protein carbonyl concentrations, indicating a possible involvement of protein oxidative damage [27].

In order to further examine the role of oxidative stress in RTT, we performed ELISA on the urinary levels of 8-OHdG, HEL, and PAO in 15 samples of ten patients with genetically confirmed RTT, in which five subjects each were aged from 4 to 22 years and 46 to 49 years. These subjects were classified into younger and older RTT patients, respectively. There were no differences in the urinary levels of any of the markers between the younger RTT patients and controls (Fig. 24.3). In contrast, the older RTT patients demonstrated significant increases in the urinary

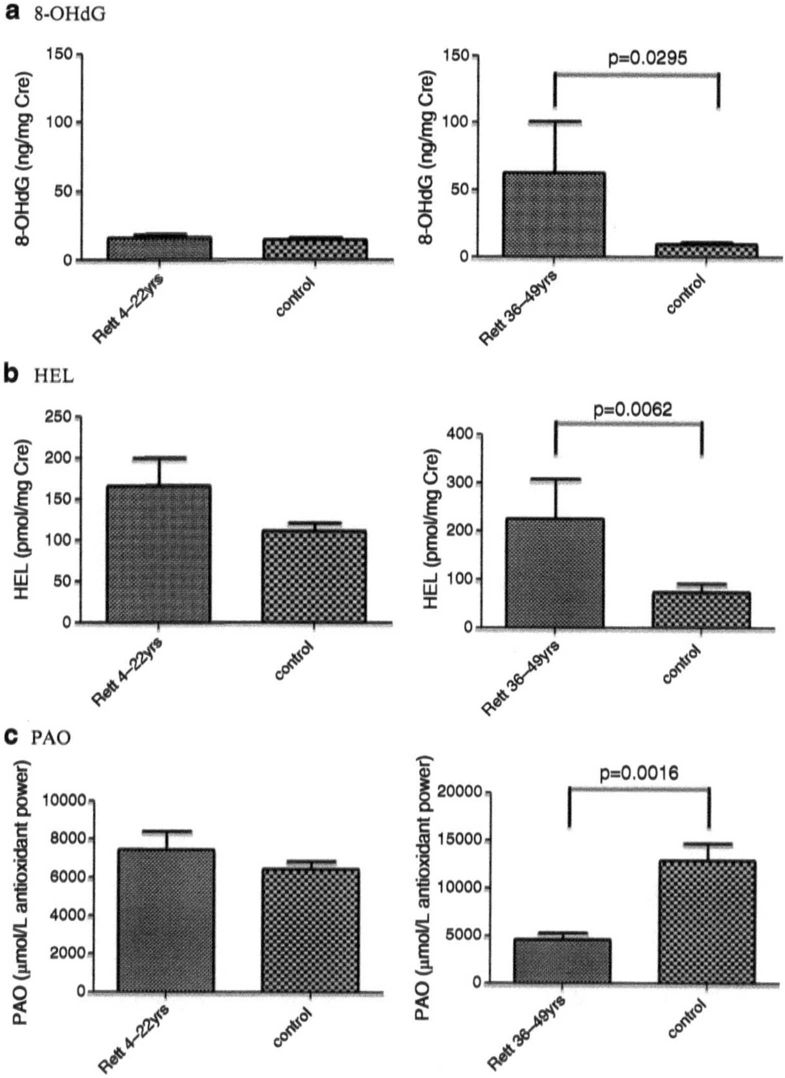

Fig. 24.3 Urinary levels of 8-OHdG (**a**), HEL (**b**), and PAO (**c**) in patients with Rett syndrome and controls

levels of both 8-OHdG and HEL, in addition to reduced urinary levels of PAO (Fig. 24.3), indicating the possibility of exaggerated oxidative stress due to impaired antioxidant mechanisms in older RTT patients. The older RTT patients who were examined in this analysis had more respiratory abnormalities and/or gastrointestinal troubles than the younger RTT patients, and such systemic complications may have been involved in the observed changes in the oxidative stress markers. Accordingly, further analyses of the CSF are required to investigate the neurological implications of oxidative stress in RTT.

24.5 Down Syndrome

Down syndrome (DS), which is the most common genetic cause of mental retardation, arises from the triplication of the entire, or even part of, chromosome 21 (trisomy21). DS is clinically characterized by craniofacial abnormalities, small brain size, accelerated aging, and cognitive defects. Increasing numbers of studies have recently shown that oxidative stress occurs in the pathogenesis and progression of DS due to a deregulation of gene/protein expression, and the increased production of ROS is accompanied by mitochondrial dysfunction [28]. The redox imbalance is likely attributed to an overexpression of Cu/ZnSOD, which is encoded by chromosome 21 and which has been investigated in in vitro, ex vivo, and animal studies [29]. Postmortem studies in DS brains have demonstrated increased levels of thiobarbituric acid-reactive substances and AGE in the cortex in fetal brain and an accumulation of 8-OHdG and nitrotyrosine in the cytoplasm of cerebral neurons [30]. Interestingly, there were no differences between DS patients and age-matched controls in the urinary levels of 8-OHdG, isoprostane, thiobarbituric acid-reactive substances, AGEs, hydrogen peroxide, and nitrite/nitrate [31]. DS patients exhibit various phenotypes, and repetitive examinations of large numbers of patients may be necessary.

24.5.1 Cerebral Palsy

Cerebral palsy (CP), which is the most common disability and which is identified in approximately 1 per 500–1,000 live births, is a nonprogressive disorder of posture and voluntary movements due to increased muscle tone. CP children often suffer from feeding difficulties and respiratory troubles, and both neurological disorders and systemic complications may cause oxidative stress in subjects with CP. However, studies on oxidative stress have rarely been performed in CP patients. Augmented lipid peroxidation with reduced antioxidant capacity in the blood has been reported in 69 CP children who were aged 1–12 years in comparison to 42 age-matched controls [32]. SOD activity in red blood cells has been shown to be reduced in CP children receiving enteral nutrition compared to CP children who are fed orally and controls [33].

The diagnosis of severe motor and intellectual disabilities (SMID) is used to designate heterogeneous developmental brain disorders that are associated with severe physical disabilities and profound mental retardation. Patients with SMID often have respiratory problems that greatly affect their life quality and expectancy [34].

Table 24.1 Preliminary immunohistochemical analysis on the oxidative stress markers in the brains in cases of congenital anomalies and perinatal hypoxic ischemic encephalopathy

Case	Age (years)/Sex	Etiology of brain diseases	Complication	8-OHdG	4-HNE
1	3/Male	Perinatal hypoxic ischemic encephalopathy	West syndrome	(–)	Cortical neurons
2	4/Male	Sturge–Weber syndrome	Neuromelamosis	(–)	(–)
3	7/Male	Congenital cytomegalovirus infection	EB-VAHS	Glial cells	(–)
4	14/Male	Perinatal hypoxic ischemic encephalopathy	HyperIgE syndrome	Glial cells	Dentate neurons
5	26/Male	Microdysgenesis in the cerebral cortex	West syndrome	Glial cells	(–)
6	40/Male	Anteriovenous malformation	Cerebral bleeding	Glial cells	(–)
7	64/Female	Brain anomaly	Liver cirrhosis	(–)	(–)
8	85/Female	Perinatal hypoxic ischemic encephalopathy	Ovarian cancer	(–)	(–)

8-OHdG 8-hydroxy-2′-deoxyguanosine, *4-HNE* 4-hydroxynonenal, *EB-VAHS* EB virus-associated hemophagocytic syndrome

We previously studied 14 SMID patients who were aged 10–48 years and 17 age-matched controls [12]. Six patients had CP and mental retardation due to perinatal hypoxic ischemic encephalopathy (HIE), and three subjects each had sequelae of acute encephalopathy and neonatal hyperbilirubinemia, respectively. The mean (SD) of the urinary levels of 8-OHdG in SMID patients, 18.8 (9.0) ng/mg Cre, was significantly increased compared to that of controls, 10.5 (2.9) ng/mg Cre. The urinary levels of 8-OHdG correlated with the severity of respiratory disturbances that were evaluated with respiratory disturbance scores but not with age or antiepileptic drug medication. However, there was no significant difference in the urinary levels of HEL between the SMID patients and controls. Respiratory disturbances may cause oxidative DNA damage in SMID patients. In addition, we performed preliminary immunohistochemical examinations of 8-OHdG and HEL in the brains of SMID autopsy cases of patients with perinatal HIE ($n = 3$) and congenital brain and/or vascular anomalies ($n = 5$) (Table 24.1). Nuclei that were immunoreactive for 8-OHdG were observed in a few glial cells in the cerebral cortex in cases 3, 5, and 6 with brain anomalies and case 4 with perinatal HIE. Cytoplasmic immunoreactivity for 4-HNE was identified in the dentate neurons of case 4 with perinatal HIE (Fig. 24.4), in addition to in the damaged neurons in the cerebral cortex of case 1 with perinatal HIE. Increased accumulations of oxidative stress markers seemed to be rare in the brains of SMID cases with brain anomalies and perinatal HIE compared to those in cases with metabolic errors and neurodegenerative disorders [1, 7, 8].

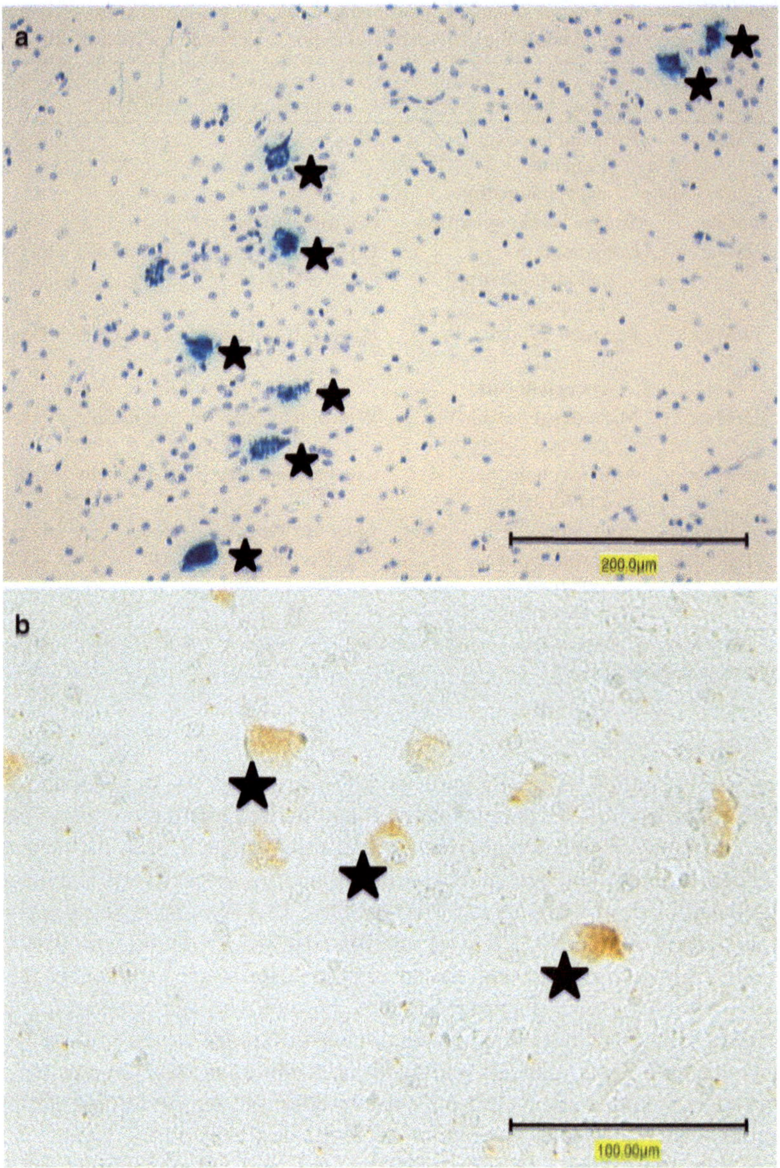

Fig. 24.4 The dentate nucleus in the case of perinatal hypoxic ischemic encephalopathy and hyperIgE syndrome (case 4 in Table 24.1) demonstrated sclerotic neurons (small stars) ((**a**) Klüver–Barrera staining, bar = 200 μm) and cytoplasmic immunoreactivity for 4-hydroxynonenal in the neurons (stars) ((**b**) bar = 100 μm)

24.6 Intractable Epilepsy

Epilepsy is the most frequent neurodegenerative disease after stroke. At least 6 % of the population is speculated to suffer one isolated seizure episode during their lifetime. Despite the increase in anticonvulsant therapies, epilepsy is medically intractable in patients with temporal lobe epilepsy, WS/LGS, progressive myoclonic epilepsy (PME), and severe myoclonic epilepsy in infants. Oxidative stress is likely to be implicated in the initiation and progression of epilepsy in animal models of epilepsy, such as genetic rat models, and in experimental seizures that are induced by kainic acid, pilocarpine, pentylenetetrazol, and trimethyltin [35]. In addition, roles of mitochondrial dysfunction and oxidative stress have been suggested as they have been shown to be acute consequences of injuries that are known to incite chronic epilepsy and to be involved in the chronic stages of epilepsy [36]. Some studies have suggested that antioxidant therapy may reduce lesions that are induced by oxidative-free radicals in some animal seizure models [37]. The antioxidant and neuroprotective properties of antiepileptic drugs such as diazepam and phenobarbital have been discussed as they may prevent seizure-induced increases in NO and peroxidation, and valproate may increase the levels of GSH [38]. A ketogenic diet, which is a high-fat, low-carbohydrate diet and which is used as a therapy for intractable epilepsy, may reduce redox signaling molecules such as hydrogen peroxide and 4-HNE and increase the production of antioxidants such as GSH in mitochondria [39].

We have examined the levels of expression of oxidative stress markers and SOD in autopsy brains of cases with PME [7, 40–42] and WS/LGS [8], and, recently, the CSF levels of oxidative markers have been determined in infants with WS [8]. Table 24.2 summarizes the results of our previous studies. The autopsy brains of patients with three disorders of different etiologies that cause PME have demonstrated exaggerated oxidative DNA damage (Fig. 24.5), lipid peroxidation, and, to a lesser extent, protein glycation. However, a predominant involvement of lipid peroxidation in WS has been indicated by both immunohistochemistry in autopsy brains (Fig. 24.6) and ELISA in the CSF. We believe that the combination of neuropathological studies and analyses of levels of oxidative stress markers in the CSF, blood, and urine will be useful for specifying the implications of oxidative stress in each disorder.

Table 24.2 Summary of data in our analysis on intratable epilepsy

Neurological disorders	Changes in oxidative stress markers/superoxide dismutase (SOD)	Regions in autopsy brains/cerebrospinal fluid	References
Progressive myoclonic epilepsy			
Neuronal ceroid-lipofuscinosis	Oxidative DNA damage ↑	Cerebral cortex	[40, 41]
	Protein glycation ↑	Cerebellar cortex	
	Lipid peroxidation ↑	Cerebral cortex, cerebeller cortex	
Dentatorubral-pallidoluysian atrophy	Oxidative DNA damage ↑	Lenticulate nucleus	[42]
	Lipid peroxidation ↑	Hippocampus, cerebellar dentate	
	Expression of MnSOD ↓	Globus pallidus, cerebellar dentate	
Lafora disease	Oxidative DNA damage ↑	Cerebral cortex, globus pallidus	[7]
	Lipid peroxidation ↑	Nuclei of cranial nerves	
West syndrome/Lennox–Gastaut syndrome			
Lissencephaly	Lipid peroxidation ↑	Midbrain tegmentum	[8]
Perinatal hypoxic ischemic encephalopathy	Lipid peroxidation ↑	Substantia nigra	
Infants with West syndrome showing poor outcome	Lipid peroxidation ↑	Cerebrospinal fluid	[8]

24.7 Conclusion

Oxidative stress, which affects nucleosides, lipids, and proteins, may be involved in the pathogenesis of ASD, RTT, DS, CP, and intractable epilepsy. Clarifying the detailed mechanisms of neurodegeneration by combining immunohistochemistry in autopsy brains and ELISA in the CSF and urine will be useful for examining the involvement of oxidative stress. In order to develop new antioxidant treatments, further analyses are required in a large number of patients.

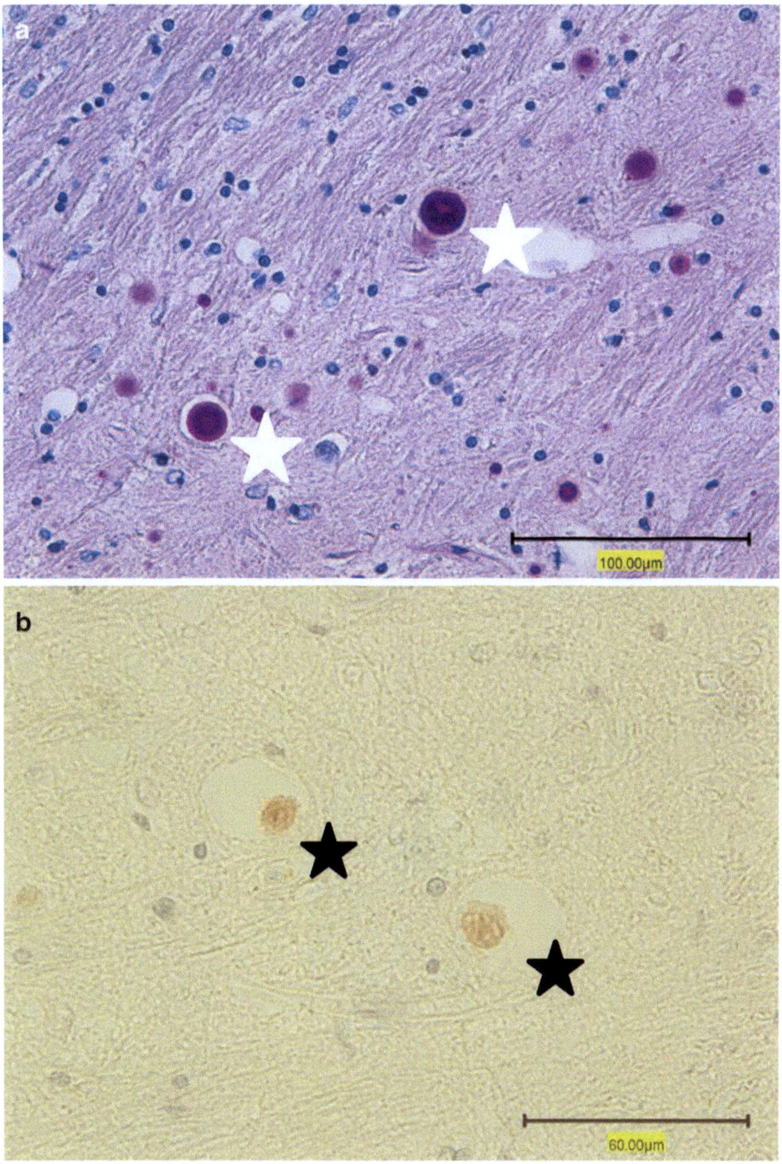

Fig. 24.5 The globus pallidus in the case of Lafora disease had Lafora bodies (white stars) ((**a**) periodic acid-Schiff staining, bar = 100 μm) and nuclear immunoreactivity for 8-hydroxy-2′-deoxyguanosine in the neurons (stars) ((**b**) bar = 60 μm)

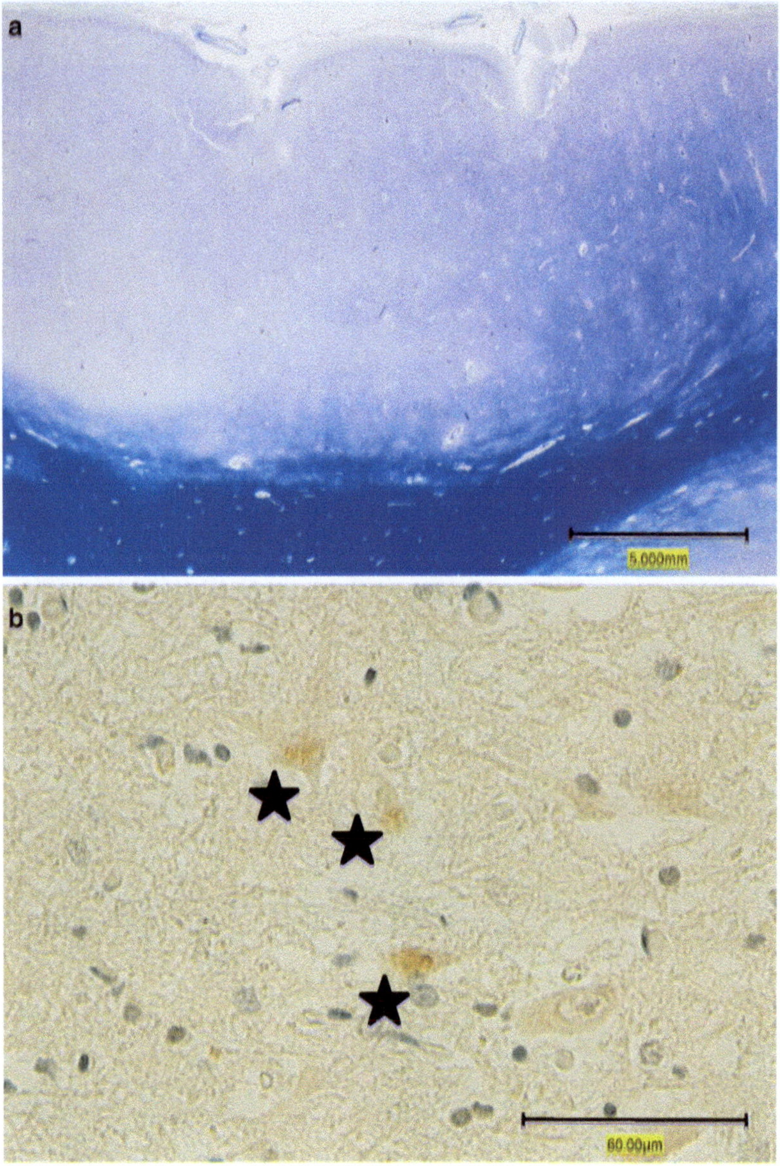

Fig. 24.6 The case of lissencephaly showed agyric frontal cortex ((**a**) Klüver–Barrera staining, bar = 5 mm) and neuronal immunoreactivity for 4-hydroxynonenal in the oculomotor nucleus (stars) ((**b**) bar = 60 μm)

References

1. Hayashi M, Tanuma N, Miyata R (2012) Oxidative stress in developmental brain disorders. In: Ahmad S (ed) Neurodegenerative diseases. Landes Bioscience, Austin, pp 278–290
2. Halliwell B (2012) Free radicals and antioxidants: updating a personal view. Nutr Rev 70:257–265. doi:10.1111/j.1753-4887.2012.00476.x

3. Halliwell B (2006) Oxidative stress and neurodegeneration: where are we now? J Neurochem 97:1634–1658. doi:10.1111/j.1471-4159.2006.03907.x
4. Toyokuni S (1999) Reactive oxygen species-induced molecular damage and its application in pathology. Pathol Int 49:91–102. doi:10.1046/j.1440-1827.1999.00829.x
5. Wang Y, Wang Y (2006) Synthesis and thermodynamic studies of oligodeoxy- ribonucleotides containing tandem lesions of thymidine glycol and 8-oxo-2′-deoxyguanosine. Chem Res Toxicol 19:837–843. doi:10.1021/tx0600321
6. Semba RD, Nicklett EJ, Ferrucci L (2010) Does accumulation of advanced glycation end products contribute to the aging phenotype? J Gerontol A Biol Sci Med Sci 65A:963–975. doi:10.1093/Gerona/glq074
7. Hayashi M (2009) Oxidative stress in developmental brain disorders. Neuropathology 29:1–8. doi:10.1111/j.1440-1789.2008.00888.x
8. Hayashi M, Tanuma N, Miyata R (2010) The involvement of oxidative stress in epilepsy. In: Kozyrev D, Slutsky V (eds) Handbook of free radicals: formation, types and effects. Nova, New York, pp 305–318
9. Hayashi M, Araki S, Kohyama J, Shioda K, Fukatsu R (2005) Oxidative nucleotide damage and superoxide dismutase expression in the brains of xeroderma pigmentosum group A and Cockayne syndrome. Brain Dev 27:34–38. doi:10.1016/j.braindev.2004.04.001
10. Izuta H, Matsunaga N, Shimazawa M, Sugiyama T, Ikeda T, Hara H (2010) Proliferative diabetic retinopathy and relations among antioxidant activity, oxidative stress, and VEGF in vitreous body. Mol Vis 16:130–136, http://www.molvis.org/molvis/v16/a16. Accessed 29 Jan 2010
11. Apak R, Guclu K, Ozyurek M, Karademir SE, Altun M (2005) Total antioxidant capacity assay of human serum using copper(ii)-neocuproine as chromogenic oxidant: the CUPRAC method. Free Radic Res 39:949–961. doi:10.1080/10715760500210145
12. Tanuma N, Miyata R, Hayashi M, Uchiyama A, Kurata K (2008) Oxidative stress as a biomarker of respiratory disturbance in patients with severe motor and intellectual disabilities. Brain Dev 30:402–409. doi:10.1016/j.braindev.2007.12.001
13. Randolph-Gips MM, Srinivasan P (2012) Modeling autism: a systems biology approach. J Clin Bioinforma 2:17. doi:10.1186/2043-9113-2-17
14. Rossignol DA, Frye RE (2012) A review of research trends in physiological abnormalities in autism spectrum disorders: immune dysregulation, inflammation, oxidative stress, mitochondrial dysfunction and environmental toxicant exposures. Mol Psychiatry 17:389–401. doi:10.1038/mp.2011.165
15. Frustaci A, Neri M, Cesario A, Adams JB, Domenici E, Bernardina BD, Bonassi S (2012) Oxidative stress-related biomarkers in autism: systematic review and meta-analysis. Free Radic Biol Med 52:2128–2141. doi:10.1016/j.freeradbiomed.2012.03.011
16. Pacorelli A, Leoncini S, De Felice C, Signorini C, Cerrone C, Valacchi G, Ciccoli L, Hayek J (2012) Non-protein-bound iron and 4-hydroxynonenal protein adducts in classic autism. Brain Dev 35:146–154. doi:10.1016/j.braindev.2012.03.011
17. Rose S, Melnyk S, Trusty TA, Pavliv O, Seidel L, Li J, Nick T, James SJ (2012) Intracellular and extracellular redox status and free radical generation in primary immune cells from children with autism. Autism Res Treat 2012:986519. doi:10.1155/2012/986519
18. Rose S, Melnyk S, Pavliv O, Bai S, Nick TG, Frye RE, James SJ (2012) Evidence of oxidative damage and inflammation associated with low glutathione redox status in the autism brain. Transl Psychiatry 2:e134. doi:10.1038/tp.2012.61
19. Chauhan A, Audhya T, Chauhan V (2012) Brain region-specific glutathione redox imbalance in autism. Neurochem Res 37:1681–1689. doi:10.1007/s11064-012-0775-4
20. Adams JB, Audhya T, McDonough-Means S, Rubin RA, Quig D, Geis E, Gehn E, Loresto M, Mitchell J, Atwood S, Bamhous S, Lee W (2011) Effect of a vitamin/mineral supplement on children and adults with autism. BMC Pediatr 11:111. doi:10.1186/1471-2431-11-111
21. Rossignol DA, Bradstreet JJ, Dyke KV, Schneider C, Freedenfeld SH, O'Hara N, Cave S, Buckley JA, Mumper EA, Frye R (2012) Hyperbaric oxygen treatment in autism spectrum disorders. Med Gas Res 2:16. doi:10.1186/2045-9912-2-16
22. Gadalla KK, Bailey ME, Cobb SR (2011) MeCP2 and Rett syndrome: reversibility and potential avenues for therapy. Biochem J 439:1–14. doi:10.1042/BJ20110648
23. Ricceri L, De Filippias B, Laviola G (2012) Rett syndrome treatment in mouse models: searching for effective targets and strategies. Neuropharmacology 68:105–115. doi:10.1016/j.neuropham.2012.08.010

24. De Felice C, Signorini C, Leoncini S, Pecorelli A, Durand T, Valacchi G, Ciccoli L, Hayek J (2012) The role of oxidative stress in Rett syndrome: an overview. Ann N Y Acad Sci 1259:121–135. doi:10.1111/j.1749-6632.2012.06611.x
25. Sofic E, Riederer P, Killian W, Rett A (1987) Reduced concentrations of ascorbic acid and glutathione in a single case of Rett syndrome: a postmortem brain study. Brain Dev 9:529–531
26. Sierra C, Vilaseca MA, Brandi N, Artuch R, Mira A, Nieto M, Pineda M (2001) Oxidative stress in Rett syndrome. Brain Dev 23(Suppl 1):S236–S239
27. De Felice C, Ciccoli L, Leoncini S, Signorini C, Rossi M, Vannuccini L, Guazzi G, Latini G, Comporti M, Valacchi G, Hayek J (2009) Systemic oxidative stress in classic Rett syndrome. Free Radic Biol Med 47:440–448. doi:10.1016/j.freeradbiomed.2009.05.016
28. Perluigi M, Butterfield DA (2012) Oxidative stress and Down syndrome: a route toward Alzheimer-like dementia. Curr Gerontol Geriatr Res 2012:724904. doi:10.1155/2012/724904
29. Pagano G, Castello G (2012) Oxidative stress and mitochondrial dysfunction in Down syndrome. In: Ahmad S (ed) Neurodegenerative diseases. Landes Bioscience, Austin, pp 291–299
30. Perluigi M, Butterfield DA (2011) The identification of protein biomarkers for oxidative stress in Down syndrome. Expert Rev Proteomics 8:427–429. doi:10.1586/EPR.11.36
31. Campos C, Guzman R, Lopez-Femandez E, Casado A (2011) Evaluation of urinary biomarkers of oxidative/nitrosative stress in children with Down syndrome. Life Sci 89:655–661. doi:10.1016/j.lfs.2011.08.006
32. Aycicek A, Iscan A (2006) Oxidative and antioxidative capacity in children with cerebral palsy. Brain Res Bull 69:666–668. doi:10.1016/j.brainresbull.2006.03.014
33. Schoendorfer NC, Vitetta L, Sharp N, DiGeronimo M, Wilson G, Coombes JS, Boyd R, Davies PS (2012) Micronutrient, antioxidant, and oxidative stress status in children with severe cerebral palsy. J Parenter Enteral Nutr 37:97–101. doi:10.1177/0148607112447200
34. Otsuka E, Hayashi M, Hamano K, Kumada S, Uchiyama A, Kurata K, Osawa M (2005) Pathological study of bronchospasms/tracheomalasia in patients with severe motor and intellectual disabilities. Brain Dev 27:70–72. doi:10.1016/j.braindev.2004.04.003
35. Shin EJ, Jeong JH, Chung YH, Kim WK, Ko KH, Bach JH, Hong JS, Yoneda Y, Kim HC (2011) Role of oxidative stress in epileptic seizures. Neurochem Int 59:122–137. doi:10.1016/j.neuint.2011:03.025
36. Waldbaum S, Patel M (2010) Mitochondrial dysfunction and oxidative stress: a contributing link to acquired epilepsy? J Bioenerg Biomembr 42:449–455. doi:10.1007/s10863-010-9320-9
37. Auiar CC, Almeida AB, Araujo PV, de Abreu RN, Chaves EM, do Vale OC, Macedo DS, Woods DJ, Fonteies MM, Vasconceios SM (2012) Oxidative stress and epilepsy: literature review. Oxid Med Cell Longev 2012:795259. doi:10.1155/2012/795259
38. Azam F, Prasad MV, Thangavel N (2012) Targeting oxidative stress component in the therapeutics of epilepsy. Curr Top Med Chem 12:994–1007, ISSN (Online): 1873–5294
39. Milder J, Patel M (2012) Modulation of oxidative stress and mitochondrial function by the ketogenic diet. Epilepsy Res 100:295–303. doi:10.1016/j.epilepsyres.2011.09.021
40. Hachiya Y, Hayashi M, Kumada S, Uchiyama A, Tsuchiya K, Kurata K (2006) Mechanisms of neurodegeneration in neuronal ceroid-lipofuscinosis. Acta Neuropathol 111:168–177. doi:10.1007/s00401-005-0024-x
41. Anzai Y, Hayashi M, Fueki N, Kurata K, Ohya T (2006) Protracted juvenile neuronal ceroid lipofuscinosis—an autopsy report and immunohistochemical analysis. Brain Dev 28:462–465. doi:10.1016/j.braindev.2005.12.004
42. Miyata R, Hayashi M, Tanuma N, Shioda K, Fukatsu R, Mizutani S (2008) Oxidative stress in neurodegeneration in dentatorubral-pallidoluysian atrophy. J Neurol Sci 264:133–139. doi:10.1016/j.jns.2007.08.025

Chapter 25
Oxidative Stress in Kidney Diseases

Kazunari Kaneko

Abbreviations

AKI	Acute kidney injury
CKD	Chronic kidney disease
CPB	Cardiopulmonary bypass
EMT	Epithelial-to-mesangial transition
EPO	Erythropoietin
ERK	Extracellular signal-regulated kinase
ESKD	End-stage kidney disease
GBM	Glomerular basement membrane
GFR	Glomerular filtration rate
GN	Glomerulonephritis
HD	Hemodialysis
HIF-1α	Hypoxia-induced factor-1α
HO	Heme-oxygenase
ICU	Intensive care units
LP	Lipid peroxide
L/P ratio	Lactate/pyruvate ratio
MAPK	Mitogen-activated protein kinase
MCP-1	Monocyte chemotactic protein-1
MDA	Malondialdehyde
MPO	Myeloperoxidase
NAC	*N*-acetylcysteine

K. Kaneko, M.D., Ph.D. (✉)
Department of Pediatrics, Kansai Medical University,
2-5-1 Shin-machi, Hirakata-shi, Osaka 573-1010, Japan
e-mail: kanekok@hirakata.kmu.ac.jp

H. Tsukahara and K. Kaneko (eds.), *Studies on Pediatric Disorders*, Oxidative Stress in Applied Basic Research and Clinical Practice, DOI 10.1007/978-1-4939-0679-6_25,

NF-κB	Nuclear factor-κB
OR	Odds ratio
OS	Oxidative stress
OSI	Oxidative stress index
PAN	Puromycin aminonucleoside
PICU	Pediatric intensive care units
ROS	Reactive oxygen species
rHuEPO	Recombinant human erythropoietin
TAC	Total antioxidant capacity
TGF-β	Transforming growth factor-β
TNF-α	Tumor necrosis factor-α
TPX	Total peroxide
VEGF	Vascular endothelial growth factor
ZnPP	Protoporphyrin IX zinc (II)

25.1 Introduction

Although it is accepted that reactive oxygen species (ROS) play a role in inflammation and tissue injury and there is clear evidence pointing to the potential utility of antioxidants in the treatment and prevention of kidney disease, the role of oxidative stress (OS) in the pathogenesis of acute and chronic kidney injury is still not completely understood and is the subject of much ongoing investigation.

In this chapter, several lines of experimental and clinical studies suggesting an important role of OS in the development of renal damage will be reviewed. The role of reactive nitrogen species in kidney disease, although an important subject, is not considered here due to space limitations.

25.2 Role of OS in Human Acute Kidney Injury

Acute kidney injury (AKI) is a clinical syndrome in which there is a rapid decrease in the glomerular filtration rate (GFR) and change in the homeostasis of the body. This is a significant problem in patients with critical illness and, indeed, approximately 5–6 % of all hospitalized adults suffer from varying degrees of AKI [1]. AKI is known to worsen mortality rates, increase duration of mechanical ventilation, and prolong hospital stays in critically ill adults and children [2, 3]. It is well known that AKI is associated with increased mortality and morbidity in critically ill children [4, 5]. Alkandari et al. recently reported that approximately 20 % of children admitted to pediatric intensive care units (PICU) developed AKI during admission and that AKI was associated with increased mortality (adjusted odds ratio (OR) = 3.7) [6]. AKI survivors are also at risk for progression to chronic kidney disease (CKD) [7]. The common causes of AKI are renal ischemia, nephrotoxic medications, and sepsis followed by primary renal diseases and hemolytic uremic syndrome.

Evidence has accumulated incriminating ROS as the causative agent of renal damage in AKI, whatever the etiology may be [8, 9]. Cell injury occurs during reperfusion of ischemic tissues when molecular oxygen is introduced into the tissues [10]. ROS cause lipid peroxidation of cell and organelle membranes, leading to the disruption of structural integrity and capacity for cell transport and energy production, especially in proximal tubular cells within the kidney. These free radicals are short-lived and cannot be measured directly, but their activity can be measured by estimating the by-products and substances involved in defense against the oxidant injury. Biomarkers of ROS include malondialdehyde (MDA; a by-product of lipid peroxidation), protein carbonyl, nitrite, and trace metals such as copper, while defenses against ROS can be substances such as ascorbic acid, ceruloplasmin, and zinc [10]. Lipid peroxide (LP) and certain enzymes, such as superoxide dismutase and glutathione peroxidase, are increased in AKI patients, and LP has predictive ability in determining the outcome of these patients [11]. The role of ROS in the pathogenesis of AKI has been demonstrated mainly in experimental and animal studies [12, 13], while studies in humans, and especially in children, are scarce [14, 15].

Mishra et al. studied OS status in 40 patients with AKI aged 0–10 years in comparison with 20 age- and gender-matched healthy children [16]. In that study, plasma MDA, protein carbonyl, nitrite, copper, ascorbic acid, zinc, and ceruloplasmin levels were measured. They found that the plasma MDA, copper, ascorbic acid, and ceruloplasmin levels were significantly raised in AKI patients. Furthermore, the levels of plasma MDA, nitrite, copper, and ceruloplasmin were significantly higher in AKI nonsurvivors in comparison with survivors. The cutoff levels of plasma nitrite ($\geq$3.6 μmol/L) and ceruloplasmin ($\geq$127 mg/dL) were most accurate in predicting mortality in AKI patients and had maximum sensitivity (100 %) and specificity (60.7 %) among the parameters studied. Thus, they concluded that the increased levels of oxidants and antioxidants suggest a possible role of OS in AKI pathogenesis.

25.3 Mechanisms of AKI Development Involving OS

It appears that there are several mechanisms by which OS causes AKI. The following are postulated mechanisms in the development of AKI by OS.

25.3.1 Ischemia–Reperfusion

The healthy kidney generates physiologically moderate amount of ROS in the course of renal oxidative metabolism. The relatively low amount of ROS generated by the kidneys is tolerated without any apparent adverse effects. In contrast, in renal ischemia, which is the most common cause of AKI in children, there is excessive production of ROS during reperfusion of the ischemic kidneys, and these ROS engender further renal injury by lipid peroxidation [13].

As it has long been known that ischemia–reperfusion of the myocardium also leads to a tremendous generation of ROS and cellular injury [17], the setting of elective surgeries such as cardiopulmonary bypass (CPB), in which the renal ischemia–reperfusion insult can be quantified prospectively, has been studied extensively. The incidence of AKI is high in patients undergoing cardiac surgery, reaching 50 % by some definitions [18]. Another study revealed that AKI after CPB is common and is associated with increases in morbidity, length of stay in the ICU and hospital, and mortality [19]. Based on these findings, ischemia–reperfusion is accepted as one of the important players in this type of renal injury, while it is clear that the pathogenesis of CPB-related AKI is multifactorial.

25.3.2 Macroscopic Glomerular Hematuria

Macroscopic hematuria is a common finding in various glomerular diseases, such as IgA nephropathy, Alport syndrome, and thin basement membrane disease [20], which do not necessarily show poor prognosis in terms of renal function. Although glomerular hematuria has been considered a clinical manifestation of glomerular diseases without real consequences on renal function and long-term prognosis, many of the studies performed have shown a relationship between macroscopic glomerular hematuria and AKI and have suggested that macroscopic hematuria-associated AKI is related to adverse long-term outcomes: up to 25 % of patients with macroscopic hematuria-associated AKI do not recover baseline renal function [21]. In cases with macroscopic hematuria-associated AKI, it has been speculated that pathophysiologic mechanisms account for the tubular injury found on renal biopsy specimens, i.e., mechanical obstruction by red blood cell casts was thought to play a role [22]. However, recent evidence points to cytotoxic effects of OS induced by hemoglobin, heme, or iron released from red blood cells, as shown in Fig. 25.1 [20]: the kidney can be damaged by large amount of heme resulting both from extrarenal heme-containing proteins (myoglobin in rhabdomyolysis and hemoglobin in hemolysis) [23, 24] and from renal heme-proteins, as occurs after ischemic or toxic insults [25, 26]. Intratubular, cell-free hemoglobin induces severe oxidative damage as a consequence of heme redox cycling between ferric and ferryl states, which generates radical species and promotes lipid peroxidation [27, 28]. Lipid peroxidation is responsible for the intense vasoconstriction and oxidative injuries observed in disorders associated with renal accumulation of hemoproteins [29]. Heme permeates plasma and organellar membranes, thereby entering the cells and facilitating cytotoxicity [30, 31]. Additional mechanisms of heme toxicity include impairment of the activity of certain glycolytic enzymes, such as glucose-6-phosphate dehydrogenase and glutathione reductase; oxidative DNA denaturation; and mitochondrial toxicity [32–35]. In addition to its direct cytotoxicity, heme can also indirectly promote

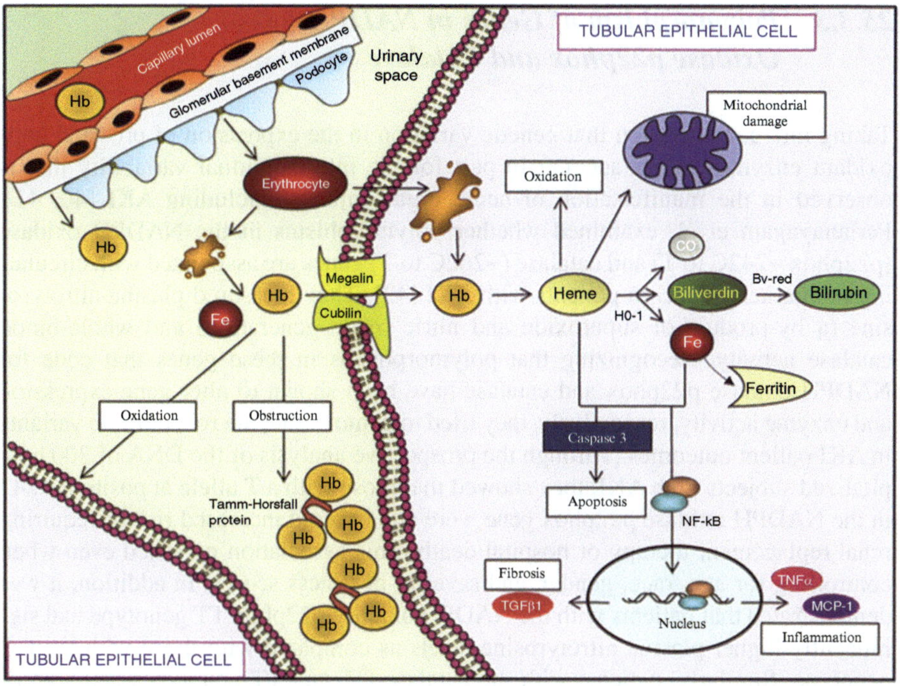

Fig. 25.1 Pathophysiologic pathways of hematuria-induced kidney damage. Hemoglobin (Hb) released by intratubular degradation of red blood cells or hemoglobin directly filtered by the glomerulus may be incorporated into proximal tubules through the megalin–cubilin receptor system or degraded in the tubular lumen, releasing heme-containing molecules and eventually free iron. Cell-free hemoglobin promotes lipid peroxidation and physical obstruction of the renal tubules by hemoglobin precipitation in association with Tamm–Horsfall protein under acidic conditions, which leads to intraluminal casts, increased intratubular pressure, and subsequent decreased GFR. Hemoglobin/heme/iron (Fe) accumulation within tubular cells generates reactive oxygen species, mitochondrial damage, caspase activation and apoptosis, upregulation of vascular adhesion molecules, and pro-inflammatory/profibrotic cytokines through activation of NF-κB transcription factor. Reproduced from [20] with permission from © 2012 by the American Society of Nephrology

chronic renal damage by inducing inflammation and fibrosis [36, 37]. Exposure to heme-proteins increases the renal expression of tumor necrosis factor (TNF)-β, monocyte chemotactic protein (MCP)-1, and transforming growth factor (TGF)-β via nuclear factor (NF)-κB transcription factor [23, 24]. The activation of these cytokine cascades serves as a "positive-feedback loop" that perpetuates renal damage beyond the initial injury phase [38] and contributes to a chronic inflammatory response, as occurs in recurrent hemolytic episodes [26]. These mechanisms of injury may be shared with hemoglobinuria or myoglobinuria-induced AKI.

Heme-oxygenase (HO) catalyzes the conversion of heme to biliverdin and is protective in animal models of heme toxicity. CD163, the recently identified scavenger receptor on the surface of the tissue macrophages, promotes the activation of anti-inflammatory pathways, opening the gates for novel therapeutic approaches [39].

25.3.3 Polymorphism of Genes in NADPH Oxidase p22phox and Catalase

Taking into consideration that genetic variation in the expression of pro- and antioxidant enzymes might account in part for the interindividual variability that is observed in the manifestation of acute organ injuries, including AKI [40, 41], Perianayagam et al. examined whether polymorphisms in the NADPH oxidase (p22phox +242C to T) and catalase (−262C to T) genes are associated with circulating biomarkers of OS in patients with AKI [42]. They measured plasma nitrotyrosine (a by-product of superoxide and nitric oxide generation) and whole-blood catalase activity. Recognizing that polymorphisms in these genes that code for NADPH oxidase p22phox and catalase have been shown to alter gene expression and enzyme activity, respectively, they tried to demonstrate the role of these variants in AKI patient outcomes. Through the prospective analysis of the DNA of 200 hospitalized subjects with AKI, they showed that those with a T allele at position +242 in the NADPH oxidase p22phox gene were at a twofold increased risk of requiring renal replacement therapy or hospital death. This association persisted even when controlling for age, race, gender, and severity of illness scores. In addition, it was demonstrated that patients with the NADPH oxidase p22phox TT genotype had significantly higher plasma nitrotyrosine levels as compared with those with the CC genotype. Similarly, patients with the catalase CT and TT genotypes had significantly lower whole-blood catalase activity as compared with those with the CC genotype. These results further lend support to the hypothesis that OS is involved in the pathogenesis of AKI.

25.3.4 Decreased Antioxidant Levels

Exercise-induced AKI in patients with idiopathic renal hypouricemia was first reported by Erley et al. in 1989 [43]. Since then, more than 50 cases have been reported, mostly from Japan [44]. It is speculated that patients with idiopathic renal hypouricemia have a 200-fold greater predisposition to exercise-induced AKI than those without renal hypouricemia [45]. Although the mechanism of exercise-induced AKI in renal hypouricemia is unclear, Murakami et al. emphasized the role of uric acid as an antioxidant [46]: renal handling of uric acid seems to be a protective mechanism against renal injury by free radicals. As both the uric acid pool and the total amount of uric acid mobilized into the proximal tubular cells are very small in patients with renal hypouricemia, renal perfusion is diminished during exercise, and reperfusion after exercise mimics ischemia–reperfusion renal injury.

Idiopathic renal hypouricemia is a rare disorder with the highest incidence of 0.15 % in Japan [47]. These patients have increased uric acid excretion due to an isolated defect in the renal tubular transport of uric acid arising from defects in a gene (*SLC22A12*) encoding the urate transporter 1 [48, 49]. As mentioned above, an important complication of this disorder is exercise-induced AKI [44, 45]. The mechanism of AKI in this setting still remains unknown, though it is speculated that OS is involved [46, 50].

We first demonstrated the oxidative imbalance by a concomitant assessment of ROS production and antioxidant system capability in a 15-year-old girl with idiopathic renal hypouricemia caused by a mutation in the urate transporter 1 gene [51]. Her serum level of ROS increased with decreasing antioxidant potential capacity soon after the initiation of anaerobic stress due to treadmill exercise. Thereafter, the serum levels of ROS and antioxidant potential showed a parallel course, returning to the baseline values at 240 min after exercise. Therefore, it appears that some patients with idiopathic renal hypouricemia demonstrate oxidative imbalance soon after exercise with a predisposition to exercise-induced acute renal failure.

25.4 AKI Therapies Involving OS Pathway Modification

While much of the OS pathway has been well documented and its role in cell injury is becoming increasingly evident, further studies will be needed to improve the prophylactic and therapeutic interventions in the setting of AKI. In this section, the clinical and experimental trials of therapies involving alterations in the OS pathway are presented.

25.4.1 Ameliorating OS in Cardiac Surgeries

Haase et al. conducted a randomized multi-blind placebo-controlled trial of high-dose *N*-acetylcysteine (NAC) in high-risk patients undergoing CPB [52]. NAC can directly scavenge ROS and therefore is expected to reduce OS during CPB; it also regenerates the glutathione pool. In this trial, 61 subjects were randomized to administration of either 24 h of high-dose NAC (300 mg/kg body weight in 5 % glucose, 1.7 L) or placebo (5 % glucose, 1.7 L), with the primary end point being absolute change in serum creatinine within the first 5 postoperative days. The study failed to demonstrate a difference in the serum creatinine levels in this early postoperative period. Nouri-Majalan et al. performed a randomized trial investigating the role of supplemental antioxidant vitamin E and the inhibitor of xanthine oxidase allopurinol in reducing ischemia–reperfusion injury after coronary artery bypass graft surgery: patients were randomized to receive 100 units vitamin E four times per day and 100 mg allopurinol twice daily for 3–5 days prior to elective surgery, or to no treatment [53]. Their results showed that prophylactic treatment with vitamin E and allopurinol had no renoprotective effects in patients with preexisting renal failure undergoing surgery. Treatment with these agents, however, reduced the duration of the ICU stay.

In summary, although several studies have been conducted, no antioxidant agent has been shown to be of benefit for the prevention or treatment of AKI in patients undergoing cardiac surgeries [54].

25.4.2 Ameliorating OS in Drug-Induced AKI

Cisplatin is an important antineoplastic agent for the treatment of solid tumors, but its clinical use is limited because of its dose-dependent renal toxicity. Cisplatin nephrotoxicity involves OS, apoptosis, inflammation, and fibrogenesis. Ajith et al. conducted a comparative study of the effects of vitamin C and vitamin E on cisplatin-induced nephrotoxicity in mice [55]. Both vitamins have well-known antioxidant properties, and the experiment was designed to expose the animals to high and low doses of both vitamins C and E (250 and 500 mg/kg of each). The vitamins were administered 1, 24, and 48 h after cisplatin injection. Higher doses of both vitamins were effective in protecting against oxidative renal damage, as measured by increasing superoxide dismutase and reduced glutathione activity.

Several other agents have also been proposed as providing effective protection against cisplatin nephrotoxicity. In a rat model, the iron chelator deferoxamine has been shown to provide functional and histological protection [56]. Satoh et al. demonstrated that edaravone, a free radical scavenger, attenuated the cisplatin-induced mitochondrial membrane potential loss, decreased the concentrations of carbonylated proteins, and exhibited cytoprotective properties in murine proximal tubular cells [57].

Much as in the case of cisplatin nephrotoxicity, gentamicin-induced renal failure has been shown to respond to prophylactic and therapeutic measures in experimental animal studies. In vivo and in vitro, gentamicin enhances the generation of ROS by altering mitochondria respiration, leading to the generation of H_2O_2 [58]. Additionally, gentamicin enhances the generation of superoxide anion and hydroxyl radical by renal cortical mitochondria. Furthermore, gentamicin induces the release of iron from these renal cortical mitochondria; this causes lipid peroxidation in vitro, with iron serving as a potent catalyst for free radical formation. Kopple et al. investigated the role of L-carnitine, an antioxidant that prevents the accumulation of end products of lipid peroxidation, in the setting of a rat model of gentamicin-induced AKI and demonstrated that L-carnitine improved renal function and ameliorated the severity of renal pathologic findings in a dose-dependent manner [59].

Despite this success in experimental models and the likely tolerability of the treatments in humans, neither of these interventions has been translated into a clinical practice.

25.4.3 Ameliorating OS in Ischemia–Reperfusion-Induced AKI by Heme-Oxygenase Induction

While the strategy has not been tested in humans to date, there are compelling data from experimental animal models regarding the role of heme-oxygenase (HO) induction in preventing AKI by decreasing OS [60]. HO is an enzyme that is capable of converting heme into carbon monoxide, iron, and biliverdin, which is in turn reduced to bilirubin by biliverdin reductase (Fig. 25.1) [20]. This degradation of heme into carbon monoxide and bilirubin is considered to be protective against

OS-related tissue injury as both the reaction products are antioxidants. HO consists of two isoforms, HO-1 (inducible) and HO-2 (constitutive): HO-1 is induced in various cells by diverse stimuli that provoke OS, including heme, heat shock, pro-inflammatory cytokines, and toxins. Given these facts, HO-1 has been examined for its potential to reduce the impact of ischemia–reperfusion-induced AKI [61]. Wei et al. employed a mouse model of rhabdomyolysis, another important clinical cause of AKI due to the release of nephrotoxins (e.g., heme) from disrupted muscles, and demonstrated that the induction of HO-1 by granulocyte colony-stimulating factor led to a reduction in AKI, and these findings were reversed in the presence of an HO-1 inhibitor, protoporphyrin IX zinc (II) (ZnPP) [62].

25.5 Role of OS in CKD

CKD is a worldwide public health problem that affects approximately 10 % of the US adult population [63] and is associated with a high prevalence of cardiovascular disease [64] and high economic cost [65]. CKD, once established, tends to progress to end-stage kidney disease (ESKD). Given the notion that oxidants are one of the important players in inflammation, a strong association of ROS with CKD is easily conceivable. While evidence exists for the presence of increased OS in CKD in adults [66–69], such evidence is scarce in children.

Recently, Hamed et al. measured plasma hypoxia-induced factor-1α (HIF-1α), vascular endothelial growth factor (VEGF), total antioxidant capacity (TAC), total peroxide (TPX), pyruvate, and lactate in 40 pediatric patients with CKD on hemodialysis (HD) and 20 healthy children [70]. They demonstrated that TAC was significantly lower while TPX, the oxidative stress index (OSI) defined as the percent ratio of TPX to the TAC level, and VEGF were higher in patients both before and after a dialysis session than in the controls. Meanwhile, before dialysis, lactate/pyruvate (L/P) ratio was lower than after dialysis. In the data collected before the dialysis session, VEGF was positively correlated with pyruvate and HIF-1α, while OSI was positively correlated with TPX, but negatively correlated with TAC. After the dialysis session, HIF-1α was negatively correlated with TPX and OSI. From these findings, the authors postulated that CKD patients experience considerable tissue hypoxia in response to OS. HD ameliorated hypoxia but lowered antioxidant levels, as evidenced by the finding that the levels of HIF-1α and TAC were lower before than after dialysis.

25.6 Mechanisms of OS-Related Renal Damage in the Progression of CKD

A sufficient body of in vitro and in vivo information exists to postulate that oxidants are important mediators in progressive kidney disease. Although there is little information on the mechanisms of OS-related renal damage in CKD in humans, we discuss a possible mechanism in the following section.

25.6.1 Renal Damage Due to ROS Produced by Underlying Conditions for CKD

One reason for OS in patients with CKD is the underlying disease itself [70]. That is, renal toxicity and immunological disorders causing kidney diseases result in an elevated formation of ROS which further deteriorate the renal function. In addition, treatment procedures for patients with CKD have also been shown to induce OS. During HD, for example, incomplete correction of the uremic toxicity together with the untoward effects of dialysis and malnutrition and the progressive worsening of the clinical condition can lead to OS. The bioincompatibility of dialysis membranes represents an important source of ROS. Losses of antioxidants via dialysis are the factors that may be responsible for the imbalance between prooxidative and antioxidative mechanisms in HD patients [71]. Furthermore, TAC in patients with renal failure was shown to be greatly diminished due to antioxidant exhaustion and inhibition [72]. The activation of neutrophils and the complement pathway during an HD session as the result of interactions of the blood with the dialysis membrane and endotoxin-contaminated dialysate, iron overload, the presence of advanced glycation end products, high homocysteine levels, intradialytic cytokine activation, or other causes could also play a role [73].

25.6.2 Excessive Protein Trafficking Through the Glomerulus

It is well known that the severity of renal tubulointerstitial injury is a major determinant of the degree and rate of progression of CKD. There has been increasing interest in the possible link between excessive protein trafficking through the glomerulus and progressive tubulointerstitial inflammation that leads to CKD, as shown in Fig. 25.2 [74]. A candidate key molecule for chemokines induced by enhanced protein uptake by renal tubular cells is NF-κB: it has been shown that albumin is a strong stimulus for H_2O_2 production, which leads to activation of NF-κB-dependent pathways, resulting in increased expression of MCP-1 and interleukin-8, which are important in the inflammatory response [75, 76]. Morigi et al. have shown that human proximal tubular cells incubated with human albumin and IgG lead to a significant and rapid increase in H_2O_2 and activation of NF-κB. Furthermore, inhibitors of protein kinase C significantly prevented H_2O_2 production and consequent NF-κB activation [75].

Tubular epithelial-to-mesangial transition (EMT) is a process in which renal tubular cells lose their epithelial phenotype and acquire new characteristic features of mesenchyme. There is growing evidence to implicate this process as a major pathway that leads to generation of interstitial myofibroblasts in the diseased kidney [77]. It has been reported that ROS mediate TGF-β1-induced EMT in renal tubular epithelial cells directly through the activation of mitogen-activated protein kinase (MAPK) and indirectly through extracellular signal-regulated kinase

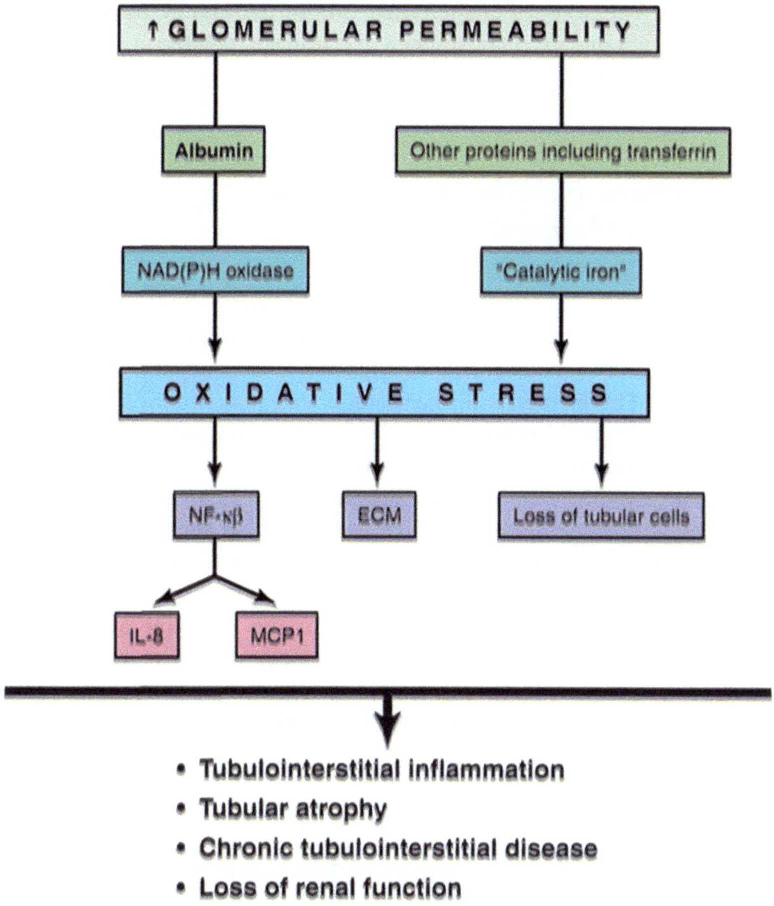

Fig. 25.2 Relation of proteinuria to oxidant stress and tubulointerstitial injury. Reproduced from [74] with permission from © 2007 by the American Society of Nephrology

(ERK)-directed Smad 2 phosphorylation, and it has been suggested that antioxidants and MAPK inhibitors may prevent EMT through these pathways and ameliorate subsequent tubulointerstitial fibrosis [78].

Another pathway to OS is speculated to be related to iron (Fig. 25.2). The data support the role of iron in models of progressive renal disease consisting of demonstration of increased iron in the kidney; enhanced oxidant generation, which provides a mechanism by which iron can be mobilized; and the beneficial effect of iron-deficient diets and iron chelators. Rats with proteinuria have increased iron content in proximal tubular cells, and iron accumulation was the only independent predictor of both functional and structural damage in this model [79]. Similarly, it has been shown that there is a substantial iron accumulation associated with increased cortical MDA in proximal tubular cells in the remnant kidney, suggesting

ROS generation. The sources of increased iron in the kidney have not been well delineated, although one group postulated that urinary transferrin provides a potential source of iron [80, 81].

25.6.3 Evidence of the Involvement of OS in Renal Injury in Animal Models

Diverse pathogenetic roles of oxidants in progressive kidney disease have been demonstrated by experimental studies [74]. Experimental glomerulonephritis (GN) can be classified into two categories, i.e., leukocyte-dependent GN and leukocyte-independent GN. The former is further categorized into two well-studied models, the anti-glomerular basement membrane (anti-GBM) antibody model and the anti-Thy 1.1 model mimicking mesangial proliferative GN, while the latter is further categorized into a puromycin aminonucleoside (PAN)-induced nephrosis model, a model of minimal-change disease, and passive Heymann nephritis induced by anti-tubular brush-border antiserum (anti-Fx1A), a model of membranous nephropathy. Although leukocytes have not been considered to be important in animal models of membranous nephropathy, there is evidence for the potential participation of a myeloperoxidase (MPO)–H_2O_2–halide system in membranous nephropathy [82, 83]. Thus, it appears that leukocytes or resident glomerular cells serve as sources for oxidants in this model. The reported roles of ROS in these experimental GN are summarized in Tables 25.1 and 25.2 [74].

25.7 Therapeutic Interventions of CKD by Modulating OS

Recently, a meta-analysis on the efficacy of antioxidant therapy in patients with CKD was reported [84]. This study included ten randomized controlled trials consisting of 1979 participants, and assessed the beneficial effects of antioxidant therapy on cardiovascular events and development of ESKD. The results showed that antioxidant therapy with vitamin E, multiple antioxidant therapy, or treatment with coenzyme Q, acetylcysteine, bardoxolone methyl, or human recombinant superoxide dismutase showed no clear overall effect on cardiovascular events such as mortality. However, some of the treatments were of significant benefit for reducing the development of ESKD. The authors concluded that antioxidant therapy in predialysis CKD patients may prevent progression to ESKD, although it does not reduce the risk of cardiovascular events. An appropriately powered study with a longer follow-up is awaited.

A novel agent that gives us hope is erythropoietin (EPO). EPO is the principal hematopoietic hormone produced by the kidney and the liver, and has been shown to regulate mammalian erythropoiesis and to exhibit diverse cellular effects in

Table 25.1 Experimental evidence for the role of oxidants in leukocyte-dependent glomerulonephritis

Leukocytes as a source of oxidants for glomerular injury
• A wide variety of soluble and particulate stimuli, including immune complexes, complement components [90], and ANCA[a] [91]
• Anti-GBM[b] enhances generation of oxidants by neutrophils in vitro
• Cytochemical detection of the presence of superoxide- and H_2O_2-generating leukocytes in anti-Thy 1.1 and anti-GBM-induced glomerulonephritis [92]
• Enhanced superoxide and hydroxyl radical are generated by macrophages that are isolated from glomeruli of rabbits with anti-GBM antibody disease [93]
• Enhanced superoxide generation by macrophages that are isolated from nephritic glomeruli (anti-thymocyte serum) [94]
Effects of oxidants that are relevant to occurrence of proteinuria in glomerular injury
• Oxidants participate in GBM[b] degradation [95]
• Infusion of myeloperoxidase-H_2O_2 induces proteinuria [96]
Evidence for the role of oxidants in animal models
• Catalase markedly reduces proteinuria, whereas superoxide dismutase has no protective effect in anti-GBM[b] antibody disease [97]
• A hydroxyl radical scavenger and an iron chelator significantly attenuate anti-GBM antibody-induced proteinuria in anti-GBM[b] antibody disease [98]
• Alpha-lipoic acid is protective in an anti-Thy 1.1 model [99]
• Liver-type fatty acid-binding protein protects mice with anti-GBM glomerulonephritis from progression of both tubulointerstitial and glomerular injury by acting as an antioxidant [100]

Reproduced with slight modification from [74] with permission from © 2007 by the American Society of Nephrology

[a]ANCA, antineutrophil cytoplasmic antibody

[b]GBM, glomerular basement membrane

non-hematopoietic tissues [85]. EPO significantly protects multiple organs in both acute and chronic diseases [85, 86]. Patel et al. reported that EPO was able to significantly attenuate the renal dysfunction and injury associated with ischemia/reperfusion in mice [87]: the increase in renal MPO activity (as a surrogate for polymorphonuclear leukocyte infiltration) and tissue MDA levels (as a surrogate for tissue lipid peroxidation) were also significantly reduced in EPO-treated mice. Bahlmann et al. demonstrated that chronic treatment with the long-acting recombinant human EPO (rHuEPO) analogue darbepoetin alpha conferred renal vascular and tissue protection and preserved renal function in the established 5/6 nephrectomy remnant kidney model in the rat, a model that features progressive injury leading to glomerular sclerosis and ischemia-induced tubulointerstitial damage [88]. Treatment with rHuEPO not only reduced renal dysfunction but also significantly improved the survival of uremic rats. In this experimental setting, the authors observed persistent activation of the Akt pathway in endothelial and epithelial glomerular cells, and reduced apoptotic cell death in renal tissue. Importantly, they used a hematologically noneffective dose of darbepoetin which did not affect hematocrit levels in treated animals. This

Table 25.2 Experimental evidence for the role of oxidants in leukocyte-independent glomerulonephritis

In a PAN[a] model of minimal-change disease
• Cultured glomerular epithelial cells exhibit an enhanced generation of H_2O_2 [101]
• Administration of scavengers of oxidants and antioxidants results in reduction in proteinuria [102–105]
• A novel free radical scavenger, edaravone, delays and ameliorates the urinary protein excretion in rats [106, 107]
• Glomerular catalytic iron is increased [108], and cytochrome P450 is an important source of the catalytic iron [109–111]
• Feeding with a selenium-deficient diet results in a marked diminution of glutathione peroxidase accompanied by an increase in proteinuria [112]
• Inhibition of superoxide dismutase by diethyldithiocarbamate results in increase in PAN[a]-induced proteinuria [113]
• Induction of antioxidant enzymes by ischemia–reperfusion injury protects against H_2O_2-induced proteinuria [114]
• Induction of antioxidant enzymes by glucocorticoids protects against PAN[a]-induced proteinuria [115]
• Apocynin, an inhibitor of NADPH oxidase, decreases superoxide production in podocytes, and inhibits endocytosis and urinary albumin excretion [116]
Evidence for the role of oxidants in passive Heymann nephritis
• There is an increased generation of H_2O_2 in passive Heymann nephritis [117]
• In a passive Heymann nephritis, a model of membranous nephropathy, hydroxyl radical scavengers and an iron chelator and probucol significantly reduce proteinuria [118, 119]
• Feeding with an iron-deficient diet results in a reduction in proteinuria [120]
• Feeding with a selenium-deficient diet results in marked diminution of glutathione peroxidase in anti-Fx1A-induced proteinuria [112]

[a]PAN, puromycin aminonucleoside (Reproduced with slight modification from [74] with permission from © 2007 by the American Society of Nephrology)

could be of considerable clinical relevance, since "low-dose" rHuEPO treatment may be a safe strategy to avoid potential adverse effects of "high-dose" rHuEPO with a risk of an increase in hematocrit with concomitant hyper-viscosity and activation of thrombocytes [88].

Fujiwara et al. recently observed that 15 CKD patients with anemia treated with rHuEPO (12,000 U administered subcutaneously once every 2 weeks) showed increased serum hemoglobin levels and decreased urinary levels of protein, liver-type fatty acid-binding protein (a biomarker of renal injury), and 8-hydroxydeoxyguanosine (a biomarker of OS) after 6 months, while serum creatinine and estimated GFR showed little difference throughout the experimental period [89]. The authors concluded that rHuEPO may ameliorate renal injury, OS, and progression of atherosclerosis in addition to improving anemia in CKD patients. It is therefore conceivable that therapy with rHuEPO may join the arsenal of approaches directed against progressive CKD.

Acknowledgments This study was partly funded by grants from the Mami Mizutani Foundation and the Human Resource Chair for Neonatologists funded by Izumi-Ohtsu City, Osaka, Japan, and by the Chair for Children with Developmental Disorders funded by Nabari City, Mie, Japan.

References

1. Uchino S, Kellum JA, Bellomo R, Doig GS, Morimatsu H, Morgera S, Schetz M, Tan I, Bouman C, Macedo E, Gibney N, Tolwani A, Ronco C, Beginning, Ending Supportive Therapy for the Kidney (BEST Kidney) Investigators (2005) Acute renal failure in critically ill patients: a multinational, multicenter study. JAMA 294:813–818
2. Hoste EAJ, Schurgers M (2008) Epidemiology of acute kidney injury: how big is the problem? Crit Care Med 36:S146–S151
3. Basu RK, Devarajan P, Wong H, Wheeler DS (2011) An update and review of acute kidney injury in pediatrics. Pediatr Crit Care Med 12:339–347
4. Merouani A, Flechelles O, Jouvet P (2012) Acute kidney injury in children. Minerva Pediatr 64:121–133
5. Kumar G, Vasudevan A (2012) Management of acute kidney injury. Indian J Pediatr 79:1069–1075
6. Alkandari O, Eddington KA, Hyder A, Gauvin F, Ducruet T, Gottesman R, Phan V, Zappitelli M (2011) Acute kidney injury is an independent risk factor for pediatric intensive care unit mortality, longer length of stay and prolonged mechanical ventilation in critically ill children: a two-center retrospective cohort study. Crit Care 15:R146
7. Chawla LS, Amdur RL, Amodeo S, Kimmel PL, Palant CE (2011) The severity of acute kidney injury predicts progression to chronic kidney disease. Kidney Int 79:1361–1369
8. Baud L, Ardaillou R (1993) Involvement of reactive oxygen species in kidney damage. Br Med Bull 49:621–629
9. Cheeseman KH, Slater TF (1993) An introduction to free radical biochemistry. Br Med Bull 49:481–493
10. Greene E, Paller M (1991) Oxygen free radicals in acute renal failure. Miner Electrolyte Metab 17:124–132
11. Dubey N, Yadav P, Dutta A, Kumar V, Ray G, Batra S (2000) Free oxygen radicals in acute renal failure. Indian Pediatr 37:153–158
12. Nath KA, Norby SM (2000) Reactive oxygen species and acute renal failure. Am J Med 109:665–678
13. Paller MS, Hoidal J, Ferris TF (1984) Oxygen free radicals in ischemic acute renal failure in the rat. J Clin Invest 74:1156–1164
14. Devasena T, Lalitha S, Padma K (2001) Lipid peroxidation, osmotic fragility and antioxidant status in children with acute post-streptococcal glomerulonephritis. Clin Chim Acta 308:155–161
15. Ferraris V, Acquier A, Ferraris JR, Vallejo G, Paz C, Mendez CF (2011) Oxidative stress status during the acute phase of haemolytic uraemic syndrome. Nephrol Dial Transplant 26:858–864
16. Mishra OP, Pooniya V, Ali Z, Upadhyay RS, Prasad R (2008) Antioxidant status of children with acute renal failure. Pediatr Nephrol 23:2047–2051
17. Hearse DJ, Humphrey SM, Bullock GR (1978) The oxygen paradox and the calcium paradox: two facets of the same problem? J Mol Cell Cardiol 10:641–668
18. Dasta JF, Kane-Gill SL, Durtschi AJ, Pathak DS, Kellum JA (2008) Costs and outcomes of acute kidney injury (AKI) following cardiac surgery. Nephrol Dial Transplant 23:1970–1974
19. Chertow GM, Levy EM, Hammermeister KE, Grover F, Daley J (1998) Independent association between acute renal failure and mortality following cardiac surgery. Am J Med 104:343–348
20. Moreno JA, Martín-Cleary C, Gutiérrez E, Toldos O, Blanco-Colio LM, Praga M, Ortiz A, Egido J (2012) AKI associated with macroscopic glomerular hematuria: clinical and pathophysiologic consequences. Clin J Am Soc Nephrol 7:175–184
21. Gutiérrez E, González E, Hernández E, Morales E, Martínez MÁ, Usera G, Praga M (2007) Factors that determine an incomplete recovery of renal function in macrohematuria-induced acute renal failure of IgA nephropathy. Clin J Am Soc Nephrol 2:51–57

22. Heyman S, Brezis M (1995) Acute renal failure in glomerular bleeding: a puzzling phenomenon. Nephrol Dial Transplant 10:591–593
23. Nath KA, Vercellotti GM, Grande JP, Miyoshi H, Paya CV, Manivel JC, Haggard JJ, Croatt AJ, Payne WD, Alam J (2001) Heme protein-induced chronic renal inflammation: suppressive effect of induced heme oxygenase-1. Kidney Int 59:106–117
24. Nath KA, Croatt AJ, Haggard JJ, Grande JP (2000) Renal response to repetitive exposure to heme proteins: chronic injury induced by an acute insult. Kidney Int 57:2423–2433
25. Hill-Kapturczak N, Chang SH, Agarwal A (2002) Heme oxygenase and the kidney. DNA Cell Biol 21:307–321
26. Zager RA (1996) Rhabdomyolysis and myohemoglobinuric acute renal failure. Kidney Int 49:314–326
27. Patel RP, Svistunenko DA, Darley-usmar VM, Symons MCR, Wilson MT (1996) Redox cycling of human methaemoglobin by H_2O_2 yields persistent ferryl iron and protein based radicals. Free Radic Res 25:117–123
28. Hogg N, Rice-Evans C, Darley-Usmar V, Wilson MT, Paganga G, Bourne L (1994) The role of lipid hydroperoxides in the myoglobin-dependent oxidation of LDL. Arch Biochem Biophys 314:39–44
29. Moore KP, Holt SG, Patel RP, Svistunenko DA, Zackert W, Goodier D, Reeder BJ, Clozel M, Anand R, Cooper CE (1998) A causative role for redox cycling of myoglobin and its inhibition by alkalinization in the pathogenesis and treatment of rhabdomyolysis-induced renal failure. J Biol Chem 273:31731–31737
30. Hebbel RP, Eaton JW (1989) Pathobiology of heme interaction with the erythrocyte membrane. Semin Hematol 26:136–149
31. Muller-Eberhard U, Fraig M (1993) Bioactivity of heme and its containment. Am J Hematol 42:59–62
32. Gonzalez-Michaca L, Farrugia G, Croatt AJ, Alam J, Nath KA (2004) Heme: a determinant of life and death in renal tubular epithelial cells. Am J Physiol Renal Physiol 286:F370–F377
33. Zager RA, Johnson AC, Hanson SY (2004) Proximal tubular cytochrome *c* efflux: determinant, and potential marker, of mitochondrial injury. Kidney Int 65:2123–2134
34. Nath KA, Balla J, Croatt AJ, Vercellotti GM (1995) Heme protein-mediated renal injury: a protective role for 21-aminosteroids in vitro and in vivo. Kidney Int 47:592–602
35. Huerta-Alardin AL, Varon J, Marik PE (2005) Bench-to-bedside review: rhabdomyolysis—an overview for clinicians. Crit Care 9:158–169
36. Tracz MJ, Alam J, Nath KA (2007) Physiology and pathophysiology of heme: implications for kidney disease. J Am Soc Nephrol 18:414–420
37. Tracz MJ, Juncos JP, Croatt AJ, Ackerman AW, Grande JP, Knutson KL, Kane GC, Terzic A, Griffin MD, Nath KA (2007) Deficiency of heme oxygenase-1 impairs renal hemodynamics and exaggerates systemic inflammatory responses to renal ischemia. Kidney Int 72:1073–1080
38. Shulman LM, Yuhas Y, Frolkis I, Gavendo S, Knecht A, Eliahou HE (1993) Glycerol induced ARF in rats is mediated by tumor necrosis factor-alpha. Kidney Int 43:1397–1401
39. Kristiansen M, Graversen JH, Jacobsen C, Sonne O, Hoffman HJ, Law SK, Moestrup SK (2001) Identification of the haemoglobin scavenger receptor. Nature 409:198–201
40. Jaber BL, Pereira BJG, Bonventre JV, Balakrishnan VS (2005) Polymorphism of host response genes: implications in the pathogenesis and treatment of acute renal failure. Kidney Int 67:14–33
41. Forsberg L, de Faire U, Morgenstern R (2001) Oxidative stress, human genetic variation, and disease. Arch Biochem Biophys 389:84–93
42. Perianayagam MC, Liangos O, Kolyada AY, Wald R, MacKinnon RW, Li L, Rao M, Balakrishnan VS, Bonventre JV, Pereira BJ, Jaber BL (2007) NADPH oxidase p22phox and catalase gene variants are associated with biomarkers of oxidative stress and adverse outcomes in acute renal failure. J Am Soc Nephrol 18:255–263
43. Erley CMM, Hirschberg R, Hoefer W, Schaefer K (1989) Acute renal failure due to uric acid nephropathy in a patient with renal hypouricemia. Klin Wochenschr 67:308–312

44. Ohta T, Sakano T, Igarashi T, Itami N, Ogawa T, ARF associated with renal Hypouricemia Research Group (2004) Exercise-induced acute renal failure associated with renal hypouricaemia: results of a questionnaire-based survey in Japan. Nephrol Dial Transplant 19:1447–1453
45. Ito O, Hasegawa Y, Sato K, Mitsui H, Yuda F, Sato H, Ito S, Kudo K (2003) A case of exercise-induced acute renal failure in a patient with idiopathic renal hypouricemia developed during antihypertensive therapy with losartan and trichlormethiazide. Hypertens Res 26:509–513
46. Murakami T, Kawakami H, Fukuda M, Furukawa S (1995) Patients with renal hypouricemia are prone to develop acute renal failure: why? Clin Nephrol 43:207–208
47. Hisatome I, Ogino K, Kotake H, Ishiko R, Saito M, Hasegawa J, Mashiba H, Nakamoto S (1989) Cause of persistent hypouricemia in outpatients. Nephron 51:13–16
48. Sperling O (1992) Renal hypouricemia: classification, tubular defect and clinical consequences. Contrib Nephrol 100:1–14
49. Enomoto A, Kimura H, Chairoungdua A, Shigeta Y, Jutabha P, Cha SH, Hosoyamada M, Takeda M, Sekine T, Igarashi T, Matsuo H, Kikuchi Y, Oda T, Ichida K, Hosoya T, Shimokata K, Niwa T, Kanai Y, Endou H (2002) Molecular identification of a renal urate–anion exchanger that regulates blood urate levels. Nature 417:447–452
50. Ames BN, Cathcart R, Schwiers E, Hochstein P (1981) Uric acid provides an antioxidant defense in humans against oxidant-and radical-caused aging and cancer: a hypothesis. Proc Natl Acad Sci U S A 78:6858–6862
51. Kaneko K, Taniguchi N, Tanabe Y, Nakano T, Hasui M, Nozu K (2009) Oxidative imbalance in idiopathic renal hypouricemia. Pediatr Nephrol 24:869–871
52. Haase M, Haase-Fielitz A, Bagshaw SM, Reade MC, Morgera S, Seevenayagam S, Matalanis G, Buxton B, Doolan L, Bellomo R (2007) Phase II, randomized, controlled trial of high-dose N-acetylcysteine in high-risk cardiac surgery patients. Crit Care Med 35:1324–1331
53. Nouri-Majalan N, Ardakani EF, Forouzannia K, Moshtaghian H (2009) Effects of allopurinol and vitamin E on renal function in patients with cardiac coronary artery bypass grafts. Vasc Health Risk Manag 5:489–494
54. Park M, Coca SG, Nigwekar SU, Garg AX, Garwood S, Parikh CR (2010) Prevention and treatment of acute kidney injury in patients undergoing cardiac surgery: a systematic review. Am J Nephrol 31:408–418
55. Ajith T, Usha S, Nivitha V (2007) Ascorbic acid and alpha-tocopherol protect anticancer drug cisplatin induced nephrotoxicity in mice: a comparative study. Clin Chim Acta 375:82–86
56. Baliga R, Zhang Z, Baliga M, Ueda N, Shah SV (1998) In vitro and in vivo evidence suggesting a role for iron in cisplatin-induced nephrotoxicity. Kidney Int 53:394–401
57. Satoh M, Kashihara N, Fujimoto S, Horike H, Tokura T, Namikoshi T, Sasaki T, Makino H (2003) A novel free radical scavenger, edarabone, protects against cisplatin-induced acute renal damage in vitro and in vivo. J Pharmacol Exp Ther 305:1183–1190
58. Walker PD, Barri Y, Shah SV (1999) Oxidant mechanisms in gentamicin nephrotoxicity. Ren Fail 21:433–442
59. Kopple JD, Ding H, Letoha A, Ivanyi B, Qing DP, Dux L, Wang HY, Sonkodi S (2002) L-carnitine ameliorates gentamicin-induced renal injury in rats. Nephrol Dial Transplant 17:2122–2131
60. Salom MG, Ceron SN, Rodriguez F, Lopez B, Hernandez I, Martinez JG, Losa AM, Fenoy FJ (2007) Heme oxygenase-1 induction improves ischemic renal failure: role of nitric oxide and peroxynitrite. Am J Physiol Heart Circ Physiol 293:H3542–H3549
61. Akagi R, Takahashi T, Sassa S (2002) Fundamental role of heme oxygenase in the protection against ischemic acute renal failure. Jpn J Pharmacol 88:127–132
62. Wei Q, Hill WD, Su Y, Huang S, Dong Z (2011) Heme oxygenase-1 induction contributes to renoprotection by G-CSF during rhabdomyolysis-associated acute kidney injury. Am J Physiol Renal Physiol 301:F162–F170
63. Eknoyan G, Levin NW (2002) K/DOQI clinical practice guidelines for chronic kidney disease: evaluation, classification, and stratification. Am J Kidney Dis 39:S1–S266

64. Sarnak MJ, Levey AS, Schoolwerth AC, Coresh J, Culleton B, Hamm LL, McCullough PA, Kasiske BL, Kelepouris E, Klag MJ (2003) Kidney disease as a risk factor for development of cardiovascular disease. Circulation 108:2154–2169
65. Xue JL, Ma JZ, Louis TA, Collins AJ (2001) Forecast of the number of patients with end-stage renal disease in the united states to the year 2010. J Am Soc Nephrol 12:2753–2758
66. Vaziri ND (2004) Roles of oxidative stress and antioxidant therapy in chronic kidney disease and hypertension. Curr Opin Nephrol Hypertens 13:93–99
67. Himmelfarb J, Stenvinkel P, Ikizler TA, Hakim RM (2002) The elephant in uremia: oxidant stress as a unifying concept of cardiovascular disease in uremia. Kidney Int 62:1524–1538
68. Ikizler T, Morrow J, Roberts L, Evanson J, Becker B, Hakim R, Shyr Y, Himmelfarb J (2002) Plasma F2-isoprostane levels are elevated in chronic hemodialysis patients. Clin Nephrol 58:190–197
69. Oberg BP, McMenamin E, Lucas FL, McMonagle E, Morrow J, Ikizler TA, Himmelfarb J (2004) Increased prevalence of oxidant stress and inflammation in patients with moderate to severe chronic kidney disease. Kidney Int 65:1009–1016
70. Hamed EA, El-Abaseri TB, Mohamed AO, Ahmed AR, El-Metwally TH (2012) Hypoxia and oxidative stress markers in pediatric patients undergoing hemodialysis: cross section study. BMC Nephrol 13:136
71. Dursun E, Dursun B, Suleymanlar G, Ozben T (2005) Effect of haemodialysis on the oxidative stress and antioxidants in diabetes mellitus. Acta Diabetol 42:123–128
72. Pavlova EL, Lilova MI, Savov VM (2005) Oxidative stress in children with kidney disease. Pediatr Nephrol 20:1599–1604
73. Schouten WEM, Grooteman MPC, Van Houte AJ, Schoorl M, Van Limbeek J, Nubé MJ (2000) Effects of dialyser and dialysate on the acute phase reaction in clinical bicarbonate dialysis. Nephrol Dial Transplant 15:379–384
74. Shah SV, Baliga R, Rajapurkar M, Fonseca VA (2007) Oxidants in chronic kidney disease. J Am Soc Nephrol 18:16–28
75. Morigi M, Macconi D, Zoja C, Donadelli R, Buelli S, Zanchi C, Ghilardi M, Remuzzi G (2002) Protein overload-induced NF-κB activation in proximal tubular cells requires H_2O_2 through a PKC-dependent pathway. J Am Soc Nephrol 13:1179–1189
76. Tang S, Leung JCK, Abe K, Chan KW, Chan LYY, Chan TM, Lai KN (2003) Albumin stimulates interleukin-8 expression in proximal tubular epithelial cells in vitro and in vivo. J Clin Invest 111:515–528
77. Liu Y (2004) Epithelial to mesenchymal transition in renal fibrogenesis: pathologic significance, molecular mechanism, and therapeutic intervention. J Am Soc Nephrol 15:1–12
78. Rhyu DY, Yang Y, Ha H, Lee GT, Song JS, Uh S, Lee HB (2005) Role of reactive oxygen species in TGF-β1-induced mitogen-activated protein kinase activation and epithelial-mesenchymal transition in renal tubular epithelial cells. J Am Soc Nephrol 16:667–675
79. Harris DCH, Tay C, Nankivell BJ (1994) Lysosomal iron accumulation and tubular damage in rat puromycin nephrosis and ageing. Clin Exp Pharmacol Physiol 21:73–81
80. Alfrey AC (1992) Toxicity of tubule fluid iron in the nephrotic syndrome. Am J Physiol 263:F637–F641
81. Cooper MA, Buddington B, Miller NL, Alfrey AC (1995) Urinary iron speciation in nephrotic syndrome. Am J Kidney Dis 25:314–319
82. Malle E, Buch T, Grone HJ (2003) Myeloperoxidase in kidney disease. Kidney Int 64:1956–1967
83. Gröne HJ, Gröne EF, Malle E (2002) Immunohistochemical detection of hypochlorite-modified proteins in glomeruli of human membranous glomerulonephritis. Lab Invest 82:5–14
84. Jun M, Venkataraman V, Razavian M, Cooper B, Zoungas S, Ninomiya T, Webster AC, Perkovic V (2012) Antioxidants for chronic kidney disease. Cochrane Database Syst Rev 10: CD008176
85. Joyeux-Faure M (2007) Cellular protection by erythropoietin: new therapeutic implications? J Pharmacol Exp Ther 323:759–762

86. Gluhovschi G, Gluhovschi C, Bob F, Velciov S, Trandafirescu V, Petrica L, Bozdog G (2008) Multiorgan-protective actions of blockers of the renin-angiotensin system, statins and erythropoietin: common pleiotropic effects in reno-, cardio- and neuroprotection. Acta Clin Belg 63:152–169
87. Patel NS, Sharples EJ, Cuzzocrea S, Chatterjee PK, Britti D, Yaqoob MM, Thiemermann C (2004) Pretreatment with EPO reduces the injury and dysfunction caused by ischemia/reperfusion in the mouse kidney in vivo. Kidney Int 66:983–989
88. Bahlmann FH, Song R, Boehm SM, Mengel M, von Wasielewski R, Lindschau C, Kirsch T, de Groot K, Laudeley R, Niemczyk E, Guler F, Menne J, Haller H, Fliser D (2004) Low-dose therapy with the long-acting erythropoietin analogue darbepoetin alpha persistently activates endothelial akt and attenuates progressive organ failure. Circulation 110:1006–1012
89. Fujiwara N, Nakamura T, Sato E, Kawagoe Y, Hikichi Y, Ueda Y, Node K (2011) Renovascular protective effects of erythropoietin in patients with chronic kidney disease. Intern Med 50:1929–1934
90. Shah SV (1989) Role of reactive oxygen metabolites in experimental glomerular disease. Kidney Int 35:1093–1106
91. Falk RJ, Terrell RS, Charles LA, Jennette JC (1990) Anti-neutrophil cytoplasmic autoantibodies induce neutrophils to degranulate and produce oxygen radicals in vitro. Proc Natl Acad Sci U S A 87:4115–4119
92. Poelstra K, Hardonk MJ, Koudstaal J, Bakker WW (1990) Intraglomerular platelet aggregation and experimental glomerulonephritis. Kidney Int 37:1500–1508
93. Boyce NW, Tipping PG, Holdsworth SR (1989) Glomerular macrophages produce reactive oxygen species in experimental glomerulonephritis. Kidney Int 35:778–782
94. Oberle GP, Niemeyer J, Thaiss F, Schoeppe W, Stahl R (1992) Increased oxygen radical and eicosanoid formation in immune-mediated mesangial cell injury. Kidney Int 42:69–74
95. Shah SV, Baricos WH, Basci A (1987) Degradation of human glomerular basement membrane by stimulated neutrophils. activation of a metalloproteinase (s) by reactive oxygen metabolites. J Clin Invest 79:25–31
96. Johnson RJ, Couser WG, Chi E, Adler S, Klebanoff S (1987) New mechanism for glomerular injury. myeloperoxidase-hydrogen peroxide-halide system. J Clin Invest 79:1379–1387
97. Rehan A, Johnson K, Wiggins R, Kunkel R, Ward P (1984) Evidence for the role of oxygen radicals in acute nephrotoxic nephritis. Lab Invest 51:396–403
98. Boyce NW, Holdsworth SR (1986) Hydroxyl radical mediation of immune renal injury by desferrioxamine. Kidney Int 30:813–817
99. Budisavljevic MN, Hodge LA, Barber K, Fulmer JR, Durazo-Arvizu RA, Self SE, Kuhlmann M, Raymond JR, Greene EL (2003) Oxidative stress in the pathogenesis of experimental mesangial proliferative glomerulonephritis. Am J Physiol Renal Physiol 285:F1138–F1148
100. Kanaguchi Y, Suzuki Y, Osaki K, Sugaya T, Horikoshi S, Tomino Y (2011) Protective effects of L-type fatty acid-binding protein (L-FABP) in proximal tubular cells against glomerular injury in anti-GBM antibody-mediated glomerulonephritis. Nephrol Dial Transplant 26:3465–3473
101. Kawaguchi M, Yamada M, Wada H, Okigaki T (1992) Roles of active oxygen species in glomerular epithelial cell injury in vitro caused by puromycin aminonucleoside. Toxicology 72:329–340
102. Diamond JR, Bonventre JV, Karnovsky MJ (1986) A role for oxygen free radicals in aminonucleoside nephrosis. Kidney Int 29:478–483
103. Beaman M, Birtwistle R, Howie A, Michael J, Adu D (1987) The role of superoxide anion and hydrogen peroxide in glomerular injury induced by puromycin aminonucleoside in rats. Clin Sci 73:320–332
104. Thakur V, Walker PD, Shah SV (1988) Evidence suggesting a role for hydroxyl radical in puromycin aminonucleoside-induced proteinuria. Kidney Int 34:494–499
105. Ricardo SD, Bertram JF, Ryan GB (1994) Antioxidants protect podocyte foot processes in puromycin aminonucleoside-treated rats. J Am Soc Nephrol 4:1974–1986
106. Someya T, Kaneko K, Yamada T, Yamashiro Y (2005) Effect of a novel free radical scavenger, edaravone, on puromycin aminonucleoside induced nephrosis in rats. Pediatr Nephrol 20:1430–1434

107. Matsumura H, Ashida A, Hirano K, Nakakura H, Tamai H (2006) Protective effect of radical scavenger edaravone against puromycin nephrosis. Clin Nephrol 66:405–410
108. Ueda N, Baliga R, Shah SV (1996) Role of 'catalytic' iron in an animal model of minimal change nephrotic syndrome. Kidney Int 49:370–373
109. Liu H, Shah SV, Baliga R (2001) Cytochrome P-450 as a source of catalytic iron in minimal change nephrotic syndrome in rats. Am J Physiol Renal Physiol 280:F88–F94
110. Liu H, Bigler SA, Henegar JR, Baliga R (2002) Cytochrome P450 2B1 mediates oxidant injury in puromycin-induced nephrotic syndrome. Kidney Int 62:868–876
111. Liu H, Baliga M, Bigler SA, Baliga R (2003) Role of cytochrome P450 2B1 in puromycin aminonucleoside-induced cytotoxicity to glomerular epithelial cells. Nephron Exp Nephrol 94:e17–e24
112. Baliga R, Baliga M, Shah SV (1992) Effect of selenium-deficient diet in experimental glomerular disease. Am J Physiol Renal Physiol 263:F56–F61
113. Hara T, Miyai H, Iida T, Futenma A, Nakamura S, Kato K (1990) Aggravation of puromycin aminonucleoside (PAN) nephrosis by the inhibition of endogenous superoxide dismutase (SOD). [Abstract]. Proceedings of the 11th international congress of nephrology, June 15–20, 1990 Springer, Tokyo, p442A
114. Yoshioka T, Bills T, Moore-Jarrett T, Greene HL, Burr IM, Ichikawa I (1990) Role of intrinsic antioxidant enzymes in renal oxidant injury. Kidney Int 38:282–288
115. Kawamura T, Yoshioka T, Bills T, Fogo A, Ichikawa I (1991) Glucocorticoid activates glomerular antioxidant enzymes and protects glomeruli from oxidant injuries. Kidney Int 40:291–301
116. Kinugasa S, Tojo A, Sakai T, Tsumura H, Takahashi M, Hirata Y, Fujita T (2011) Selective albuminuria via podocyte albumin transport in puromycin nephrotic rats is attenuated by an inhibitor of NADPH oxidase. Kidney Int 80:1328–1338
117. Neale TJ, Ullrich R, Ojha P, Poczewski H, Verhoeven AJ, Kerjaschki D (1993) Reactive oxygen species and neutrophil respiratory burst cytochrome b558 are produced by kidney glomerular cells in passive heymann nephritis. Proc Natl Acad Sci U S A 90:3645–3649
118. Shah SV (1988) Evidence suggesting a role for hydroxyl radical in passive heymann nephritis in rats. Am J Physiol 254:F337–F344
119. Brownlee M (2001) Biochemistry and molecular cell biology of diabetic complications. Nature 414:813–820
120. Baliga R, Ueda N, Shah SV (1996) Kidney iron status in passive heymann nephritis and the effect of an iron-deficient diet. J Am Soc Nephrol 7:1183–1188

Chapter 26
Trauma

Fukashi Serizawa, Eric Patterson, Gediminas Cepinskas, and Douglas D. Fraser

Abbreviations

ER	Endoplasmic reticulum
GCS	Glasgow coma scale
GPx	Glutathione peroxidase
HO-1	Heme oxygenase-1
4-HNE	4-Hyroxynonenal
L-NAME	L-NG-nitroarginine methyl ester
LP	Lipid peroxidation
MDA	Malondialdehyde
NADPH	Nicotinamide adenine dinucleotide phosphate
NOS	Nitric oxide synthase
NQO1	NAD(P)H quinonereductase-1
Nrf2/ARE	Nuclear factor E2-related factor2/antioxidant response element
3-NT	3-Nitrotyrosine
$ONOO^-$	Peroxynitrite
ORP	Oxidation–reduction potential

F. Serizawa, M.D., Ph.D. • E. Patterson, Ph.D. • G. Cepinskas, D.V.M., Ph.D.
Centre for Critical Illness Research, London, ON, Canada

D.D. Fraser, M.D., Ph.D. (✉)
Pediatric Critical Care Medicine, Translational Research Centre,
Children's Health Research Institute and Centre for Critical Illness Research,
Western University, London, Ontario, Canada
e-mail: douglas.fraser@lhsc.on.ca

H. Tsukahara and K. Kaneko (eds.), *Studies on Pediatric Disorders*, Oxidative Stress in Applied Basic Research and Clinical Practice, DOI 10.1007/978-1-4939-0679-6_26,

PBI	Primary brain injury
PBN	Alpha-phenyl-*tert*-butyl nitrone
PCOOH	Phosphatidylcholine hydroperoxide
PEG-SOD	Polyethylene glycol-conjugated superoxide dismutase
Prx	Peroxiredoxin
RNS	Reactive nitrogen species
ROS	Reactive oxygen species
SBI	Secondary brain injury
SFN	Sulforaphane
SOD	Superoxide dismutase
TBI	Traumatic brain injury
UDP	Uridine 5′-diphospho
XJB-5-131	4-Hydroxy-2,2,6,6-tetramethyl piperidine-1-oxyl nitroxide

26.1 Introduction

Trauma is a leading cause of death in children and adolescents [1]. Blunt injury is the most frequent cause of pediatric trauma, with penetrating injury accounting for only a minority of trauma admissions in most pediatric centers. Trauma causes a systemic inflammatory response, in addition to specific organ injuries [2]. Among pediatric trauma patients, brain injury is the leading cause of death (Table 26.1) and therefore a major determinant of outcomes [3, 4].

26.1.1 Trauma Resuscitation and Oxidative Stress

Rapid trauma resuscitation is imperative for survival [5]. The aims of trauma resuscitation are to restore circulating blood volume and optimize organ perfusion and tissue oxygen delivery. Post-traumatic intravascular expansion with crystalloid is often critical for survival, but can be associated with unintended consequences such as reperfusion injury and over-resuscitation syndrome, with subsequent widespread oxidative stress to both the vascular endothelium and end organs [6, 7].

26.1.2 Pediatric Traumatic Brain Injury

Traumatic brain injury (TBI) is classified according to mechanism, severity, and anatomical features on neuroimaging. A common pathophysiological finding after closed TBI is diffuse axonal injury, secondary to coup and contra-coup mechanisms, and shearing stress due to rotational and acceleration/deceleration forces [8]. The Glasgow coma scale (GCS) is the most common classification system for TBI

Table 26.1 Pediatric severe trauma—patient demographics and epidemiology of injuries

Male: Female	72 %: 28 %
Age	
1 year	4 %
1–4 years	5 %
5–9 years	17 %
10–15 years	40 %
16–17 years	34 %
Mechanism of injury	
Motor vehicle collision	49 %
Pedestrian (vs. motor vehicle)	20 %
Fall from height	12 %
Sporting	10 %
Pedal bike	7 %
Assault	2 %
% of patients with injuries	
Brain	61 %
Thoracic	51 %
Abdominal	46 %
Orthopedic	46 %
Facial	17 %
Spinal cord	3 %

Data compiled from a North American Level 1 Trauma Centre [3, 5, 79]
Injury severity scores ≥ 12

Table 26.2 Glasgow coma scale

Score	Best motor response	Best verbal response	Eye opening	Severity	Total score
1	None	None	None	Severe	3–8
2	Decorticate posturing	Mutters unintelligibly	Opens to pain		
3	Decerebrate posturing	Inappropriate speech	Opens to command	Moderate	9–12
4	Withdraws to pain	Confused	Opens spontaneously		
5	Localizing response to pain	Alert and oriented	N/A	Mild	13–15
6	Obeys commands	N/A	N/A		
Total	1–6	1–5	1–4	–	–

severity and is used to assess a patient's level of consciousness and level of neurologic functioning [9, 10]. The GCS assesses three components including best motor response, best verbal response, and eye opening (Table 26.2). Ominous features visualized on brain imaging include the presence of hemorrhage, mass lesions, midline shifts, edema, and/or herniation syndromes.

After TBI, there is an increased risk of brain ischemia. Dysregulation of cerebral blood flow can reduce blood supply, placing brain tissue in peril of post-traumatic ischemic injury. Two major mechanisms may contribute to this risk of brain

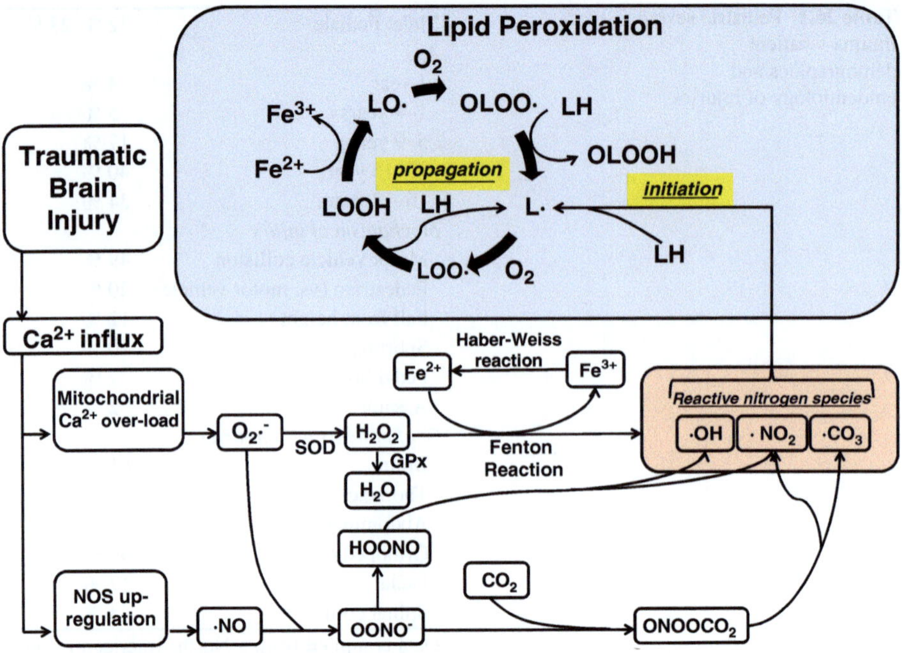

Fig. 26.1 The oxidative stress cascade in traumatic brain injury. TBI induces influx of Ca^{2+} into the mitochondria that activates the oxidative stress cascade. The formation of superoxide and up-regulation of NOS leads to further ROS/RNS production, resulting in excessive oxidative stress and subsequent lipid peroxidation. These oxidative molecules have positive feedback effects on the cascade itself, resulting in continuous ROS/RNS generation and massive neuron death

ischemia: first, direct mechanical impact and structural damage to the intracerebral arteries, and second, local vasospasms of cerebral arteries due to regional extravasations of blood into the brain parenchyma.

Mechanical damage caused by external forces at the time of trauma is the primary brain injury (PBI). Secondary brain injury (SBI) ensues with time, lasting days to weeks, and involves increases in nonspecific plasma membrane permeability, the generation of free radicals, calpain activation, glutamate-mediated excitotoxicity, disruptions in ion homeostasis, failure of mitochondrial respiration, and microvascular damage [11].

Oxidative stress is an important mediator in SBI, caused by the chemical imbalance between the generation and breakdown of reactive oxygen species (ROS) and reactive nitrogen species (RNS). The generation of excessive ROS/RNS starts immediately after mechanical impact, and the key reaction contributing to brain injury is lipid peroxidation (LP) caused by peroxynitrite ($ONOO^-$) [12, 13]. Fatty acid peroxides derived from LP are relatively stable and easily diffuse into the intracellular space due to their hydrophobic nature, causing irreversible damage to neuronal membranes and proteins. Oxidative stress cascades have positive feedback effects on the LP cascade itself and could result in endless ROS/RNS generation and massive neuronal death (Fig. 26.1).

Table 26.3 Reactive oxygen species

ROS/RNS	Image	Explanation
Oxygen (O_2)	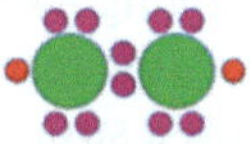	Each molecule has one unpaired electron. Oxygen is not included in the free-radicals. ORP=0.94
Superoxide ($^{\bullet}O_2^-$)	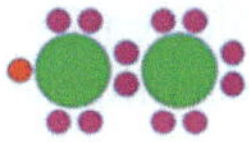	Free radical. One oxygen atom has one unpaired electron. Most likely produced in mitochondria and acts as an oxidant and reductant
Hydrogen peroxide (H_2O_2)	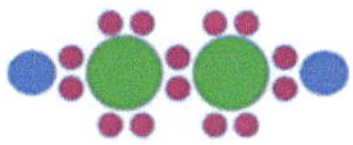	SOD catalyzes the dismutation of superoxide into hydrogen peroxide ORP=1.30
Hydroxy radical ($^{\bullet}OH$)	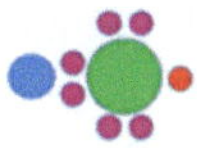	Very potent free-radical. The oxygen atom has one unpaired electron and is formed through the Fenton reaction. ORP=2.05
Hydro perradical ($HOO^{\bullet}$)	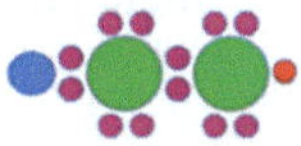	Potential for free-radical generation
Nitric oxide ($^{\bullet}NO$)	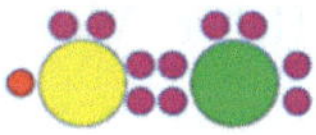	Generated by NOS

oxygen atom, hydrogen atom, nitrogen atom, paired electron, unpaired electron

ORP oxygen–reduction potential, *SOD* superoxide dismutase, *NOS* nitric oxide synthase

Typically, mammalian cells have well-organized anti-oxidant systems to balance ROS/RNS production [14, 15], such as the superoxide scavenger superoxide dismutase (SOD) and the H_2O_2 scavengers glutathione peroxidase (GPx) and peroxiredoxin (Prx). The generation of ROS/RNS following TBI can potentially overwhelm these defense systems. Recently, anti-oxidative therapeutic approaches have been studied in TBI and include scavengers of superoxide, hydrogen peroxide-decomposing molecules/catalysts, carbonyl compounds, inhibitors of LP, and pharmacological boosting of the endogenous anti-oxidant system.

26.1.3 Reactive Oxygen Species in TBI

ROS is a general term for chemically reactive molecules derived from oxygen. ROS that have an unpaired electron are "radicals" and those that do not have an unpaired electron are "non-radicals" (Table 26.3). Radicals include the superoxide ($O_2^{\bullet -}$), hydroxyl ($^{\bullet}OH$), hydroperoxyl ($HOO^{\bullet}$), lipid alkoxyl ($LO^{\bullet}$), lipid peroxyl ($LOO^{\bullet}$) radicals, and nitric oxide ($^{\bullet}NO$); while non-radicals include hydrogen peroxide (H_2O_2) and peroxynitrite ($ONOO^-$).

TBI initiates the oxygen radical reaction cascade with a sharp rise in neuronal intracellular Ca^{2+} concentration, followed by mitochondrial Ca^{2+} overload. Following this event, O_2 becoming $O_2^{\bullet-}$ is accepting a single electron and working as either an oxidant or reductant. SOD catalyzes the reaction of $O_2^{\bullet-}$ with H^+ forming H_2O_2 and O_2. However, H_2O_2 does not have a strong oxidizability (oxidation–reduction potential [ORP] = 1.30); it is stable, easily diffuses through the cell membrane and has a long lifetime. H_2O_2 is decomposed into $^{\bullet}OH$ via the Fenton reaction and $^{\bullet}OH$ has a very short lifetime (10^{-6} s) but strong oxidizability (ORP = 2.05); it is highly reactive with enzymes, cytoskeletal proteins, lipids, and nucleic acids.

$$Fe^{2+} + H_2O_2 \rightarrow Fe^{3+} + \cdot OH + OH^- \left(\text{Fenton reaction}\right)$$

$O_2^{\bullet-}$ also acts as a reductant and donates a single electron to Fe^{3+} in the Haber–Weiss reaction.

$$O_2 \cdot^- + Fe^{3+} \rightarrow Fe^{2+} + O_2 \left(\text{Haber - Weiss reaction}\right)$$

Through this reaction, Fe^{3+} is reduced to Fe^{2+} and back again due to the Fenton reaction and, as a result, $^{\bullet}OH$ production is increased. In the oxidation–reduction reactions of free radicals, iron ions play an important role by donating or accepting an electron. To prevent this damage, mammals restrict the amount of free iron using reversible binding proteins such as ferritin and transferrin. The binding rates between iron and ferritin/transferrin decrease under conditions of tissue acidosis (i.e. trauma) however, resulting in an increase in iron release from extravasated iron-binding proteins to the damaged tissues and subsequent oxygen radical production. In addition, mechanical damage causes hemoglobin accumulation, derived from hematoma formation, releasing iron ions, which also drives oxidative damage.

26.1.4 Reactive Nitrogen Species in TBI

Oxidative molecules derived from $^{\bullet}NO$ and $O_2^{\bullet-}$ are called RNS. Three isoforms of nitric oxide synthase (NOS; endothelial, neuronal, and inducible) are up-regulated after TBI for at least 24 h [16, 17], and the reaction between $^{\bullet}NO$ and $O_2^{\bullet-}$ forms the highly reactive peroxynitrite ($ONOO^-$) (Fig. 26.1) [13].

$$O_2^{\cdot -} + {}^{\cdot}NO \rightarrow ONOO^-$$

Administration of a NOS inhibitor showed a neuroprotective effect in a rodent head injury model [18–20], suggesting that an up-regulation of NOS contributes to neuronal damage. $ONOO^-$ has two metabolic pathways and forms three metabolic by-products. First, after peroxynitrous acid (ONOOH) formation via $ONOO^-$ protonation, $^{\bullet}NO_2$ and $^{\bullet}OH$ are produced through its dissociation process.

Second, the reaction of $ONOO^-$ and carbon dioxide forms nitrosoperoxocarbonate ($ONOOCO_2$), which is resolved into $^{\bullet}NO_2$ and $^{\bullet}CO_3$.

$$ONOOH \rightarrow {}^{\cdot}NO_2 + {}^{\cdot}OH$$

$$ONOOCO_2 \rightarrow {}^{\cdot}NO_2 + {}^{\cdot}CO_3$$

These three by-products ($^{\bullet}OH/^{\bullet}NO_2/^{\bullet}CO_3$) are highly reactive cytotoxic free radicals and induce LP. LP causes oxidative stress and generates irreversible damage to proteins, cell membrane lipids, and nuclear proteins. For example, NO_2 derived from $ONOO^-$ causes tyrosine nitrosation to form 3-nitrotyrosine (3-NT), a nitrosative stress marker. This nitrosative stress marker has a cytotoxic effect, which is weakened by the NOS inhibitor L-NG-Nitroarginine Methyl Ester (L-NAME) through decreasing the accumulation of 3-NT, in a brain injury model [21].

26.1.5 Lipid Peroxidation

LP is the oxidative degradation of lipids by ROS/RNS; major lipid peroxides are shown in Table 26.4. Polyunsaturated fatty acids, such as arachidonic acid, linoleic acid, eicosapentaenoic acid, and docosahexaenoic acid, are enriched in brain. Typically, brain injury by an oxidative radical chain reaction takes place in three steps. In the first step (initiation), a ROS causes protonation of a polyunsaturated fatty acid and forms a lipid radical ($L^{\bullet}$)

$$LH + ROS \rightarrow L^{\cdot} + ROS - H$$

In the second step (propagation), the lipid radical reacts with oxygen and forms a $LOO^{\bullet}$.

$$L^{\cdot} + O_2 \rightarrow LOO^{\cdot}$$

$LOO^{\bullet}$ in turn causes protonation of polyunsaturated fatty acids to form lipid hydroperoxides (LOOH) and regenerate the lipid radicals again, thus creating a vicious circle of lipid peroxidation.

$$LOO^{\cdot} + LH \rightarrow LOOH + L^{\cdot}$$

In this radical chain reaction, Fe^{2+} and Fe^{3+} enhance oxidative stress through the reactions shown below, where iron behaves as both oxidant and reductant.

$$LOOH + Fe^{2+} \rightarrow LO^{\cdot} + {}^{\cdot}OH + Fe^{3+}$$

$$LOOH + Fe^{3+} \rightarrow LOO^{\cdot} + Fe^{2+}$$

The $LO^{\bullet}$ and $LOO^{\bullet}$, by-products of these reactions, cause further lipid peroxidation and increase oxidative stress.

Table 26.4 Lipid peroxides

Lipid peroxide	Symbol	Explanation
Lipid	LH	Polyunsaturated fatty acid.
Lipid radical	$L^{\cdot}$	Protonated lipid.
Lipid alkoxyl radical	$LO^{\cdot}$	One oxygen atom bound to a lipid radical.
Lipid peroxyl radical	$LOO^{\cdot}$	Two oxygen atoms bound to a lipid radical. This is called the peroxide bond (–O–O–).
Lipid hydro peroxide	LOOH	LOOH is more stable than the other free-radicals but still has oxidizability and is classified as a free-radical.

$$LO^{\cdot} + LH \rightarrow LOH + L^{\cdot}$$

$$LOO^{\cdot} + LH \rightarrow LOOH + L^{\cdot}$$

In the third step (termination), the oxidative stress cascade ends with material depletion and the lipid radical changes into a non-radical product. Acrolein and 4-hyroxynonenal (4-HNE) are end products of LP in TBI animal models [22], initiating neurotoxicity by binding to lysine or histidine and changing the structures and functions of affected proteins.

26.1.6 *Biomarkers of Oxidative Stress in TBI*

Several compounds are useful as oxidative stress biomarkers in TBI. F2-isoprostanes are bioactive, cyclopentanone, prostaglandin-like compounds generated from the peroxidation of arachidonyl-containing lipids and are known lipid biomarkers of oxidative stress [23]. The concentration of thiobarbituric acid-reactive substances in cerebrospinal fluid, which reflects LP, is associated with the severity of head injury in adults [24]. Additionally, in adult TBI patients, increased levels of nitrotyrosine, which is a marker of oxidative/nitrosative damage to proteins, correlated with poor neurological outcomes. In adult severe TBI patients (GCS $\leq$ 8), increased levels of 3-NT, which is a marker of oxidative/nitrosative damage to proteins, correlated with poor neurological outcomes [25]. Protein thiols, markers of oxidative damage to proteins, were also used as injury biomarkers after TBI [26].

26.2 Antioxidant Therapy Strategies

Oxidative stress contributes to the initiation and propagation of SBI, suggesting that anti-oxidative therapies targeting ROS molecules may have therapeutic utility. One approach is to scavenge superoxide, which is the starting point for the oxidative stress cascade. Therapeutic scavenging of superoxide after mechanical impact, however, is limited by the brief window of time to block the initial ROS production. An alternative therapeutic approach is to inhibit LP, with a focus on various target molecules. This latter approach benefits from a longer treatment window and thus, is more realistic as a therapeutic intervention.

26.2.1 Superoxide Scavengers

Superoxide, as a by-product of the arachidonic acid cascade, is the first hyperactivated ROS and causes post-traumatic microvascular dysfunction. Administration of SOD immediately after the TBI reduced microvascular dysfunction. Consequently, a small clinical trial was conducted using an improved metabolically stable SOD (polyethylene glycol-conjugated superoxide dismutase, PEG-SOD) administered to patients within 8 h of TBI and a neuroprotective tendency was shown. In contrast, a subsequent multi-center phase II trial on severe TBI patients (GCS≤8) showed no improvement in survival rate or neuroprotective effect [27, 28]. The study concluded that scavenging superoxide effectively by the administration of agents was unrealistic due to the brevity of the time window. In addition, PEG-SOD is rather a large compound and might not be able to pass the blood–brain barrier with sufficient concentration to scavenge superoxide. On the other hand, transgenic mice over-expressing Cu/Zn SOD showed a significant neuroprotective effect against TBI [29–31]. These results support the idea that superoxide plays an important role in the oxidative stress cascade following TBI. However, therapeutic targeting has now shifted from oxidants to the down-stream, post-traumatic free radical chain reaction cascade.

26.2.2 Lipid Peroxidation Inhibitors

Tirilazad: Tirilazad (U-74006F) acts as a LOO• scavenger and inhibitor of the reaction between LOO• and polyunsaturated fatty acids through a membrane stabilizing effect. After Tirilazad showed neuroprotective effects in an animal TBI model [32, 33], two multi-center phase III trials were conducted in moderate (GCS 9-12) and severe (GCS ≤ 8) TBI patients in North America and Europe. The treatment, 2.5 mg/kg of either Tirilazad or a placebo agent administered every 6 h for 5 days, was started in these two trials within 4 h of the TBI. While the results of the trial performed in North America remain unpublished due to an inappropriate randomization procedure, the data obtained from the European trial was reported in 1998 [34]. According to the results of the European trial, Tirilazad provided no beneficial effects to moderate (GCS 9-12) or severe (GCS ≤ 8) TBI patients. However, post hoc analysis following the European trial revealed a significant reduction in mortality in male moderate (GCS 9-12) or severe (GCS ≤ 8) TBI patients with the complication of traumatic subarachnoid hemorrhage as compared to the placebo group (moderate, 6 % vs. 24 %; severe, 34 % vs. 43 %). The reasons for failure in this trial may be that Tirilazad did not have sufficient permeability through the blood–brain barrier into the brain parenchyma, and/or differences in metabolism of the drug and its outcomes between genders [35].

U83836E: U83836E is a second-generation lazaroid (21-amino steroid) that has a non-steroidal structure. Neuronal calpain activation occurs within minutes to hours following either contusive or diffuse brain trauma in animals [36, 37], suggesting calpains are an early mediator of neuronal damage. U83836E suppressed calpain-mediated cytoskeletal degradation in the same animal model [38]. In addition, U83836E also acts as a LOO• scavenger, maintaining mitochondrial respiration and calcium buffering functions [39]. Based on these studies, U83836E exerts neuroprotective effects in multiple steps of the oxidative stress cascade.

Resveratrol: Resveratrol (*trans*-3,5,4′-trihydroxystilbene) came to scientific attention as a possible explanation for the "French Paradox"—the low incidence of heart disease among French people, who eat a relatively high-fat diet [40]. In a spinal cord injury model, resveratrol decreased oxidative stress [41] and improved post-traumatic edema. In addition, resveratrol activated Na^+-K^+-ATPase [42] and showed neuroprotective and neurorecovery effects [43]. These effects are due to up-regulation of SOD and down-regulation of malondialdehyde (MDA), inflammatory cytokines, and pro-apoptotic proteins [44]. Upon publication of these results, the effects of resveratrol in a TBI animal model were studied and decreases in brain edema, MDA, xanthine oxidase, and NO were shown [45]. In addition, acute treatment with resveratrol has a neuroprotective role against trauma-induced hippocampal neuron loss and associated cognitive impairment in rats [46].

26.2.3 Scavengers of Peroxynitrite ($ONOO^-$) and $ONOO^-$-Derived Free Radicals

Alpha-phenyl-*tert*-butyl nitrone (PBN), NXY-059 (a thiol analog of PBN) and tempol are scavengers of peroxynitrite and peroxynitrite-derived free radicals. PBN was reported to decrease the lesion volume and hemispheric loss of brain tissue in a rodent TBI model [47]. Tempol has the most potent anti-oxidative effect and scavenges many ROS and RNS ($O_2^{\bullet-}$/H_2O_2/$ONOO^-$-derived $^{\bullet}NO_2$/$^{\bullet}CO_3$) [48, 49]. In a mouse TBI model, tempol decreased oxidative damage through the inhibition of post-traumatic LP and protein nitrification, which resulted in maintaining mitochondrial bioenergetics, decreasing calpain-mediated cytoskeletal damage [50], and thus suppressing neurodegradation. In addition, tempol suppressed post-traumatic brain edema and improved neurological recovery [50]. Tempol also had a mitochondrial protective effect within 1 h after TBI [50], while others described tempol as improving motor function after spinal cord contusion over days [51].

26.2.4 Carbonyl Scavenging: Penicillamine and Phenelzine

As described above, aldehydes derived from lipid peroxidation (4-HNE, acrolein, and MDA) are highly reactive with amino acids, thereby causing disruption of protein function via covalent modification of amino acids. Compounds that covalently bind to aldehydes, act as carbonyl scavengers and are neuroprotective in TBI models. For example, D-penicillamine irreversibly binds to primary aldehydes and scavenges peroxynitrite [52]. In isolated rat brain mitochondria, D-penicillamine reduced peroxynitrite-induced mitochondrial respiratory failure, accompanied by a decrease in 4-HNE level [53]. In addition, administration of penicillamine in the acute phase of a mouse TBI model aided neurologic recovery [54].

More recently, it was revealed that compounds with a specific structure (–NH–NH_2), including hydralazine (an anti-hypertensive drug), phenelzine (an anti-depressant) and iproniazid (an anti-tuberculosis drug) have neuroprotective effects via chemical scavenging of 4-HNE and acrolein in a spinal cord injury model [55, 56]. Neuroprotection obtained with hydralazine is still effective even after 4-HNE binds to proteins, hence this drug is thought to be useful due to the duration of its therapeutic window [57]. On the other hand, clinical application of hydralazine to TBI or spinal cord injury might be unrealistic due to concerns about its potential vaso-dilation effect in the context of the hypotension that typically follows central nervous system injury. Phenelzine and iproniazid do not have hypotensive effects but, in spite of their long-term clinical use, almost no studies have examined their neuroprotective profiles.

26.2.5 Nrf2/ARE Signaling Activation

PKR-like endoplasmic reticulum (ER) kinase is known as a stress sensor [58] and regulates the nuclear factor E2-related factor2/antioxidant response element (Nrf2/ARE) via phosphorylation. Nrf2/ARE regulates the expression of anti-oxidant enzymes such as glutathione *S*-transferase, heme oxygenase-1 (HO-1), NAD(P)H quinonereductase-1 (NQO1), and γ-glutamylcysteine synthetase [59, 60]. Chemical compounds that activate Nrf2/ARE, include DL-sulforaphane, tertbutylhydroquinone, lipoic acid, fumaric acid, and curcumin. These compounds have shown promise in studies of cerebral infarction and neurodegradation [61, 62]. In addition, Nrf2 knockout mice are more sensitive to oxidative stress toxicity following TBI and exhibit neurologic deficits compared to their wild-type counterparts [63]. Sulforaphane (SFN) is known to induce antioxidant genes such as HO-1, glutathione transferases, Uridine 5′-diphospho (UDP)-glucuronyltransferase and NAD(P)H:quinone oxidoreductase I via Nrf2/ARE activation [64]. Although up-regulation of Nrf2 and HO-1 was reported previously in rat brain injury models [65, 66], administration of SFN in various TBI animal models showed further up-regulation of Nrf2, HO-1, NQO1, and glutathione *S*-transferase-α1, resulting in the reduction of brain edema, decreased oxidative stress, strengthened blood–brain barrier integrity, and improved cognition [67]. Another Nfr2 activator, tertbutylhydroquinone, showed neuroprotective effects via inactivation of NF-κB and decreased inflammatory cytokines (TNF-α and IL-1β) in a mouse TBI model [68]. Taken together, the anti-oxidative effects triggered by Nrf2 activation appear due to up-regulation of anti-oxidant enzymes and the reduction of inflammation following TBI.

26.2.6 A Cardiolipin Oxygenation Inhibitor

Very recently, it was reported that cardiolipin oxygenation generates neuronal death signals and that these can be prevented by XJB-5-131 (4-hydroxy-2,2,6,6-tetramethyl piperidine-1-oxyl nitroxide), which is a mitochondria-targeted small molecule and bifunctional antioxidant, that scavenges radicals [69]. In a rat TBI model, XJB-5-131 blocked cardiolipin oxidation and improved behavior. In addition, XJB-5-131 attenuated apoptotic neuronal death. The assumed mechanisms of XJB-5-131 are an SOD mimetic-action and an electron scavenging effect. In the former case, XJB-5-131 would retain its radical state and in the latter case, XJB-5-131 is reduced to hydroxylamine.

26.2.7 Hypothermia Therapy

The precise mechanisms behind hypothermia therapy's effects on oxidative stress in SBI are still unknown, but some beneficial effects and outcomes have been obtained [70–72]. Hypothermia reduces calcium influx into cells, a key early event in the oxidative stress cascade after TBI (Fig. 26.1), and inhibits excitotoxicity [73, 74]. Furthermore, hypothermia inhibits early gene expression and stress responses via altered microRNAs [75]. Previous studies showed that NOSs are up-regulated immediately following a mechanical impact, where they contribute to the oxidative stress cascade [16, 17]. Hypothermia therapy might inhibit up-regulation of NOSs and thereby attenuate oxidative stress. Recently, therapeutic hypothermia was shown to preserve antioxidant defenses in cerebrospinal fluid obtained from children with TBI by measuring oxidative biomarkers, including protein-thiols and F2-isoprostane [76]. On the other hand, the largest randomized controlled trial using therapeutic hypothermia after severe TBI ($GCS \leq 8$) in children failed to demonstrate a benefit [77] and thus, therapeutic cooling remains controversial [78].

26.3 Conclusion

Trauma remains the number one killer of children and adolescents, and the brain is the most frequently injured organ. At present, there are no direct therapies to reduce brain cell injury. One approach to improve trauma outcomes is the administration of rationally designed therapeutic agents to interrupt or inhibit oxidative/nitrosative stress during SBI. A therapeutic reduction in systemic and/or brain oxidative stress represents potential hope for injured children.

References

1. Krug SE, Tuggle DW (2008) Management of pediatric trauma. Pediatrics 121:849–854
2. Wetzel RC, Burns RC (2002) Multiple trauma in children: critical care overview. Crit Care Med 30:S468–S477
3. Forward K, Chan M, Stewart TC, Gilliland J, Campbell C, Fraser DD (2010) Injury analyses in rural children: comparison of old-order Anabaptists and non-Anabaptists. J Trauma 69:1294–1299
4. Morrison G, Fraser DD, Cepinskas G (2013) Mechanisms and consequences of acquired brain injury during development. Pathophysiology 20:49–57
5. Al-Sharif A, Thakur V, Al-Farsi S, Singh RN, Kornecki A, Seabrook JA, Fraser DD (2012) Resuscitation volume in paediatric non-haemorrhagic blunt trauma. Injury 43:2078–2082
6. Laplace C, Huet O, Vicaut E, Ract C, Martin L, Benhamou D, Duranteau J (2005) Endothelial oxidative stress induced by serum from patients with severe trauma hemorrhage. Intensive Care Med 31:1174–1180
7. van Golen RF, van Gulik TM, Heger M (2012) Mechanistic overview of reactive species-induced degradation of the endothelial glycocalyx during hepatic ischemia/reperfusion injury. Free Radic Biol Med 52:1382–1402
8. Meythaler JM, Peduzzi JD, Eleftheriou E, Novack TA (2001) Current concepts: diffuse axonal injury-associated traumatic brain injury. Arch Phys Med Rehabil 82:1461–1471
9. Teasdale G, Jennett B (1974) Assessment of coma and impaired consciousness. A practical scale. Lancet 2:81–84
10. McDonald CM, Jaffe KM, Fay GC, Polissar NL, Martin KM, Liao S, Rivara JB (1994) Comparison of indices of traumatic brain injury severity as predictors of neurobehavioral outcome in children. Arch Phys Med Rehabil 75:328–337
11. Pettus EH, Christman CW, Giebel ML, Povlishock JT (1994) Traumatically induced altered membrane permeability: its relationship to traumatically induced reactive axonal change. J Neurotrauma 11:507–522
12. Gutteridge JM (1995) Lipid peroxidation and antioxidants as biomarkers of tissue damage. Clin Chem 41:1819–1828
13. Beckman JS (1991) The double-edged role of nitric oxide in brain function and superoxide-mediated injury. J Dev Physiol 15:53–59
14. Pompella A, Visvikis A, Paolicchi A, De Tata V, Casini AF (2003) The changing faces of glutathione, a cellular protagonist. Biochem Pharmacol 66:1499–1503
15. Rhee SG, Chae HZ, Kim K (2005) Peroxiredoxins: a historical overview and speculative preview of novel mechanisms and emerging concepts in cell signaling. Free Radic Biol Med 38:1543–1552
16. Rao VL, Dogan A, Bowen KK, Dempsey RJ (1999) Traumatic injury to rat brain upregulates neuronal nitric oxide synthase expression and L-[3H]nitroarginine binding. J Neurotrauma 16:865–877
17. Gahm C, Holmin S, Mathiesen T (2000) Temporal profiles and cellular sources of three nitric oxide synthase isoforms in the brain after experimental contusion. Neurosurgery 46:169–177
18. Wada K, Chatzipanteli K, Kraydieh S, Busto R, Dietrich WD (1998) Inducible nitric oxide synthase expression after traumatic brain injury and neuroprotection with aminoguanidine treatment in rats. Neurosurgery 43:1427–1436
19. Mesenge C, Verrecchia C, Allix M, Boulu RR, Plotkine M (1996) Reduction of the neurological deficit in mice with traumatic brain injury by nitric oxide synthase inhibitors. J Neurotrauma 13:209–214

20. Wallis RA, Panizzon KL, Girard JM (1996) Traumatic neuroprotection with inhibitors of nitric oxide and ADP-ribosylation. Brain Res 710:169–177
21. Bringold U, Ghafourifar P, Richter C (2000) Peroxynitrite formed by mitochondrial NO synthase promotes mitochondrial Ca2+ release. Free Radic Biol Med 29:343–348
22. Shao C, Roberts KN, Markesbery WR, Scheff SW, Lovell MA (2006) Oxidative stress in head trauma in aging. Free Radic Biol Med 41:77–85
23. Morrow JD, Awad JA, Boss HJ, Blair IA, Roberts LJ 2nd (1992) Non-cyclooxygenase-derived prostanoids (F2-isoprostanes) are formed in situ on phospholipids. Proc Natl Acad Sci U S A 89:10721–10725
24. Kasprzak HA, Wozniak A, Drewa G, Wozniak B (2001) Enhanced lipid peroxidation processes in patients after brain contusion. J Neurotrauma 18:793–797
25. Darwish RS, Amiridze N, Aarabi B (2007) Nitrotyrosine as an oxidative stress marker: evidence for involvement in neurologic outcome in human traumatic brain injury. J Trauma 63:439–442
26. Tyurin VA, Tyurina YY, Borisenko GG, Sokolova TV, Ritov VB, Quinn PJ, Rose M, Kochanek P, Graham SH, Kagan VE (2000) Oxidative stress following traumatic brain injury in rats: quantitation of biomarkers and detection of free radical intermediates. J Neurochem 75:2178–2189
27. Muizelaar JP, Kupiec JW, Rapp LA (1995) PEG-SOD after head injury. J Neurosurg 83:942
28. Muizelaar JP, Marmarou A, Young HF, Choi SC, Wolf A, Schneider RL, Kontos HA (1993) Improving the outcome of severe head injury with the oxygen radical scavenger polyethylene glycol-conjugated superoxide dismutase: a phase II trial. J Neurosurg 78:375–382
29. Mikawa S, Kinouchi H, Kamii H, Gobbel GT, Chen SF, Carlson E, Epstein CJ, Chan PH (1996) Attenuation of acute and chronic damage following traumatic brain injury in copper, zinc-superoxide dismutase transgenic mice. J Neurosurg 85:885–891
30. Lewen A, Matz P, Chan PH (2000) Free radical pathways in CNS injury. J Neurotrauma 17:871–890
31. Xiong Y, Shie FS, Zhang J, Lee CP, Ho YS (2005) Prevention of mitochondrial dysfunction in post-traumatic mouse brain by superoxide dismutase. J Neurochem 95:732–744
32. Dimlich RV, Tornheim PA, Kindel RM, Hall ED, Braughler JM, McCall JM (1990) Effects of a 21-aminosteroid (U-74006F) on cerebral metabolites and edema after severe experimental head trauma. Adv Neurol 52:365–375
33. McIntosh TK, Thomas M, Smith D, Banbury M (1992) The novel 21-aminosteroid U74006F attenuates cerebral edema and improves survival after brain injury in the rat. J Neurotrauma 9:33–46
34. Marshall LF, Maas AI, Marshall SB, Bricolo A, Fearnside M, Iannotti F, Klauber MR, Lagarrigue J, Lobato R, Persson L, Pickard JD, Piek J, Servadei F, Wellis GN, Morris GF, Means ED, Musch B (1998) A multicenter trial on the efficacy of using tirilazad mesylate in cases of head injury. J Neurosurg 89:519–525
35. Farin A, Deutsch R, Biegon A, Marshall LF (2003) Sex-related differences in patients with severe head injury: greater susceptibility to brain swelling in female patients 50 years of age and younger. J Neurosurg 98:32–36
36. Kampfl A, Posmantur R, Nixon R, Grynspan F, Zhao X, Liu SJ, Newcomb JK, Clifton GL, Hayes RL (1996) mu-calpain activation and calpain-mediated cytoskeletal proteolysis following traumatic brain injury. J Neurochem 67:1575–1583
37. Saatman KE, Bozyczko-Coyne D, Marcy V, Siman R, McIntosh TK (1996) Prolonged calpain-mediated spectrin breakdown occurs regionally following experimental brain injury in the rat. J Neuropathol Exp Neurol 55:850–860
38. Mustafa AG, Wang JA, Carrico KM, Hall ED (2011) Pharmacological inhibition of lipid peroxidation attenuates calpain-mediated cytoskeletal degradation after traumatic brain injury. J Neurochem 117:579–588
39. Mustafa AG, Singh IN, Wang J, Carrico KM, Hall ED (2010) Mitochondrial protection after traumatic brain injury by scavenging lipid peroxyl radicals. J Neurochem 114:271–280
40. Kopp P (1998) Resveratrol, a phytoestrogen found in red wine. A possible explanation for the conundrum of the 'French paradox'? Eur J Endocrinol 138:619–620

41. Kiziltepe U, Turan NN, Han U, Ulus AT, Akar F (2004) Resveratrol, a red wine polyphenol, protects spinal cord from ischemia–reperfusion injury. J Vasc Surg 40:138–145
42. Yang YB, Piao YJ (2003) Effects of resveratrol on secondary damages after acute spinal cord injury in rats. Acta Pharmacol Sin 24:703–710
43. Ates O, Cayli S, Altinoz E, Gurses I, Yucel N, Kocak A, Yologlu S, Turkoz Y (2006) Effects of resveratrol and methylprednisolone on biochemical, neurobehavioral and histopathological recovery after experimental spinal cord injury. Acta Pharmacol Sin 27:1317–1325
44. Liu C, Shi Z, Fan L, Zhang C, Wang K, Wang B (2011) Resveratrol improves neuron protection and functional recovery in rat model of spinal cord injury. Brain Res 1374:100–109
45. Ates O, Cayli S, Altinoz E, Gurses I, Yucel N, Sener M, Kocak A, Yologlu S (2007) Neuroprotection by resveratrol against traumatic brain injury in rats. Mol Cell Biochem 294:137–144
46. Sonmez U, Sonmez A, Erbil G, Tekmen I, Baykara B (2007) Neuroprotective effects of resveratrol against traumatic brain injury in immature rats. Neurosci Lett 420:133–137
47. Marklund N, Clausen F, Lewen A, Hovda DA, Olsson Y, Hillered L (2001) alpha-Phenyl-tert-N-butyl nitrone (PBN) improves functional and morphological outcome after cortical contusion injury in the rat. Acta Neurochir (Wien) 143:73–81
48. Carroll RT, Galatsis P, Borosky S, Kopec KK, Kumar V, Althaus JS, Hall ED (2000) 4-Hydroxy-2,2,6,6-tetramethylpiperidine-1-oxyl (tempol) inhibits peroxynitrite-mediated phenol nitration. Chem Res Toxicol 13:294–300
49. Wilcox CS (2010) Effects of tempol and redox-cycling nitroxides in models of oxidative stress. Pharmacol Ther 126:119–145
50. Deng-Bryant Y, Singh IN, Carrico KM, Hall ED (2008) Neuroprotective effects of tempol, a catalytic scavenger of peroxynitrite-derived free radicals, in a mouse traumatic brain injury model. J Cereb Blood Flow Metab 28:1114–1126
51. Hillard VH, Peng H, Zhang Y, Das K, Murali R, Etlinger JD, Zeman RJ (2004) Tempol, a nitroxide antioxidant, improves locomotor and histological outcomes after spinal cord contusion in rats. J Neurotrauma 21:1405–1414
52. Althaus JS, Oien TT, Fici GJ, Scherch HM, Sethy VH, VonVoigtlander PF (1994) Structure activity relationships of peroxynitrite scavengers: an approach to nitric oxide neurotoxicity. Res Commun Chem Pathol Pharmacol 83:243–254
53. Singh IN, Sullivan PG, Hall ED (2007) Peroxynitrite-mediated oxidative damage to brain mitochondria: protective effects of peroxynitrite scavengers. J Neurosci Res 85:2216–2223
54. Hall ED, Kupina NC, Althaus JS (1999) Peroxynitrite scavengers for the acute treatment of traumatic brain injury. Ann N Y Acad Sci 890:462–468
55. Hamann K, Durkes A, Ouyang H, Uchida K, Pond A, Shi R (2008) Critical role of acrolein in secondary injury following ex vivo spinal cord trauma. J Neurochem 107:712–721
56. Hamann K, Shi R (2009) Acrolein scavenging: a potential novel mechanism of attenuating oxidative stress following spinal cord injury. J Neurochem 111:1348–1356
57. Galvani S, Coatrieux C, Elbaz M, Grazide MH, Thiers JC, Parini A, Uchida K, Kamar N, Rostaing L, Baltas M, Salvayre R, Negre-Salvayre A (2008) Carbonyl scavenger and antiatherogenic effects of hydrazine derivatives. Free Radic Biol Med 45:1457–1467
58. Zhang K, Kaufman RJ (2008) From endoplasmic-reticulum stress to the inflammatory response. Nature 454:455–462
59. Kensler TW, Wakabayashi N, Biswal S (2007) Cell survival responses to environmental stresses via the Keap1-Nrf2-ARE pathway. Annu Rev Pharmacol Toxicol 47:89–116
60. Zhang DD (2006) Mechanistic studies of the Nrf2-Keap1 signaling pathway. Drug Metab Rev 38:769–789
61. Baird L, Dinkova-Kostova AT (2011) The cytoprotective role of the Keap1-Nrf2 pathway. Arch Toxicol 85:241–272
62. Shih AY, Johnson DA, Wong G, Kraft AD, Jiang L, Erb H, Johnson JA, Murphy TH (2003) Coordinate regulation of glutathione biosynthesis and release by Nrf2-expressing glia potently protects neurons from oxidative stress. J Neurosci 23:3394–3406

63. Shih AY, Imbeault S, Barakauskas V, Erb H, Jiang L, Li P, Murphy TH (2005) Induction of the Nrf2-driven antioxidant response confers neuroprotection during mitochondrial stress in vivo. J Biol Chem 280:22925–22936
64. Misiewicz I, Skupinska K, Kowalska E, Lubinski J, Kasprzycka-Guttman T (2004) Sulforaphane-mediated induction of a phase 2 detoxifying enzyme NAD(P)H:quinone reductase and apoptosis in human lymphoblastoid cells. Acta Biochim Pol 51:711–721
65. Chen G, Fang Q, Zhang J, Zhou D, Wang Z (2011) Role of the Nrf2-ARE pathway in early brain injury after experimental subarachnoid hemorrhage. J Neurosci Res 89:515–523
66. Yan W, Wang HD, Hu ZG, Wang QF, Yin HX (2008) Activation of Nrf2-ARE pathway in brain after traumatic brain injury. Neurosci Lett 431:150–154
67. Dash PK, Zhao J, Orsi SA, Zhang M, Moore AN (2009) Sulforaphane improves cognitive function administered following traumatic brain injury. Neurosci Lett 460:103–107
68. Jin W, Kong J, Wang H, Wu J, Lu T, Jiang J, Ni H, Liang W (2011) Protective effect of tert-butylhydroquinone on cerebral inflammatory response following traumatic brain injury in mice. Injury 42:714–718
69. Ji J, Kline AE, Amoscato A, Samhan-Arias AK, Sparvero LJ, Tyurin VA, Tyurina YY, Fink B, Manole MD, Puccio AM, Okonkwo DO, Cheng JP, Alexander H, Clark RS, Kochanek PM, Wipf P, Kagan VE, Bayir H (2012) Lipidomics identifies cardiolipin oxidation as a mitochondrial target for redox therapy of brain injury. Nat Neurosci 15:1407–1413
70. Adelson PD, Ragheb J, Kanev P, Brockmeyer D, Beers SR, Brown SD, Cassidy LD, Chang Y, Levin H (2005) Phase II clinical trial of moderate hypothermia after severe traumatic brain injury in children. Neurosurgery 56:740–754; discussion 740–754
71. Bernard SA, Gray TW, Buist MD, Jones BM, Silvester W, Gutteridge G, Smith K (2002) Treatment of comatose survivors of out-of-hospital cardiac arrest with induced hypothermia. N Engl J Med 346:557–563
72. Shankaran S, Laptook AR, Ehrenkranz RA, Tyson JE, McDonald SA, Donovan EF, Fanaroff AA, Poole WK, Wright LL, Higgins RD, Finer NN, Carlo WA, Duara S, Oh W, Cotten CM, Stevenson DK, Stoll BJ, Lemons JA, Guillet R, Jobe AH (2005) Whole-body hypothermia for neonates with hypoxic-ischemic encephalopathy. N Engl J Med 353:1574–1584
73. Colbourne F, Grooms SY, Zukin RS, Buchan AM, Bennett MV (2003) Hypothermia rescues hippocampal CA1 neurons and attenuates down-regulation of the AMPA receptor GluR2 subunit after forebrain ischemia. Proc Natl Acad Sci U S A 100:2906–2910
74. Dietrich WD, Bramlett HM (2010) The evidence for hypothermia as a neuroprotectant in traumatic brain injury. Neurotherapeutics 7:43–50
75. Truettner JS, Alonso OF, Bramlett HM, Dietrich WD (2011) Therapeutic hypothermia alters microRNA responses to traumatic brain injury in rats. J Cereb Blood Flow Metab 31:1897–1907
76. Bayir H, Adelson PD, Wisniewski SR, Shore P, Lai Y, Brown D, Janesko-Feldman KL, Kagan VE, Kochanek PM (2009) Therapeutic hypothermia preserves antioxidant defenses after severe traumatic brain injury in infants and children. Crit Care Med 37:689–695
77. Hutchison JS, Ward RE, Lacroix J, Hebert PC, Barnes MA, Bohn DJ, Dirks PB, Doucette S, Fergusson D, Gottesman R, Joffe AR, Kirpalani HM, Meyer PG, Morris KP, Moher D, Singh RN, Skippen PW (2008) Hypothermia therapy after traumatic brain injury in children. N Engl J Med 358:2447–2456
78. Fraser DD, Morrison G (2009) Brain oxidative stress after traumatic brain injury … cool it? Crit Care Med 37:787–788
79. Sangha GS, Pepelassis D, Buffo-Sequeira I, Seabrook JA, Fraser DD (2012) Serum troponin-I as an indicator of clinically significant myocardial injury in paediatric trauma patients. Injury 43:2046–2050

Chapter 27
Oxidative Stress and Pulmonary Vascular Disorders

Christine M. Gross, Saurabh Aggarwal, Ruslan Rafikov, and Stephen M. Black

Abbreviations

ADMA	Asymmetrical dimethylarginine
BH_4	Tetrahydrobiopterin
BPD	Bronchopulmonary dysplasia
CHD	Congenital heart disease
CPAP	Continuous positive airway pressure
CrAT	Carnitine acetyltransferase
DUOX	Dual oxidase
eNOS	Endothelial NOS
ET-1	Endothelin-1
FAD	Flavin adenine dinucleotide
GPx	Glutathione peroxidase
GSH	Glutathione
GS-SG	Glutathione disulfide
iNOS	Inducible NOS
IPAH	Idiopathic pulmonary arterial hypertension
NAC	*N*-acetylcysteine
NAD+	Nicotinamide adenine dinucleotide
NADH	Nicotinamide adenine dinucleotide, reduced form
NADPH	Nicotinamide adenine dinucleotide phosphate
NO	Nitric oxide

C.M. Gross • S. Aggarwal • R. Rafikov • S.M. Black, Ph.D. (✉)
The Pulmonary Disease Program, Vascular Biology Center, Georgia Regents University, 1459 Laney Walker Boulevard, CB 3211-B, Augusta, GA 30912, USA
e-mail: SBLACK@gru.edu

H. Tsukahara and K. Kaneko (eds.), *Studies on Pediatric Disorders*, Oxidative Stress in Applied Basic Research and Clinical Practice, DOI 10.1007/978-1-4939-0679-6_27,

NOS	Nitric oxide synthase
NOX	Nicotinamide adenine dinucleotide phosphate oxidase
$p47^{phox}$	Protein 47 kD phagocyte oxidase
$p67^{phox}$	Protein 67 kD phagocyte oxidase
PAEC	Pulmonary artery endothelial cells
PAH	Pulmonary arterial hypertension
PAP	Pulmonary artery pressure
PASMC	Pulmonary artery smooth muscle cells
PBF	Pulmonary blood flow
PEG	Polyethylene glycol
PH	Pulmonary hypertension
PPHN	Persistent pulmonary hypertension of the newborn
PVR	Pulmonary vascular resistance
RDS	Respiratory distress syndrome
ROP	Retinopathy of prematurity
ROS	Reactive oxygen species
SCD	Sickle cell disease
SO_2	Sulfur dioxide
SOD	Superoxide dismutase
UCP-2	Uncoupling protein-2
VEGF	Vascular endothelial growth factor
XO	Xanthine oxidase
XOR	Xanthine oxidoreductase

27.1 Introduction

Life in an oxygen-containing environment inevitably leads to the production of oxygen-derived radicals. The survival of organisms from bacteria to mammals under these "noxious" conditions is facilitated by the evolutionary development of specific enzyme systems that sense and inactivate these species. However, when these primary protection mechanisms are overwhelmed by excessive reactive oxygen species (ROS) production, the situation of "oxidative stress" occurs resulting in the hyper-activation of cellular signaling pathways, as well as the modification of cellular macromolecules. ROS include radicals such as superoxide (O_2^-), hydroxyl ($OH^·$), peroxyl ($RO_2^·$), lipid peroxyl ($LOO^·$), as well as non-radical species, such as hydrogen peroxide (H_2O_2), hypochloric acid (HOCl), ozone (O_3), and lipid peroxide (LOOH). Although initially believed to be toxic species, it is now clear that certain ROS act as signaling intermediates. The main ROS involved in signaling are O_2^- and H_2O_2. ROS are generated in multiple cell types in the vasculature and are generated both from enzymatic (nicotinamide adenine dinucleotide phosphate (NADPH) oxidase, xanthine oxidase (XO), uncoupled nitric oxide synthase (NOS)) and nonenzymatic (mitochondria) sources. The following sections will describe in detail the sources of ROS and the antioxidant systems present within the cell that are designed to maintain ROS within a physiologic range required for their signaling properties. In addition where possible we will tie the dysregulation of these systems to the development of pediatric disorders with a major focus on those responsible for neonatal lung disease.

27.2 Sources of Reactive Oxygen Species

Multiple cell types are sources of ROS in the vasculature. ROS can be generated by enzymatic sources located in multiple cellular compartments as well as cellular organelles. The major sources of ROS in the vasculature are NADPH oxidase, xanthine oxidase, uncoupled NOS, and dysfunctional mitochondria. The mechanisms underlying the upregulation of ROS from these systems are complex and will be dealt with in the following sections.

27.2.1 NADPH Oxidase

One of the most important cellular ROS generators is NADPH oxidase, which is located on the cell membrane, and other organelles, of multiple cell types in the vasculature [1–4]. The NADPH oxidase complex is a transmembrane multimeric protein first discovered in phagocytes where it consumes NADPH to produce O_2^-, which in turn is utilized to kill invading bacteria captured in phagosomes [5]. Intensive study over the last two decades has led to the identification of a multi-gene family consisting of seven isoforms that are expressed in multiple different tissues and undergo complex regulation (Fig. 27.1). The prototype NADPH oxidase produces O_2^- by catalyzing a one-electron reduction of molecular O_2, using nicotinamide adenine dinucleotide (NADH) or NADPH as a donor [6]. However, it is now clear that only nicotinamide adenine dinucleotide phosphate oxidase (NOX) 1–3 and NOX5 produce O_2^-, while NOX4 and dual oxidase (DUOX) 1/2 predominantly generate H_2O_2 (Fig. 27.1). Vascular cells express all the NOX isoforms except NOX3. The physiological role of the NOX enzymes in vascular cells is still unclear although it is usually attributed to redox-regulated signal transduction. The O_2^- and H_2O_2 produced by the NADPH oxidase complexes can oxidize cysteine residues within proteins, which alter their activity, either positively or negatively, leading to either physiological or pathological signaling.

Newborns are very susceptible to oxidative stress conditions and this can result in complications, such as pulmonary hypertension (PH) or hyperoxia-induced lung injury. The exposure of newborn mice to 95 % oxygen results in a distinct phenotype of compromised alveolar and vascular development in the lung [7]. This condition is accompanied with increased activity of the NOX1 and NOX2 isoforms [8].

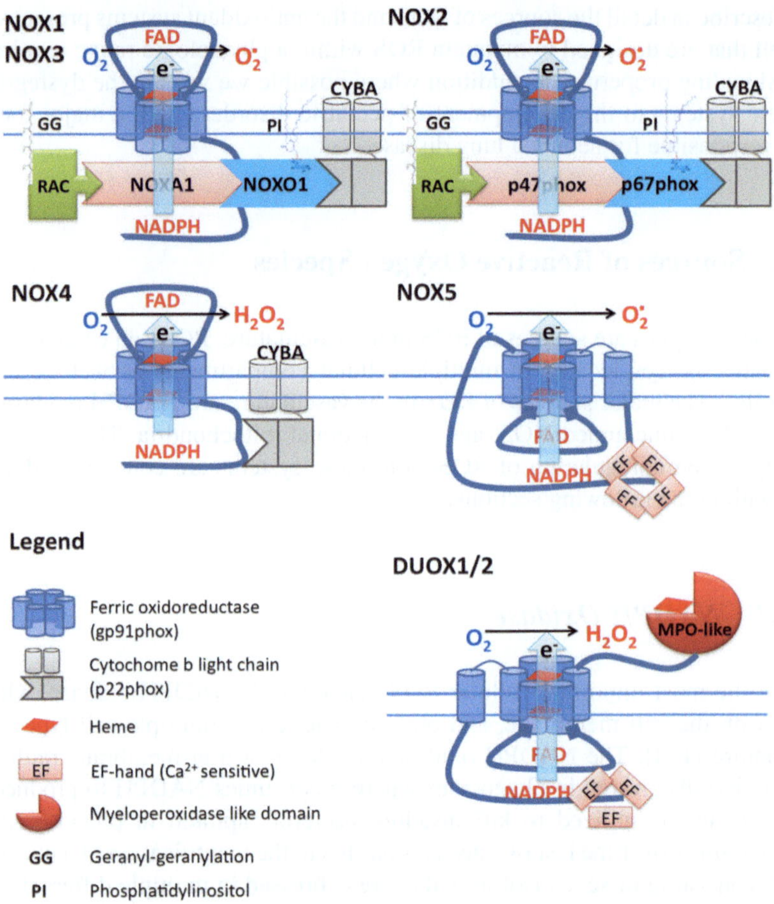

Fig. 27.1 Schematic representation of the NADPH oxidase (NOX) multi-gene family. Data from the Universal Protein Resource (UNIPROT) protein database was obtained. These data allow for both the prediction of transmembrane sequences and for the identification of potential interactions between the protein subunits of the NOX enzymes. All NOX family proteins have six highly conserved transmembrane domains that together compose the ferric oxidoreductase core. The ferric oxidoreductase core has two heme moieties, the NAD binding pocket, and the NADPH binding site. NOX5 and DUOX1/2 have an additional transmembrane domain that links to a calcium-sensitive EF-hand in NOX5 or a myeloperoxidase-like domain in DUOX1/2. NOX isoforms 1–4 require the cytochrome b light chain ($p22^{phox}$) for activation. NOX1/3 and NOX2 require the formation of a multi-protein complex to become activated. NOX1/3 require the binding of NOXA1 and NOXO1. NOX2 requires the binding of $p47^{phox}$ and $p67^{phox}$. In addition, NOX1/3 and NOX2 all require the binding of Rac1. In turn, Rac1 is modified by geranyl-geranylation, while NOXO1 and $p67^{phox}$ bind to phosphatidylinositol. These modifications allow the translocation of the regulatory complex to the plasma membrane resulting in NOX activation

Further, hyperoxia-dependent ROS production, and the accompanying cellular damage, is significantly reduced in the lungs of NOX1-deficient mice confirming the importance of the NOX1 isoform in hyperoxia-induced lung injury [8]. Hypoxia, counterintuitively, can also result in oxidative damage. It is well documented that

NOX4 and other NOX isoforms contribute to the hypoxia-induced ROS generation in the lung [9]. For example, ROS generated by NADPH oxidase contribute to the abnormal responses in pulmonary resistance arteries (PRAs) of piglets exposed to chronic hypoxia for 3 days [10]. Another study has explored the contribution of NOX-derived ROS at a more advanced stage of PH, after 10 days of chronic hypoxia. In this case, NOX1 isoform expression and the accumulation of protein 67kd phagocyte oxidase ($p67^{phox}$) in the membrane fraction were observed, while the expression of NOX4 was unchanged [11]. In lambs with endothelial dysfunction either associated with increased pulmonary blood flow (PBF) or induced through chronic ductal constriction NADPH oxidase-derived O_2^- is also enhanced [12]. Endothelin-1 (ET-1) released from endothelial cells appears to be an important mediator of vascular diseases that require ROS [13, 14]. Plasma levels of ET-1 are increased in a number of animal models, including angiotensin II-induced hypertension [15], spontaneously hypertensive stroke-prone rats, and deoxycorticosterone acetate-salt-induced hypertension [16], and in a lamb model of persistent pulmonary hypertension of the newborn (PPHN) [17]. Adults with idiopathic PH also have higher lung ET-1 levels [18]. It has been recently reported that in the SU5416/hypoxia model of PH that NADPH oxidase-derived O_2^- generation and the vascular remodeling are attenuated by treatment with the combined ET receptor antagonist, bosentan [18]. ET-1 has been shown to increase NADPH oxidase activity and O_2^- levels in cultured pulmonary artery smooth muscle cells (PASMC) [19], while the increased O_2^- in lambs with both acute and chronic ductal constriction correlates with increased ET-1 levels [20].

27.2.2 *Xanthine Oxidase*

Xanthine oxidoreductase (XOR) is a homodimer with an N-terminal iron sulfur cluster, a central flavin adenine dinucleotide (FAD) binding region, and a C-terminal active site containing molybdopterin (Fig. 27.2). Xanthine dehydrogenase (XD) is the primary form of XOR, which can be converted to xanthine oxidase (XO) through posttranslational modifications (Fig. 27.2). During purine catabolism, most of the ROS are produced by XO.

Ischemia–reperfusion injury is associated with significant mortality and long-term morbidity in asphyxiated term and preterm infants. The re-oxygenation of hypoxic tissues has classically been associated with the conversion of XD to XO and is a significant source of ROS production [21]. This phenomenon has also been observed in adult human patients undergoing reconstructive surgery. After ischemia and reperfusion (tourniquet release), plasma XO activity and uric acid concentrations increase [22]. In addition, preterm infants classified as having poor outcome had significantly higher plasma XO activity and lipid hydroperoxide levels than term infants and normal preterm infants [23]. Similarly, the concentration of uric acid is higher in fetuses with well-defined criteria of intrapartum hypoxia, while elevated hypoxanthine and xanthine levels are correlated with fetal heart rate tracings indicative of fetal stress [24]. Xanthine levels in premature newborns are higher

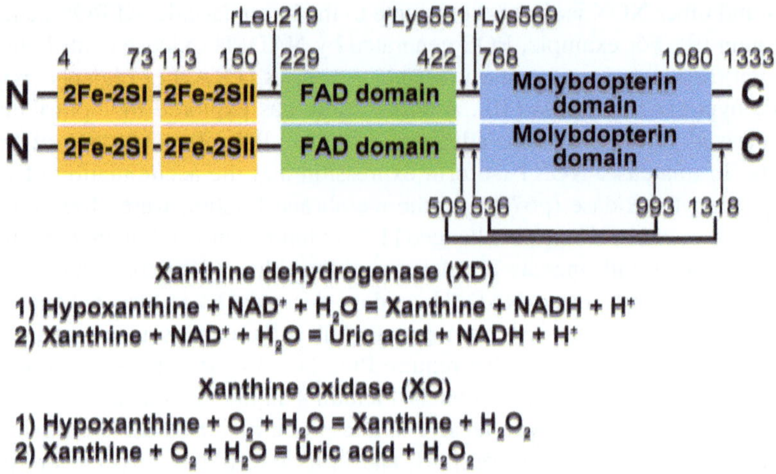

Fig. 27.2 Xanthine oxidoreductase (XOR) is a hydroxylase involved in the oxidative metabolism of purines. XOR is a homodimer with an N-terminal iron sulfur domain containing two [2Fe–2S] clusters (2Fe–2SI and 2Fe–2SII), a central FAD binding region, and a C-terminal active site containing molybdopterin. In humans, each monomer is a 1,333-amino acid peptide [141], and each acts as independent catalytic units [142]. XOR is primarily in the form of xanthine dehydrogenase (XD); however, XD can be switched to xanthine oxidase (XO) through several posttranslational modifications. The cleavage of rat XD by trypsin after Lys551 [143] or by pancreatin after Leu219 and Lys569 [144] will irreversibly convert XD to XO. Similarly, intra-disulfide bridges between Cys535 and Cys992 and Cys1316 and Cys1324 in bovine [142] and rat XD [145] will reversibly switch XD to XO, while in humans, intra-disulfide bonds between Cys536 and Cys993 and Cys509 and Cys1318 have been identified [146]. During the hydroxylation of hypoxanthine to uric acid, substrate oxidation occurs at the molybdenum center, and then the electrons travel across the two [2Fe–2S] clusters to the FAD domain where the re-oxidation of the enzyme occurs by reducing either NAD^+ or oxygen

than in mature newborns, while the highest levels in this study were found in premature newborns with respiratory distress syndrome (RDS) [25]. Another study has shown that XO activity increases in newborns with RDS and in critically ill children; however, XO activity later decreases in the reparative phase [26], presumably by an adaptive process. In premature infants with RDS and bronchopulmonary dysplasia (BPD), XO activity in endotracheal tube aspirates rises even higher than in infants with simple RDS [27]. Interestingly, high concentrations of elastase correlated with XO activity and enabled the subsequent prediction of BPD [27]. Indeed, proteases, such as neutrophil elastase, have been known to induce the switch from XD to XO [28]. XO activity is also increased in infants with septic shock syndrome during early, pro-inflammatory stages, while lower XO activity is found in later, anti-inflammatory stages [29]. Plasma XO activity is also increased in sickle cell disease (SCD) [30]. In addition, it has also been suggested that premature infants with RDS could be predisposed to developing retinopathy of prematurity (ROP) because increased levels of hypoxanthine in the vitreous humor of the eye have been found [31]. Increased hypoxanthine [32] and uric acid [33] levels are also found in infants with cavitating periventricular leukomalacia, a significant precursor of cerebral palsy. In addition, hyperuricemia is a common finding in patients with PH and is indicative of increased XO activity [34–38]. Elevated uric acid levels in patients with PPH correlate with pulmonary vascular resistance (PVR) and mortality and negatively correlate with cardiac output, suggesting that high serum uric acid content is associated with the severity of PPH [34, 35]. High serum uric acid levels also correlate with increased clinical severity of idiopathic pulmonary arterial hypertension (IPAH), congenital heart disease (CHD)-related PAH, and connective tissue disease-related PAH [36, 37]. A similar study in children with PAH shows that pulmonary artery pressure (PAP) and PVR are positively associated with, while cardiac index is negatively associated with, serum uric acid levels [38]. In adult patients with IPAH, flow-mediated vasodilation is impaired and plasma malondialdehyde and XO activity are increased [39, 40].

27.2.3 Endothelial Nitric Oxide Synthase

A dramatic drop in PVR and increases in PBF accompany the transition from fetal to postnatal life. Nitric oxide (NO) generated in the pulmonary vascular endothelium by endothelial NO synthase (eNOS) plays a pivotal role in the adaptation of the perinatal circulation to air breathing life. NO is generated by the five-electron oxidation of the guanidino nitrogen of L-arginine (Fig. 27.3). eNOS differs from inducible NOS (iNOS) in that its activity is dependent on the presence of the calcium/calmodulin complex. Thus, eNOS activity can be regulated by changes in intracellular calcium concentrations and can be activated by agonists that increase calcium levels, such as bradykinin [41], estradiol [42], and vascular endothelial growth factor (VEGF) [43]. Mutations in the eNOS gene increase PAP, impair the relaxation to acetylcholine [44], and promote DNA synthesis and PASMC proliferation [45]. NO has also been demonstrated to induce apoptosis [46] in PASMC and inhibit the production of growth factors, including ET-1 and platelet-derived growth factor (PDGF), in pulmonary artery endothelial cells (PAEC) [47]. Rather than being a constitutive enzyme as was first suggested, eNOS is dynamically regulated at the transcriptional, posttranscriptional, and posttranslational levels. However, the most complex regulation of eNOS appears to be through its phosphorylation (Fig. 27.3). The diminished expression of eNOS [48] and the generation of O_2^- instead of NO, referred to as NOS uncoupling [49] (Fig 27.4), have been identified as major contributors to decreased NO signaling and impaired vascular function. The formation of O_2^- can in turn decrease the bioavailability of NO by combining rapidly to form $ONOO^-$. Exposure to high levels of oxidative stress during the neonatal period can also lead to NOS uncoupling and vascular pathologies in adulthood. In the transition from fetal to neonatal life, an abrupt increase in blood O_2 content and O_2 availability within first few minutes after birth exposes the neonate to a burst of O_2^- generation. In premature infants, the lack of antioxidant defense mechanisms coupled with the need for O_2 supplementation, because of lung immaturity, can lead

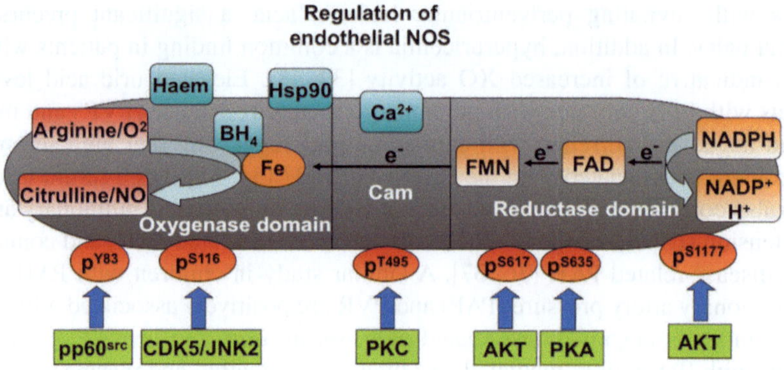

Fig. 27.3 Endothelial nitric oxide synthase (eNOS) produces NO by catalyzing a five-electron oxidation of a guanidino nitrogen of L-arginine. The oxidation of arginine to citrulline occurs via two successive monooxygenation reactions during which 2 mol of O_2 and 1.5 mol of nicotinamide adenine dinucleotide phosphate (NADPH) are consumed per mole of NO formed. The enzyme is a dimer consisting of two identical monomers, which can be divided into two major domains: a C-terminal reductase domain and an N-terminal oxygenase domain. The N-terminal oxygenase domain contains cysteine-ligated heme and tetrahydrobiopterin (BH_4), the C-terminal reductase domain binds flavin mononucleotide (FMN) and flavin adenine dinucleotide (FAD), and an intervening calmodulin (Cam) binding region plays an important role in both the structure and function of the enzyme. In addition, the binding of heat shock protein 90 (Hsp90) to the reductase domain is important for the activation of eNOS. Phosphorylation of eNOS is a major mechanism for posttranslational regulation of eNOS. Phosphorylation at residues serine 617 and serine1177 by Akt, serine 645 and serine 1177 by protein kinase A (PKA), and tyrosine 83 by $pp60^{Src}$ activates eNOS. However, the phosphorylation of threonine 495 by protein kinase C (PKC) and serine 116 by cyclin-dependent kinase 5 (CDK5) and c-jun n-terminal kinase 2 (JNK2) is associated with eNOS inhibition

to significant oxidative stress in the immediate neonatal period and predispose them to vascular abnormalities in adulthood.

The inability of a newborn to successfully achieve and sustain the decline in PVR at birth results in PPHN, a condition associated with significant morbidity and mortality. The inhibition of NOS activity in utero reduces NO production and leads to hemodynamic alterations that are similar to those seen in PPHN [50]. In addition, early administration of exogenous (inhaled) NO causes potent and selective pulmonary vasodilation in neonates with respiratory failure and PH [51]. Impaired NO bioavailability is also associated with the development of PH in a subset of children with SCD. The decompartmentalization of hemoglobin into the plasma, as the result of hemolysis in SCD, scavenges NO and impairs vascular function [52]. Treatments that inactivate plasma hemoglobin restore NO bioavailability and improve PBF [52]. In a mouse model of SCD, the development of PH is associated with a reduction in eNOS activity as a result of the disruption of the eNOS dimer, hemolysis-related NO scavenging by free hemoglobin in plasma, and increased arginase activity [53]. In a randomized clinical trial, oral administration of L-arginine produced a 15 % reduction in pulmonary artery systolic pressure in patients with SCD [54]. In a small pilot study, the administration of inhaled NO to children suffering

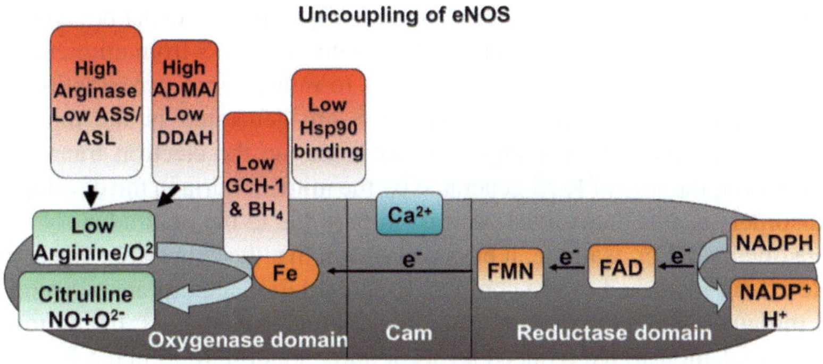

Fig. 27.4 The cartoon depicts the mechanisms by which endothelial nitric oxide synthase (eNOS) is uncoupled, switching from the physiological production of NO to a predominant generation of superoxide (O_2^-). The major pathways involved are the decrease in the NOS cofactor tetrahydrobiopterin (BH_4) and the enzyme GTP cyclohydrolase 1 (GCH-1), which is required for the de novo generation of BH_4, the pathological accumulation of the endogenous NOS inhibitor, asymmetrical dimethylarginine (ADMA), and the decreased expression or the activity of the enzyme, dimethylarginine dimethylaminohydrolase (DDAH), responsible for the degradation of ADMA. In addition, the depletion of the substrate L-arginine is secondary to the increased activity of the enzyme, arginase, which is required for the degradation of arginine or the decreased activity of the enzymes, argino-succinate synthase (ASS) and arginosuccinate lyase (ASL), which recycle L-arginine from L-citrulline. The decreased interaction of eNOS with heat shock protein 90 (hsp90) also plays an important role in the uncoupling of the enzyme

from sickle cell pain crises suggested that the treatment lowered the amount of analgesic therapy [55]. Sildenafil, which inhibits phosphodiesterase 5 and therefore enhances the biologic effect of NO on guanylate cyclase, has also been demonstrated to benefit SCD patients by reducing their PAP [56]. In premature infants, the development of PH is also associated with BPD. The pathophysiology of BPD appears to be caused by a combination of factors including inflammatory injury, mechanical ventilation, oxygen therapy, and lung prematurity. Although the direct role of NOS uncoupling in BPD has not yet been established, animal models of BPD exhibit reduced pulmonary eNOS [57], while premature infants with BPD have high levels of 3-nitrotyrosine [58], suggesting that eNOS may be uncoupled. The role of inhaled NO therapy in the amelioration of BPD is inconclusive and is dependent upon the initial severity of the disease where infants with milder pathology respond better than the ones with more severe disease [59].

27.2.4 Mitochondrial Dysfunction

The disruption of mitochondrial function is a critical event in a number of pathologic conditions including hypoxia-ischemic injuries [60], stroke [61], and diabetes [62–66]. Mitochondrial dysfunction also appears to be important in the development of PH

[67–69]. Mitochondria are a significant source of O_2^- and H_2O_2 under both physiologic and pathologic conditions. The increased production of ROS from mitochondria can be deleterious to the cell due to their ability to induce lipid peroxidation, protein oxidation, and DNA damage [70, 71]. ROS are generated in the mitochondria through the activities of the complexes that make up the electron transport chain [72]. However, the role of ROS generated by the mitochondria in the development of lung disorders is still unresolved, at least in part due to the contradictory nature of the data obtained. For example, it has been shown that the expression and activity of complex II are increased, while complex II-derived ROS are elevated, in a rat model of PH initiated using the injection of monocrotaline (MCT) [73]. In contrast, other studies have demonstrated reduced levels of mitochondrial ROS in the fawn-hooded (FH) rat [74] that spontaneously develops PH, and in models of PH induced by hypoxia and MCT [75]. Low levels of mitochondrial-derived ROS have also been found in PASMC isolated from patients with PH [76]. However, others have found that under hypoxic conditions, mitochondrial-derived ROS are elevated [77–79], possibly due to derangements in complex II [80]. In PAEC, asymmetrical dimethylarginine (ADMA)-mediated eNOS uncoupling induces mitochondrial dysfunction by increasing both uncoupling protein-2 (UCP-2) and mitochondrial ROS generation [81]. Similarly, lambs with PH secondary to increased PBF [82] also have endothelial mitochondrial dysfunction, as evidenced by increases in UCP-2, decreases in superoxide dismutase (SOD)-2, and increased lactate levels [83].

Two other mitochondrial metabolic pathways, acyl-CoA and carnitine acetyltransferase (CrAT), appear to be of critical importance in maintaining normal mitochondrial function. Recent data have shown that under conditions of metabolic stress, mitochondria accumulate acyl-CoA (Fig. 27.5), which is normally maintained in homeostasis with carnitine. High acyl-CoA levels inhibit multiple enzymatic processes involved in oxidative metabolism [84]. Acyl-CoA also inhibits multiple other mitochondrial enzymes downstream from its own synthesis resulting in a metabolic roadblock within the mitochondrial matrix [85–88]. Also, cell-based and in vivo studies from Ames' group implicate oxidative damage to key mitochondrial enzymes as important mediators of the decrease in affinity (K_m) for carnitine, the enzyme substrate [1–3]. The principal enzyme affected by this process is CrAT (Fig. 27.4) that catalyzes an equilibrium reaction between short-chain acyl-CoA and CoA, and acyl-carnitine and carnitine (Fig. 27.5). In aging rats, this process is associated with a loss of the membrane phospholipid cardiolipin, reduction in mitochondrial membrane potential, and a decrease in cellular oxygen uptake [89, 90]. Recent studies have begun to evaluate the contribution of these integral mitochondrial processes to the pathophysiology of PH associated with increased PBF in the neonatal period. Data indicate that carnitine homeostasis is disrupted in these lambs and this is a multifactorial process involving both the decreased expression of key carnitine homeostasis enzymes and the posttranslational inhibition of CrAT through increased nitration [83]. Further, it has been shown that the loss of carnitine homeostasis correlates with increased mitochondrial ROS generation and the disruption of mitochondrial bioenergetics [83, 91, 92]. This process appears to be driven by ADMA-mediated eNOS uncoupling and the redistribution of eNOS from the plasma

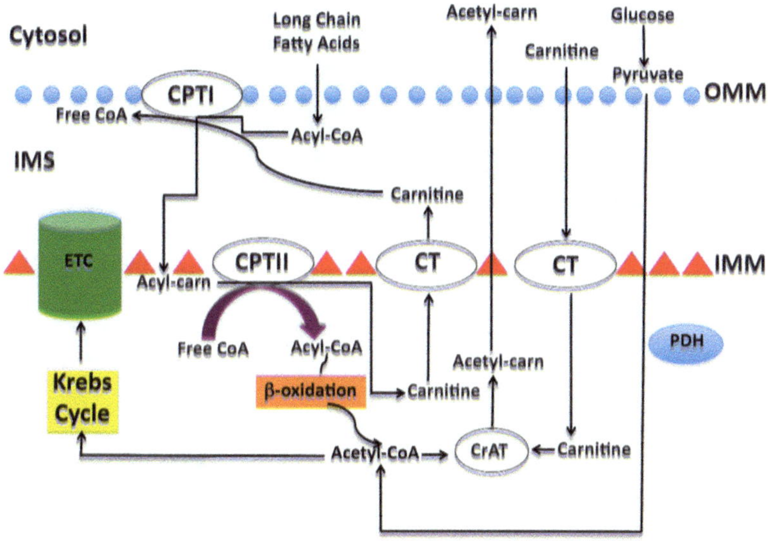

Fig. 27.5 Carnitine-dependent enzymes and transporters associated with the mitochondrial metabolism of acyl moieties. *CPTI* carnitine palmitolyl transferase 1, *CT* carnitine translocase, *CrAT* carnitine acetyltransferase, *CoA* coenzyme A, *Carn* carnitine, *IMM* inner mitochondrial membrane, *OMM* outer mitochondrial membrane, *IMS* intermembrane space, *PDH* pyruvate dehydrogenase

membrane to the mitochondria through the nitration-mediated phosphorylation of eNOS via Akt1. This has been demonstrated both in vitro [81, 93] and in vivo [92]. The disruption of mitochondrial bioenergetics leads to a decrease in ATP levels which in turn reduces hsp90 activity [81, 82]. As eNOS is chaperoned by hsp90 [94] this results in a "feed-forward" signaling cascade in which eNOS becomes progressively more uncoupled leading to less NO—and more $ONOO^-$—being produced driving pulmonary endothelial dysfunction. Further, recent intervention studies have shown that chronic L-carnitine supplementation preserves both NO signaling and pulmonary endothelial function [91, 95]. It is also tempting to speculate that agents that help to maintain eNOS in its coupled state and/or prevent its mitochondrial translocation could have therapeutic benefit for the treatment of children born with CHD that results in increased PBF.

27.3 Antioxidant Enzyme Systems

The redox balance in cells is partially maintained by antioxidant enzymes, and their dysregulation is involved in the pathology of several vascular disorders [96]. SOD, catalase, and glutathione peroxidase (GPx) are major vascular enzymatic antioxidants (Fig. 27.6). It is generally thought that increases in oxidative stress induce a compensatory increase

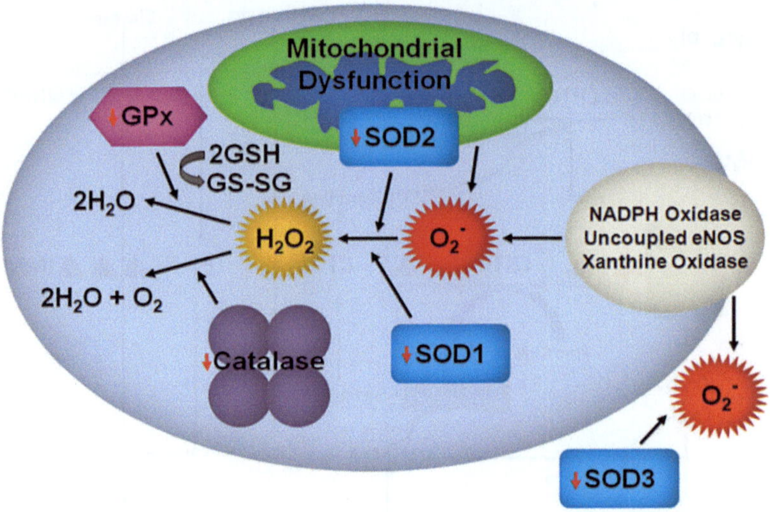

Fig. 27.6 The scavenging of reactive oxygen species (ROS) by antioxidant enzymes is necessary to maintain the redox homeostasis of the cell. The three superoxide dismutase (SOD) isotypes, Cu/ZnSOD (SOD1), MnSOD (SOD2), and EC-SOD (SOD3), are located in the cytosol, mitochondria, and extracellular space, respectively, where they reduce O_2^- to H_2O_2. Tetrameric catalase converts two H_2O_2 molecules to two H_2O molecules and one O_2 molecule in one of the fastest enzymatic reactions known. Similarly, to decompose one H_2O_2 molecule to two H_2O molecules, glutathione peroxidase (GPx) oxidizes two glutathione (GSH) to glutathione disulfide (GS-SG). In many pediatric vascular disorders, the levels of these antioxidant enzymes fail to respond to increases in ROS, thereby further compounding the cellular injury induced by oxidative stress. Several clinical trials now indicate that antioxidant supplementation can ameliorate disease progression in respiratory distress syndrome (RDS), bronchopulmonary dysplasia (BPD), retinopathy of prematurity (ROP), and patients with sickle cell disease (SCD)

in antioxidant enzymes; failure to do so predisposes the vasculature to vasoconstrictive, vasoproliferative, and edematous disorders. CuZnSOD (SOD1) exists in the cytosol, MnSOD (SOD2) is present in the mitochondria, and EC-SOD (SOD3) is extracellular in nature (Fig. 27.6). SOD catalyzes the conversion of O_2^- into H_2O_2, whereas catalase reduces H_2O_2 into oxygen and water [96] (Fig. 27.6). In addition, GPx decomposes H_2O_2 into water and glutathione disulfide (GSSG) (Fig. 27.6). The following sections will discuss in some detail the latest information regarding the role of these antioxidant enzyme systems in the neonatal period.

27.3.1 Superoxide Dismutase

There is abundant data in several animal models of PH that SOD expression and/or activity are decreased [97–100]. However, few studies have been carried out in humans and even fewer in children with PH. One such study, in infants with CHD with PH, has shown that there are increased levels of oxidative stress both before and after corrective open heart surgery, and this correlates with reductions in plasma and lung SOD and lung glutathione (GSH) [101]. However, in adults with primary PH, GSH levels have been shown to be higher than control patients, while SOD and GPx activities appear to be unchanged [102]. In contrast, reports have suggested that oxidative stress markers, such as 5-oxo-eicosatetraenoic acid, 5-hydroxyeicosatetraenoic acid, and malondialdehyde, increase, while SOD levels and/or activity decrease, in adult patients with IPAH [39, 103, 104]. Interestingly, decreases in SOD and GPx activity have been shown to positively correlate with increases in PAP and PVR [105]. Further, pulmonary arteries from patients with PAH have decreased SOD2 staining in the media and adventitia [106]. Similarly, PAEC isolated from patients with IPAH also have decreased expression of SOD2 [107].

SOD is important for determining outcome in pediatric vascular disorders with high levels of SOD associated with increased survival and decreased morbidity. The activities of major antioxidant enzymes, SOD, catalase, and GPx, increase toward the end of term gestation in several animal models, which is thought to prepare the neonate for life in an O_2-rich environment [100, 108, 109]. In humans, SOD activity in the lungs of premature infants is substantially lower than in the lungs of full-term newborns [110, 111]. There also appears to be gender differences in SOD expression and activity, as a prospective study has shown that owing to greater antioxidant capabilities and reduced markers of oxidative stress, female preterm infants have better outcomes in conditions such as BPD, intra-periventricular hemorrhage, or ROP than males [112]. SOD activity also appears to be involved in the development of severe RDS. SOD activity increases in premature infants given oxygen therapy; however, in patients that later develop RDS, SOD activity does not increase [113, 114]. Similarly, in another study in premature infants with RDS, initial SOD levels are less than in infants without RDS [115]. This study also found that erythrocyte SOD content inversely correlates with the development of BPD and mortality [115]. A prospective double-blind controlled study found that the administration of SOD to neonates with RDS decreased the development of BPD, attenuated clinical presentations of the disease, and reduced the need for continuous positive airway pressure (CPAP), suggesting that SOD therapy might be effective in alleviating the symptoms of BPD in infants with RDS [116]. In addition to BPD, preterm infants given intensive neonatal care and supplemental oxygen are also at risk of developing ROP. ROP is a major cause of blindness in children and involves the abnormal growth of retinal blood vessels possibly leading to retinal detachment. Oxidative stress is also implicated in this disease process. Preterm neonates that later developed ROP have increased markers of oxidative stress [117]. SOD and GPx activities, while higher than normal preterm infants, decrease over time compared to preterm infants who do not develop ROP and who appear to be able to increase in antioxidant activities over time [117]. Again, the key role of SOD in the development of ROP has been demonstrated by a recent clinical trial in which SOD administration reduced the severity and incidence of ROP by 53 % in preterm infants (<25 weeks) [118].

27.3.2 Catalase

Catalase is a tetrameric antioxidant enzyme that is located in the cytoplasm and in peroxisomes and converts H_2O_2 into water and oxygen. The available data relating to the regulation of catalase levels and/or activity in humans is confusing, likely due to the lack of published studies. For example, catalase activity is up-regulated in preterm infants with RDS [119], while catalase levels are lower in the lungs of neonates who die after RDS [120]. Catalase activity is increased in infants that eventually develop BPD compared with infants with RDS alone [27], while premature lambs with RDS treated with intravenous SOD and catalase have less lung damage and hemorrhaging [121]. Furthermore, SOD and catalase supplementation ameliorate hyperoxic lung injury in rats [122]. Catalase levels are also reduced in other pediatric disorders, such as SCD [123, 124]. In adults with COPD and PH, catalase and SOD activities do not appear to be different than control non-smoking patients [125], although lower catalase activities in patients with COPD have been reported [126]. Similarly, catalase activity in the lungs of patients with IPAH was reported to be unchanged [104], while another study has indicated that catalase activity may actually vary during the development of PH [100]. In juvenile lambs with increased PBF and PH, catalase levels and activity are decreased at 2 weeks of age, but return to normal levels at 4 and 8 weeks [100], suggesting that the down-regulation of catalase may play an important role in early oxidative stress in the pathogenesis of the disease. In the lamb model of PPHN induced by in utero ductal ligation, H_2O_2 concentrations are elevated in pulmonary arteries, although catalase and GPx expression and activities remain unaffected [17]. These studies highlight a central theme that a delay in or the failure to up-regulate antioxidant enzymes in the presence of increased oxidative stress predisposes the vasculature to oxidative damage and dysfunction.

Although the therapeutic benefits of catalase supplementation in humans with PH are still unclear, in animals, catalase supplementation has been shown to have protective effects on pulmonary function in PH. Catalase activity in the MCT model of PH was increased, while treatment with a sulfur dioxide (SO_2) donor further increased catalase and SOD activities which normalized pulmonary vessel morphology [127]. In a rat model of PH induced by hyperoxia, a daily intratracheal injection of liposome-encapsulated catalase prevented the vascular and lung tissue damage associated with oxygen toxicity [128]. Interestingly, liposome-encapsulated SOD or SOD+catalase did not have an effect [128]. Similarly, a combined treatment of both polyethylene glycol (PEG)-catalase and PEG-SOD did not attenuate group B streptococcus-induced PH [129], suggesting that the efficacy of antioxidants may be specific to the disease itself. PEG-catalase improves the vasodilator responses to exogenous NO in pulmonary arteries isolated from lambs after in utero ductal ligation [17]. The administration of intratracheal PEG-catalase in this model has also been shown to elevate SOD3 activity, decrease O_2^- levels, and improve arterial oxygenation when the lambs were ventilated with 100 % O_2 [98]. This study is important, as it suggests that the H_2O_2 generated during hyperoxic ventilation may be capable of stimulating O_2^- production through the oxidative modification and inhibition of SOD3 and that part of the protective mechanism of intratracheal catalase is through its ability to preserve SOD3 activity and thereby improve vascular function [98].

27.3.3 Glutathione Peroxidase (GPx)

The most prevalent isoform, GPx1, contains selenium and is located in the cytosol. GPx requires GSH as a co-substrate. Thus, one must evaluate these systems in tandem. Several studies indicate that GPx activity is decreased in pediatric disorders. For example, erythrocyte GPx activity is reduced in premature infants with, compared to those without, BPD [130]. In fact, preterm infants with RDS who later recover can be distinguished from those who later develop BPD using variables, such as gestational age, Apgar scores, erythrocyte selenium, GSH, GPx, and SOD [131]. Similarly, decreased GPx activity is considered a determinant for ROP [132]. Infants with ROP have less GSH, more GSSG, and the highest ratio of GSSG/GSH compared with control and ROP patients in remission [133]. Erythrocyte GSH content is also reduced in SCD patients with PH, thereby possibly contributing to oxidative hemolysis [134]. Similarly, GSH levels [135], SOD, and GPx activities [136] are lower in pediatric thalassemia patients with vascular endothelial dysfunction and increased arterial stiffness [135]. Similarly, in adults with primary PH [137] or IPAH [105], GPx activities are lower than healthy patients. In contrast, GPx activity in erythrocytes from these primary PH patients was greater than control. This difference was most pronounced in patients with severe cardiac insufficiency and PH [137]. Another study found that, while SOD and GPx activities were decreased in the lungs of IPAH patients, catalase activity and GSH levels were unchanged [104]. Interestingly, PAP in these patients could be predicted by utilizing regression analysis of SOD, GPx, and exhaled NO levels [104]. Decreased erythrocyte GPx activity is also present in patients with COPD and secondary PH, while the erythrocyte activities of SOD and catalase are unchanged [125]. The GSH precursor, *N*-acetylcysteine (NAC), has been shown to reduce oxidative stress [138] and vaso-occlusive episodes [139] in sickle cell patients and has shown benefits in adults with RDS [140]. This suggests that targeting the GSH/GPx system may have therapeutic potential that should be evaluated.

27.4 Conclusion

It is clear that oxidative stress plays a major role in the development of a number of pediatric vascular disorders especially in the developing lung. Further, given the success of targeting these enzymes in animal models it is likely that there is the potential for therapeutic interventions in children. The availability of specific inhibitors, both pharmacologic and peptide based, makes NOX a tempting therapeutic target. However, the success of L-carnitine in attenuating mitochondrial and endothelial dysfunction in prevention studies warrants further studies to evaluate its ability to reverse underlying disease states. In conclusion, we are in an exciting time for pediatric investigations as we start to move from animal-based studies into human-based clinical trials.

Acknowledgments The authors would like to acknowledge research support from the National Institutes of Health (HL60190, HL67841, HL084739, R21HD057406, HL61284, P01HL0101902) and the Foundation Leducq. Christine Gross is supported, in part, by a pre-doctoral fellowship from the Southeast Affiliates of the American Heart Association (12PRE12060224). Dr. Black also thanks the efforts of all past and present laboratory members.

References

1. Mittal M, Roth M, Konig P, Hofmann S, Dony E, Goyal P, Selbitz AC, Schermuly RT, Ghofrani HA, Kwapiszewska G, Kummer W, Klepetko W, Hoda MA, Fink L, Hanze J, Seeger W, Grimminger F, Schmidt HH, Weissmann N (2007) Hypoxia-dependent regulation of nonphagocytic NADPH oxidase subunit NOX4 in the pulmonary vasculature. Circ Res 101(3):258–267
2. Babior BM (2000) The NADPH oxidase of endothelial cells. IUBMB Life 50(4–5):267–269
3. Vignais PV (2002) The superoxide-generating NADPH oxidase: structural aspects and activation mechanism. Cell Mol Life Sci 59(9):1428–1459
4. Li S, Tabar SS, Malec V, Eul BG, Klepetko W, Weissmann N, Grimminger F, Seeger W, Rose F, Hanze J (2008) NOX4 regulates ROS levels under normoxic and hypoxic conditions, triggers proliferation, and inhibits apoptosis in pulmonary artery adventitial fibroblasts. Antioxid Redox Signal 10(10):1687–1698
5. McRipley RJ, Sbarra AJ (1967) Role of the phagocyte in host-parasite interactions. XI. Relationship between stimulated oxidative metabolism and hydrogen peroxide formation, and intracellular killing. J Bacteriol 94(5):1417–1424
6. Griendling KK, Sorescu D, Ushio-Fukai M (2000) NAD(P)H oxidase: role in cardiovascular biology and disease. Circ Res 86(5):494–501
7. Auten RL, Mason SN, Auten KM, Brahmajothi M (2009) Hyperoxia impairs postnatal alveolar epithelial development via NADPH oxidase in newborn mice. Am J Physiol Lung Cell Mol Physiol 297(1):L134–L142
8. Carnesecchi S, Deffert C, Pagano A, Garrido-Urbani S, Metrailler-Ruchonnet I, Schappi M, Donati Y, Matthay MA, Krause KH, Barazzone Argiroffo C (2009) NADPH oxidase-1 plays a crucial role in hyperoxia-induced acute lung injury in mice. Am J Respir Crit Care Med 180(10):972–981
9. Nisbet RE, Graves AS, Kleinhenz DJ, Rupnow HL, Reed AL, Fan TH, Mitchell PO, Sutliff RL, Hart CM (2009) The role of NADPH oxidase in chronic intermittent hypoxia-induced pulmonary hypertension in mice. Am J Respir Cell Mol Biol 40(5):601–609
10. Fike CD, Slaughter JC, Kaplowitz MR, Zhang Y, Aschner JL (2008) Reactive oxygen species from NADPH oxidase contribute to altered pulmonary vascular responses in piglets with chronic hypoxia-induced pulmonary hypertension. Am J Physiol Lung Cell Mol Physiol 295(5):L881–L888
11. Dennis KE, Aschner JL, Milatovic D, Schmidt JW, Aschner M, Kaplowitz MR, Zhang Y, Fike CD (2009) NADPH oxidases and reactive oxygen species at different stages of chronic hypoxia-induced pulmonary hypertension in newborn piglets. Am J Physiol Lung Cell Mol Physiol 297(4):L596–L607

12. Oishi PE, Sharma S, Datar SA, Kumar S, Aggarwal S, Lu Q, Raff G, Azakie A, Hsu JH, Sajti E, Fratz S, Black SM, Fineman JR (2013) Rosiglitazone preserves pulmonary vascular function in lambs with increased pulmonary blood flow. Pediatr Res 73(1):54–61
13. Pollock DM (2005) Endothelin, angiotensin, and oxidative stress in hypertension. Hypertension 45(4):477–480
14. Remuzzi G, Perico N, Benigni A (2002) New therapeutics that antagonize endothelin: promises and frustrations. Nat Rev Drug Discov 1(12):986–1001
15. Alexander BT, Cockrell KL, Rinewalt AN, Herrington JN, Granger JP (2001) Enhanced renal expression of preproendothelin mRNA during chronic angiotensin II hypertension. Am J Physiol Regul Integr Comp Physiol 280(5):R1388–R1392
16. Abdel-Sayed S, Nussberger J, Aubert JF, Gohlke P, Brunner HR, Brakch N (2003) Measurement of plasma endothelin-1 in experimental hypertension and in healthy subjects. Am J Hypertens 16(7):515–521
17. Wedgwood S, Steinhorn RH, Bunderson M, Wilham J, Lakshminrusimha S, Brennan LA, Black SM (2005) Increased hydrogen peroxide downregulates soluble guanylate cyclase in the lungs of lambs with persistent pulmonary hypertension of the newborn. Am J Physiol Lung Cell Mol Physiol 289(4):L660–L666
18. Rafikova O, Rafikov R, Kumar S, Sharma S, Aggarwal S, Schneider F, Jonigk D, Black SM, Tofovic SP (2013) Bosentan inhibits oxidative and nitrosative stress and rescues occlusive pulmonary hypertension. Free Radic Biol Med 56:28–43
19. Wedgwood S, Lakshminrusimha S, Czech L, Schumacker PT, Steinhorn RH (2013) Increased p22(phox)/Nox4 expression is involved in remodeling through hydrogen peroxide signaling in experimental persistent pulmonary hypertension of the newborn. Antioxid Redox Signal 18(14):1765–1776
20. Ovadia B, Bekker JM, Fitzgerald RK, Kon A, Thelitz S, Johengen MJ, Hendricks-Munoz K, Gerrets R, Black SM, Fineman JR (2002) Nitric oxide-endothelin-1 interactions after acute ductal constriction in fetal lambs. Am J Physiol Heart Circ Physiol 282(3):H862–H871
21. Bindoli A, Cavallini L, Rigobello MP, Coassin M, Di Lisa F (1988) Modification of the xanthine-converting enzyme of perfused rat heart during ischemia and oxidative stress. Free Radic Biol Med 4(3):163–167
22. Friedl HP, Smith DJ, Till GO, Thomson PD, Louis DS, Ward PA (1990) Ischemia–reperfusion in humans. Appearance of xanthine oxidase activity. Am J Pathol 136(3):491–495
23. Supnet MC, David-Cu R, Walther FJ (1994) Plasma xanthine oxidase activity and lipid hydroperoxide levels in preterm infants. Pediatr Res 36(3):283–287
24. Porter KB, O'Brien WF, Benoit R (1992) Comparison of cord purine metabolites to maternal and neonatal variables of hypoxia. Obstet Gynecol 79(3):394–397
25. Karmazsin L, Balla G (1985) Plasma hypoxanthine and xanthine levels in the early newborn period in problem-free preterm babies and those with idiopathic respiratory distress syndrome. Acta Paediatr Hung 26(1):1–9
26. Boda D, Nemeth I (1989) Measurement of urinary caffeine metabolites reflecting the "in vivo" xanthine oxidase activity in premature infants with RDS and in hypoxic states of children. Biomed Biochim Acta 48(2–3):S31–S35
27. Contreras M, Hariharan N, Lewandoski JR, Ciesielski W, Koscik R, Zimmerman JJ (1996) Bronchoalveolar oxyradical inflammatory elements herald bronchopulmonary dysplasia. Crit Care Med 24(1):29–37
28. Phan SH, Gannon DE, Ward PA, Karmiol S (1992) Mechanism of neutrophil-induced xanthine dehydrogenase to xanthine oxidase conversion in endothelial cells: evidence of a role for elastase. Am J Respir Cell Mol Biol 6(3):270–278
29. Nemeth I, Boda D (2001) Xanthine oxidase activity and blood glutathione redox ratio in infants and children with septic shock syndrome. Intensive Care Med 27(1):216–221
30. Aslan M, Ryan TM, Adler B, Townes TM, Parks DA, Thompson JA, Tousson A, Gladwin MT, Patel RP, Tarpey MM, Batinic-Haberle I, White CR, Freeman BA (2001) Oxygen radical inhibition of nitric oxide-dependent vascular function in sickle cell disease. Proc Natl Acad Sci U S A 98(26):15215–15220

31. Saugstad OD, Rognum TO (1988) High postmortem levels of hypoxanthine in the vitreous humor of premature babies with respiratory distress syndrome. Pediatrics 81(3):395–398
32. Russell GA, Jeffers G, Cooke RW (1992) Plasma hypoxanthine: a marker for hypoxic-ischaemic induced periventricular leucomalacia? Arch Dis Child 67(4 Spec No):388–392
33. Perlman JM, Risser R (1998) Relationship of uric acid concentrations and severe intraventricular hemorrhage/leukomalacia in the premature infant. J Pediatr 132(3 Pt 1):436–439
34. Nagaya N, Uematsu M, Satoh T, Kyotani S, Sakamaki F, Nakanishi N, Yamagishi M, Kunieda T, Miyatake K (1999) Serum uric acid levels correlate with the severity and the mortality of primary pulmonary hypertension. Am J Respir Crit Care Med 160(2):487–492
35. Voelkel MA, Wynne KM, Badesch DB, Groves BM, Voelkel NF (2000) Hyperuricemia in severe pulmonary hypertension. Chest 117(1):19–24
36. Jiang X, Han ZY, Wang Y, Xu XQ, Ma CR, Wu Y, Pan L, Jing ZC (2008) Hemodynamic variables and clinical features correlated with serum uric acid in patients with pulmonary arterial hypertension. Chin Med J (Engl) 121(24):2497–2503
37. Njaman W, Iesaki T, Iwama Y, Takasaki Y, Daida H (2007) Serum uric acid as a prognostic predictor in pulmonary arterial hypertension with connective tissue disease. Int Heart J 48(4):523–532
38. Van Albada ME, Loot FG, Fokkema R, Roofthooft MT, Berger RM (2008) Biological serum markers in the management of pediatric pulmonary arterial hypertension. Pediatr Res 63(3):321–327
39. Gabrielli LA, Castro PF, Godoy I, Mellado R, Bourge RC, Alcaino H, Chiong M, Greig D, Verdejo HE, Navarro M, Lopez R, Toro B, Quiroga C, Diaz-Araya G, Lavandero S, Garcia L (2011) Systemic oxidative stress and endothelial dysfunction is associated with an attenuated acute vascular response to inhaled prostanoid in pulmonary artery hypertension patients. J Card Fail 17(12):1012–1017
40. Spiekermann S, Schenk K, Hoeper MM (2009) Increased xanthine oxidase activity in idiopathic pulmonary arterial hypertension. Eur Respir J 34(1):276
41. Blatter LA, Taha Z, Mesaros S, Shacklock PS, Wier WG, Malinski T (1995) Simultaneous measurements of Ca2+ and nitric oxide in bradykinin-stimulated vascular endothelial cells. Circ Res 76(5):922–924
42. Stefano GB, Cadet P, Breton C, Goumon Y, Prevot V, Dessaint JP, Beauvillain JC, Roumier AS, Welters I, Salzet M (2000) Estradiol-stimulated nitric oxide release in human granulocytes is dependent on intracellular calcium transients: evidence of a cell surface estrogen receptor. Blood 95(12):3951–3958
43. Faehling M, Kroll J, Fohr KJ, Fellbrich G, Mayr U, Trischler G, Waltenberger J (2002) Essential role of calcium in vascular endothelial growth factor A-induced signaling: mechanism of the antiangiogenic effect of carboxyamidotriazole. FASEB J 16(13):1805–1807
44. Steudel W, Ichinose F, Huang PL, Hurford WE, Jones RC, Bevan JA, Fishman MC, Zapol WM (1997) Pulmonary vasoconstriction and hypertension in mice with targeted disruption of the endothelial nitric oxide synthase (NOS 3) gene. Circ Res 81(1):34–41
45. Ambalavanan N, Mariani G, Bulger A, Philips IJ (1999) Role of nitric oxide in regulating neonatal porcine pulmonary artery smooth muscle cell proliferation. Biol Neonate 76(5):291–300
46. Krick S, Platoshyn O, Sweeney M, McDaniel SS, Zhang S, Rubin LJ, Yuan JX (2002) Nitric oxide induces apoptosis by activating K+ channels in pulmonary vascular smooth muscle cells. Am J Physiol Heart Circ Physiol 282(1):H184–H193
47. Kourembanas S, McQuillan LP, Leung GK, Faller DV (1993) Nitric oxide regulates the expression of vasoconstrictors and growth factors by vascular endothelium under both normoxia and hypoxia. J Clin Invest 92(1):99–104
48. Giaid A, Saleh D (1995) Reduced expression of endothelial nitric oxide synthase in the lungs of patients with pulmonary hypertension. N Engl J Med 333(4):214–221
49. d'Uscio LV (2011) eNOS uncoupling in pulmonary hypertension. Cardiovasc Res 92(3):359–360
50. Villanueva ME, Zaher FM, Svinarich DM, Konduri GG (1998) Decreased gene expression of endothelial nitric oxide synthase in newborns with persistent pulmonary hypertension. Pediatr Res 44(3):338–343

51. Gonzalez A, Fabres J, D'Apremont I, Urcelay G, Avaca M, Gandolfi C, Kattan J (2010) Randomized controlled trial of early compared with delayed use of inhaled nitric oxide in newborns with a moderate respiratory failure and pulmonary hypertension. J Perinatol 30(6):420–424
52. Reiter CD, Wang X, Tanus-Santos JE, Hogg N, Cannon RO 3rd, Schechter AN, Gladwin MT (2002) Cell-free hemoglobin limits nitric oxide bioavailability in sickle-cell disease. Nat Med 8(12):1383–1389
53. Hsu LL, Champion HC, Campbell-Lee SA, Bivalacqua TJ, Manci EA, Diwan BA, Schimel DM, Cochard AE, Wang X, Schechter AN, Noguchi CT, Gladwin MT (2007) Hemolysis in sickle cell mice causes pulmonary hypertension due to global impairment in nitric oxide bioavailability. Blood 109(7):3088–3098
54. Morris CR, Morris SM Jr, Hagar W, Van Warmerdam J, Claster S, Kepka-Lenhart D, Machado L, Kuypers FA, Vichinsky EP (2003) Arginine therapy: a new treatment for pulmonary hypertension in sickle cell disease? Am J Respir Crit Care Med 168(1):63–69
55. Weiner DL, Hibberd PL, Betit P, Cooper AB, Botelho CA, Brugnara C (2003) Preliminary assessment of inhaled nitric oxide for acute vaso-occlusive crisis in pediatric patients with sickle cell disease. JAMA 289(9):1136–1142
56. Machado RF, Martyr S, Kato GJ, Barst RJ, Anthi A, Robinson MR, Hunter L, Coles W, Nichols J, Hunter C, Sachdev V, Castro O, Gladwin MT (2005) Sildenafil therapy in patients with sickle cell disease and pulmonary hypertension. Br J Haematol 130(3):445–453
57. Afshar S, Gibson LL, Yuhanna IS, Sherman TS, Kerecman JD, Grubb PH, Yoder BA, McCurnin DC, Shaul PW (2003) Pulmonary NO synthase expression is attenuated in a fetal baboon model of chronic lung disease. Am J Physiol Lung Cell Mol Physiol 284(5):L749–L758
58. Banks BA, Ischiropoulos H, McClelland M, Ballard PL, Ballard RA (1998) Plasma 3-nitrotyrosine is elevated in premature infants who develop bronchopulmonary dysplasia. Pediatrics 101(5):870–874
59. Sosenko IR, Bancalari E (2010) NO for preterm infants at risk of bronchopulmonary dysplasia. Lancet 376(9738):308–310
60. Blomgren K, Zhu C, Hallin U, Hagberg H (2003) Mitochondria and ischemic reperfusion damage in the adult and in the developing brain. Biochem Biophys Res Commun 304(3):551–559
61. Sims NR, Anderson MF (2002) Mitochondrial contributions to tissue damage in stroke. Neurochem Int 40(6):511–526
62. Duchen MR (2004) Mitochondria in health and disease: perspectives on a new mitochondrial biology. Mol Aspects Med 25(4):365–451
63. Duchen MR (2004) Roles of mitochondria in health and disease. Diabetes 53(Suppl 1):S96–S102
64. Moreira PI, Santos MS, Moreno AM, Proenca T, Seica R, Oliveira CR (2004) Effect of streptozotocin-induced diabetes on rat brain mitochondria. J Neuroendocrinol 16(1):32–38
65. Nishio Y, Kanazawa A, Nagai Y, Inagaki H, Kashiwagi A (2004) Regulation and role of the mitochondrial transcription factor in the diabetic rat heart. Ann N Y Acad Sci 1011:78–85
66. Petersen KF, Dufour S, Befroy D, Garcia R, Shulman GI (2004) Impaired mitochondrial activity in the insulin-resistant offspring of patients with type 2 diabetes. N Engl J Med 350(7):664–671
67. Barclay AR, Sholler G, Christodolou J, Shun A, Arbuckle S, Dorney S, Stormon MO (2005) Pulmonary hypertension—a new manifestation of mitochondrial disease. J Inherit Metab Dis 28(6):1081–1089
68. Venditti CP, Harris MC, Huff D, Peterside I, Munson D, Weber HS, Rome J, Kaye EM, Shanske S, Sacconi S, Tay S, DiMauro S, Berry GT (2004) Congenital cardiomyopathy and pulmonary hypertension: another fatal variant of cytochrome-c oxidase deficiency. J Inherit Metab Dis 27(6):735–739
69. Sproule DM, Dyme J, Coku J, de Vinck D, Rosenzweig E, Chung WK, De Vivo DC (2008) Pulmonary artery hypertension in a child with MELAS due to a point mutation of the mitochondrial tRNA((Leu)) gene (m.3243A>G). J Inherit Metab Dis 1–7

70. Tribe RM, Poston L (1996) Oxidative stress and lipids in diabetes: a role in endothelium vasodilator dysfunction? Vasc Med 1(3):195–206
71. Ronson RS, Nakamura M, Vinten-Johansen J (1999) The cardiovascular effects and implications of peroxynitrite. Cardiovasc Res 44(1):47–59
72. Quinlan CL, Orr AL, Perevoshchikova IV, Treberg JR, Ackrell BA, Brand MD (2012) Mitochondrial complex II can generate reactive oxygen species at high rates in both the forward and reverse reactions. J Biol Chem 287(32):27255–27264
73. Redout EM, Wagner MJ, Zuidwijk MJ, Boer C, Musters RJ, van Hardeveld C, Paulus WJ, Simonides WS (2007) Right-ventricular failure is associated with increased mitochondrial complex II activity and production of reactive oxygen species. Cardiovasc Res 75(4):770–781
74. Bonnet S, Michelakis ED, Porter CJ, Andrade-Navarro MA, Thebaud B, Bonnet S, Haromy A, Harry G, Moudgil R, McMurtry MS, Weir EK, Archer SL (2006) An abnormal mitochondrial-hypoxia inducible factor-1alpha-Kv channel pathway disrupts oxygen sensing and triggers pulmonary arterial hypertension in fawn hooded rats: similarities to human pulmonary arterial hypertension. Circulation 113(22):2630–2641
75. McMurtry MS, Bonnet S, Wu X, Dyck JR, Haromy A, Hashimoto K, Michelakis ED (2004) Dichloroacetate prevents and reverses pulmonary hypertension by inducing pulmonary artery smooth muscle cell apoptosis. Circ Res 95(8):830–840
76. Bonnet S, Rochefort G, Sutendra G, Archer SL, Haromy A, Webster L, Hashimoto K, Bonnet SN, Michelakis ED (2007) The nuclear factor of activated T cells in pulmonary arterial hypertension can be therapeutically targeted. Proc Natl Acad Sci U S A 104(27):11418–11423
77. Rathore R, Zheng YM, Niu CF, Liu QH, Korde A, Ho YS, Wang YX (2008) Hypoxia activates NADPH oxidase to increase [ROS]i and [Ca2+]i through the mitochondrial ROS-PKCepsilon signaling axis in pulmonary artery smooth muscle cells. Free Radic Biol Med 45(9):1223–1231
78. Waypa GB, Chandel NS, Schumacker PT (2001) Model for hypoxic pulmonary vasoconstriction involving mitochondrial oxygen sensing. Circ Res 88(12):1259–1266
79. Waypa GB, Marks JD, Mack MM, Boriboun C, Mungai PT, Schumacker PT (2002) Mitochondrial reactive oxygen species trigger calcium increases during hypoxia in pulmonary arterial myocytes. Circ Res 91(8):719–726
80. Paddenberg R, Ishaq B, Goldenberg A, Faulhammer P, Rose F, Weissmann N, Braun-Dullaeus RC, Kummer W (2003) Essential role of complex II of the respiratory chain in hypoxia-induced ROS generation in the pulmonary vasculature. Am J Physiol Lung Cell Mol Physiol 284(5):L710–L719
81. Sud N, Wells SM, Sharma S, Wiseman DA, Wilham J, Black SM (2008) Asymmetric dimethylarginine inhibits HSP90 activity in pulmonary arterial endothelial cells: role of mitochondrial dysfunction. Am J Physiol Cell Physiol 294(6):C1407–C1418
82. Sun X, Fratz S, Sharma S, Hou Y, Rafikov R, Kumar S, Rehmani I, Tian J, Smith A, Schreiber C, Reiser J, Naumann S, Haag S, Hess J, Catravas JD, Patterson C, Fineman JR, Black SM (2011) C-terminus of heat shock protein 70-interacting protein-dependent GTP cyclohydrolase I degradation in lambs with increased pulmonary blood flow. Am J Respir Cell Mol Biol 45(1):163–171
83. Sharma S, Sud N, Wiseman DA, Carter AL, Kumar S, Hou Y, Rau T, Wilham J, Harmon C, Oishi P, Fineman JR, Black SM (2008) Altered carnitine homeostasis is associated with decreased mitochondrial function and altered nitric oxide signaling in lambs with pulmonary hypertension. Am J Physiol Lung Cell Mol Physiol 294(1):L46–L56
84. Sharma S, Black SM (2009) Carnitine homeostasis, mitochondrial function, and cardiovascular disease. Drug Discov Today Dis Mech 6(1–4):e31–e39
85. Matsuishi T, Stumpf DA, Seliem M, Eguren LA, Chrislip K (1991) Propionate mitochondrial toxicity in liver and skeletal muscle: acyl CoA levels. Biochem Med Metab Biol 45(2):244–253
86. Pande SV, Blanchaer MC (1971) Reversible inhibition of mitochondrial adenosine diphosphate phosphorylation by long chain acyl coenzyme A esters. J Biol Chem 246(2):402–411
87. Shug AL, Shrago E, Bittar N, Folts JD, Koke JR (1975) Acyl-CoA inhibition of adenine nucleotide translocation in ischemic myocardium. Am J Physiol 228(3):689–692

88. Stumpf DA, McAfee J, Parks JK, Eguren L (1980) Propionate inhibition of succinate: CoA ligase (GDP) and the citric acid cycle in mitochondria. Pediatr Res 14(10):1127–1131
89. Liu J, Killilea DW, Ames BN (2002) Age-associated mitochondrial oxidative decay: improvement of carnitine acetyltransferase substrate-binding affinity and activity in brain by feeding old rats acetyl-l-carnitine and/or R-alpha-lipoic acid. Proc Natl Acad Sci U S A 99(4):1876–1881
90. Hagen TM, Liu J, Lykkesfeldt J, Wehr CM, Ingersoll RT, Vinarsky V, Bartholomew JC, Ames BN (2002) Feeding acetyl-l-carnitine and lipoic acid to old rats significantly improves metabolic function while decreasing oxidative stress. Proc Natl Acad Sci U S A 99(4):1870–1875
91. Sharma S, Sun X, Kumar S, Rafikov R, Aramburo A, Kalkan G, Tian J, Rehmani I, Kallarackal S, Fineman JR, Black SM (2012) Preserving mitochondrial function prevents the proteasomal degradation of GTP cyclohydrolase I. Free Radic Biol Med 53(2):216–229
92. Sun X, Sharma S, Fratz S, Kumar S, Rafikov R, Aggarwal S, Rafikova O, Lu Q, Burns T, Dasarathy S, Wright J, Schreiber C, Radman M, Fineman JR, Black SM (2013) Disruption of endothelial cell mitochondrial bioenergetics in lambs with increased pulmonary blood flow. Antioxid Redox Signal 18(14):1739–1752
93. Rafikov R, Rafikova O, Aggarwal S, Gross C, Desai J, Fulton D, Black SM (2013) Asymmetric dimethylarginine induces endothelial nitric oxide synthase mitochondrial redistribution through the nitration-mediated activation of Akt1. J Biol Chem 288(9):6212–6226
94. Rafikov R, Fonseca FV, Kumar S, Pardo D, Darragh C, Elms S, Fulton D, Black SM (2011) eNOS activation and NO function: structural motifs responsible for the posttranslational control of endothelial nitric oxide synthase activity. J Endocrinol 210(3):271–284
95. Sharma S, Aramburo A, Rafikov R, Sun X, Kumar S, Oishi PE, Datar SA, Raff G, Xoinis K, Kalkan G, Fratz S, Fineman JR, Black SM (2013) l-carnitine preserves endothelial function in a lamb model of increased pulmonary blood flow. Pediatr Res 74(1):39–47
96. Wassmann S, Wassmann K, Nickenig G (2004) Modulation of oxidant and antioxidant enzyme expression and function in vascular cells. Hypertension 44(4):381–386
97. Brennan LA, Steinhorn RH, Wedgwood S, Mata-Greenwood E, Roark EA, Russell JA, Black SM (2003) Increased superoxide generation is associated with pulmonary hypertension in fetal lambs: a role for NADPH oxidase. Circ Res 92(6):683–691
98. Wedgwood S, Lakshminrusimha S, Fukai T, Russell JA, Schumacker PT, Steinhorn RH (2011) Hydrogen peroxide regulates extracellular superoxide dismutase activity and expression in neonatal pulmonary hypertension. Antioxid Redox Signal 15(6):1497–1506
99. Fike CD, Aschner JL, Zhang Y, Salvemini D, Kaplowitz MR (2005) Superoxide and chronic hypoxia-induced pulmonary hypertension in newborn piglets. Chest 128(6 Suppl):555S–556S
100. Sharma S, Grobe AC, Wiseman DA, Kumar S, Englaish M, Najwer I, Benavidez E, Oishi P, Azakie A, Fineman JR, Black SM (2007) Lung antioxidant enzymes are regulated by development and increased pulmonary blood flow. Am J Physiol Lung Cell Mol Physiol 293(4):L960–L971
101. Asano M, Mishima A, Takeuchi Y, Usami S, Kotani H, Suzuki Y, Yura J (1993) [Lipid peroxide and free radical scavengers in congenital heart disease with pulmonary hypertension]. Nihon Geka Gakkai Zasshi 94(12):1299–1304 (in Japanese)
102. Kaneko FT, Arroliga AC, Dweik RA, Comhair SA, Laskowski D, Oppedisano R, Thomassen MJ, Erzurum SC (1998) Biochemical reaction products of nitric oxide as quantitative markers of primary pulmonary hypertension. Am J Respir Crit Care Med 158(3):917–923
103. Bowers R, Cool C, Murphy RC, Tuder RM, Hopken MW, Flores SC, Voelkel NF (2004) Oxidative stress in severe pulmonary hypertension. Am J Respir Crit Care Med 169(6):764–769
104. Masri FA, Comhair SA, Dostanic-Larson I, Kaneko FT, Dweik RA, Arroliga AC, Erzurum SC (2008) Deficiency of lung antioxidants in idiopathic pulmonary arterial hypertension. Clin Transl Sci 1(2):99–106
105. Mata M, Sarrion I, Milian L, Juan G, Ramon M, Naufal D, Gil J, Ridocci F, Fabregat-Andres O, Cortijo J (2012) PGC-1alpha induction in pulmonary arterial hypertension. Oxid Med Cell Longev 2012:236572

106. Archer SL, Marsboom G, Kim GH, Zhang HJ, Toth PT, Svensson EC, Dyck JR, Gomberg-Maitland M, Thebaud B, Husain AN, Cipriani N, Rehman J (2010) Epigenetic attenuation of mitochondrial superoxide dismutase 2 in pulmonary arterial hypertension: a basis for excessive cell proliferation and a new therapeutic target. Circulation 121(24):2661–2671
107. Fijalkowska I, Xu W, Comhair SA, Janocha AJ, Mavrakis LA, Krishnamachary B, Zhen L, Mao T, Richter A, Erzurum SC, Tuder RM (2010) Hypoxia inducible-factor1alpha regulates the metabolic shift of pulmonary hypertensive endothelial cells. Am J Pathol 176(3): 1130–1138
108. Frank L, Groseclose EE (1984) Preparation for birth into an O2-rich environment: the antioxidant enzymes in the developing rabbit lung. Pediatr Res 18(3):240–244
109. Asikainen TM, Raivio KO, Saksela M, Kinnula VL (1998) Expression and developmental profile of antioxidant enzymes in human lung and liver. Am J Respir Cell Mol Biol 19(6):942–949
110. Autor AP, Frank L, Roberts RJ (1976) Developmental characteristics of pulmonary superoxide dismutase: relationship to idiopathic respiratory distress syndrome. Pediatr Res 10(3):154–158
111. Huertas JR, Palomino N, Ochoa JJ, Quiles JL, Ramirez-Tortosa MC, Battino M, Robles R, Mataix J (1998) Lipid peroxidation and antioxidants in erythrocyte membranes of full-term and preterm newborns. Biofactors 8(1–2):133–137
112. Vento M, Aguar M, Escobar J, Arduini A, Escrig R, Brugada M, Izquierdo I, Asensi MA, Sastre J, Saenz P, Gimeno A (2009) Antenatal steroids and antioxidant enzyme activity in preterm infants: influence of gender and timing. Antioxid Redox Signal 11(12):2945–2955
113. Frank L, Autor AP, Roberts RJ (1977) Oxygen therapy and hyaline membrane disease: the effect of hyperoxia on pulmonary superoxide dismutase activity and the mediating role of plasma or serum. J Pediatr 90(1):105–110
114. Schroder A, Herting E, Speer CP (1999) [Superoxide dismutase and catalase activity in tracheobronchial secretions after surfactant treatment of newborn infants with respiratory distress syndrome]. Z Geburtshilfe Neonatol 203(5):201–206 (in German)
115. Bonta VW, Gawron ER, Warshaw JB (1977) Neonatal red cell superoxide dismutase enzyme levels: possible role as a cellular defense mechanism against pulmonary oxygen toxicity. Pediatr Res 11(6):754–757
116. Rosenfeld W, Evans H, Concepcion L, Jhaveri R, Schaeffer H, Friedman A (1984) Prevention of bronchopulmonary dysplasia by administration of bovine superoxide dismutase in preterm infants with respiratory distress syndrome. J Pediatr 105(5):781–785
117. Garg U, Jain A, Singla P, Beri S, Garg R, Saili A (2012) Free radical status in retinopathy of prematurity. Indian J Clin Biochem 27(2):196–199
118. Parad RB, Allred EN, Rosenfeld WN, Davis JM (2012) Reduction of retinopathy of prematurity in extremely low gestational age newborns treated with recombinant human Cu/Zn superoxide dismutase. Neonatology 102(2):139–144
119. Singh SK, Tandon A, Kumari S, Ravi RN, Ray GN, Batra S (1998) Changes in anti-oxidant enzymes and lipid peroxidation in hyaline membrane disease. Indian J Pediatr 65(4):609–614
120. Bazowska G, Jendryczko A (1996) [Antioxidant enzyme activities in fetal and neonatal lung: lowered activities of these enzymes in children with RDS]. Ginekol Pol 67(2):70–74 (in Polish)
121. Walther FJ, Gidding CE, Kuipers IM, Willebrand D, Bevers EM, Abuchowski A, Viau AT (1986) Prevention of oxygen toxicity with superoxide dismutase and catalase in premature lambs. J Free Radic Biol Med 2(4):289–293
122. Padmanabhan RV, Gudapaty R, Liener IE, Schwartz BA, Hoidal JR (1985) Protection against pulmonary oxygen toxicity in rats by the intratracheal administration of liposome-encapsulated superoxide dismutase or catalase. Am Rev Respir Dis 132(1):164–167
123. Beretta L, Gerli GC, Ferraresi R, Agostoni A, Gualandri V, Orsini GB (1983) Antioxidant system in sickle red cells. Acta Haematol 70(3):194–197

124. Agostoni A, Gerli GC, Beretta L, Bianchi M, Vignoli M, Bombelli F (1980) Superoxide dismutase, catalase and glutathione peroxidase activities in maternal and cord blood erythrocytes. J Clin Chem Clin Biochem 18(11):771–773
125. Joppa P, Petrasova D, Stancak B, Dorkova Z, Tkacova R (2007) Oxidative stress in patients with COPD and pulmonary hypertension. Wien Klin Wochenschr 119(13–14):428–434
126. Aksenova TA, Parkhomenko IV, Gorbunov VV (2008) [Lipid peroxidation in hypertensive disease concurrent with chronic obstructive pulmonary disease]. Klin Lab Diagn 8:17–19
127. Jin HF, Du SX, Zhao X, Wei HL, Wang YF, Liang YF, Tang CS, Du JB (2008) Effects of endogenous sulfur dioxide on monocrotaline-induced pulmonary hypertension in rats. Acta Pharmacol Sin 29(10):1157–1166
128. Thibeault DW, Rezaiekhaligh M, Mabry S, Beringer T (1991) Prevention of chronic pulmonary oxygen toxicity in young rats with liposome-encapsulated catalase administered intratracheally. Pediatr Pulmonol 11(4):318–327
129. Carpenter D, Larkin H, Chang A, Morris E, O'Neill J, Curtis J (2001) Superoxide dismutase and catalase do not affect the pulmonary hypertensive response to group B streptococcus in the lamb. Pediatr Res 49(2):181–188
130. Fu RH, Chiu TH, Chiang MC, Lien R, Chou YH, Chiang CC, Cho YH, Yang PH (2008) Lower erythrocyte glutathione peroxidase activity in bronchopulmonary dysplasia in the first week of neonatal life. Neonatology 93(4):269–275
131. Falciglia HS, Johnson JR, Sullivan J, Hall CF, Miller JD, Riechmann GC, Falciglia GA (2003) Role of antioxidant nutrients and lipid peroxidation in premature infants with respiratory distress syndrome and bronchopulmonary dysplasia. Am J Perinatol 20(2):97–107
132. Fu RH, Chiu TH, Chiang MC, Cho YH, Lien R, Chiang CC, Yang PH (2007) Erythrocyte antioxyenzyme activity in preterm infants with retinopathy of prematurity. Neonatology 92(1):59–63
133. Papp A, Nemeth I, Karg E, Papp E (1999) Glutathione status in retinopathy of prematurity. Free Radic Biol Med 27(7–8):738–743
134. Morris CR, Suh JH, Hagar W, Larkin S, Bland DA, Steinberg MH, Vichinsky EP, Shigenaga M, Ames B, Kuypers FA, Klings ES (2008) Erythrocyte glutamine depletion, altered redox environment, and pulmonary hypertension in sickle cell disease. Blood 111(1):402–410
135. Detchaporn P, Kukongviriyapan U, Prawan A, Jetsrisuparb A, Greenwald SE, Kukongviriyapan V (2012) Altered vascular function, arterial stiffness, and antioxidant gene responses in pediatric thalassemia patients. Pediatr Cardiol 33(7):1054–1060
136. Waseem F, Khemomal KA, Sajid R (2011) Antioxidant status in beta thalassemia major: a single-center study. Indian J Pathol Microbiol 54(4):761–763
137. Irodova NL, Lankin VZ, Konovalova GK, Kochetov AG, Chazova IE (2002) Oxidative stress in patients with primary pulmonary hypertension. Bull Exp Biol Med 133(6):580–582
138. Nur E, Brandjes DP, Teerlink T, Otten HM, Oude Elferink RP, Muskiet F, Evers LM, ten Cate H, Biemond BJ, Duits AJ, Schnog JJ (2012) N-acetylcysteine reduces oxidative stress in sickle cell patients. Ann Hematol 91(7):1097–1105
139. Pace BS, Shartava A, Pack-Mabien A, Mulekar M, Ardia A, Goodman SR (2003) Effects of *N*-acetylcysteine on dense cell formation in sickle cell disease. Am J Hematol 73(1):26–32
140. Soltan-Sharifi MS, Mojtahedzadeh M, Najafi A, Reza Khajavi M, Reza Rouini M, Moradi M, Mohammadirad A, Abdollahi M (2007) Improvement by *N*-acetylcysteine of acute respiratory distress syndrome through increasing intracellular glutathione, and extracellular thiol molecules and anti-oxidant power: evidence for underlying toxicological mechanisms. Hum Exp Toxicol 26(9):697–703
141. Ichida K, Amaya Y, Noda K, Minoshima S, Hosoya T, Sakai O, Shimizu N, Nishino T (1993) Cloning of the cDNA encoding human xanthine dehydrogenase (oxidase): structural analysis of the protein and chromosomal location of the gene. Gene 133(2):279–284
142. Nishino T, Okamoto K, Eger BT, Pai EF (2008) Mammalian xanthine oxidoreductase—mechanism of transition from xanthine dehydrogenase to xanthine oxidase. FEBS J 275(13):3278–3289
143. Amaya Y, Yamazaki K, Sato M, Noda K, Nishino T (1990) Proteolytic conversion of xanthine dehydrogenase from the NAD-dependent type to the O2-dependent type. Amino acid sequence

of rat liver xanthine dehydrogenase and identification of the cleavage sites of the enzyme protein during irreversible conversion by trypsin. J Biol Chem 265(24):14170–14175
144. Enroth C, Eger BT, Okamoto K, Nishino T, Pai EF (2000) Crystal structures of bovine milk xanthine dehydrogenase and xanthine oxidase: structure-based mechanism of conversion. Proc Natl Acad Sci U S A 97(20):10723–10728
145. Nishino T, Okamoto K, Kawaguchi Y, Hori H, Matsumura T, Eger BT, Pai EF (2005) Mechanism of the conversion of xanthine dehydrogenase to xanthine oxidase: identification of the two cysteine disulfide bonds and crystal structure of a non-convertible rat liver xanthine dehydrogenase mutant. J Biol Chem 280(26):24888–24894
146. Pearson AR, Godber BLJ, Eisenthal R, Taylor GL, Harrison R (2009) Human milk xanthine dehydrogenase is incompletely converted to the oxidase form in the absence of proteolysis. A structural explanation. Submitted (FEB-2009) to the PDB data bank

Index

H. Tsukahara and K. Kaneko (eds.), *Studies on Pediatric Disorders*, Oxidative Stress in Applied Basic Research and Clinical Practice, DOI 10.1007/978-1-4939-0679-6,

B

F

G

K

MIX
Papier aus verantwortungsvollen Quellen
Paper from responsible sources
FSC® C105338

If you have any concerns about our products,
you can contact us on
ProductSafety@springernature.com

In case Publisher is established outside the EU,
the EU authorized representative is:
Springer Nature Customer Service Center GmbH
Europaplatz 3, 69115 Heidelberg, Germany

Printed by Libri Plureos GmbH
in Hamburg, Germany